Handbook of Experimental Pharmacology

Volume 264

The *Handbook of Experimental Pharmacology* is one of the most authoritative and influential book series in pharmacology. It provides critical and comprehensive discussions of the most significant areas of pharmacological research, written by leading international authorities. Each volume in the series represents the most informative and contemporary account of its subject available, making it an unrivalled reference source.

HEP is indexed in PubMed and Scopus.

More information about this series at http://www.springer.com/series/164

Harald H. H. W. Schmidt • Pietro Ghezzi •
Antonio Cuadrado
Editors

Reactive Oxygen Species

Network Pharmacology and Therapeutic Applications

Springer

Editors
Harald H. H. W. Schmidt
Department of Pharmacology and Personalised Medicine
Maastricht University
Maastricht, The Netherlands

Pietro Ghezzi
Brighton and Sussex Medical School
Brighton, UK

Antonio Cuadrado
Department of Biochemistry, Faculty of Medicine
Autonomous University of Madrid
Madrid, Spain

ISSN 0171-2004 ISSN 1865-0325 (electronic)
Handbook of Experimental Pharmacology
ISBN 978-3-030-68509-6 ISBN 978-3-030-68510-2 (eBook)
https://doi.org/10.1007/978-3-030-68510-2

This Springer imprint is published by the registered company Springer Nature Switzerland AG.
The registered company address is: Gewerbestrasse 11, 6330 Cham, Switzerland

Preface

For decades, reactive oxygen species (ROS) have been suggested as a possible disease mechanism and therapeutic target. This went hand in hand with the introduction of the term 'oxidative stress' and promoting antioxidants as preventative or curative intervention. Almost every disease has been correlated with oxidative stress or markers thereof. However, neither any therapeutic nor any diagnostic has made it into the clinic and there is no accepted ROS-induced disease (ROSopathy). Is this because ROS do not play a role in disease? The chapters in this book suggest this is not the case and provide evidence. Several misconceptions have led this field in the wrong direction. Most relevant, ROS are clearly not solely triggering disease. They also fulfil important signalling and metabolic functions in our body. Thus, the term 'oxidative stress' was a bad choice as it implied ROS as being primarily a stressor and risk, and that any strategy that eliminates or scavenges them, if this were indeed possible, would be beneficial. But ROS are essential and thus substantially reducing ROS levels is bound to have side-effects by neutralising any beneficial role they may have, exactly as has been observed in numerous unsuccessful clinical trials. Moreover, not all ROS may be equal. ROS from a certain source in a specific compartment may be essential, whilst in a neighbouring compartment, cell or organ a different source is highly dysregulated and does harm. But how to identify this? In Part I, we demystify the simplistic view of the role of ROS, i.e. small amounts being beneficial, large amounts detrimental. We instead review their key physiological, at low and high concentrations, and introduce an entirely new way of identifying ROS-related disease mechanism and drug targets by systems medicine. In Part II, we then review approaches based on modulation of the endogenous antioxidant systems; in Part III, how to inhibit disease triggering sources of ROS; in Part IV, conditions where even ROS stimulation or substitution is beneficial; and, most innovative, in Part V we extend the ROS field to non-ROS enzymes, i.e. ROS targets. This is a message for all fields. Our current concept in pharmacology to look at isolated, curated networks around certain signalling molecules (cAMP, calcium) or mechanisms (tyrosine kinase, GPCR) are mindmaps that have nothing to do with disease mechanisms. Instead disease modules are more often than not hybrids from different parts of different signalling pathways. Thus, this handbook opens a new chapter in redox medicine, pharmacology and drug discovery in general. This is not just another book about redox biology and biochemistry. Outstanding experts who provided the field with landmark publications have joined forces. The editors and several of the contributing authors were networked by the European Cooperation on Science and Technology (COST) with actions dedicated to ROS and systems medicine, which is gratefully acknowledged as a major facilitator of this book and game changer in innovating our thinking and concepts of ROS, redox medicine, pharmacology and drug discovery.

Maastricht, The Netherlands Harald H. H. W. Schmidt
Brighton, UK Pietro Ghezzi
Madrid, Spain Antonio Cuadrado

Contents

Part I

ROS Revisited: Changing the Focus from Oxidative Stress and Redox Biology to Redox Medicine

Demystifying Oxidative Stress

Pietro Ghezzi and Arshag D. Mooradian

Contents

Abstract

The hypothesis that reactive oxygen species (ROS) can be not just associated with but causally implicated in disease was first made in 1956, but so far, the oxidative stress theory of disease has not led to major therapeutic breakthrough, and the use of antioxidant is now confined to the field of complementary medicine. This

P. Ghezzi (✉)
Brighton and Sussex Medical School, Brighton, UK
e-mail: p.ghezzi@bsms.ac.uk

A. D. Mooradian
Department of Medicine, University of Florida College of Medicine, Jacksonville, FL, USA
e-mail: arshag.mooradian@jax.ufl.edu

H. H. H. W. Schmidt et al. (eds.), *Reactive Oxygen Species*,
Handbook of Experimental Pharmacology 264, https://doi.org/10.1007/164_2020_379

chapter reviews the lack of high-level clinical evidence for the effectiveness of antioxidants in preventing disease and the epistemological problems of the oxidative stress theory of disease. We conclude on possible ways forward to test this hypothesis with approaches that take into account personalized medicine.

Graphical Abstract

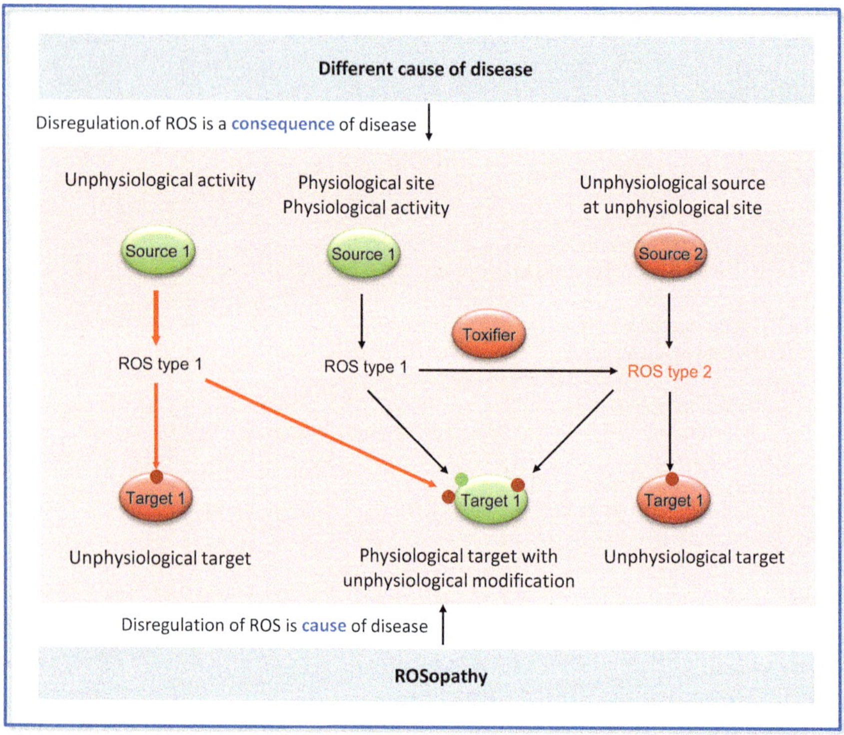

The previous oxidative stress model has helped neither to diagnose nor to treat possibly ROS-related or ROS-dependent diseases. The redox balance concept that low ROS levels are beneficial or tolerable and high levels are disease triggers and best reduced is apparently wrong. Physiological ROS signalling may become dysfunctional or a disease trigger by at least five mechanisms: a physiological source may appear at an unphysiological site, a physiological source may be underactivated (less common) or overactivated (more common), a new source may appear, a physiological source may be overactivated or underactivated, and a toxifying enzyme may convert an ROS signal molecule into a more reactive molecule. The latter three mechanisms may reach a physiological or nonphysiological target. All of these dysregulations may be the direct and essential cause of a disease (rarely the case) or just a secondary epiphenomenon, which will disappear once the non-ROS-related cause of the disease is cured (much more common). Importantly, these mechanisms are the same for almost every signalling system. Causal target validation (sources, toxifiers and targets) is essential in order to identify effective drugs and therapies for ROSopathies.

Keywords

Antioxidants · Clinical trials · Epistemology · Evidence-based medicine · Supplements

1 Oxidative Stress: The Origins

The concept that ROS can be toxic precedes the introduction of the expression "oxidative stress" (OS). This term was first used by Ernest Beutler in a 1970 paper on the effect of glutathione reductase deficiency in red blood cells (Paniker et al. 1970). In the following 10 years, OS was only mentioned 24 times (source: PubMed).

However, in considering the origins of this concept, we must bear in mind that the term became established in the scientific literature only following the pioneering studies by Helmut Sies and co-workers published in 1985.

We have used Google Ngram Viewer (https://books.google.com/ngrams) to show how the terminology in the field has evolved (Fig. 1). Probably the earliest expression related to this is "oxygen toxicity" (with a visible increase in frequency in 1970 and decline after 1985), followed by "lipid peroxidation". Oxidative stress, like "reactive oxygen species" or "reactive oxygen intermediates", appears in the late 1980s. The graph shows other expressions related to the field, such as "superoxide", "paraquat" and "oxygen radicals".

Oxygen toxicity is defined as the toxic effects of exposure to oxygen concentrations higher than ambient concentration (21%) and is known to occur in patients exposed for prolonged period to oxygen therapy, including premature infants (Clark and Lambertsen 1971).

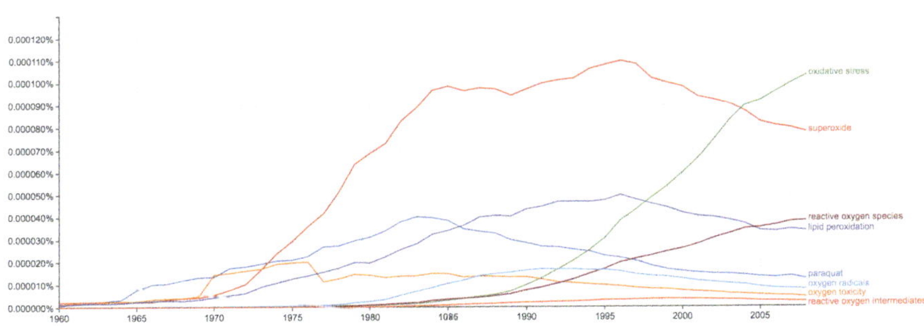

Fig. 1 Chronology of usage of terms related to oxidative stress

As early as 1954, Gershman suggested that oxygen toxicity and radiation injury have a common mechanism mediated by free radical and discussed the protective effects of antioxidants such as propyl gallate and nordihydroguaiaretic acid (Gerschman et al. 1954).

Lipid peroxidation is another term that relates to oxidative stress and has been studied in the context of oxygen and ozone toxicity (Wood and Watson 1967), carbon tetrachloride hepatotoxicity (Comporti et al. 1965) and pulmonary toxicity of paraquat (Bus et al. 1974).

2 Oxidative Stress: The Establishment of the Concept

Although the first occurrence of the expression is found in the context of glutathione in red blood cells in a 1970 paper by the laboratory of biochemist Ernest Beutler (Paniker et al. 1970), oxidative stress was formulated as a concept in a series of papers by Helmuth Sies published from 1985 on (Sies et al. 2017). The expression is often meant as a state of cellular stress caused by an imbalance between the amount of ROS produced and the ROS-degrading systems (antioxidants).

Several discoveries were instrumental to the growth of this concept, also shown in Fig. 1, and included the work by McCord and Fridovich that led to the discovery of the ability of xanthine oxidase to produce superoxide and of superoxide dismutase (SOD) to reduce them to hydrogen peroxide (McCord and Fridovich 1968, 1969). This provided two essential enzymatic tools to generate or scavenge superoxide and study its biological effects.

3 The Oxidative Stress Theory of Disease

It is generally accepted that ROS can alter the structure and function of key constituents of human organism, notably proteins, lipids, carbohydrate moieties and DNA (Kalyanaraman 2013). Thus, it is readily accepted that quenching of these radicals would reduce the likelihood of tissue damage, thereby preventing a host of ailments including cancer and cardiovascular disease.

Independent of the terminology used, the first well-known article where oxidative stress is described as a pathogenic mechanism is the free radical theory of ageing by Harman (1956) where also it was postulated that the use of antioxidants such as cysteine would improve the ageing process. After that study, oxidative stress has been associated, in thousands of research papers, with most of the existing pathological conditions, not just complications of ageing. Many of these studies have, more or less explicitly, concluded that the use of antioxidants would have positive effect on the pathological process, either therapeutic or preventative.

As we will describe in the next section, despite most of the published works reported either a strong association between "oxidative stress" and disease onset or severity or positive effect of antioxidants in a variety of disease models, 60 years after the 1956 publication, there are no antioxidant molecules approved for any indication. We discussed already that the only possible exception is edaravone (approved in Japan for stroke and ALS and in the USA for ALS).[1]

This contrasts with other theories of disease, such as the cytokine theory of disease, which was originally postulated in 1985 and led to the approval, in 1998 (just 14 years later), of anti-TNF antibodies for rheumatoid arthritis (Feldmann and Maini 2001), opening the way to subsequent approval of other antibodies to cytokines or their receptors.

As a result, there is a risk that the antioxidant approach will be confined to the field of complementary/alternative medicine and its multibillion market of nutraceutical supplements, rather than leading to significant development in drug discovery, and we will try to highlight the problems of the theory of oxidative stress to point out the problems left open and research questions.

4 The Disappointments

When antioxidants were tested in the clinical arena, this often led to major disappointments. One example was the free radical scavenger NXY-059 for which there was a body of evidence in animal models of stroke and a first clinical trial (Lees et al. 2006) was found ineffective in a larger clinical trial (Shuaib et al. 2007). This failure was then seen in connection with other failures for the same indication, and the low quality of the preclinical research was blamed, leading to discussion of guidelines for preclinical drug development in stroke (Savitz and Fisher 2007), although Slemmer et al. discussed other possible reasons for the failure such as the difficulty of increasing antioxidant defence in men and the fact that NXY-059, like many antioxidants, can easily oxidize (Slemmer et al. 2008). Another such example is the use of glutathione-replenishing therapy in acute respiratory distress syndrome which, despite positive results in animal models, including by one of the authors of this chapter (Gatti et al. 1993), was found not effective in clinical trials (Kollef and Schuster 1995).

Indeed, a number of large observational studies have concluded that consumption of antioxidant vitamin-rich food reduces the risk of cardiovascular events (reviewed

[1]However, although edaravone has been developed as a free radical scavenger, recent studies show that it also acts as an Nrf2 inducer (Li et al. 2016, 2017; Wu et al. 2017). Although one may argue that this results, in the end, in a free radical scavenger effect because Nrf2 drives the transcription of several antioxidant genes, Nrf2 has among its target genes cytoprotective factors that are independent of an antioxidant mechanism and downregulates inflammation though a direct cross-talk with NF-kB (Cuadrado et al. 2018). Furthermore, activation of Nrf2 is a common feature of pro-oxidant and electrophilic molecules (Cuadrado et al. 2018). We will discuss the off-target effects of "antioxidants" later in this chapter.

Table 1 Approaches to the study of the role of oxidative stress in disease

	Experiments/model	Outcomes/readouts
Observation	In patients' samples or cells In disease models in vivo or in vitro	ROS Biomarkers of OS Indirect evidence measuring known oxidants or antioxidants (SOD, XO, etc.) and *assuming this will have an impact*
Intervention	Addition of antioxidants/oxidants (including induction/inhibition of oxidative stress) Genetic or chemical modification of enzymes (Transgenic/KO/ inhibitors/inducers)	Disease severity Biomarkers of OS ROS

in Hasanain and Mooradian (2002), (2004), Mooradian (2006), Sheikh-Ali et al. (2011)). The advantages of these studies include the large size of the population surveyed, sound design of the studies and accepted methodology of data collection. However, like all observational studies, they fail to show causal relationship between the variables of interest. In addition, observational studies of the effect of nutrition on health are generally marred by the limitations of relying on dietary recall and the difficulty in identifying the specific nutrient that lead to the outcomes of interest. In most observational studies, the specific antioxidant vitamin is not measured, and individuals who consume antioxidant-rich foods are generally more health conscious and have a healthier lifestyle.

Traditionally, randomized placebo-controlled trials have been used to establish the safety and efficacy of a therapeutic intervention. These trials have been reviewed in past publications (Hasnain and Mooradian 2002, 2004; Mooradian 2006; Sheikh-Ali et al. 2011). The larger interventional trials (those that included more than 5,000 subjects) are summarized in Tables 1 and 2. These trials by and large have failed to show any beneficial effects of antioxidant vitamins (Table 1) (Heart Protection Study Collaborative Group 2002; Heart Outcomes Prevention Evaluation Study Investigators 2000; Dietary supplementation with n-3 polyunsaturated fatty acids and vitamin E after myocardial infarction: results of the GISSI-Prevenzione trial. Gruppo Italiano per lo Studio della Sopravvivenza nell'Infarto miocardico 1999; Albert et al. 2008; Cook 2007; Hennekens et al. 1996; Lee et al. 2005; Lin et al. 2008), and some have raised the possibility that beta-carotene can increase the risk of cardiovascular events and lung cancer in vulnerable populations, while vitamin E may increase the risk of haemorrhagic stroke (Table 2) (Omenn et al. 1996; Sesso 2008; Alpha-Tocopherol BCCPSG 1994)). One exception was a large Chinese trial in 29,584 participants aged 40–69 years that followed the subjects for an average of 5.4 years (Blot et al. 1993). In this trial, the study participants were randomized to receive one of the following four micronutrient combinations: (1) daily 5,000 IU retinol with 22.5 mg zinc, (2) 3.2 mg riboflavin with 40 mg niacin, (3) 120 mg ascorbic acid with 30 mg molybdenum and (4) a mixture of 15 mg beta-carotene with 50 mg selenium and 30 mg α-tocopherol. In this trial, there was a reduction in

Table 2 Large (over 5,000 subjects) randomized double-blind placebo-controlled trials of antioxidant vitamins where neither harm nor benefit was demonstrated

Population	Antioxidant	Outcome	Reference
22,071 male physicians at 40–84 years of age studied for 12 years	β-Carotene 50-mg QOD	β-Carotene had no significant effects on malignant neoplasms or CV disease	Hennekens et al. (1996)
11,324 Italian patients with myocardial infarction studied for 3.5 years	n-3 poly-unsaturated fatty acid 1 g daily, E 300 mg daily	Vitamin E conferred no significant benefit	Dietary supplementation with n-3 polyunsaturated fatty acids and vitamin E after myocardial infarction: results of the GISSI-Prevenzione trial. Gruppo Italiano per lo Studio della Sopravvivenza nell'Infarto miocardico (1999)
9,541 high-risk patients 55 years of age or older studied for 5 years	$E = 400$ IU/d	No change in the incidence of malignant neoplasms, cardiovascular disease or death from all causes	Heart Outcomes Prevention Evaluation Study Investigators (2000)
20,536 high-risk men and women on simvastatin or placebo at 40–80 years of age studied for 5 years	E 600 mg C 250 mg β-Carotene 20 mg daily	Antioxidants do not decrease mortality or the incidence of vascular disease, cancer or other major outcomes	Heart Protection Study Collaborative Group (2002)
39,876 healthy women aged at least 45 years studied for an average of 10.1 years	$E = 600$ IU QOD	Vitamin E did not decrease major cardiovascular events, cancer or total mortality	Lee et al. (2005)
8,171 women with CHD or 3 CHD risk factors aged 40 years or older and studied for a mean of 9.4-years	E 600 IU QOD C 500 mg/day ß-carotene 50 mg QOD	No overall effects of ascorbic acid, vitamin E or beta-carotene on cardiovascular events among women at high risk for CV disease	Cook (2007)

QOD every other day dosing, *CV* cardiovascular, *CHD* coronary heart disease

total and cancer-related mortality, especially for mortality from stomach cancer in those on the combination regimen of β-carotene, selenium and α-tocopherol (Blot et al. 1993). The generalizability of this observation is questionable as the study population was from an area that is naturally deficient in selenium and the observed favourable effect may have been the result of selenium supplementation. It is noteworthy that a meta-analysis of seven randomized trials of vitamin E treatment (81,788 patients) and of eight trials of beta-carotene treatment (138,113 patients) concluded that vitamin E in the dose range used (50 to 800 IU) did not alter

Table 3 Large (over 5,000 subjects) randomized double-blind placebo-controlled trials of antioxidant vitamins where potential harm was demonstrated

Population	Antioxidant	Outcome	Source
29,133 Finnish male smokers 50 to 69 years of age followed up for 5–8 years	E 50 mg, ß-carotene 20 mg daily dosing	E: ß-carotene increased lung cancer (by 16%) and CV mortality. Total mortality was 8% higher among those who received beta-carotene primarily because of more deaths from lung cancer and ischemic heart disease	Alpha-Tocopherol BCCPSG (1994)
18,314 men and women at high risk for lung cancer with a mean age of 57 years and followed up for a mean of 4 years	30 mg of ß-carotene and 25,000 IU of retinal palmitate daily	Intervention was stopped 21 months early because there were 28% more lung cancers and 17% more deaths in the active intervention group	Hennekens et al. (1996)
14,641 healthy male physicians 50 years or older followed up for a mean of 8 years	E 400 IU C 500 mg ß-carotene 50 mg daily dosing	No effect on CV events but vitamin E increased risk of haemorrhagic stroke (HR, 1.74 [95% CI, 1.04–2.91]; $P = 0.04$)	Sesso (2008)

CV cardiovascular

cardiovascular or overall mortality, while supplementation with beta-carotene (15 to 50 mg) led to a small but significant increase in cardiovascular deaths (3.4 vs 3.1%, $P = 0.003$) and in all-cause mortality (7.4% vs 7.0%, $P = 0.003$) (Vivekananthan et al. 2003). Another meta-analysis also found that monotherapy with vitamin E may promote the development of congestive heart failure (Thomson et al. 2009). Another meta-analysis of 68 randomized trials with 232,606 participants concluded that treatment with β-carotene, vitamin A and vitamin E may increase mortality, whereas the effects of vitamin C and selenium on mortality need further investigation (Bjelakovic et al. 2012) (Table 3).

Thus, some antioxidant vitamins, notably beta-carotene, may be harmful (Alpha-Tocopherol BCCPSG 1994; Omenn et al. 1996). There are also other smaller trials where the potential harm of antioxidants has been highlighted. The HDL-raising effect of nicotinic acid could be blunted with high-dose vitamin E (Cheung et al. 2001), and daily ingestion of antioxidant vitamins such as vitamin C and E prevents exercise-related enhancement of insulin sensitivity (Ristow et al. 2009). These observations taken together raise the spectre that antioxidants may be "worse than useless" (Gomez-Cabrera et al. 2012).

The biological plausibility of the hypothesis that reducing oxidative stress with antioxidant supplementation is so strong that the failure of antioxidants to have any beneficial effects in clinical trials seems on the surface to be paradoxical. It can be argued that the limitations with the available interventional trials preclude excluding the role of antioxidants in reducing human pathology. After all, these trials were carried out in people with well-established underlying pathology, the duration of the interventions was often short, and there are no reliable biomarkers of oxidative stress in humans to stratify the participants in the study and monitor the adequacy of the dosing regimen. Given the latter limitation, it would be impossible to know if sufficient amounts of the ingested antioxidants were absorbed or were biologically active.

5 The Epistemological Weakness of the Oxidative Stress Theory of Disease (OSTD)

We pointed out elsewhere some of the epistemological weaknesses of the theory that OS is a causal mechanism of disease, compared to other theories of disease such as the cytokine theory of disease or the bacterial theory of disease.

We will summarize below the main points to be taken into consideration when formulating a theory of disease considering the experimental evidence obtained in biological systems that supports the OSTD.

First of all, what kind of evidence can we obtain to support a role for oxidative stress in disease?

We can first distinguish between observational evidence and interventional evidence, summarized in Table 1.

5.1 Observational Evidence

Observational evidence is where, for instance, something related to oxidative stress is measured in a patient cohort and is compared to healthy subjects or subjects with an unrelated disease. We identified from the literature three types of measurements.

The first type is when levels of ROS can be directly measured in biological fluids of patients (in vivo) or ROS production is measured in vitro in cells obtained from patients (this is usually defined as ex vivo).

The second type consists in the measurement of biomarkers of oxidative stress that are, as we pointed out elsewhere, molecules formed as products of the oxidation of cellular molecules by ROS.

The third type is more an indirect and hypothetical evidence. For instance, observing that, in patients, the levels of an antioxidant (e.g. SOD, catalase, vitamin C, vitamin E) are decreased can generate the assumption that those patients will be in a state of oxidative stress because antioxidant defence is defective. Likewise, an observed increase in ROS-generating enzymes (e.g. XO, NOX) is often thought to result, de facto, in oxidative stress due to overproduction of ROS.

In these studies, researchers will be looking for associations between the oxidative stress-associated variables (ROS, biomarkers, oxidants/antioxidants), either by comparing healthy versus disease samples or by correlating the measurement with indicators of disease severity such as C-reactive protein (CRP) in inflammatory disease, the Sequential Organ Failure Assessment (SOFA) score in sepsis or a correlation of variables with the disease course.

The pros of clinical cohort studies are that they investigate a real-world patient population. A major disadvantage is the heterogeneity of the population and associated variability that would make necessary to have a large sample size for the study to be adequately powered. Other disadvantages are the difficulty to meet the requirements of timeliness and storage when labile markers are measured or the difficulty of obtaining samples from the onset of disease and/or at different time points, so that studies often only take a snapshot. The multicausal nature of many diseases is an additional problem that will be discussed later in this chapter.

This approach is not limited to clinical studies, and the same three typologies of evidence apply to experimental models of disease, in vivo or in vitro.

In vivo models are those where a specific disease is induced in animals (e.g. collagen-induced arthritis, endotoxin-induced lung injury, dextran-induced colitis).

Animal models have many advantages such as the ease of collecting samples at different times, including before induction of disease; access to tissues and organs, not just biological fluids; possibility of using sophisticated but more invasive techniques (such as chemiluminescence, transgenic animals expressing fluorescent proteins, etc.); and lack of subject variability by using inbred animals of the same age. In addition, the disease is induced in a "clean" way (with a very specific trigger), and there are no comorbidities or variabilities due, for instance, to differences in diet and lifestyle.

The disadvantages are the same as the advantages. Reductionism, the use of a model where we isolate one mechanism of disease, makes it difficult to extrapolate results and, even more, hypotheses on causes and mechanisms to the real patient. Furthermore, the mechanism by which the disease is induced in mice may not be representative of the causal mechanisms that lead to the corresponding disease in patients.

The problems associated with reductionism are even more evident in case of in vitro models of disease, often using a single-cell type. Examples are endotoxin-stimulated macrophages, used as a catch-all model of inflammation, or beta-amyloid-induced neurotoxicity in neuronal cells. On the other hand, these models allow the application of experimental techniques that would be very difficult to apply to in vivo studies.

Like in clinical studies, in vitro, ex vivo or in vivo (animal) observational studies will search for an association between disease (or an indicator of it) and oxidative stress. As in clinical trials, this can take the shape of a comparison between health and disease (normal mice versus mice with arthritis, macrophages with endotoxin compared to those without endotoxin, etc.) or a correlation in a dose-response or time-course experiment.

The other aspect of all observational studies, whether experimental or not, is the measurement related to oxidative stress used in the study.

To study a TNF-mediated disease, we want to measure TNF and not a generic marker of inflammation such as C-reactive protein; to postulate the infectious origin of a disease, we would like to be able to isolate the pathogen, not just measure fever. Likewise, when hypothesizing an ROS-mediated disease, we would like to obtain evidence of increased ROS levels. However, this is nearly impossible because, unlike cytokines or viruses, ROS have half-lives in the range of microseconds or nanoseconds.

This is the main reason why most studies (and virtually all those in patients) measure "biomarkers of oxidative stress". We have discussed elsewhere the limitation of these biomarkers (Frijhoff et al. 2015a; Ghezzi et al. 2018), but we could summarize them as follows: (1) they are not measuring ROS, but they are only an indirect indicator of their formation; (2) often, there are multiple chemical reactions leading to their formation from the initial reaction with a specific ROS; and (3) many of them are not specific as they can also be produced by other metabolic pathways (for instance, malondialdehyde can also be produced during arachidonic acid metabolism).

5.2 Limitations of Observational Studies: Drawing the Arrow

All the studies falling in the category above are about identifying an association between disease and oxidative stress (however, measured). It is worth remembering the axiom "association does not mean causation", and be cautious when drawing an arrow between "oxidative stress" and "disease x". For instance, if we find that a disease is associated with oxidative stress, that is, all patients with that disease have oxidative stress, we can say "if *disease x*, then *oxidative stress*".

Because oxidative stress is often seen as a mechanism of disease, or at least that would be the research hypothesis in such studies, the above finding will be often depicted (as it is in most review articles):

1. Oxidative Stress → Disease x

This is often read by biomedical scientists as a causal link, "oxidative stress *causes* disease x". This is, of course, fine if this is what is meant. However, bearing in mind the axiom mentioned above, the fact that all patients with a disease have (higher) oxidative stress could also be explained with the hypothesis that "disease x causes oxidative stress" and draw the arrow in the opposite direction:

2. Disease x ← Oxidative Stress

In symbolic logic, the arrow indicates a material conditional. This means that the arrow in 1 means "if oxidative stress, then disease x". It is true whenever oxidative stress is false and also when oxidative stress is true and disease x is true (i.e. when oxidative stress and disease are both present). It is only false when oxidative stress is true and disease x is false (i.e. when we have oxidative stress but not the disease). This means that the arrow in a material conditional means that all those with oxidative stress will have the disease. This is more or less we would expect from a causal link, but it may not be a causal link.

In fact, the other important concept in disease causation is that of the confounder. For instance, we could hypothesize that having observed an association between drinking and disease x, we could hypothesize that alcohol causes the disease. However, people who drink may also be more frequently smokers, and this would be the confounder. Then, the two parts of the equation (alcohol and disease) are not causally related to each other, but it could be that alcohol is just associated with smoking and that smoking causes disease x.

In our case, for instance, if we observe an association between oxidative stress and atherosclerosis, we might hypothesize that oxidative stress causes atherosclerosis. However, both are associated with inflammation, and it could be that inflammation (the confounder) causes both atherosclerosis and oxidative stress. In this case, the mechanism would be that inflammation induces atherosclerosis and oxidative stress may not necessarily be implicated in the formation of atherosclerosis.

5.3 Limitations of Observational Studies: Multicausal Diseases

Another limitation in the use of observational studies to formulate theories of disease arises from their application to multicausal disease. Many of the diseases that are the challenge of our time are multicausal diseases. Unlike a thrombotic stroke, which is caused by a blood clot that forms in an artery, or an infectious disease, which is due to an infectious virus or bacterium, chronic inflammatory diseases, neurodegenerative diseases or diabetes are likely to be due to a mixture of causes, at least until proven differently.

This concept of component causes has been well described by Rothman (1976). According to that concept, a disease can be caused by a mixture of causes that add up to form a "sufficient cause". This model is normally depicted with what is now called

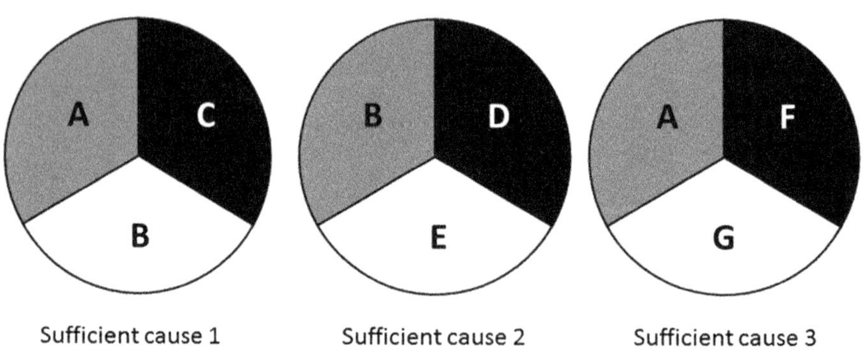

Fig. 2 The Rothman pie model of disease causation, where different combinations of component causes (**a**–**g**) that, alone, would not cause disease form a sufficient cause. A multifactorial disease can result from different sufficient causes. Modified from Rothman (1976)

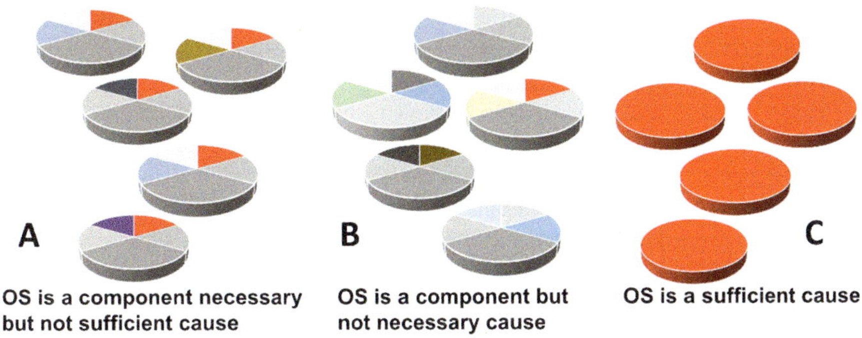

Fig. 3 Possible roles of oxidative stress in multifactorial disease. (**a**) OS is always necessary but not sufficient; (**b**) oxidative stress is necessary only in some component causes; (**c**) oxidative stress is sufficient, and does not need any other component cause. From Ghezzi et al. (2018)

the "Rothman pie". The scheme in Fig. 2, from the original study by Rothman, shows a disease that can be caused by three "sufficient causes" (the pies), each made up of five different "component causes" (the slices). Each of those component causes by itself cannot produce the disease. Of note, there are two types of component causes. For instance, from Fig. 2, we could understand that cause A is necessary to induce the disease (along with other component causes), while B or D is necessary only when in the presence of a specific mixture of component causes.

This concept, although now part of any textbook of epidemiology, is often forgotten not only by laboratory-based researchers but also by clinical ones. Let us consider the specific case of oxidative stress as a cause in disease x. Figure 3 shows three possibilities for the role of oxidative stress. In the first case (right panel), oxidative stress (in red) must be present for the disease to be induced but needs to be present in combination with other specific component causes; in this case, oxidative stress is a necessary component and will be present in all patients. In the second case (middle panel), the disease can develop with different component causes, in some cases comprising oxidative stress but not always; in this case, oxidative stress will only be present in some patients, not all. The third possibility (right panel) is the simplest case, where oxidative stress alone can cause the disease, but even in this case, one cannot exclude that the same disease can also be induced by another cause, independent of oxidative stress. This means that unless we hypothesize that oxidative stress is a necessary component cause for a multicausal disease, we should not expect it to be present in every patient.

5.4 Limitations of Observational Studies: Mechanistic Biomarkers Versus Surrogate Biomarkers

Several publications have suggested that oxidative stress has a key role in multicausal diseases such as cardiovascular disease (Heitzer et al. 2001), diabetes (Evans et al. 2002) and neurodegenerative diseases (Barnham et al. 2004). These hypotheses are supported by many studies reporting increased levels of various biomarkers of oxidative stress in patients.

In fact, Stoker and Keaney (2004) already discussed the possibility that the unsuccessful trials of antioxidants in atherosclerosis could be due to the fact that oxidative stress is not a cause of atherosclerosis but that both are caused by inflammation (the concept of confounder discussed above).

However, there is another issue with measurements of oxidative stress biomarkers in patients. If we consider the Rothman model, we should expect a biomarker of oxidative stress to be elevated, and pass the statistical significance test, only in those patients where oxidative stress is present. If oxidative stress was present in all patients with a disease, there are two possibilities.

The first possibility is that oxidative stress is a necessary component cause that needs to be present in all patients with the disease. In this case, one might expect that antioxidant therapy will be effective, with the caveat of its specificity and off-target effects, which we will discuss later.

The second explanation for observing oxidative stress in all patients with a disease is that oxidative stress is a consequence, not a cause. In that case, the disease would be induced by one or several mechanisms and will then result in oxidative stress. In this case, we may not expect that a therapy targeting oxidative stress would be effective more than we would expect an antipyretic drug to cure an infection. In this second case, however, biomarkers of oxidative stress could correlate with the disease and its severity and be very good surrogate biomarkers of disease.

5.5 Interventional Studies and Their Purpose

We discussed above how it is not straightforward to demonstrate the role of a mechanism of disease with observational studies, whether these are in patient cohorts or animal models.

The other approach is to use a modifier to intervene in the experimental model, in vitro or in vivo. In our case, this can be done by either increasing or decreasing the amount of ROS in the system.

The first approach, usually done in vitro, is used to see whether ROS can induce a disease or activate a pathway known to be a causal mechanism of the disease. Examples of this approach could be testing whether ROS induce the production of inflammatory cytokines in macrophages or if they cause neuronal death. A practical problem of these studies is that ROS have very short half-lives (nanoseconds to milliseconds), while the in vitro assays may require 24 h or more. In this case, researchers often use ROS-generating system, such as adding xanthine oxidase or glucose oxidase and their substrates, to achieve a continuous production of ROS. Redox cycling compounds that generate ROS, such as menadione, are also used.

One major limitation of these experiments is that it is difficult to expose the target cells to the same concentrations of ROS that would be produced in vivo during the development of the disease and with the same temporal pattern. This may lead to an overestimation of the role of ROS in disease.

The other limitation is not specific to the theory of oxidative stress but is implicit in most experimental models that are necessarily reductive, that is, they are designed to isolate one pathway avoiding interference from any other variable. This is a well-known problem, described by Illari and Russo as the problem of external validity or "extrapolation problem" (Illari et al. 2011) and others defined as "inferential reproducibility" (Goodman et al. 2016).

To return to the problem of multicausal disease, oxidative stress has been implicated in many of them, but the induction of disease needs other component causes, and other inducers are seldom present in the experimental model where ROS are added as a modifier.

The other limitation of the approach using addition of ROS to an experimental model is that this is hardly applicable to in vivo models due to the lack of accessibility of the target tissue. Indirect ways of increasing endogenous ROS are also possible, such as knocking out genes encoding antioxidant enzymes (e.g. SOD) or decreasing the levels of endogenous antioxidants (e.g. glutathione). However, we should not assume that blocking an antioxidant will always result in increased ROS concentrations. In fact, the endogenous antioxidant defence system is robust and redundant; blocking one peroxide-detoxifying enzyme such as glutathione peroxidase may not necessarily increase the concentration of hydrogen peroxide as this is also eliminated by several other enzymes (catalase, peroxiredoxins, etc.).

5.6 Using Antioxidants as Tools and the Problem of Their Specificity

The second approach is to block ROS production in a disease-relevant model, in vitro or in vivo. For instance, to study the role of oxidative stress in inflammation, we could test an ROS scavenger in vitro in a model where macrophages, or mice, are exposed to endotoxin (a standard inflammatory stimulus) and production of inflammatory cytokine is measured, or a disease indicator. For instance, in the past, we have administered a thiol antioxidant to mice in a model of endotoxin-induced lung injury and found that it inhibited both the production of inflammatory cytokines and pulmonary oedema (Gatti et al. 1993) and have concluded that "data strongly support the hypothesis that ROS may play a key role" in that model of disease. Sometimes we are led to consider this sort of evidence, where an antioxidant ameliorates a disease model at the highest level, clear-cut evidence of a role of oxidative stress, but how strong is this evidence, and how much it can be extrapolated to draw more general conclusions on the mechanism of acute respiratory distress syndrome? These approaches are not unambiguous. Most endogenous "antioxidants" have other functions than scavenging ROS (for instance, we found that in the context of inflammation, glutathione has signalling functions independent of its ROS-scavenging action (Diotallevi et al. 2017). Thus, evidence obtained in vivo with an antioxidant has not the same value for the oxidative stress theory of disease that an experiment with anti-TNF antibodies has for the cytokine theory of disease. A problem with many so-called antioxidants is that they are seldom specific, unlike, for instance, anti-cytokine molecules. For instance, many of them are also reducing agents. All thiol-based antioxidant will reduce labile disulphides (Laragione et al. 2003).

Redox-active compounds can act both as anti- or pro-oxidants. For instance, GSH is a major defence against H2O2, serving as the electron donor for GSH peroxidase, that catalyses the reaction:

$$2GSH + H_2O_2 - > GSSG + 2H_2O$$

However, as pointed out by Gilbert (Gilbert et al. 1957; Gilbert 1963), it can also act as an oxidant and generate oxygen radicals:

$$GSH + O2 - > HO_2. + GS$$

$$HO_2. + RH - > R. + H_2O_2$$

Redox reactions are complex, and other antioxidants can have pro-oxidant activities, depending on the experimental model, including vitamin C (Podmore et al. 1998), polyphenols (Halliwell 2008), curcumin (Ahsan and Hadi 1998), quercetin (Rahman et al. 1990), carotenoids (El-Agamey et al. 2004) and resveratrol (de la Lastra and Villegas 2007).

While in oversimplified models of ROS-mediated damage in vitro or in animal models it is possible to isolate the nonspecific effects and put the antioxidant action in evidence, this may not reflect the complex clinical setting of disease.

It should also be noted that in an exemplary adaptive response of the organism, oxidants activate the transcription factor Nrf2 that induces the production of several antioxidant genes (as discussed in another chapter in this book), which might explain why, in certain models, pro-oxidants such as flavonoids can result in an antioxidant phenotype (Lee-Hilz et al. 2006).

All these off-target effects of antioxidants may contribute to the lack of translational success, which one would expect from drugs lacking specificity.

Finally, the concept of oxidative stress has recently evolved due to the identification of a regulatory, signalling role of ROS (particularly hydrogen peroxide) at the physiological level that led to the use of the terms "redox regulation" and "oxidative eustress" (Sies 2017, 2019; Sies and Jones 2020). This implies that a blanket removal of ROS will not only eliminate a potential cause of disease but also affect physiological processes.

6 The Shortcut: Vitamins and Supplements

Despite the problems discussed earlier in this chapter and the lack of success of many clinical trials, the notion that "antioxidants are good" has become one of the "science myths that will never die" (Scudellari 2015), and the concept is popular in the lay public.

It is no surprise that while there are no antioxidant drugs, many have taken the shortcut of antioxidant supplements, which represent a huge market where nutritional advice and basic research findings are often extrapolated to support the use of supplements to promote healthy ageing and prevent and cure disease.

Although there is strong evidence that appropriate intake of fruit is essential in a healthy diet, and this has been taken up by most public health guidelines, there is no conclusive evidence that this is due to antioxidants and vitamins, above the doses needed to avoid deficiency, and Choi et al. pose the interesting question "if combinations with other food components are needed for effective protection, or if Vitamins C and E are largely surrogate biomarkers of a 'healthy' diet, but not the key protective agents" (Choi et al. 2004).

7 The Way Forward

The tone of this chapter was in line with the title, demystify – trying to make clear that the fact that oxidative stress is a cause of disease is still a theory. Because this is considered not a theory but just plain truth by many, the tone might seem pessimistic.

In fact, the current evidence from the clinical trials indicate that there is no justification for prescribing antioxidant supplementation to people (Hasanain and Mooradian 2002, 2004; Mooradian 2006; Sheikh-Ali et al. 2011). There are no known benefits, and, in some cases, antioxidants may have unfavourable effects on human health. However, the consumption of fresh food rich in antioxidants may have benefits as suggested by observational studies (Sheikh-Ali et al. 2011). This may be the result of having the antioxidant vitamins in association with other nutritional components that optimize the biological effects of these vitamins. In general, vitamin and mineral supplementation should be reserved for those who are at risk of nutritional deficiency. This leaves open the possibility of evidence-based dietary recommendation. In this case, the use of exogenous antioxidants would not be recommended unless there was a documented deficiency, as it happens with vitamin C for scurvy or vitamin D for Northern countries such as the UK (Choices 2016). We considered nutritional recommendations a topic that does not fit a *Handbook of Pharmacology*.

Future studies should focus on developing novel and clinically useful biomarkers of oxidative stress to identify people who are better candidates for antioxidant therapy. Novel organelle-specific antioxidants that are known to protect against oxidative damage rather than quenching of free radicals should be tested in clinical trials. In addition, since oxidative stress is one of many stresses that promote cellular dysfunction, agents that have pan stress modification capabilities should be developed and tested (Mooradian 2016).

Several observations made in the last two decades reveal the complexity of the role of free radicals. Multiple biochemical pathways generate free radicals, and it is possible that different antioxidants with different mechanisms of action should be combined to study the role of antioxidant therapy. Perhaps inhibiting specific oxygen-toxicity pathways would be more effective intervention than relying on nonspecific ROS scavengers (Ghezzi et al. 2017, 2020). It is noteworthy that evidence in cell culture and animal models suggest that reactive oxygen species generation in response to a signal undergoes rapid adaption. Thus, dextrose-induced ROS generated in endothelial cells is normalized after 6 h of continuous treatment with excess dextrose (27.5 Mm) (Horani et al. 2004). These adaptive responses are likely the result of short- and long-term cellular changes including induction of genes expressing enzymes critical to quenching of free radicals.

It is noteworthy that free radicals not only are agents of disruption but also have important roles in signal transduction and maintenance of optimal redox state of the various compartments of the cell notably the endoplasmic reticulum (ER) (Kalyanaraman 2013; Mooradian et al. 2016). This organelle, the site of protein folding, has oxidative redox potential necessary to form disulphide bonds to fold proteins properly. Thus, it is conceivable that antioxidants in high doses may alter the internal oxidative milieu of this organelle and thereby cause ER stress. Such cross talk between oxidative load and ER stress has been described in coronary artery endothelial cell cultures (Mooradian et al. 2016). This phenomenon could be an example of "reductive stress". The latter describes the dysfunction in a biologic environment where the antioxidative defences outweigh the oxidative load required for the optimal cellular signalling and metabolism. Since this phenomenon may well be compartmentalized, it is possible that antioxidant supplementation may be protective in one compartment while it has deleterious effects on another compartment. This speculation is supported by the observation that in the worm *Caenorhabditis elegans*, elevated reactive oxygen species (ROS) in the mitochondria acts to increase lifespan, while elevated ROS in the cytoplasm decreases lifespan (Schaar et al. 2015). The importance of an oxidative environment for promoting insulin sensitivity is suggested by studies of patients with mutations impairing the production of antioxidant selenoproteins (Schoenmakers et al. 2010). These patients have signs of oxidative damage in tissues but maintain high level of insulin sensitivity even in the context of obesity. In addition, a small clinical trial found that physiological amounts of the antioxidants vitamin C and vitamin E abrogated the capacity of physical exercise to enhance insulin sensitivity (Ristow et al. 2009). Similarly, the infusion with the antioxidant N-acetylcysteine or treatment with resveratrol blunts the positive effects of exercise on human skeletal muscle and cardiovascular health (Gliemann et al. 2013; Petersen et al. 2012). Thus, the optimal redox state for various biologic functions varies considerably, and this may pose a challenge in identifying therapeutic antioxidants that target specific cellular functions.

The vast literature on oxidative stress, ROS and antioxidants in disease might be telling us something. Despite the inadequacy of many biomarkers of oxidative stress, there is good evidence that increase ROS production may really take place in some disease, although maybe not all of them. Also, despite the many off-target effects of most antioxidants used experimentally, for some disease models, the evidence for a protective effect is certainly strong enough to deserve further study. This may require, however, to consider the issues described above.

From the analysis of the recent literature, we can identify some strategies being considered to move forward.

8 Specific Targeting

If in disease x the mechanistic hypothesis is that activation of enzyme y causes overproduction of ROS in a given tissue, then we should target that enzyme rather than administering large doses of antioxidants and reductants.

We should look at the example of the treatment of chronic myelogenous leukaemia (CML). The finding, in 1990, that CML associated with the Philadelphia chromosome results in the constitutive activation of Bcr-Abl tyrosine kinase led to the development of successful inhibitors of specific Bcr-Abl inhibitors to treat CML (Capdeville et al. 2002). Although Bcr-Abl is a protein kinase that, as such, transfers a phosphate group from ATP to specific protein substrates, nobody ever proposed to "scavenge" ATP to cure CML or to block all protein kinases.

As mentioned in other chapters in this book, a number of enzymes evolved to produce ROS, for instance, xanthine oxidase (XO) or various isoforms of NADPH oxidases (NOX). If, in a disease, a specific enzyme is activated and produces higher than normal quantities of ROS in a specific tissue, it would make sense to inhibit the enzyme or its isoform rather than nonspecifically lower ROS levels in the organism.

Example for this is the use of NOX inhibitors in stroke or XO inhibitors, as discussed in other chapters in this book.

9 Personalized Medicine and Biomarkers

Personalized medicine is the flavour of the month, but we often forget the implications. Stating that "each patient is different" could imply that not all of them need the same treatment. If, according to the Rothman model, we think that oxidative stress may not necessarily be present as a causal component in all patients with a disease, as shown earlier in Fig. 3, then we should not expect antioxidants to be effective in clinical trials on the entire patient population. We should first identify those patients where higher levels of ROS are present.

This requires being able to measure ROS in disease the way we can measure inflammatory cytokines such as TNF or IL-6. However, as we discussed elsewhere, it is not realistic to measure ROS in disease because of their short half-lives, ranging from seconds to microseconds (Ghezzi et al. 2017). We therefore resort to rely on indirect evidence of ROS production. There are many biomarkers of oxidative stress, usually products of the reaction of different ROS with biological molecules, including DNA oxidation products (e.g. 8-oxo guanine), lipid oxidation products (e.g. malondialdehyde) or proteins (e.g. carbonylated proteins) (Frijhoff et al. 2015b). These biomarkers often lack specificity, as they can also be produced by other biochemical pathways or, like malondialdehyde, and their production is the result of a series of complex reactions.

To apply a personalized medicine approach, we will need to establish the significance and specificity of these biomarkers and what they really measure. It is possible that new biomarkers that are indicators of the response of the organism to oxidative stress, including transcription factors such as Nrf2 (Cuadrado et al. 2018), redox-dependent genes (Diotallevi et al. 2017) or micro-RNA (Bedreag et al. 2016), will help.

References

Ahsan H, Hadi SM (1998) Strand scission in DNA induced by curcumin in the presence of Cu(II). Cancer Lett 124(1):23–30

Albert CM, Cook NR, Gaziano JM, Zaharris E, MacFadyen J, Danielson E, Buring JE, Manson JE (2008) Effect of folic acid and B vitamins on risk of cardiovascular events and total mortality among women at high risk for cardiovascular disease. JAMA 299(17):2027. https://doi.org/10.1001/jama.299.17.2027

Alpha-Tocopherol BCCPSG (1994) The effect of vitamin E and beta carotene on the incidence of lung cancer and other cancers in male smokers. N Engl J Med 330(15):1029–1035. https://doi.org/10.1056/NEJM199404143301501

Barnham KJ, Masters CL, Bush AI (2004) Neurodegenerative diseases and oxidative stress. Nat Rev Drug Discov 3(3):205–214. https://doi.org/10.1038/nrd1330

Bedreag OH, Sandesc D, Chiriac SD, Rogobete AF, Cradigati AC, Sarandan M, Dumache R, Nartita R, Papurica M (2016) The use of circulating miRNAs as biomarkers for oxidative stress in critically ill polytrauma patients. Clin Lab 62(3):263–274

Bjelakovic G, Nikolova D, Gluud LL, Simonetti RG, Gluud C (2012) Antioxidant supplements for prevention of mortality in healthy participants and patients with various diseases. Cochrane Database Syst Rev 3:CD007176. https://doi.org/10.1002/14651858.CD007176.pub2

Blot WJ, Li JY, Taylor PR, Guo W, Dawsey S, Wang GQ, Yang CS, Zheng SF, Gail M, Li GY et al (1993) Nutrition intervention trials in Linxian, China: supplementation with specific vitamin/mineral combinations, cancer incidence, and disease-specific mortality in the general population. J Natl Cancer Inst 85(18):1483–1492

Bus JS, Aust SD, Gibson JE (1974) Superoxide- and singlet oxygen-catalyzed lipid peroxidation as a possible mechanism for paraquat (methyl viologen) toxicity. Biochem Biophys Res Commun 58(3):749–755

Capdeville R, Buchdunger E, Zimmermann J, Matter A (2002) Glivec (STI571, imatinib), a rationally developed, targeted anticancer drug. Nat Rev Drug Discov 1(7):493–502. https://doi.org/10.1038/nrd839

Cheung MC, Zhao XQ, Chait A, Albers JJ, Brown BG (2001) Antioxidant supplements block the response of HDL to simvastatin-niacin therapy in patients with coronary artery disease and low HDL. Arterioscler Thromb Vasc Biol 21(8):1320–1326

Choi SW, Benzie IF, Collins AR, Hannigan BM, Strain JJ (2004) Vitamins C and E: acute interactive effects on biomarkers of antioxidant defence and oxidative stress. Mutat Res 551(1–2):109–117. https://doi.org/10.1016/j.mrfmmm.2004.03.006

Choices N (2016) The new guidelines on vitamin D – what you need to know. https://www.nhs.uk/news/food-and-diet/the-new-guidelines-on-vitamin-d-what-you-need-to-know/; https://web.

archive.org/web/20180312142230/https://www.nhs.uk/news/food-and-diet/the-new-guidelines-on-vitamin-d-what-you-need-to-know/. Accessed 12 Mar 2018

Clark JM, Lambertsen CJ (1971) Pulmonary oxygen toxicity: a review. Pharmacol Rev 23 (2):37–133

Comporti M, Saccocci C, Dianzani MU (1965) Effect of CCl-4 in vitro and in vivo on lipid peroxidation of rat liver homogenates and subcellular fractions. Enzymologia 29(3):185–204

Cook NR (2007) A randomized factorial trial of vitamins C and E and beta carotene in the secondary prevention of cardiovascular events in women. Arch Intern Med 167(15):1610. https://doi.org/10.1001/archinte.167.15.1610

Cuadrado A, Manda G, Hassan A, Alcaraz MJ, Barbas C, Daiber A, Ghezzi P, Leon R, Lopez MG, Oliva B, Pajares M, Rojo AI, Robledinos-Anton N, Valverde AM, Guney E, Schmidt H (2018) Transcription factor NRF2 as a therapeutic target for chronic diseases: a systems medicine approach. Pharmacol Rev 70(2):348–383. https://doi.org/10.1124/pr.117.014753

de la Lastra CA, Villegas I (2007) Resveratrol as an antioxidant and pro-oxidant agent: mechanisms and clinical implications. Biochem Soc Trans 35(Pt 5):1156–1160. https://doi.org/10.1042/BST0351156

Dietary supplementation with n-3 polyunsaturated fatty acids and vitamin E after myocardial infarction: results of the GISSI-Prevenzione trial. Gruppo Italiano per lo Studio della Sopravvivenza nell'Infarto miocardico (1999) Lancet 354(9177):447–455

Diotallevi M, Checconi P, Palamara AT, Celestino I, Coppo L, Holmgren A, Abbas K, Peyrot F, Mengozzi M, Ghezzi P (2017) Glutathione fine-tunes the innate immune response toward antiviral pathways in a macrophage cell line independently of its antioxidant properties. Front Immunol 8:1239. https://doi.org/10.3389/fimmu.2017.01239

El-Agamey A, Lowe GM, McGarvey DJ, Mortensen A, Phillip DM, Truscott TG, Young AJ (2004) Carotenoid radical chemistry and antioxidant/pro-oxidant properties. Arch Biochem Biophys 430(1):37–48. https://doi.org/10.1016/j.abb.2004.03.007

Evans JL, Goldfine ID, Maddux BA, Grodsky GM (2002) Oxidative stress and stress-activated signaling pathways: a unifying hypothesis of type 2 diabetes. Endocr Rev 23(5):599–622. https://doi.org/10.1210/er.2001-0039

Feldmann M, Maini RN (2001) Anti-TNF alpha therapy of rheumatoid arthritis: what have we learned? Annu Rev Immunol 19:163–196. https://doi.org/10.1146/annurev.immunol.19.1.163

Frijhoff J, Winyard PG, Zarkovic N, Davies SS, Stocker R, Cheng D, Knight AR, Taylor EL, Oettrich J, Ruskovska T, Cipak Gasparovic A, Cuadrado A, Weber D, Poulsen HE, Grune T, Schmidt HH, Ghezzi P (2015a) Clinical relevance of biomarkers of oxidative stress. Antioxid Redox Signal 23:1144. https://doi.org/10.1089/ars.2015.6317

Frijhoff J, Winyard PG, Zarkovic N, Davies SS, Stocker R, Cheng D, Knight AR, Taylor EL, Oettrich J, Ruskovska T, Gasparovic AC, Cuadrado A, Weber D, Poulsen HE, Grune T, Schmidt HH, Ghezzi P (2015b) Clinical relevance of biomarkers of oxidative stress. Antioxid Redox Signal 23(14):1144–1170. https://doi.org/10.1089/ars.2015.6317

Gatti S, Faggioni R, Echtenacher B, Ghezzi P (1993) Role of tumour necrosis factor and reactive oxygen intermediates in lipopolysaccharide-induced pulmonary oedema and lethality. Clin Exp Immunol 91(3):456–461

Gerschman R, Gilbert DL, Nye SW, Dwyer P, Fenn WO (1954) Oxygen poisoning and x-irradiation: a mechanism in common. Science 119(3097):623–626

Ghezzi P, Jaquet V, Marcucci F, Schmidt HH (2017) The oxidative stress theory of disease: levels of evidence and epistemological aspects. Br J Pharmacol 174:1784–1796. https://doi.org/10.1111/bph.13544

Ghezzi P, Davies K, Delaney A, Floridi L (2018) Theory of signs and statistical approach to big data in assessing the relevance of clinical biomarkers of inflammation and oxidative stress. Proc Natl Acad Sci U S A 115(10):2473–2477. https://doi.org/10.1073/pnas.1719807115

Ghezzi P, Ghiara V, Davies K (2020) Epistemological challenges of the oxidative stress theory of disease and the problem of biomarkers. In: Oxidative stress. Academic Press, London, pp 13–27

Gilbert DL (1963) The role of pro-oxidants and antioxidants in oxygen toxicity. Radiat Res Suppl 3:44–53

Gilbert DL, Gerschman R, Cohen J, Sherwood W (1957) The influence of high oxygen pressures on the viscosity of solutions of sodium desoxyribonucleic acid and of sodium Alginate1. J Am Chem Soc 79(21):5677–5680

Gliemann L, Schmidt JF, Olesen J, Bienso RS, Peronard SL, Grandjean SU, Mortensen SP, Nyberg M, Bangsbo J, Pilegaard H, Hellsten Y (2013) Resveratrol blunts the positive effects of exercise training on cardiovascular health in aged men. J Physiol 591(20):5047–5059. https://doi.org/10.1113/jphysiol.2013.258061

Gomez-Cabrera MC, Ristow M, Vina J (2012) Antioxidant supplements in exercise: worse than useless? Am J Physiol Endocrinol Metab 302(4):E476–E477; author reply E478-479. https://doi.org/10.1152/ajpendo.00567.2011

Goodman SN, Fanelli D, Ioannidis JP (2016) What does research reproducibility mean? Sci Transl Med 8(341):341ps312. https://doi.org/10.1126/scitranslmed.aaf5027

Halliwell B (2008) Are polyphenols antioxidants or pro-oxidants? What do we learn from cell culture and in vivo studies? Arch Biochem Biophys 476(2):107–112. https://doi.org/10.1016/j.abb.2008.01.028

Harman D (1956) Aging: a theory based on free radical and radiation chemistry. J Gerontol 11 (3):298–300

Hasanain B, Mooradian AD (2002) Antioxidant vitamins and their influence in diabetes mellitus. Curr Diab Rep 2(5):448–456. https://doi.org/10.1007/s11892-002-0110-6

Hasnain BI, Mooradian AD (2004) Recent trials of antioxidant therapy: what should we be telling our patients? Cleve Clin J Med 71(4):327–334. https://doi.org/10.3949/ccjm.71.4.327

Heart Outcomes Prevention Evaluation Study Investigators (2000) Vitamin E supplementation and cardiovascular events in high-risk patients. N Engl J Med 342(3):154–160. https://doi.org/10.1056/NEJM200001203420302

Heart Protection Study Collaborative Group (2002) MRC/BHF heart protection study of antioxidant vitamin supplementation in 20,536 high-risk individuals: a randomised placebo-controlled trial. Lancet 360(9326):23–33. https://doi.org/10.1016/S0140-6736(02)09328-5

Heitzer T, Schlinzig T, Krohn K, Meinertz T, Munzel T (2001) Endothelial dysfunction, oxidative stress, and risk of cardiovascular events in patients with coronary artery disease. Circulation 104 (22):2673–2678

Hennekens CH, Buring JE, Manson JE, Stampfer M, Rosner B, Cook NR, Belanger C, LaMotte F, Gaziano JM, Ridker PM, Willett W, Peto R (1996) Lack of effect of long-term supplementation with Beta carotene on the incidence of malignant neoplasms and cardiovascular disease. N Engl J Med 334(18):1145–1149. https://doi.org/10.1056/nejm199605023341801

Horani MH, Haas MJ, Mooradian AD (2004) Rapid adaptive down regulation of oxidative burst induced by high dextrose in human umbilical vein endothelial cells. Diabetes Res Clin Pract 66 (1):7–12. https://doi.org/10.1016/j.diabres.2004.02.013

Illari P, Russo F, Williamson J (2011) Causality in the sciences. OUP, Oxford

Kalyanaraman B (2013) Teaching the basics of redox biology to medical and graduate students: oxidants, antioxidants and disease mechanisms. Redox Biol 1(1):244–257. https://doi.org/10.1016/j.redox.2013.01.014

Kollef MH, Schuster DP (1995) The acute respiratory distress syndrome. N Engl J Med 332 (1):27–37. https://doi.org/10.1056/NEJM199501053320106

Laragione T, Bonetto V, Casoni F, Massignan T, Bianchi G, Gianazza E, Ghezzi P (2003) Redox regulation of surface protein thiols: identification of integrin alpha-4 as a molecular target by using redox proteomics. Proc Natl Acad Sci U S A 100(25):14737–14741. https://doi.org/10.1073/pnas.2434516100

Lee IM, Cook NR, Gaziano JM, Gordon D, Ridker PM, Manson JE, Hennekens CH, Buring JE (2005) Vitamin E in the primary prevention of cardiovascular disease and Cancer. JAMA 294 (1):56. https://doi.org/10.1001/jama.294.1.56

Lee-Hilz YY, Boerboom AM, Westphal AH, Berkel WJ, Aarts JM, Rietjens IM (2006) Pro-oxidant activity of flavonoids induces EpRE-mediated gene expression. Chem Res Toxicol 19 (11):1499–1505. https://doi.org/10.1021/tx060157q

Lees KR, Zivin JA, Ashwood T, Davalos A, Davis SM, Diener HC, Grotta J, Lyden P, Shuaib A, Hardemark HG, Wasiewski WW, Stroke-Acute Ischemic NXYTTI (2006) NXY-059 for acute ischemic stroke. N Engl J Med 354(6):588–600. https://doi.org/10.1056/NEJMoa052980

Li Z, Ma QQ, Yan Y, Xu FD, Zhang XY, Zhou WQ, Feng ZC (2016) Edaravone attenuates hippocampal damage in an infant mouse model of pneumococcal meningitis by reducing HMGB1 and iNOS expression via the Nrf2/HO-1 pathway. Acta Pharmacol Sin 37 (10):1298–1306. https://doi.org/10.1038/aps.2016.71

Li Y, Liu H, Zeng W, Wei J (2017) Edaravone protects against hyperosmolarity-induced oxidative stress and apoptosis in primary human corneal epithelial cells. PLoS One 12(3):e0174437. https://doi.org/10.1371/journal.pone.0174437

Lin J, Cook NR, Albert C, Zaharris E, Gaziano JM, Van Denburgh M, Buring JE, Manson JE (2008) Vitamins C and E and Beta carotene supplementation and cancer risk: a randomized controlled trial. JNCI J Natl Cancer Inst 101(1):14–23. https://doi.org/10.1093/jnci/djn438

McCord JM, Fridovich I (1968) The reduction of cytochrome c by milk xanthine oxidase. J Biol Chem 243(21):5753–5760

McCord JM, Fridovich I (1969) Superoxide dismutase. An enzymic function for erythrocuprein (hemocuprein). J Biol Chem 244(22):6049–6055

Mooradian AD (2006) Antioxidants and diabetes. In: Nestlé nutrition workshop series: clinical and performance program. KARGER. https://doi.org/10.1159/000094429

Mooradian AD (2016) Targeting select cellular stress pathways to prevent hyperglycemia-related complications: shifting the paradigm. Drugs 76(11):1081–1091. https://doi.org/10.1007/s40265-016-0609-9

Mooradian AD, Onstead-Haas L, Haas MJ (2016) Asymmetrical cross-talk between the endoplasmic reticulum stress and oxidative stress caused by dextrose. Life Sci 144:37–48. https://doi.org/10.1016/j.lfs.2015.11.016

Omenn GS, Goodman GE, Thornquist MD, Balmes J, Cullen MR, Glass A, Keogh JP, Meyskens FL, Valanis B, Williams JH, Barnhart S, Cherniack MG, Brodkin CA, Hammar S (1996) Risk factors for lung Cancer and for intervention effects in CARET, the Beta-carotene and retinol efficacy trial. JNCI J Natl Cancer Inst 88(21):1550–1559. https://doi.org/10.1093/jnci/88.21.1550

Paniker NV, Srivastava SK, Beutler E (1970) Glutathione metabolism of the red cells. Effect of glutathione reductase deficiency on the stimulation of hexose monophosphate shunt under oxidative stress. Biochim Biophys Acta 215(3):456–460

Petersen AC, McKenna MJ, Medved I, Murphy KT, Brown MJ, Della Gatta P, Cameron-Smith D (2012) Infusion with the antioxidant N-acetylcysteine attenuates early adaptive responses to exercise in human skeletal muscle. Acta Physiol (Oxf) 204(3):382–392. https://doi.org/10.1111/j.1748-1716.2011.02344.x

Podmore ID, Griffiths HR, Herbert KE, Mistry N, Mistry P, Lunec J (1998) Vitamin C exhibits pro-oxidant properties. Nature 392(6676):559. https://doi.org/10.1038/33308

Rahman A, Hadi SM, Parish JH (1990) Complexes involving quercetin, DNA and Cu(II). Carcinogenesis 11(11):2001–2003

Ristow M, Zarse K, Oberbach A, Kloting N, Birringer M, Kiehntopf M, Stumvoll M, Kahn CR, Bluher M (2009) Antioxidants prevent health-promoting effects of physical exercise in humans. Proc Natl Acad Sci U S A 106(21):8665–8670. https://doi.org/10.1073/pnas.0903485106

Rothman KJ (1976) Causes. Am J Epidemiol 104(6):587–592

Savitz SI, Fisher M (2007) Future of neuroprotection for acute stroke: in the aftermath of the SAINT trials. Ann Neurol 61(5):396–402. https://doi.org/10.1002/ana.21127

Schaar CE, Dues DJ, Spielbauer KK, Machiela E, Cooper JF, Senchuk M, Hekimi S, Van Raamsdonk JM (2015) Mitochondrial and cytoplasmic ROS have opposing effects on lifespan. PLoS Genet 11(2):e1004972. https://doi.org/10.1371/journal.pgen.1004972

Schoenmakers E, Agostini M, Mitchell C, Schoenmakers N, Papp L, Rajanayagam O, Padidela R, Ceron-Gutierrez L, Doffinger R, Prevosto C, Luan J, Montano S, Lu J, Castanet M, Clemons N, Groeneveld M, Castets P, Karbaschi M, Aitken S, Dixon A, Williams J, Campi I, Blount M, Burton H, Muntoni F, O'Donovan D, Dean A, Warren A, Brierley C, Baguley D, Guicheney P, Fitzgerald R, Coles A, Gaston H, Todd P, Holmgren A, Khanna KK, Cooke M, Semple R, Halsall D, Wareham N, Schwabe J, Grasso L, Beck-Peccoz P, Ogunko A, Dattani M, Gurnell M, Chatterjee K (2010) Mutations in the selenocysteine insertion sequence-binding protein 2 gene lead to a multisystem selenoprotein deficiency disorder in humans. J Clin Invest 120(12):4220–4235. https://doi.org/10.1172/JCI43653

Scudellari M (2015) The science myths that will not die. Nature 528(7582):322–325

Sesso HD (2008) Vitamins E and C in the prevention of cardiovascular disease in men. JAMA 300 (18):2123. https://doi.org/10.1001/jama.2008.600

Sheikh-Ali M, Chehade JM, Mooradian AD (2011) The antioxidant paradox in diabetes mellitus. Am J Ther 18(3):266–278. https://doi.org/10.1097/mjt.0b013e3181b7badf

Shuaib A, Lees KR, Lyden P, Grotta J, Davalos A, Davis SM, Diener HC, Ashwood T, Wasiewski WW, Emeribe U, Investigators SIT (2007) NXY-059 for the treatment of acute ischemic stroke. N Engl J Med 357(6):562–571. https://doi.org/10.1056/NEJMoa070240

Sies H (2017) Hydrogen peroxide as a central redox signaling molecule in physiological oxidative stress: oxidative eustress. Redox Biol 11:613–619

Sies H (2019) Oxidative stress: eustress and distress. Academic Press, London

Sies H, Jones DP (2020) Reactive oxygen species (ROS) as pleiotropic physiological signalling agents. Nat Rev Mol Cell Biol:1–21

Sies H, Berndt C, Jones DP (2017) Oxidative stress. Annu Rev Biochem 86:715–748. https://doi.org/10.1146/annurev-biochem-061516-045037

Slemmer JE, Shacka JJ, Sweeney MI, Weber JT (2008) Antioxidants and free radical scavengers for the treatment of stroke, traumatic brain injury and aging. Curr Med Chem 15(4):404–414

Stocker R, Keaney JF Jr (2004) Role of oxidative modifications in atherosclerosis. Physiol Rev 84 (4):1381–1478. https://doi.org/10.1152/physrev.00047.2003

Thomson MJ, Frenneaux MP, Kaski JC (2009) Antioxidant treatment for heart failure: friend or foe? QJM 102(5):305–310. https://doi.org/10.1093/qjmed/hcn160

Vivekananthan DP, Penn MS, Sapp SK, Hsu A, Topol EJ (2003) Use of antioxidant vitamins for the prevention of cardiovascular disease: meta-analysis of randomised trials. Lancet 361 (9374):2017–2023. https://doi.org/10.1016/S0140-6736(03)13637-9

Wood JD, Watson WJ (1967) Lipid peroxidation (thiobarbituric acid reacting material) and enzyme inhibition in rat brain homogenates exposed to oxygen at high pressure. Can J Physiol Pharmacol 45(4):752–755

Wu J, Ren J, Yao S, Wang J, Huang L, Zhou P, Yun D, Xu Q, Wu S, Wang Z, Qiu P (2017) Novel antioxidants' synthesis and their anti-oxidative activity through activating Nrf2 signaling pathway. Bioorg Med Chem Lett 27(7):1616–1619. https://doi.org/10.1016/j.bmcl.2017.02.006

Oxidants in Physiological Processes

Ulla G. Knaus

Contents

Abstract

A number of diseases and conditions have been associated with prolonged or persistent exposure to non-physiological levels of reactive oxygen species (ROS). Similarly, ROS underproduction due to loss-of-function mutations in superoxide or hydrogen peroxide (H_2O_2)-generating enzymes is a risk factor or causative for certain diseases. However, ROS are required for basic cell functions; in particular the diffusible second messenger H_2O_2 that serves as signaling molecule in redox processes. This activity sets H_2O_2 apart from highly reactive oxygen radicals and influences the approach to drug discovery, clinical utility, and therapeutic intervention. Here we review the chemical and biological fundamentals of ROS with emphasis on H_2O_2 as a signaling conduit and initiator of redox relays and propose an integrated view of physiological versus non-physiological reactive species.

U. G. Knaus (✉)
Conway Institute, School of Medicine, University College Dublin, Dublin, Ireland
e-mail: ulla.knaus@ucd.ie

H. H. H. W. Schmidt et al. (eds.), *Reactive Oxygen Species*,
Handbook of Experimental Pharmacology 264, https://doi.org/10.1007/164_2020_380

Therapeutic interventions that target persistently altered ROS levels should include both selective inhibition of a specific source of primary ROS and careful consideration of a targeted pro-oxidant approach, an avenue that is still underdeveloped. Both strategies require attention to redox dynamics in complex cellular systems, integration of the overall spatiotemporal cellular environment, and target validation to yield effective and safe therapeutics.

Graphical Abstract

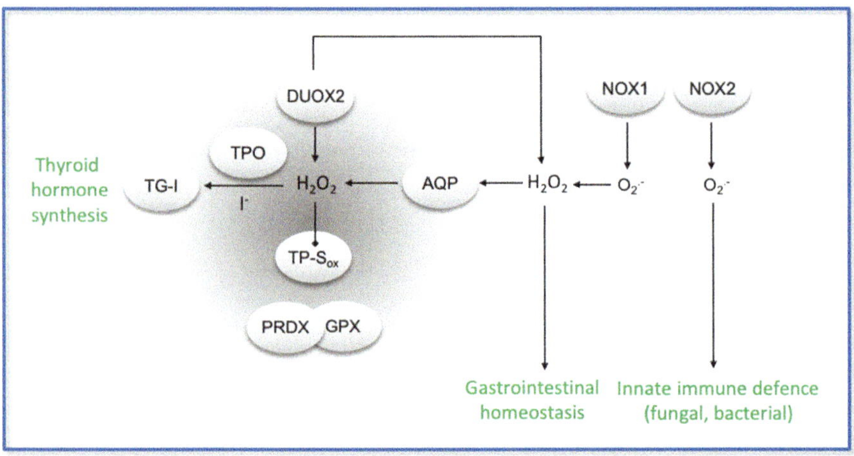

The only professional primary ROS producers are NADPH oxidases (NOX1-5, DUOX1-2). Many other enzymes, e.g., xanthine oxidase (XO), monoamine oxidases (MAO), lysyl oxidases (LO), lipoxygenase (LOX), and cyclooxygenase (COX), produce superoxide and H_2O_2 secondary to their primary metabolic function. Superoxide is too reactive to disseminate, but H_2O_2 is diffusible, only limited by adjacent PRDXs or GPXs, and can be apically secreted and imported into cells through aquaporin (AQP) channels. H_2O_2 redox signaling includes oxidation of the active site thiol in protein tyrosine phosphatases, which will inhibit their activity and thereby increase tyrosine phosphorylation on target proteins. Essential functions include the oxidative burst by NOX2 as antimicrobial innate immune response; gastrointestinal NOX1 and DUOX2 generating low H_2O_2 concentrations sufficient to trigger antivirulence mechanisms; and thyroidal DUOX2 essential for providing H_2O_2 reduced by TPO to oxidize iodide to an iodinating form which is then attached to tyrosyls in TG. Loss-of-function (LoF) variants in TPO or DUOX2 cause congenital hypothyroidism and LoF variants in the NOX2 complex chronic granulomatous disease.

Keywords

Chronic granulomatous disease (CGD) · Congenital hypothyroidism · Hydrogen peroxide (H_2O_2) · Inflammatory bowel disease (IBD) · Mitochondrial electron transport chain · NADPH oxidases (NOX) · Reactive oxygen species (ROS) · Redox relay · Redox signaling · Superoxide ($O_2^{\bullet-}$)

1 Reactive Oxygen Species (ROS)

1.1 Chemistry and Biological Context

All aerobic organisms need molecular oxygen (atmospheric O_2 at 21%) for efficient energy production in mitochondria (Halliwell 2006). Partial reduction of O_2 leads to formation of reactive oxygen species (ROS). The term ROS needs to be used cautiously as it includes radicals, such as superoxide anion radical ($O_2^{\bullet-}$) and hydroxyl radical ($HO^{\bullet}$), non-radical species hydrogen peroxide (H_2O_2), and adducts such as hypochlorous acid (HOCl) or nitrogen-containing species including peroxynitrite ($ONOO^-$) and nitrogen dioxide ($^{\bullet}NO_2$) formed by the reaction of superoxide with nitric oxide radicals ($^{\bullet}NO$). A microenvironment containing H_2O_2, a peroxidase and a halide (e.g., Cl^-) or pseudohalide (e.g., SCN^-) in sufficient concentration will promote the generation of hypohalous acids. The superoxide radical as a primary species is formed by the one-electron reduction of molecular oxygen. Often interconversion of oxygen-derived species will take place; for example, in a hydrophilic environment, two $O_2^{\bullet-}$ molecules will interact spontaneously or catalyzed by superoxide dismutase (SOD) enzymes, in a coupled oxidation-reduction reaction termed dismutation which will generate H_2O_2. A number of enzymes will produce directly H_2O_2 when O_2 is available. In the presence of transition metal ions such as ferrous ion, the Fenton or Haber-Weiss reaction occurs, resulting in the decomposition of H_2O_2 to hydroxyl radical ($^{\bullet}OH$) and hydroxide ion (OH^-) (Winterbourn 2013).

The reactivity and half-life of these various oxygen-derived species differ greatly, with $HO^{\bullet}$ being highly reactive with a very short half-life (10^{-9} s), followed by $O_2^{\bullet-}$ (10^{-5} s) and H_2O_2 that is a relatively weak oxidant but fairly stable (10^{-2}–10^{-3} s) (Pryor 1986; Sies 1993). The charge of $O_2^{\bullet-}$ prevents its diffusion through membranes, permitting only oxidation of adjacent targets, whereas H_2O_2 is diffusible, transverses membranes via aquaporin channels, and reacts preferentially with thiol-containing proteins, thereby initiating redox signaling. Due to its high, non-targeted reactivity, the $^{\bullet}OH$ species is likely responsible for genomic instability by oxidizing DNA, lipids, and proteins leading to DNA damage, including single- and double-strand breaks, DNA-DNA intrastrand adducts, and DNA-protein cross-links. To protect susceptible targets and to maintain cellular integrity, all reactive species will be converted into secondary species or will be eliminated. Preferentially, elimination of H_2O_2 is by intracellular scavengers, often called antioxidant systems, including peroxiredoxins (PRDX) and glutathione peroxidase (GPX). PRDXs are cysteine-dependent peroxide reductases characterized by an active site Pxxx(T/S)

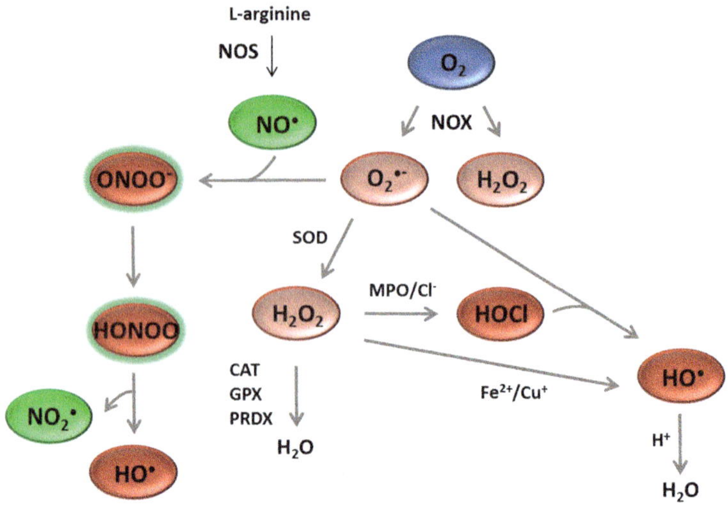

Fig. 1 Scheme of the RONS network. Generation, conversions, and adduct products of reactive oxygen and nitrogen species (RONS), depicting molecular oxygen (O_2), primary ROS (superoxide radical ($O_2^{\bullet -}$), hydrogen peroxide (H_2O_2)), examples of secondary ROS (hydroxyl radical ($HO^{\bullet}$), hypochlorous acid (HOCl)), nitrogen derived species (nitric oxide radical ($NO^{\bullet}$), nitrogen dioxide radical ($NO_2^{\bullet}$)), and oxygen-nitrogen adduct products (peroxynitrite ($OONO^{-}$), peroxynitrous acid (HOONO)), including some of the enzymes involved in RONS generation, conversion, and degradation processes (NADPH oxidase (NOX), nitric oxide synthetase (NOS), superoxide dismutase (SOD), catalase (CAT), glutathione peroxidase (GPX), peroxiredoxin (PRDX), myeloperoxidase(MPO))

xxC sequence in combination with a conserved arginine. Typical PRDXs containing a second "resolving" cysteine recycle by using thioredoxins or glutaredoxins and show isoform-dependent cell compartment-specific localization (Park et al. 2014; Poole and Nelson 2016). High abundance and high reactivity (10^4–10^7 M^{-1} s^{-1}) enable PRDXs to react promptly with peroxide or redox-sensitive targets by transmitting oxidative equivalents to another target via a redox relay mechanism (Winterbourn and Peskin 2016; Stocker et al. 2018b). GPX enzymes reduce H_2O_2 and organic hydroperoxides and use as their reductants either glutathione (GSH) or redoxins (Brigelius-Flohe and Maiorino 2013). The slow decomposition of H_2O_2 to water and O_2 can be also accelerated by the enzyme catalase, mainly expressed in peroxisomes. The continuous formation, interconversion, and breakdown of oxygen-derived species assure a rapid, dynamic response to changes in the cellular environment (Fig. 1).

1.2 Enzymatic ROS Sources

Many enzymes produce $O_2^{\bullet -}$ or H_2O_2, here referred to as primary ROS, as by-product (i.e., secondary enzymatic product) of their primary metabolic function. Examples include xanthine oxidase, monoamine oxidases, lysyl oxidases, lipoxygenase, and cyclooxygenase. Mitochondrial redox signaling relies mainly on $O_2^{\bullet-}$ generated by complex I and III of the electron transport chain (ETC), although other mitochondrial proteins such as monoaminooxidase B, glycerol-3-phosphate dehydrogenase, and the electron transferring flavoprotein ubiquinone oxidoreductase will contribute to mitochondrial ROS (mtROS). The extent of mitochondrial $O_2^{\bullet-}$ production correlates with the mitochondrial ultrastructure, the mitochondrial status (e.g., alterations of the membrane potential, changes in OXPHOS), and oxygen availability. Under hypoxic conditions, superoxide anion produced by mitochondrial complex III undergoes SOD-catalyzed conversion to H_2O_2 that diffuses into the cytosol and stabilizes hypoxia-inducible factor-1α (HIF-1α) (McElroy and Chandel 2017). HIF-1α translocation into the nucleus initiates transcription of genes that promote adaptation to the hypoxic environment. In most conditions, H_2O_2 serves as the main mitochondrial ROS signal. Conversion of $O_2^{\bullet-}$ to H_2O_2 is accomplished by manganese superoxide dismutase (SOD2) in the mitochondrial matrix or by Cu, Zn superoxide dismutase (SOD1) in the intermembrane space (Collins et al. 2012). The overall extent of the mitochondrial redox signal depends on production by the ETC, degradation by peroxiredoxins and glutathione peroxidase, and release into the cytosol. Mitochondrial redox signaling has been connected to immune system regulation, metabolic functions, the circadian clock, and cellular lifespan (Schieber and Chandel 2014). While it is not yet fully understood if and by which means enzymes can regulate the timing and rate of $O_2^{\bullet-}$ as a side product, a second messenger (H_2O_2) will be produced that will participate in redox signaling and alter physiological functions.

Some flavin-containing proteins may generate superoxide in a specific context while engaging with partner proteins. For example, MICAL (molecules interacting with CasL) proteins are scaffolding proteins that contain a flavomonooxygenase domain which generates a redox potential. MICALs can either oxidize directly methionines to methionine sulfoxides (e.g., in actin) via a hydroperoxyflavin intermediate or generate indirectly H_2O_2 for oxidation when interacting with partner proteins such as semaphorin 3A (Nadella et al. 2005; Hung et al. 2011; Giridharan and Caplan 2014). MICAL-induced oxidation targets actin polymerization or microtubule assembly, thereby inducing cytoskeletal rearrangements or growth cone collapse.

The only professional primary ROS producers are NADPH oxidases, a family of seven mammalian enzymes (NOX1-5, DUOX1-2) (Bedard and Krause 2007; Kawahara et al. 2007). NOX/DUOX enzymes are expressed in every cell type and tissue, often with multiple oxidases present in the same cell, where they generate $O_2^{\bullet-}$ (or in some instances H_2O_2) in a spatially and temporally controlled manner when a particular stimulus triggers activation. The members of this family contain a conserved catalytic core that permits the transfer of two electrons from cytoplasmic NADPH to oxidized FAD, followed by one electron transport across the cell membrane, via two inequivalent low potential hemes, to molecular oxygen (Cross and Segal 2004). The reduction of O_2 generates $O_2^{\bullet-}$ in the extracellular space, endosomal lumen or phagosome, depending on the type of membrane that incorporates the oxidase. NOX1-3 and NOX5 follow this established paradigm, while for NOX4 and DUOX1-2 enzymes, only H_2O_2 can be detected, likely due to a yet undetermined modification of the membrane incorporated three-dimensional structure of the active oxidase complex (von Lohneysen et al. 2010; Augsburger et al. 2019). Signals emanating from $O_2^{\bullet-}$ will be short range, affecting only susceptible proteins in the immediate vicinity, as $O_2^{\bullet-}$ is short-lived and cannot traverse membranes. H_2O_2 on the other hand contributes to intracellular redox signaling by diffusion or aquaporin-mediated reentry of H_2O_2 through membranes. H_2O_2 initiates redox relays, protein modifications, and cell to cell communication. Adduct products of $O_2^{\bullet-}$ (i.e., $ONOO^-$) alter signal transduction by thiol oxidation, tyrosine nitration, and lipid peroxidation, while the H_2O_2 adduct HOCl serves in a highly specialized immune defense function by killing microorganisms after phagocytic uptake (Randall et al. 2014; Radi 2018; Nauseef 2019). The regulation of NOX/DUOX enzymes is multifaceted, requiring stimulus-dependent posttranslational modifications, assembly of several cytosolic partner proteins at a membrane-bound heterodimeric complex, and GDP to GTP exchange on the small molecular weight GTPase RAC, leading to formation of an active multimeric complex (Sumimoto 2008). In some cases transcriptional upregulation, calcium binding or epigenetic changes need to occur to support catalytic activity or location-specific biological function of a particular oxidase. Additional regulatory features of oxidases are discrete subcellular localization in different cell types and stimulus-dependent translocation, for example, movement from the endoplasmatic reticulum to the plasma membrane. Mechanisms for turning off the catalytic activity of oxidases are not yet completely understood. RAC GTP hydrolysis and dephosphorylation of cytosolic components are considered the dominant factors in termination of oxidase activity, but other mechanisms such as citrullination of cytosolic components by protein arginase deiminase 4 (PAD4), leading to complex disassembly, have been reported (Decoursey and Ligeti 2005; Zhou et al. 2018). Primary ROS generated by NADPH oxidases undergo the same conversion mechanisms (e.g., SOD1-3), adduct reactions with nitric oxide ($^{\bullet}NO$) or (pseudo)halides (Cl^-, SCN^-), and degradation processes (catalase, radical scavengers, antioxidant systems) as ROS produced by other enzymes.

2 Hydrogen Peroxide and Redox Signaling

2.1 Models for Signal Transmission

The key role of H_2O_2 in driving signaling pathways by oxidation-reduction reactions is undisputed, and recent studies have shed light on how initial oxidation transmits signals inside the cell (Stocker et al. 2018b). Estimates of the basal H_2O_2 concentration in the cytosol are in the low nanomolar range (1–10 nM), rising transiently up to 50-fold or more (500–700 nM) when NADPH oxidases generate $O_2^{\bullet-}$ or H_2O_2 (Stocker et al. 2018a). The current models suggests that cytosolic H_2O_2 is limited to a 0.3 μm gradient radiating from the enzymatic ROS source due to the high reactivity of adjacently localized cytoplasmic PRDXs and GPXs (Travasso et al. 2017). In these conditions reduced PRDXs or GPXs which are present in abundance will be oxidized and will facilitate indirectly thiol oxidation of target proteins. Quantitative modeling supports that often not H_2O_2 itself but sulfenic or disulfide forms of 2-Cys peroxiredoxins may oxidize targets in a localized redox relay (Travasso et al. 2017). Redox reactive cysteines in proteins will be oxidized not directly by H_2O_2 but indirectly by PRDXs. Rhee and coworkers reported that PRDX3-SO_2/sulfiredoxin oscillations enable rhythmic release of H_2O_2 from the mitochondria (Rhee and Kil 2016). Released mtH_2O_2 has been connected to immune signaling, autophagy, and cell cycle regulation among other cellular functions. At higher H_2O_2 concentrations, the process of sensing H_2O_2 and transferring oxidative equivalents via a redox relay can collapse and H_2O_2 will accumulate, diffuse, and oxidize less reactive targets located further away (Travasso et al. 2017). Posttranslational modifications such as phosphorylations can inactivate PRDXs, thereby permitting local accumulation of H_2O_2 that induces cysteine oxidation (Woo et al. 2010). These scenarios were modeled as outside-in H_2O_2 flux, but they may also be accurate for intracellular release of H_2O_2 from redoxosomes or other compartments (Oakley et al. 2009; Zana et al. 2018). In plants and model organisms coordinated calcium and H_2O_2 waves have been postulated as communication system, transmitting information inside the cell or between cells (Gilroy et al. 2016; Vestergaard et al. 2012), but this has not been demonstrated convincingly in mammalian systems.

This model cannot account for gradient disturbances by concomitant or sequential H_2O_2 generation by multiple ROS sources. In those conditions H_2O_2 levels will dynamically fluctuate, and localized gradients will be reinforced or collapse. PRDX family members or other peroxidases in oxidized or non-oxidized form may associate with different signaling complexes, thereby inducing, stabilizing, or hindering translocation of signaling platforms. Posttranslational modifications that rely not only on oxidation but also on reactive nitrogen or sulfur species will likely be altered in such complex systems, and oxidation of suitable targets by highly reactive $O_2^{\bullet-}$ and/or $^{\bullet}OH$ may take place simultaneously. Cells in tissues will also integrate signals generated by cell-cell contact, by stimulated or apoptotic cells nearby, as well as signals transmitted by the physical microenvironment such as changes in the extracellular matrix or shear stress. The overall reaction to multiple stimuli in a complex system will necessitate the seamless integration of various oxygen species and other signaling modifiers. The robustness of such complex systems is usually ensured by redundancy and compensation. Perturbations of this system will not be tolerated if they occur persistently or exceed a certain threshold of physiological tolerability. In a first scenario, disturbances will have detrimental outcome for cellular integrity, viability, or both and will result in cell death, tissue injury, and disease. In other circumstances the response to vital signaling inputs will be muted or enhanced on a permanent basis, yet again leading to compromised cell function and disease. This second scenario occurs when inherited or de novo mutations in primary ROS-generating or ROS-degrading enzymes are present. Examples are loss-of-function or reduced function variants in NADPH oxidase isoforms (see 3.1–3.3). While gain-of-function variants in ROS sources are not yet identified, variants in other genes are coupled to continuously augmented primary ROS production.

Examples are *CCM1* linked to cerebral cavernous malformations, and *PTPN22* or *PTPN11*, which are associated with autoimmune disease and myeloproliferative disorders, respectively (Bayley et al. 2015; Xu et al. 2013; Li et al. 2015; Goitre et al. 2014). Thus, ROS levels in a certain, predefined physiological range are required to maintain health.

2.2 Redox Signaling Pathways

The first step in H_2O_2-elicited redox signaling involves oxidation of a thiol group to a disulfide group on the target protein, which will often change the function of the protein (Winterbourn 2008; Finkel 2011). The reaction of H_2O_2 with cysteines is dependent on the overall microenvironment of the thiol and on the pK_a, as the reaction occurs exclusively with the thiolate anion (Fig. 2). These parameters together with accessibility and reaction rate contribute to the specificity of modifying particular thiols. The initially formed product is usually sulfenic acid (or in some cases thiyl radical), which can react with glutathione (GSH) to form disulfide bonds or can modify proteins by glutathionylation or sulfenylamide formation. The reversibility of thiol modifications is accomplished by PRDX, GPX, or thioredoxin reductase (TRX). While direct oxidation of protein thiols by high concentrations

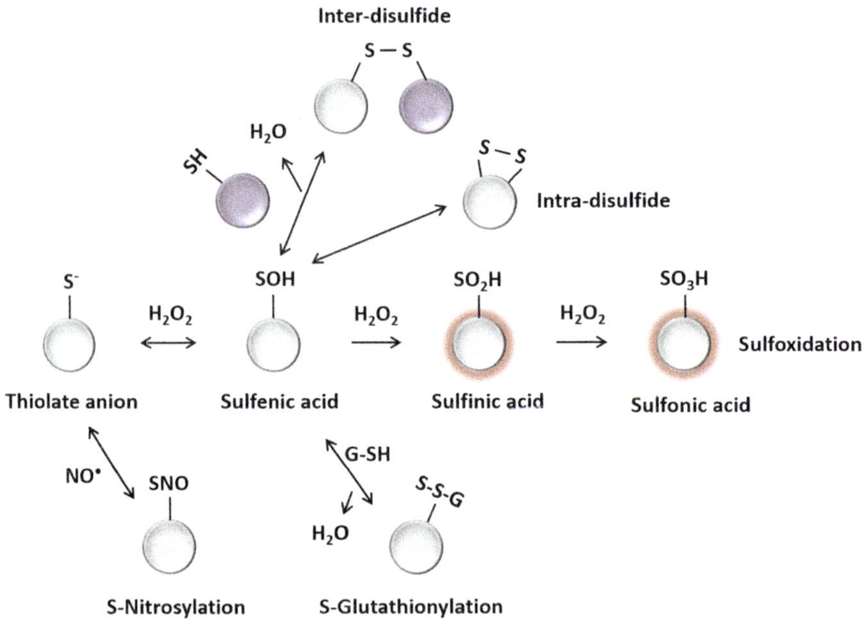

Fig. 2 Redox-dependent cysteine modifications as cellular signal transducers. Reversible and irreversible (red glow) cysteine modifications by hydrogen peroxide or nitric oxide, leading to sulfoxidation, S-nitrosylation, S-glutathionylation, and disulfide bond formation of target proteins. Modified from J. Cell Biol. 2011, 194:7

of H_2O_2 can occur, the favored model for physiological signaling is the redox relay, whereby H_2O_2 reacts readily with cytosolic 2-Cys peroxiredoxins, which transmit oxidizing equivalents to target proteins (Stocker et al. 2018a; Travasso et al. 2017). The reaction rate of H_2O_2 with the active site cysteine in PRDX2 is a million-fold higher than with the phosphatase PTP1B, a target in oxidant signaling pathways (Paulsen and Carroll 2013). This indirect PRDX-mediated thiol oxidation was also reported for the oxidation of the apoptosis signal-regulating kinase 1 (ASK1) and the transcription factor STAT3 (Jarvis et al. 2012; Sobotta et al. 2015). Similarly, oxidation of a redox-sensitive cysteine close to the phosphate binding P-loop of RAC1 might be mediated by PRDX6, which binds to the NOX2 complex and enhances its catalytic activity (Hobbs et al. 2014; Ambruso et al. 2012).

A well-known example for H_2O_2 redox signaling is oxidation of the active site thiol in protein tyrosine phosphatases, which will inhibit their activity and thereby increase tyrosine phosphorylation on target receptors, protein kinases, and other proteins (Tonks 2013). Typically, such deactivation of phosphatases is transient and facilitates phosphorylation/dephosphorylation switches that trigger spatiotemporally controlled signaling. For epidermal growth factor receptor, vascular endothelial growth factor receptor, fibroblast growth factor receptor, protein kinase B (PKB, Akt), ASK1, c-Src, and inhibitory κB kinase (IKK), all receptor and non-receptor tyrosine kinases with redox-sensitive cysteine residues, activity depends on redox regulation (Truong and Carroll 2013). All of these kinases are subject to many regulatory inputs that control their context-dependent activation state. For example, H_2O_2-induced oxidation of specific c-Src cysteine residues promotes sulfenylation, glutathionylation, and disulfide bond formation, which will alter c-Src conformation, tyrosine phosphorylation, and kinase activity (Heppner et al. 2018; Zhang and Forman 2014; Giannoni et al. 2005). The redox activation of ASK1 is a multistep process that depends on oxidation of cysteines in TRX1 and ASK1, leading to dissociation of the inactive TRX1-ASK1 complex and conformational changes in ASK1 (Liu et al. 2000; Nadeau et al. 2007). Thiol oxidation will alter not only protein folding and interfere with protein-protein interactions of protein kinase cascades, but will modify also the activity of transcription factors or ion channels (Brigelius-Flohe and Flohe 2011; Bogeski and Niemeyer 2014).

3 Contribution of Physiological ROS to Health

Adequate generation of H_2O_2 for redox signaling and specialized biological processes is essential for maintaining cellular functions and, when disrupted permanently, will cause disease or greatly increase susceptibility for disease. We outline here only cell functions that when disturbed by permanently altered $O_2^{\bullet-}/H_2O_2$ generation cause clearly defined, clinical human disease phenotypes or are associated with increased susceptibility for human disease. These examples are based on genetic variants in the NADPH oxidase family of proteins, as variants in other ROS sources have not yet been identified in patients.

3.1 Innate Immune Response

Innate immune cells such as neutrophils, macrophages, and dendritic cells are critical for host defense. These cells exert antimicrobial effects, connect to other immune cell types, and are required for resolution of infectious and inflammatory events. Many activities of innate immune cells are redox-regulated, but in contrast to other cell types, ROS fulfill also specialized functions such as killing of microbes by products of stimulated NOX2. Mutational inactivation of NOX2 (*CYBB;* X-linked CGD) or in genes required for NOX2 complex formation (*CYBA*, *NCF1*, *NCF2*, *NCF4*; Autosomal Recessive-CGD) causes chronic granulomatous disease (CGD), an inherited immunodeficiency that leads to recurrent life-threatening fungal and bacterial infections in patients due to superoxide deficiency in innate immune cells (Roos 2016). The primary function of the NOX2-dependent phagocyte oxidative burst is pathogen control, in particular toward certain fungi and bacteria (*A. fumigatus*, *S. aureus*, *S. marcescens*, *Nocardia* spp., *B. cepacia*). In addition, NOX2-generated H_2O_2 participates in redox signaling, serving a protective, anti-inflammatory role that is essential for resolution of inflammation (Campbell et al. 2014). The immune cells of CGD patients produce significantly increased levels of chemokines and pro-inflammatory cytokines (IL-8, IL-6, TNF-α), resulting in inflammatory conditions such as granulomas and colitis (O'Neill et al. 2015; Nauseef 2019; Dinauer 2019). Lack of infection control, hyperinflammation, and development of skin granulomas are also present in Nox2-deficient mice. How NOX2-derived $O_2^{\bullet -}/H_2O_2$ alters specific transcriptional pathways, dampens inflammation, and contributes to resolution of inflammatory processes is not yet completely defined.

Current therapy includes aggressive diagnostic evaluation of clinical complaints, antibacterial and antifungal prophylaxis, and IFNγ-1b (ACTIMMUNE®). Curative treatment such as allogeneic hematopoietic stem cell transplantation (HSCT) has been used for certain CGD patients. Gene therapy trials with improved lentiviral vectors show promising results, and targeted repair studies using CRISPR gene editing technology on iPSC have commenced (De Ravin et al. 2017; Merling et al. 2017; Sweeney et al. 2019; Kohn et al. 2020). Finding new avenues for adjunctive CGD therapy may include activating a distinct primary ROS source which can partially compensate for NOX2 complex inactivation. A modest increase in primary ROS levels is likely sufficient to reach the threshold required for increased patient survival, because female X-CGD carriers retaining at least 10% of neutrophils oxidase functional are phenotypically normal (Kuhns et al. 2010; Marciano et al. 2018). Increasing primary ROS in CGD was recently undertaken by using the PPARγ agonist pioglitazone, a drug approved for Type 2 diabetes, in an animal model of CGD (*Cybb*$^{-/-}$ mice). Pioglitazone enhanced the oxidative environment in neutrophils, monocytes, and macrophages by increasing mtROS generation, enabling the killing of *Staphylococcus aureus* ex vivo and in vivo (Fernandez-Boyanapalli et al. 2015b). Pioglitazone treatment improved also efferocytosis, the engulfment and removal of dying cells, in monocytes derived from CGD patients and may thus suppress the enhanced production of pro-inflammatory mediators observed in CGD macrophages (Fernandez-Boyanapalli et al. 2015a). In Nox2-deficient mice, pioglitazone reversed sterile inflammation by enhancing clearance of neutrophils and reducing cytokine production (Fernandez-Boyanapalli et al. 2010). A Phase1/2 clinical trial investigating efficacy and safety of pioglitazone in children with CGD will be completed in 2020. Even if the frequency of infections declines, it will be too early to assess if pioglitazone or other PPARγ agonists will provide long-term benefit in CGD, but the overall approach seems feasible and promising.

3.2 Gastrointestinal Homeostasis

Effective homeostasis in the intestine provides the initial host defense against intestinal pathogens; this includes colonization resistance provided by microbiota, the physical epithelial barrier reinforced by selective permeability and mucus, and epithelial secretions that include proteinaceous substances, organic substances, and chemicals. One of these chemicals is H_2O_2 that is released into the intestinal lumen by the epithelial NADPH oxidases NOX1 and DUOX2 after sensing pathogens, pathogen attachment, or invasion (Aviello and Knaus 2018). These oxidases are less efficient in primary ROS generation than the phagocyte oxidase NOX2, but their activity is often sustained over a longer period. Even though the achievable H_2O_2 concentration in the intestine is too low to exert direct microbicidal activity, diffusion of nanomolar H_2O_2 into bacteria triggers several antivirulence mechanisms such as inhibiting transcription of the LEE pathogenicity island encoding the Type 3 secretion system of enteropathogenic *E. coli* or globally downregulating phosphotyrosine signaling, which governs polysaccharide biosynthesis, metabolic pathways, and virulence determinants (Corcionivoschi et al. 2012; Alvarez et al. 2016; Pircalabioru et al. 2016). Reducing virulence and weakening pathogen fitness facilitates more efficient clearance of the invading microorganisms. In the low oxygen environment of the colon (~3% oxygen at the tip of the villi), nanomolar H_2O_2 concentrations seem to be sufficient to provide initial protection against certain microorganisms.

In homeostatic conditions and, in particular, when the function of epithelial NADPH oxidases is downregulated or disrupted by mutation (Hayes et al. 2015; Parlato et al. 2017; Schwerd et al. 2018), supplementation with physiological concentrations of H_2O_2 will provide host benefit in the gastrointestinal tract. One can envision various approaches to provide H_2O_2 to the mucosa, but commonly certain strains of commensal bacteria, in particular lactobacilli, have been used as treatment for intestinal infections (Isolauri et al. 2002; Lievin-Le Moal and Servin 2014). Lactobacilli produce lactic acid, bacteriocins, and inorganic substances and secrete H_2O_2. With such an array of bioactive compounds, it is difficult to assess which of these compounds individually or in combination provides the observed health benefits in infections. Antibacterial activity of *Lactobacillus*-generated H_2O_2 has been observed when using culture supernatants with or without catalase addition (Reid 2008; Atassi and Servin 2010). Identification and deletion of the H_2O_2-

generating enzymes in *L. johnsonii* NCC533 linked H_2O_2 production conclusively to anti-infective mechanisms in mice (Hertzberger et al. 2014; Pircalabioru et al. 2016; Knaus et al. 2017). These observations further support the prophylactic intake of lactobacilli to strengthen colonization resistance and antivirulence mechanisms.

Intestinal inflammatory diseases have long been associated with oxidative damage, but substances with "antioxidant" activity show mixed performance in clinical trials of inflammatory bowel diseases (IBD) (Moura et al. 2015). The notion of increased superoxide levels initiating or perpetuating intestinal inflammation needs to be reevaluated as 40–50% of CGD patients with loss-of-function mutations in the NOX2 complex will develop CGD-IBD (Falcone and Holland 2019), and reduced function of the NADPH oxidases NOX1, NOX2, or DUOX2 predisposes to very early onset IBD (Huang et al. 2016; Hayes et al. 2015; Parlato et al. 2017; Dhillon et al. 2014; Stenke et al. 2019; Schwerd et al. 2018). Physiological levels of H_2O_2 are essential for epithelial barrier maintenance (e.g., mucus layer, autophagy, wound healing) and for communication with the microbiota (Leoni et al. 2013; Aviello and Knaus 2018; Aviello et al. 2019). Hence, enhancing luminal H_2O_2 concentrations provides not only benefit in intestinal infections, but also in inflammatory disease. Administration of *Lactobacillus reuteri* increased mucus thickness in mice by an unidentified mechanism (Ahl et al. 2016), while *Lactobacillus johnsonii*-derived H_2O_2 was required for accelerated tissue restitution and recovery from chemically induced colitis (Singh et al. 2018). However, lactobacilli have not always performed well in clinical studies, possibly due to the changes in mucus quantity and quality triggered by active inflammation, thereby decreasing mucus-associated attachment sites required for colonization. Additionally, in immunocompromised patients or UC patients with increased intestinal barrier permeability, high doses of lactobacilli have been connected to bacteremia (Vahabnezhad et al. 2013; Sherid et al. 2016). Another pro-oxidant therapeutic approach in IBD is administering the cytokine GM-CSF, which will stimulate and strengthen the host's immune system (Bilsborough et al. 2016). GM-CSF promotes differentiation and proliferation of myeloid cells, but also acts as priming agent for neutrophils, enhancing their oxidative burst (Egea et al. 2010; El-Benna et al. 2016). Secondary ROS generation by neutrophils is often considered destructive by leading to tissue injury, but neutrophils perform also protective and restitutive functions (Parkos 2016). Further, as mentioned earlier, 40–50% of CGD patients suffer from intestinal inflammation duc to reduced superoxide production. These conflicting observations clearly indicate that more mechanistic insight is needed.

3.3 Thyroid Hormone Synthesis

Thyroid hormone synthesis is dependent on iodine availability, basolateral iodide uptake by active transport via the sodium/iodide symporter (NIS), thyroperoxidase (TPO), H_2O_2, and the iodine acceptor protein thyroglobulin (TG). The production and storage of thyroid hormones take place in thyroid follicles, with TPO reducing H_2O_2 to H_2O, thereby oxidizing iodide to an iodinating form and attaching it to tyrosyl residues in TG. Initial iodination of TG leads to monoiodotyrosine (MIT) and diiodotyrosine (DIT), which then react with each other in a second TPO-H_2O_2-catalyzed coupling reaction to form T3 (triiodothyronine, MIT + DIT) and T4 (thyroxine, DIT+DIT). TG-bound T3 and T4 are stored and released into the bloodstream after TG cleavage. The essential H_2O_2 for iodide oxidation steps is provided by the calcium-activated NADPH oxidase DUOX2, which co-localizes with TPO. Apical thyrocyte membrane localization and catalytic activity of DUOX2 is dependent on dimerization with DUOXA2, glycosylation, and other posttranslational modifications (Rousset et al. 2000; Carre et al. 2015).

Loss-of-function variants in TPO or DUOX2 cause transient or permanent congenital hypothyroidism (CH), confirming the requirement of these enzymes in catalyzing thyroid hormone synthesis. A well-defined CH phenotype-genotype correlation, as suggested by initial characterization of *DUOX2* variants (Moreno et al. 2002), has not been observed after additional homozygous, heterozygous, and compound heterozygous mutations were identified and characterized (Ohye and Sugawara 2010; O'Neill et al. 2015; De Deken and Miot 2019; Peters et al. 2019; Dufort et al. 2019). The variety of clinical phenotypes associated with DUOX2 (and rarely with DUOXA2) mutations suggests that the related, thyrocyte-expressed DUOX1/DUOXA1 oxidase may compensate in some circumstances and that possibly other environmental factors modify the outcome. Mutational inactivation of Duox2 in mice ($Duox2^{thyd}$) or deletion of both dimerization partners ($Duoxa^{-/-}$ mice) caused severe CH with undetectable serum T4 levels (Grasberger et al. 2012; Donko et al. 2014), reinforcing the critical role of DUOX2 in thyroid hormone production. Recently, retrospective analysis of CH patients revealed a higher overall IBD prevalence, especially when transient CH was present, which may correlate with the presence of DUOX2 mutations (Grasberger et al. 2018). In contrast, very early onset IBD patients with rare inactivating *DUOX2* variants had normal thyroid hormone levels at birth (Hayes et al. 2015; Parlato et al. 2017), suggesting incomplete penetrance or tissue-specific factors that regulate DUOX2 expression.

4 Outlook

Cells in tissues exist in context with several primary ROS sources being triggered, in parallel or sequentially, by chemical, biological, and physical inputs. In these circumstances multiple H_2O_2 signals of different strength and duration will be generated, transmitted, received, converted, and removed. One can assume that physiological levels of H_2O_2 will persist in this cellular environment, not in a steady-state but fluctuating across a homeostatic range (Fig. 3). We propose that persistent ROS over- or underproduction permits homeostatic fluctuations only across a confined range. Short exposure to high ROS levels will be tolerated due to the actions of various protection, removal, and repair systems, but persistent exposure will trigger inappropriate signals and conversion to more damaging secondary ROS. In permanent low ROS scenarios, some redox signals will be sustained due to proximity or high affinity of targets, while others are disrupted. As the

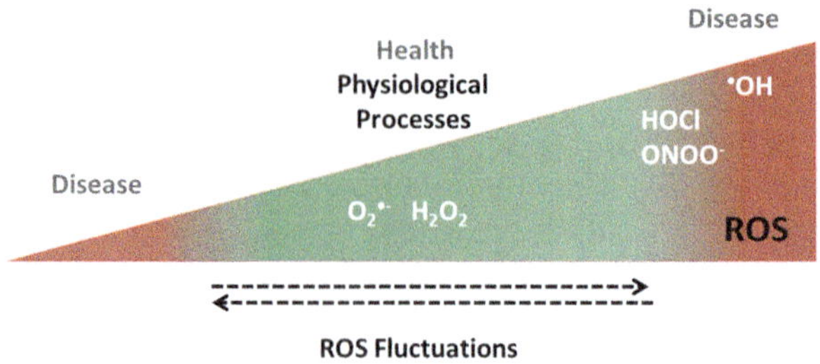

Fig. 3 Homeostatic ROS fluctuations in a multi-ROS source cellular setting. In physiological conditions, ROS levels will continuously change over a wide range (oxidative fluctuations), while disease states are characterized by permanently reduced ROS generation due to loss-of-function gene variants or drugs or by persistently high ROS levels with increased production of secondary ROS

remaining H_2O_2 still fulfills certain signaling functions, the system will have less pressure for counter regulation and compensation. Hence, persistently low ROS levels will result in deficiencies and physiological stress responses that may give rise to damaging long-term effects. Therapeutic intervention in diseases with a well-defined ROS over-/underproduction will necessitate strategies that safeguard physiological ROS generation and redox signaling. In certain diseases the identification of a critical enzymatic ROS source generating a particular unwanted signal will be required for therapeutic targeting, but in other settings the activation of alternative primary ROS-generating enzymes may compensate when H_2O_2 signaling is permanently downregulated. We propose that therapeutics for certain disease conditions in the future will include not only specific enzyme inhibitors but also drugs or biologicals that induce or supply H_2O_2 at physiological concentrations.

Acknowledgments Our work has been primarily supported by the National Institutes of Health (USA), Science Foundation Ireland, and the National Children's Research Center (both Ireland). I would like to thank William M. Nauseef, University of Iowa, for critical review, and colleagues in the NOX field and in EU-ROS, a European Network for Redox Biology Research, for stimulating discussions over the years.

References

Ahl D, Liu H, Schreiber O, Roos S, Phillipson M, Holm L (2016) Lactobacillus reuteri increases mucus thickness and ameliorates dextran sulphate sodium-induced colitis in mice. Acta Physiol 217(4):300–310. https://doi.org/10.1111/apha.12695

Alvarez LA, Kovacic L, Rodriguez J, Gosemann JH, Kubica M, Pircalabioru GG, Friedmacher F, Cean A, Ghise A, Sarandan MB, Puri P, Daff S, Plettner E, von Kriegsheim A, Bourke B, Knaus UG (2016) NADPH oxidase-derived H2O2 subverts pathogen signaling by oxidative phosphotyrosine conversion to PB-DOPA. Proc Natl Acad Sci U S A 113(37):10406–10411. https://doi.org/10.1073/pnas.1605443113

Ambruso DR, Ellison MA, Thurman GW, Leto TL (2012) Peroxiredoxin 6 translocates to the plasma membrane during neutrophil activation and is required for optimal NADPH oxidase activity. Biochim Biophys Acta 1823(2):306–315. https://doi.org/10.1016/j.bbamcr.2011.11.014

Atassi F, Servin AL (2010) Individual and co-operative roles of lactic acid and hydrogen peroxide in the killing activity of enteric strain Lactobacillus johnsonii NCC933 and vaginal strain Lactobacillus gasseri KS120.1 against enteric, uropathogenic and vaginosis-associated pathogens. FEMS Microbiol Lett 304(1):29–38. https://doi.org/10.1111/j.1574-6968.2009.01887.x

Augsburger F, Filippova A, Rasti D, Seredenina T, Lam M, Maghzal G, Mahiout Z, Jansen-Durr P, Knaus UG, Doroshow J, Stocker R, Krause KH, Jaquet V (2019) Pharmacological characterization of the seven human NOX isoforms and their inhibitors. Redox Biol 26:101272. https://doi.org/10.1016/j.redox.2019.101272

Aviello G, Knaus UG (2018) NADPH oxidases and ROS signaling in the gastrointestinal tract. Mucosal Immunol 11(4):1011–1023. https://doi.org/10.1038/s41385-018-0021-8

Aviello G, Singh AK, O'Neill S, Conroy E, Gallagher W, D'Agostino G, Walker AW, Bourke B, Scholz D, Knaus UG (2019) Colitis susceptibility in mice with reactive oxygen species deficiency is mediated by mucus barrier and immune defense defects. Mucosal Immunol 12(6):1316–1326. https://doi.org/10.1038/s41385-019-0205-x

Bayley R, Kite KA, McGettrick HM, Smith JP, Kitas GD, Buckley CD, Young SP (2015) The autoimmune-associated genetic variant PTPN22 R620W enhances neutrophil activation and function in patients with rheumatoid arthritis and healthy individuals. Ann Rheum Dis 74(8):1588–1595. https://doi.org/10.1136/annrheumdis-2013-204796

Bedard K, Krause KH (2007) The NOX family of ROS-generating NADPH oxidases: physiology and pathophysiology. Physiol Rev 87(1):245–313. 87/1/245 [pii]. https://doi.org/10.1152/physrev.00044.2005

Bilsborough J, Targan SR, Snapper SB (2016) Therapeutic targets in inflammatory bowel disease: current and future. Am J Gastroenterol Suppl 3(3):27–37. https://doi.org/10.1038/ajgsup.2016.18

Bogeski I, Niemeyer BA (2014) Redox regulation of ion channels. Antioxid Redox Signal 21(6):859–862. https://doi.org/10.1089/ars.2014.6019

Brigelius-Flohe R, Flohe L (2011) Basic principles and emerging concepts in the redox control of transcription factors. Antioxid Redox Signal 15(8):2335–2381. https://doi.org/10.1089/ars.2010.3534

Brigelius-Flohe R, Maiorino M (2013) Glutathione peroxidases. Biochim Biophys Acta 1830(5):3289–3303. https://doi.org/10.1016/j.bbagen.2012.11.020

Campbell EL, Bruyninckx WJ, Kelly CJ, Glover LE, McNamee EN, Bowers BE, Bayless AJ, Scully M, Saeedi BJ, Golden-Mason L, Ehrentraut SF, Curtis VF, Burgess A, Garvey JF, Sorensen A, Nemenoff R, Jedlicka P, Taylor CT, Kominsky DJ, Colgan SP (2014) Transmigrating neutrophils shape the mucosal microenvironment through localized oxygen depletion to influence resolution of inflammation. Immunity 40(1):66–77. https://doi.org/10.1016/j.immuni.2013.11.020

Carre A, Louzada RA, Fortunato RS, Ameziane-El-Hassani R, Morand S, Ogryzko V, de Carvalho DP, Grasberger H, Leto TL, Dupuy C (2015) When an intramolecular disulfide bridge governs the interaction of DUOX2 with its partner DUOXA2. Antioxid Redox Signal 23(9):724–733. https://doi.org/10.1089/ars.2015.6265

Collins Y, Chouchani ET, James AM, Menger KE, Cocheme HM, Murphy MP (2012) Mitochondrial redox signalling at a glance. J Cell Sci 125(Pt 4):801–806. https://doi.org/10.1242/jcs.098475

Corcionivoschi N, Alvarez LA, Sharp TH, Strengert M, Alemka A, Mantell J, Verkade P, Knaus UG, Bourke B (2012) Mucosal reactive oxygen species decrease virulence by disrupting campylobacter jejuni phosphotyrosine signaling. Cell Host Microbe 12(1):47–59. https://doi.org/10.1016/j.chom.2012.05.018

Cross AR, Segal AW (2004) The NADPH oxidase of professional phagocytes--prototype of the NOX electron transport chain systems. Biochim Biophys Acta 1657(1):1–22. https://doi.org/10.1016/j.bbabio.2004.03.008. S0005272804000556 [pii]

De Deken X, Miot F (2019) DUOX defects and their roles in congenital hypothyroidism. Methods Mol Biol 1982:667–693. https://doi.org/10.1007/978-1-4939-9424-3_37

De Ravin SS, Li L, Wu X, Choi U, Allen C, Koontz S, Lee J, Theobald-Whiting N, Chu J, Garofalo M, Sweeney C, Kardava L, Moir S, Viley A, Natarajan P, Su L, Kuhns D, Zarember KA, Peshwa MV, Malech HL (2017) CRISPR-Cas9 gene repair of hematopoietic stem cells from patients with X-linked chronic granulomatous disease. Sci Transl Med 9(372):eaah3480. https://doi.org/10.1126/scitranslmed.aah3480

Decoursey TE, Ligeti E (2005) Regulation and termination of NADPH oxidase activity. Cell Mol Life Sci 62(19–20):2173–2193. https://doi.org/10.1007/s00018-005-5177-1

Dhillon SS, Fattouh R, Elkadri A, Xu W, Murchie R, Walters T, Guo C, Mack D, Huynh HQ, Baksh S, Silverberg MS, Griffiths AM, Snapper SB, Brumell JH, Muise AM (2014) Variants in nicotinamide adenine dinucleotide phosphate oxidase complex components determine susceptibility to very early onset inflammatory bowel disease. Gastroenterology 147(3):680–689 e682. https://doi.org/10.1053/j.gastro.2014.06.005

Dinauer MC (2019) Inflammatory consequences of inherited disorders affecting neutrophil function. Blood 133(20):2130–2139. https://doi.org/10.1182/blood-2018-11-844563

Donko A, Morand S, Korzeniowska A, Boudreau HE, Zana M, Hunyady L, Geiszt M, Leto TL (2014) Hypothyroidism-associated missense mutation impairs NADPH oxidase activity and intracellular trafficking of Duox2. Free Radic Biol Med 73:190–200. https://doi.org/10.1016/j.freeradbiomed.2014.05.006

Dufort G, Larrivee-Vanier S, Eugene D, De Deken X, Seebauer B, Heinimann K, Levesque S, Gravel S, Szinnai G, Van Vliet G, Deladoey J (2019) Wide spectrum of DUOX2 deficiency: from life-threatening compressive goiter in infancy to lifelong euthyroidism. Thyroid 29(7):1018–1022. https://doi.org/10.1089/thy.2018.0461

Egea L, Hirata Y, Kagnoff MF (2010) GM-CSF: a role in immune and inflammatory reactions in the intestine. Expert Rev Gastroenterol Hepatol 4(6):723–731. https://doi.org/10.1586/egh.10.73

El-Benna J, Hurtado-Nedelec M, Marzaioli V, Marie JC, Gougerot-Pocidalo MA, Dang PM (2016) Priming of the neutrophil respiratory burst: role in host defense and inflammation. Immunol Rev 273(1):180–193. https://doi.org/10.1111/imr.12447

Falcone EL, Holland SM (2019) Gastrointestinal complications in chronic granulomatous disease. Methods Mol Biol 1982:573–586. https://doi.org/10.1007/978-1-4939-9424-3_34

Fernandez-Boyanapalli R, Frasch SC, Riches DW, Vandivier RW, Henson PM, Bratton DL (2010) PPARgamma activation normalizes resolution of acute sterile inflammation in murine chronic granulomatous disease. Blood 116(22):4512–4522. https://doi.org/10.1182/blood-2010-02-272005

Fernandez-Boyanapalli RF, Falcone EL, Zerbe CS, Marciano BE, Frasch SC, Henson PM, Holland SM, Bratton DL (2015a) Impaired efferocytosis in human chronic granulomatous disease is reversed by pioglitazone treatment. J Allergy Clin Immunol 136(5):1399–1401 e1393. https://doi.org/10.1016/j.jaci.2015.07.034

Fernandez-Boyanapalli RF, Frasch SC, Thomas SM, Malcolm KC, Nicks M, Harbeck RJ, Jakubzick CV, Nemenoff R, Henson PM, Holland SM, Bratton DL (2015b) Pioglitazone restores phagocyte mitochondrial oxidants and bactericidal capacity in chronic granulomatous disease. J Allergy Clin Immunol 135(2):517–527 e512. https://doi.org/10.1016/j.jaci.2014.10.034

Finkel T (2011) Signal transduction by reactive oxygen species. J Cell Biol 194(1):7–15. https://doi.org/10.1083/jcb.201102095

Giannoni E, Buricchi F, Raugei G, Ramponi G, Chiarugi P (2005) Intracellular reactive oxygen species activate Src tyrosine kinase during cell adhesion and anchorage-dependent cell growth. Mol Cell Biol 25(15):6391–6403. https://doi.org/10.1128/MCB.25.15.6391-6403.2005

Gilroy S, Bialasek M, Suzuki N, Gorecka M, Devireddy AR, Karpinski S, Mittler R (2016) ROS, calcium, and electric signals: key mediators of rapid systemic signaling in plants. Plant Physiol 171(3):1606–1615. https://doi.org/10.1104/pp.16.00434

Giridharan SS, Caplan S (2014) MICAL-family proteins: complex regulators of the actin cytoskeleton. Antioxid Redox Signal 20(13):2059–2073. https://doi.org/10.1089/ars.2013.5487

Goitre L, De Luca E, Braggion S, Trapani E, Guglielmotto M, Biasi F, Forni M, Moglia A, Trabalzini L, Retta SF (2014) KRIT1 loss of function causes a ROS-dependent upregulation of c-Jun. Free Radic Biol Med 68:134–147. https://doi.org/10.1016/j.freeradbiomed.2013.11.020

Grasberger H, De Deken X, Mayo OB, Raad H, Weiss M, Liao XH, Refetoff S (2012) Mice deficient in dual oxidase maturation factors are severely hypothyroid. Mol Endocrinol 26 (3):481–492. https://doi.org/10.1210/me.2011-1320

Grasberger H, Noureldin M, Kao TD, Adler J, Lee JM, Bishu S, El-Zaatari M, Kao JY, Waljee AK (2018) Increased risk for inflammatory bowel disease in congenital hypothyroidism supports the existence of a shared susceptibility factor. Sci Rep 8(1):10158. https://doi.org/10.1038/s41598-018-28586-5

Halliwell B (2006) Reactive species and antioxidants. Redox biology is a fundamental theme of aerobic life. Plant Physiol 141(2):312–322. https://doi.org/10.1104/pp.106.077073

Hayes P, Dhillon S, O'Neill K, Thoeni C, Hui KY, Elkadri A, Guo CH, Kovacic L, Aviello G, Alvarez LA, Griffiths AM, Snapper SB, Brant SR, Doroshow JH, Silverberg MS, Peter I, McGovern DP, Cho J, Brumell JH, Uhlig HH, Bourke B, Muise AA, Knaus UG (2015) Defects in NADPH oxidase genes NOX1 and DUOX2 in very early onset inflammatory bowel disease. Cell Mol Gastroenterol Hepatol 1(5):489–502. https://doi.org/10.1016/j.jcmgh.2015.06.005

Heppner DE, Dustin CM, Liao C, Hristova M, Veith C, Little AC, Ahlers BA, White SL, Deng B, Lam YW, Li J, van der Vliet A (2018) Direct cysteine sulfenylation drives activation of the Src kinase. Nat Commun 9(1):4522. https://doi.org/10.1038/s41467-018-06790-1

Hertzberger R, Arents J, Dekker HL, Pridmore RD, Gysler C, Kleerebezem M, de Mattos MJ (2014) H(2)O(2) production in species of the Lactobacillus acidophilus group: a central role for a novel NADH-dependent flavin reductase. Appl Environ Microbiol 80(7):2229–2239. https://doi.org/10.1128/AEM.04272-13

Hobbs GA, Zhou B, Cox AD, Campbell SL (2014) Rho GTPases, oxidation, and cell redox control. Small GTPases 5:e28579. https://doi.org/10.4161/sgtp.28579

Huang C, De Ravin SS, Paul AR, Heller T, Ho N, Wu Datta L, Zerbe CS, Marciano BE, Kuhns DB, Kader HA, Holland SM, Malech HL, Brant SR, Consortium NIG (2016) Genetic risk for inflammatory bowel disease is a determinant of crohn's disease development in chronic granulomatous disease. Inflamm Bowel Dis 22(12):2794–2801. https://doi.org/10.1097/MIB.0000000000000966

Hung RJ, Pak CW, Terman JR (2011) Direct redox regulation of F-actin assembly and disassembly by Mical. Science 334(6063):1710–1713. https://doi.org/10.1126/science.1211956

Isolauri E, Kirjavainen PV, Salminen S (2002) Probiotics: a role in the treatment of intestinal infection and inflammation? Gut 50(Suppl 3):III54–III59

Jarvis RM, Hughes SM, Ledgerwood EC (2012) Peroxiredoxin 1 functions as a signal peroxidase to receive, transduce, and transmit peroxide signals in mammalian cells. Free Radic Biol Med 53 (7):1522–1530. https://doi.org/10.1016/j.freeradbiomed.2012.08.001

Kawahara T, Quinn MT, Lambeth JD (2007) Molecular evolution of the reactive oxygen-generating NADPH oxidase (Nox/Duox) family of enzymes. BMC Evol Biol 7:109. 1471-2148-7-109 [pii]. https://doi.org/10.1186/1471-2148-7-109

Knaus UG, Hertzberger R, Pircalabioru GG, Yousefi SP, Branco Dos Santos F (2017) Pathogen control at the intestinal mucosa – H2O2 to the rescue. Gut Microbes 8(1):67–74. https://doi.org/10.1080/19490976.2017.1279378

Kohn DB, Booth C, Kang EM, Pai SY, Shaw KL, Santilli G, Armant M, Buckland KF, Choi U, De Ravin SS, Dorsey MJ, Kuo CY, Leon-Rico D, Rivat C, Izotova N, Gilmour K, Snell K, Dip JX, Darwish J, Morris EC, Terrazas D, Wang LD, Bauser CA, Paprotka T, Kuhns DB, Gregg J,

Raymond HE, Everett JK, Honnet G, Biasco L, Newburger PE, Bushman FD, Grez M, Gaspar HB, Williams DA, Malech HL, Galy A, Thrasher AJ, Net CGDc (2020) Lentiviral gene therapy for X-linked chronic granulomatous disease. Nat Med 26(2):200–206. https://doi.org/10.1038/s41591-019-0735-5

Kuhns DB, Alvord WG, Heller T, Feld JJ, Pike KM, Marciano BE, Uzel G, DeRavin SS, Priel DA, Soule BP, Zarember KA, Malech HL, Holland SM, Gallin JI (2010) Residual NADPH oxidase and survival in chronic granulomatous disease. N Engl J Med 363(27):2600–2610. https://doi.org/10.1056/NEJMoa1007097

Leoni G, Alam A, Neumann PA, Lambeth JD, Cheng G, McCoy J, Hilgarth RS, Kundu K, Murthy N, Kusters D, Reutelingsperger C, Perretti M, Parkos CA, Neish AS, Nusrat A (2013) Annexin A1, formyl peptide receptor, and NOX1 orchestrate epithelial repair. J Clin Invest 123 (1):443–454. https://doi.org/10.1172/JCI65831

Li XJ, Goodwin CB, Nabinger SC, Richine BM, Yang Z, Hanenberg H, Ohnishi H, Matozaki T, Feng GS, Chan RJ (2015) Protein-tyrosine phosphatase Shp2 positively regulates macrophage oxidative burst. J Biol Chem 290(7):3894–3909. https://doi.org/10.1074/jbc.M114.614057

Lievin-Le Moal V, Servin AL (2014) Anti-infective activities of lactobacillus strains in the human intestinal microbiota: from probiotics to gastrointestinal anti-infectious biotherapeutic agents. Clin Microbiol Rev 27(2):167–199. https://doi.org/10.1128/CMR.00080-13

Liu H, Nishitoh H, Ichijo H, Kyriakis JM (2000) Activation of apoptosis signal-regulating kinase 1 (ASK1) by tumor necrosis factor receptor-associated factor 2 requires prior dissociation of the ASK1 inhibitor thioredoxin. Mol Cell Biol 20(6):2198–2208. https://doi.org/10.1128/mcb.20.6.2198-2208.2000

Marciano BE, Zerbe CS, Falcone EL, Ding L, DeRavin SS, Daub J, Kreuzburg S, Yockey L, Hunsberger S, Foruraghi L, Barnhart LA, Matharu K, Anderson V, Darnell DN, Frein C, Fink DL, Lau KP, Long Priel DA, Gallin JI, Malech HL, Uzel G, Freeman AF, Kuhns DB, Rosenzweig SD, Holland SM (2018) X-linked carriers of chronic granulomatous disease: illness, lyonization, and stability. J Allergy Clin Immunol 141(1):365–371. https://doi.org/10.1016/j.jaci.2017.04.035

McElroy GS, Chandel NS (2017) Mitochondria control acute and chronic responses to hypoxia. Exp Cell Res 356(2):217–222. https://doi.org/10.1016/j.yexcr.2017.03.034

Merling RK, Kuhns DB, Sweeney CL, Wu X, Burkett S, Chu J, Lee J, Koontz S, Di Pasquale G, Afione SA, Chiorini JA, Kang EM, Choi U, De Ravin SS, Malech HL (2017) Gene-edited pseudogene resurrection corrects p47(phox)-deficient chronic granulomatous disease. Blood Adv 1(4):270–278. https://doi.org/10.1182/bloodadvances.2016001214

Moreno JC, Bikker H, Kempers MJ, van Trotsenburg AS, Baas F, de Vijlder JJ, Vulsma T, Ris-Stalpers C (2002) Inactivating mutations in the gene for thyroid oxidase 2 (THOX2) and congenital hypothyroidism. N Engl J Med 347(2):95–102. https://doi.org/10.1056/NEJMoa012752

Moura FA, de Andrade KQ, dos Santos JC, Araujo OR, Goulart MO (2015) Antioxidant therapy for treatment of inflammatory bowel disease: does it work? Redox Biol 6:617–639. https://doi.org/10.1016/j.redox.2015.10.006

Nadeau PJ, Charette SJ, Toledano MB, Landry J (2007) Disulfide bond-mediated multimerization of Ask1 and its reduction by thioredoxin-1 regulate H(2)O(2)-induced c-Jun NH(2)-terminal kinase activation and apoptosis. Mol Biol Cell 18(10):3903–3913. https://doi.org/10.1091/mbc.e07-05-0491

Nadella M, Bianchet MA, Gabelli SB, Barrila J, Amzel LM (2005) Structure and activity of the axon guidance protein MICAL. Proc Natl Acad Sci U S A 102(46):16830–16835. https://doi.org/10.1073/pnas.0504838102

Nauseef WM (2019) The phagocyte NOX2 NADPH oxidase in microbial killing and cell signaling. Curr Opin Immunol 60:130–140. https://doi.org/10.1016/j.coi.2019.05.006

O'Neill S, Brault J, Stasia MJ, Knaus UG (2015) Genetic disorders coupled to ROS deficiency. Redox Biol 6:135–156. https://doi.org/10.1016/j.redox.2015.07.009

Oakley FD, Abbott D, Li Q, Engelhardt JF (2009) Signaling components of redox active endosomes: the redoxosomes. Antioxid Redox Signal 11(6):1313–1333. https://doi.org/10.1089/ARS.2008.2363

Ohye H, Sugawara M (2010) Dual oxidase, hydrogen peroxide and thyroid diseases. Exp Biol Med (Maywood) 235(4):424–433. https://doi.org/10.1258/ebm.2009.009241

Park J, Lee S, Lee S, Kang SW (2014) 2-cys peroxiredoxins: emerging hubs determining redox dependency of mammalian signaling networks. Int J Cell Biol 2014:715867. https://doi.org/10.1155/2014/715867

Parkos CA (2016) Neutrophil-epithelial interactions: a double-edged sword. Am J Pathol 186 (6):1404–1416. https://doi.org/10.1016/j.ajpath.2016.02.001

Parlato M, Charbit-Henrion F, Hayes P, Tiberti A, Aloi M, Cucchiara S, Begue B, Bras M, Pouliet A, Rakotobe S, Ruemmele F, Knaus UG, Cerf-Bensussan N (2017) First identification of biallelic inherited DUOX2 inactivating mutations as a cause of very early onset inflammatory bowel disease. Gastroenterology 153(2):609–611 e603. https://doi.org/10.1053/j.gastro.2016.12.053

Paulsen CE, Carroll KS (2013) Cysteine-mediated redox signaling: chemistry, biology, and tools for discovery. Chem Rev 113(7):4633–4679. https://doi.org/10.1021/cr300163e

Peters C, Nicholas AK, Schoenmakers E, Lyons G, Langham S, Serra EG, Sebire NJ, Muzza M, Fugazzola L, Schoenmakers N (2019) DUOX2/DUOXA2 mutations frequently cause congenital hypothyroidism that evades detection on newborn screening in the United Kingdom. Thyroid 29(6):790–801. https://doi.org/10.1089/thy.2018.0587

Pircalabioru G, Aviello G, Kubica M, Zhdanov A, Paclet MH, Brennan L, Hertzberger R, Papkovsky D, Bourke B, Knaus UG (2016) Defensive mutualism rescues NADPH oxidase inactivation in gut infection. Cell Host Microbe 19(5):651–663. https://doi.org/10.1016/j.chom.2016.04.007

Poole LB, Nelson KJ (2016) Distribution and features of the six classes of peroxiredoxins. Mol Cells 39(1):53–59. https://doi.org/10.14348/molcells.2016.2330

Pryor WA (1986) Oxy-radicals and related species: their formation, lifetimes, and reactions. Annu Rev Physiol 48:657–667. https://doi.org/10.1146/annurev.ph.48.030186.003301

Radi R (2018) Oxygen radicals, nitric oxide, and peroxynitrite: redox pathways in molecular medicine. Proc Natl Acad Sci U S A 115(23):5839–5848. https://doi.org/10.1073/pnas.1804932115

Randall LM, Manta B, Hugo M, Gil M, Batthyany C, Trujillo M, Poole LB, Denicola A (2014) Nitration transforms a sensitive peroxiredoxin 2 into a more active and robust peroxidase. J Biol Chem 289(22):15536–15543. https://doi.org/10.1074/jbc.M113.539213

Reid G (2008) Probiotic lactobacilli for urogenital health in women. J Clin Gastroenterol 42(Suppl 3 Pt 2):S234–S236. https://doi.org/10.1097/MCG.0b013e31817f1298

Rhee SG, Kil IS (2016) Mitochondrial H2O2 signaling is controlled by the concerted action of peroxiredoxin III and sulfiredoxin: linking mitochondrial function to circadian rhythm. Free Radic Biol Med 99:120–127. https://doi.org/10.1016/j.freeradbiomed.2016.07.029

Roos D (2016) Chronic granulomatous disease. Br Med Bull 118(1):50–63. https://doi.org/10.1093/bmb/ldw009

Rousset B, Dupuy C, Miot F, Dumont J (2000) Chapter 2 thyroid hormone synthesis and secretion. In: Feingold KR, Anawalt B, Boyce A et al (eds) Endotext. MDtext, South Dartmouth

Schieber M, Chandel NS (2014) ROS function in redox signaling and oxidative stress. Curr Biol 24 (10):R453–R462. https://doi.org/10.1016/j.cub.2014.03.034

Schwerd T, Bryant RV, Pandey S, Capitani M, Meran L, Cazier JB, Jung J, Mondal K, Parkes M, Mathew CG, Fiedler K, McCarthy DJ, Consortium WGS, Sullivan PB, Rodrigues A, Travis SPL, Moore C, Sambrook J, Ouwehand WH, Roberts DJ, Danesh J, Study I, Russell RK, Wilson DC, Kelsen JR, Cornall R, Denson LA, Kugathasan S, Knaus UG, Serra EG, Anderson CA, Duerr RH, McGovern DP, Cho J, Powrie F, Li VS, Muise AM, Uhlig HH, Oxford IBDcsi, investigators CiIg, Consortium UIG (2018) NOX1 loss-of-function genetic variants in patients with inflammatory bowel disease. Mucosal Immunol 11(2):562–574. https://doi.org/10.1038/mi.2017.74

Sherid M, Samo S, Sulaiman S, Husein H, Sifuentes H, Sridhar S (2016) Liver abscess and bacteremia caused by lactobacillus: role of probiotics? Case report and review of the literature. BMC Gastroenterol 16(1):138. https://doi.org/10.1186/s12876-016-0552-y

Sies H (1993) Strategies of antioxidant defense. Eur J Biochem 215(2):213–219

Singh AK, Hertzberger RY, Knaus UG (2018) Hydrogen peroxide production by lactobacilli promotes epithelial restitution during colitis. Redox Biol 16:11–20. https://doi.org/10.1016/j.redox.2018.02.003

Sobotta MC, Liou W, Stocker S, Talwar D, Oehler M, Ruppert T, Scharf AN, Dick TP (2015) Peroxiredoxin-2 and STAT3 form a redox relay for H2O2 signaling. Nat Chem Biol 11 (1):64–70. https://doi.org/10.1038/nchembio.1695

Stenke E, Bourke B, Knaus UG (2019) NAPDH oxidases in inflammatory bowel disease. Methods Mol Biol 1982:695–713. https://doi.org/10.1007/978-1-4939-9424-3_38

Stocker S, Maurer M, Ruppert T, Dick TP (2018a) A role for 2-Cys peroxiredoxins in facilitating cytosolic protein thiol oxidation. Nat Chem Biol 14(2):148–155. https://doi.org/10.1038/nchembio.2536

Stocker S, Van Laer K, Mijuskovic A, Dick TP (2018b) The conundrum of hydrogen peroxide signaling and the emerging role of peroxiredoxins as redox relay hubs. Antioxid Redox Signal 28(7):558–573. https://doi.org/10.1089/ars.2017.7162

Sumimoto H (2008) Structure, regulation and evolution of Nox-family NADPH oxidases that produce reactive oxygen species. FEBS J 275(13):3249–3277. https://doi.org/10.1111/j.1742-4658.2008.06488.x

Sweeney CL, Merling RK, De Ravin SS, Choi U, Malech HL (2019) Gene editing in chronic granulomatous disease. Methods Mol Biol 1982:623–665. https://doi.org/10.1007/978-1-4939-9424-3_36

Tonks NK (2013) Protein tyrosine phosphatases--from housekeeping enzymes to master regulators of signal transduction. FEBS J 280(2):346–378. https://doi.org/10.1111/febs.12077

Travasso RDM, Sampaio Dos Aidos F, Bayani A, Abranches P, Salvador A (2017) Localized redox relays as a privileged mode of cytoplasmic hydrogen peroxide signaling. Redox Biol 12:233–245. https://doi.org/10.1016/j.redox.2017.01.003

Truong TH, Carroll KS (2013) Redox regulation of protein kinases. Crit Rev Biochem Mol Biol 48 (4):332–356. https://doi.org/10.3109/10409238.2013.790873

Vahabnezhad E, Mochon AB, Wozniak LJ, Ziring DA (2013) Lactobacillus bacteremia associated with probiotic use in a pediatric patient with ulcerative colitis. J Clin Gastroenterol 47 (5):437–439. https://doi.org/10.1097/MCG.0b013e318279abf0

Vestergaard CL, Flyvbjerg H, Moller IM (2012) Intracellular signaling by diffusion: can waves of hydrogen peroxide transmit intracellular information in plant cells? Front Plant Sci 3:295. https://doi.org/10.3389/fpls.2012.00295

von Lohneysen K, Noack D, Wood MR, Friedman JS, Knaus UG (2010) Structural insights into Nox4 and Nox2: motifs involved in function and cellular localization. Mol Cell Biol 30 (4):961–975. https://doi.org/10.1128/MCB.01393-09

Winterbourn CC (2008) Reconciling the chemistry and biology of reactive oxygen species. Nat Chem Biol 4(5):278–286. https://doi.org/10.1038/nchembio.85

Winterbourn CC (2013) The biological chemistry of hydrogen peroxide. Methods Enzymol 528:3–25. https://doi.org/10.1016/B978-0-12-405881-1.00001-X

Winterbourn CC, Peskin AV (2016) Kinetic approaches to measuring peroxiredoxin reactivity. Mol Cells 39(1):26. https://doi.org/10.14348/molcells.2016.2325

Woo HA, Yim SH, Shin DH, Kang D, Yu DY, Rhee SG (2010) Inactivation of peroxiredoxin I by phosphorylation allows localized H(2)O(2) accumulation for cell signaling. Cell 140 (4):517–528. https://doi.org/10.1016/j.cell.2010.01.009

Xu D, Zheng H, Yu WM, Qu CK (2013) Activating mutations in protein tyrosine phosphatase Ptpn11 (Shp2) enhance reactive oxygen species production that contributes to myeloproliferative disorder. PLoS One 8(5):e63152. https://doi.org/10.1371/journal.pone.0063152
Zana M, Peterfi Z, Kovacs HA, Toth ZE, Enyedi B, Morel F, Paclet MH, Donko A, Morand S, Leto TL, Geiszt M (2018) Interaction between p22(phox) and Nox4 in the endoplasmic reticulum suggests a unique mechanism of NADPH oxidase complex formation. Free Radic Biol Med 116:41–49. https://doi.org/10.1016/j.freeradbiomed.2017.12.031
Zhang H, Forman HJ (2014) TGFbeta1 rapidly activates Src through a non-canonical redox mechanism. Free Radic Biol Med 75(Suppl 1):S4. https://doi.org/10.1016/j.freeradbiomed.2014.10.831
Zhou Y, An LL, Chaerkady R, Mittereder N, Clarke L, Cohen TS, Chen B, Hess S, Sims GP, Mustelin T (2018) Evidence for a direct link between PAD4-mediated citrullination and the oxidative burst in human neutrophils. Sci Rep 8(1):15228. https://doi.org/10.1038/s41598-018-33385-z

Network Medicine-Based Unbiased Disease Modules for Drug and Diagnostic Target Identification in ROSopathies

Cristian Nogales, Alexander G. B. Grønning, Sepideh Sadegh, Jan Baumbach, and Harald H. H. W. Schmidt

Contents

Cristian Nogales and Alexander G. B. Grønning contributed equally to this work.

C. Nogales (✉) · H. H. H. W. Schmidt
Department of Pharmacology and Personalised Medicine, Maastricht University, Maastricht, The Netherlands
e-mail: cnogales@ppmlab.net

A. G. B. Grønning
Department of Mathematics and Computer Science, University of Southern Denmark, Odense, Denmark

S. Sadegh
Chair of Experimental Bioinformatics, TUM School of Life Sciences, Technical University of Munich, Freising, Germany

J. Baumbach
Department of Mathematics and Computer Science, University of Southern Denmark, Odense, Denmark

Chair of Experimental Bioinformatics, TUM School of Life Sciences, Technical University of Munich, Freising, Germany

H. H. H. W. Schmidt et al. (eds.), *Reactive Oxygen Species*,
Handbook of Experimental Pharmacology 264, https://doi.org/10.1007/164_2020_386

Abstract

Most diseases are defined by a symptom, not a mechanism. Consequently, therapies remain symptomatic. In reverse, many potential disease mechanisms remain in arbitrary search for clinical relevance. Reactive oxygen species (ROS) are such an example. It is an attractive hypothesis that dysregulation of ROS can become a disease trigger. Indeed, elevated ROS levels of various biomarkers have been correlated with almost every disease, yet after decades of research without any therapeutic application. We here present a first systematic, non-hypothesis-based approach to transform this field as a proof of concept for biomedical research in general. We selected as seed proteins 9 families with 42 members of clinically researched ROS-generating enzymes, ROS-metabolizing enzymes or ROS targets. Applying an unbiased network medicine approach, their first neighbours were connected, and, based on a stringent subnet participation degree (SPD) of 0.4, hub nodes excluded. This resulted in 12 distinct human interactome-based ROS signalling modules, while 8 proteins remaining unconnected. This ROSome is in sharp contrast to commonly used highly curated and integrated KEGG, HMDB or WikiPathways. These latter serve more as mind maps of possible ROS signalling events but may lack important interactions and often do not take different cellular and subcellular localization into account. Moreover, novel non-ROS-related proteins were part of these forming functional hybrids, such as the NOX5/sGC, NOX1,2/NOS2, NRF2/ENC-1 and MPO/SP-A modules. Thus, ROS sources are not interchangeable but associated with distinct disease processes or not at all. Module members represent leads for precision diagnostics to stratify patients with specific ROSopathies for precision intervention.

Graphical Abstract

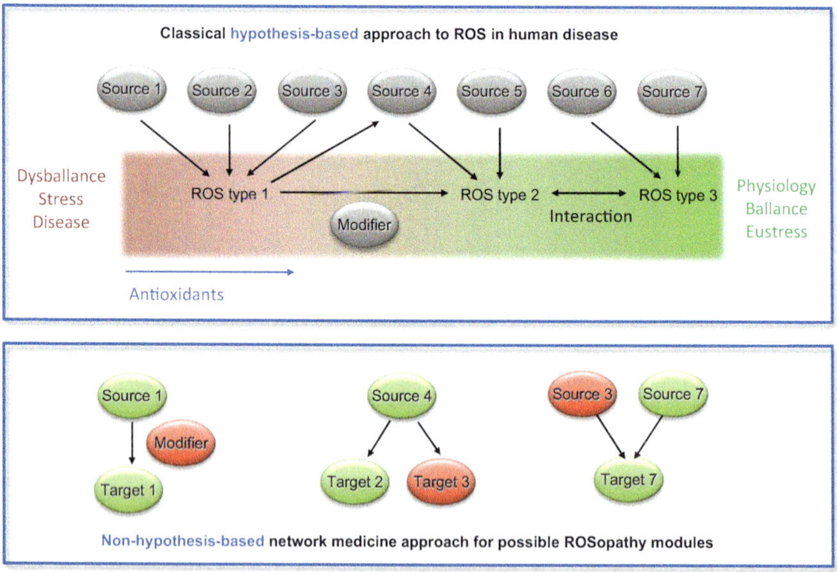

The upper panel shows the classical approach to generate hypotheses for a role of ROS in a given disease by focusing on ROS levels and to some degree the ROS type or metabolite. Low levels are considered physiological; higher amounts are thought to cause a redox imbalance, oxidative stress and eventually disease. The source of ROS is less relevant; there is also ROS-induced ROS formation, i.e. by secondary sources (see upwards arrow). The non-hypothesis-based network medicine approach uses genetically or otherwise validated risk genes to construct disease-relevant signalling modules, which will contain also ROS targets. Not all ROS sources will be relevant for a given disease; some may not be disease relevant at all. The three examples show (from left to right) the disease-relevant appearance of an unphysiological ROS modifier/toxifier protein, ROS target or ROS source.

Keywords

ROS · Systems medicine · Network pharmacology · Precision medicine · Precision diagnostics

There is a conceptual problem with the way most diseases are defined and consequently treated. Taxonomies are based on either (1) organs, e.g. heart failure diseases, chronic kidney disease or polyneuropathy; (2) the name of a doctor, e.g. Parkinson's disease, Alzheimer's disease or Crohn's disease; or (3) a symptom, e.g. hypertension, hyperthyroidism or asthma. One reason for this is that most diseases are mechanistically not understood. One result is that we have to treat symptoms, often chronically and not curative, and with low precision (i.e. high numbers needed to treat) when it comes to patient-relevant outcomes (i.e. mortality).

Having a closer look at hypertension, i.e. a blood pressure above a certain threshold level, one could even argue that it is not a disease but a risk factor. It is associated with 60% of all strokes and 50% of all cases of coronary heart disease, the primary cause of death and the second of disability worldwide (He and MacGregor 2007). Nevertheless, antihypertensive drugs, even if they normalize blood pressure – or remove the symptom – need high numbers needed to treat (NNT), namely, up to 81, to prevent a stroke or myocardial infarction. Thus, none of pharmacotherapeutic options are curative, effective or precise (Ogden et al. 2000). Moreover, some of the most popular antihypertensive drugs have a race bias. Hypertensive blacks treated with angiotensin-converting enzyme (ACE) inhibitors have poorer cardiovascular outcomes than whites (Ogedegbe et al. 2015). Precision medicine instead focuses on the response to therapy of the individual, not the group, reducing NNT by understanding and treating the precise pathological mechanisms (Schork 2015). Consequently current disease definitions will need to be replaced by mechanistic ones. Hypertension, like many other current "disease" definitions, will most likely be split up according to different mechanistic endotypes.

1 ROSopathies

One such possible mechanism involves reactive oxygen species (ROS). They are important for the maintenance of cellular homeostasis. If endogenous redox signalling is changed in an unphysiological manner, e.g. with respect quantity or type of ROS or its subcellular location, cellular and bodily dysfunctions can result (Frijhoff et al. 2015; Casas et al. 2015). Such ROS-associated disease states (ROSopathies) may range from hypertension and other cardiovascular pathologies (Gracia et al. 2017) to neurological disorders (Gray et al. 2013; Jha et al. 2014; Wilkinson-Berka et al. 2014) and chronic inflammation and cancer (Dröge 2002; Thanan et al. 2014), to name a few. These diverse diseases have in common aberrant redox signalling as a component in their development. However, correlation means not necessarily causality and establishing specific and the search for predictive ROS-associated biomarkers with clinical and therapeutic relevance has proven a fruitless endeavour (Frijhoff et al. 2015). Furthermore, therapeutic approaches with antioxidants, based on the concept that a little ROS is tolerable and above a certain threshold needs to be scavenged, have failed to provide patient benefit (Schmidt et al. 2015; Lapchak 2010). The conceptual mistake here was to ignore the fact that ROS at all levels of concentration can also have beneficial effects. Thus, in the ROS field, there is a massive need for change in our understanding of the roles of ROS in health and disease and how to treat ROS-associated conditions beyond the failed antioxidant approach.

A precision medicine approach to ROSopathies would be to identify more classical pharmacological protein targets, e.g. different enzymatic sources of ROS, to validate which of these is in a given condition a disease trigger, treat it but leave all other ROS sources untouched. Thereby dysfunctional ROS signalling would be cured, and physiological signalling could continue (Casas et al. 2015). In line with most recent therapeutic concepts are increasingly mechanistically orientated and target-specific ROS-forming, ROS-toxifying or ROS target enzymes (Altenhöfer et al. 2012; Hochman et al. 2007).

2 From Single Therapeutic Targets to Biased Pathways to Unbiased Modules

Another emerging concept in defining disease mechanisms is that with the exception of monogenic diseases, most common and chronic diseases are not caused by a single protein but a network around one or more candidates. Pathways, however, are currently defined in a highly curated manner resembling more mind maps than actually validated subcellular entities. Even though they are perpetuated from one review article to the next or because of that, they are biased, are highly curated, lack important interactions and do not integrate sufficient information about subcellular localizations of molecules. In general, biological networks can suffer from two types of biases: technical biases and study biases (Schaefer et al. 2015). Some technical biases are caused by experimental procedures which tend to detect interactions between highly abundant proteins (Björklund et al. 2008; von Mering et al. 2002; Ivanic et al. 2009), while others are inclined to detect interaction between proteins located in the nucleus (Jensen and Bork 2008). Study biases stem from overlapping interests of researchers, meaning that some proteins are studied more than others since they have shown higher biomedical and pharmaceutical relevance (Alanis-Lobato and Andrade-Navarro 2017; Schaefer et al. 2015). In line with this, it is becoming increasingly evident that high node degree (interaction partners) of, e.g. disease proteins is caused by study biases and not caused by them being essential elements in the interactome (Alanis-Lobato and Andrade-Navarro 2017). A recent study showed that a strong correlation exists between human protein-protein interaction (PPI) network centrality (number of physical interactions, betweenness, closeness) and number of times scientific publications mention protein representing genes (Alvarez-Ponce et al. 2017). Each protein can be regulated to suppress unphysiological formation of ROS. However, it remains unknown which proteins are relevant to target for the different ROSopathies and which of them are involved in the development of the disease. To locate potential drug targets, or combinations of drug targets, in an unbiased manner and to assess the impact of drug effects on molecular pathways, biological networks that accurately reflect cellular and pathological signalling are needed.

3 Hybrid Modules and Not All ROS-Related Proteins Will Be Disease Relevant

Moreover, pathways have again an artificial taxonomy, i.e. ROS signalling, cAMP signalling, Ca^{2+} signalling and tyrosine kinase signalling, as if they would exist in isolation, hardly interacting with each other, and in that totality, i.e. all components strictly interact with all other components. Typically, these sources, toxifiers and targets are summarized in mind map-like ROS signalling pathways (Bigarella et al. 2014; Bickers and Athar 2006; Yang et al. 2018; Zhao et al. 2017; Moloney and Cotter 2018) and databases such as KEGG, HMDB or WikiPathways. These often consider more or less one common cellular ROS pool or a cellular redox balance (see Fig. 1).

This assumption, however, is unlikely. Already, differential subcellular localization or differential expression in different cells will result in the fact that many protein components of one classical signalling pathway will never interact with each other; in reverse co-expression in specific subcellular compartments will result in the fact that signalling components from two or more pathways form a local hybrid interaction module, not captured in any curated pathway but possibly key to understand a disease mechanism. Thus, we need a new, unbiased manner how to construct these signalling modules. Network medicine does so by constructing them from the interactome, beginning with clinically highly validated seed proteins and extending them by their first neighbours, but with a limit to not exclude highly connected (unspecific) proteins, i.e. hubs. As a result, ROSopathy modules are highly likely to contain components of other classical signalling pathways.

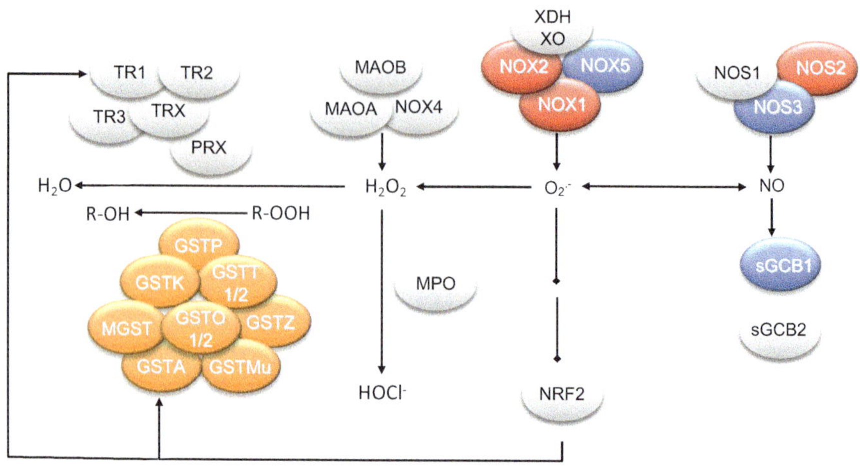

Fig. 1 Classical ROS signalling pathway. Perpetuated schematic representation of ROS signalling pathways. ROS enzymatic sources and targets are usually drawn interacting next to each other

Moreover, not all proteins qualify as disease proteins; some are too essential, and some are not essential at all. According to the Pareto principle, all networks can be modulated by 20% of its components, assuming that 20% of all proteins are critical and qualify as disease proteins. In reverse, we hypothesize that approximately 80% of all ROS-related proteins will not appear in a module or will not be disease relevant.

4 In Silico-Based Prediction of Modules for ROSopathies

To construct one or more ROS modules from protein-protein interaction (PPI) networks of relevant interactions, we selected as seed proteins 9 families with 42 members of clinically trialled or, ideally, validated ROS-generating enzymes, ROS-metabolizing enzymes or ROS target proteins. These included the ROS-generating enzymes nitric oxide synthase (NOS), monoamine oxidase (MAO), xanthine oxidase (XO), NADPH oxidase (NOX) and the ROS toxifier myeloperoxidase (MPO) (see Table 1). Because of lack of therapeutic relevance, mitochondria, for example, were excluded.

PPI experimental information was extracted from the Integrative Interaction Database (IID), which integrates human interaction data from different sources (Kotlyar et al. 2019). Selected disease-relevant ROS enzymes were used as seeds and all first neighbour interacting proteins in IID added to build an initial network with 765 nodes or proteins and 12,193 edges or PPIs. In order to extract disease-relevant signalling modules, a subnet participation degree (SPD) score was calculated for each protein in the network. The SPD score measures how enriched and specific are the interactions of a protein for the given subnetwork; it is defined as the

Table 1 List of seed proteins selected from 9 families with 42 members of clinically relevant researched ROS-generating enzymes, ROS-metabolizing enzymes or ROS targets

ROS-forming enzyme	UniProt entry name
NRF2	NF2L2_HUMAN
NOX1	NOX1_HUMAN
NOX2	CY24B_HUMAN
NOX3	NOX3_HUMAN
NOX4	NOX4_HUMAN
NOX5	NOX5_HUMAN
NOS1	NOS1_HUMAN
NOS2	NOS2_HUMAN
NOS3	NOS3_HUMAN
XO	XDH_HUMAN
MAOA	AOFA_HUMAN
MAOB	AOFB_HUMAN
MPO	PERM_HUMAN
ThR1	TRXR1_HUMAN
ThR2	TRXR2_HUMAN
ThR3	TRXR3_HUMAN
sGC	GCYA1_HUMAN
	GCYB1_HUMAN
	GCYA2_HUMAN
	GCYB2_HUMAN
GSTA1	GSTA1_HUMAN
GSTA2	GSTA2_HUMAN
GSTA3	GSTA3_HUMAN
GSTA4	GSTA4_HUMAN
GSTA5	GSTA5_HUMAN
GSTK1	GSTK1_HUMAN
GSTM1	GSTM1_HUMAN
GSTM2	GSTM2_HUMAN
GSTM3	GSTM3_HUMAN
GSTM4	GSTM4_HUMAN
GSTM5	GSTM5_HUMAN
GSTO1	GSTO1_HUMAN
GSTO2	GSTO2_HUMAN
GSTP1	GSTP1_HUMAN
GSTT1	GSTT1_HUMAN
GSTT2	GST2_HUMAN
GSTT2B?	GSTT2_HUMAN
GSTT4	GSTT4_HUMAN
GSTZ1	MAAI_HUMAN
MGST1	MGST1_HUMAN
MGST2	MGST2_HUMAN
MGST3	MGST3_HUMAN

degree or number of interactions of the protein nodes within the subnetwork normalized by the number of interactions of the node in the full interactome, i.e. IID. A stringent SPD cut-off value of 0.4 was selected, corresponding to 95% of the cumulative sum of the percentage of the protein nodes. Proteins nodes with a SPD score below 0.4 were excluded from the final subnetwork, isolating most module-specific interactions while excluding non-specific proteins. The final

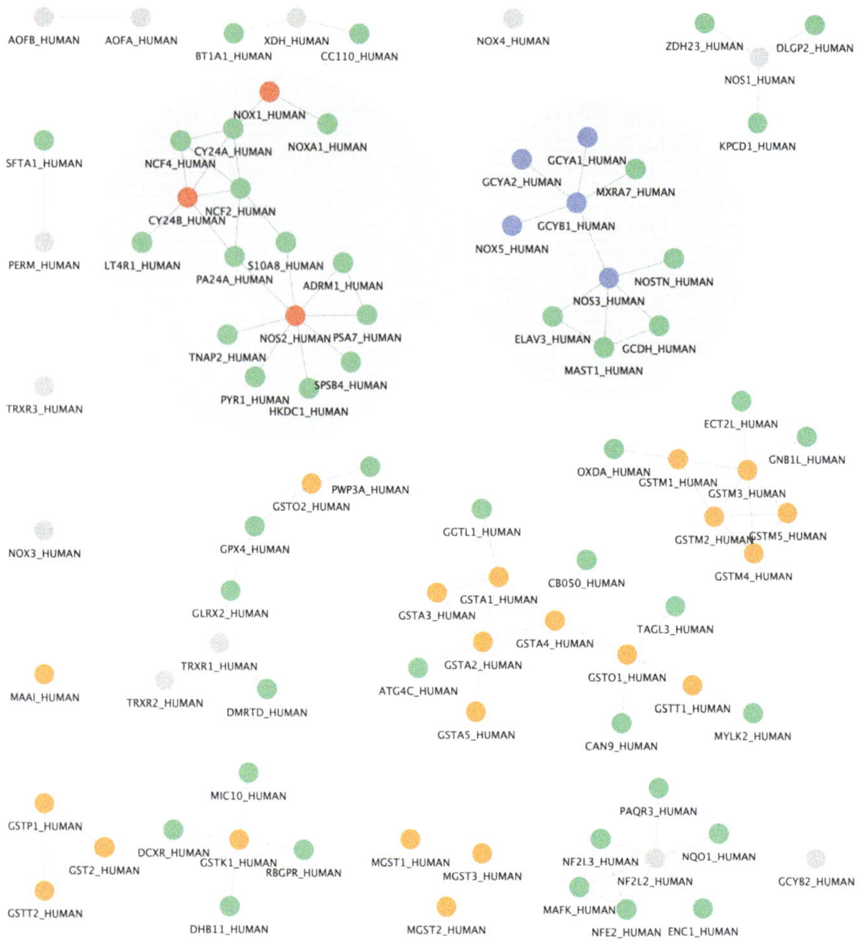

Fig. 2 PPI network and isolated ROS disease-relevant modules. 42 clinically researched ROS-generating enzymes, ROS-metabolizing enzymes or ROS targets were used as seeds to start building the PPI network. After pruning the network using a 0.4 SPD cut-off, 12 potential ROS disease-relevant modules were extracted, and 8 proteins remained isolated with no interactions. First neighbour proteins appear coloured green, and different seed proteins are coloured grey, yellow, blue or red, following the previous colour scheme in Fig. 1

subnetwork (Fig. 2) includes 88 nodes or proteins and 86 edges or PPIs; 12 different ROS-related disease modules were isolated, and 8 proteins appeared without any connections (Table 2).

Table 2 Disease-relevant ROS modules extracted from the PPI network shown in Fig. 2

UniProt name	Gene name	Common name
Module 1		
CY24B_HUMAN	CYBB	NADPH oxidase 2*
PYR1_HUMAN	CAD	CAD protein
HKDC1_HUMAN	HKDC1	Hexokinase HKDC1*
NOX1_HUMAN	NOX1	NADPH oxidase 1*
ADRM1_HUMAN	ADRM1	Proteasomal ubiquitin receptor ADRM1*
NOS2_HUMAN	NOS2	Nitric oxide synthase, inducible*
NOXA1_HUMAN	NOXA1	NADPH oxidase activator 1*
TNAP2_HUMAN	TNFAIP2	Tumour necrosis factor alpha-induced protein 2*
LT4R1_HUMAN	LTB4R	Leukotriene B4 receptor 1*
PA24A_HUMAN	PLA2G4A	Cytosolic phospholipase A2*
SPSB4_HUMAN	SPSB4	SPRY domain-containing SOCS box protein 4*
NCF4_HUMAN	NCF4	Neutrophil cytosol factor 4*
PSA7_HUMAN	PSMA7	Proteasome subunit alpha type-7*
NCF2_HUMAN	NCF2	Neutrophil cytosol factor 2*
CY24A_HUMAN	CYBA	Cytochrome b-245 light chain*
S10A8_HUMAN	S100A8	Protein S100-A8*
Module 2		
ATG4C_HUMAN	ATG4C	Cysteine protease ATG4C*
CAN9_HUMAN	CAPN9	Calpain-9*
GSTA2_HUMAN	GSTA2	Glutathione S-transferase A2*
GSTT1_HUMAN	GSTT1	Glutathione S-transferase theta-1*
GSTA1_HUMAN	GSTA1	Glutathione S-transferase A1*
GGTL1_HUMAN	GGTLC1	Glutathione hydrolase light chain 1*
GSTO1_HUMAN	GSTO1	Glutathione S-transferase omega-1*
CB050_HUMAN	C2orf50	Uncharacterized protein C2orf50
GSTA5_HUMAN	GSTA5	Glutathione S-transferase A5*
TAGL3_HUMAN	TAGLN3	Transgelin-3*
GSTA4_HUMAN	GSTA4	Glutathione S-transferase A4*
GSTA3_HUMAN	GSTA3	Glutathione S-transferase A3*
MYLK2_HUMAN	MYLK2	Myosin light chain kinase 2*
Module 3		
GCDH_HUMAN	GCDH	Glutaryl-CoA dehydrogenase*
ELAV3_HUMAN	ELAVL3	ELAV-like protein 3 *
GCYB1_HUMAN	GUCY1B1	Guanylate cyclase soluble subunit beta-1*
GCYA2_HUMAN	GUCY1A2	Guanylate cyclase soluble subunit alpha-2*
MXRA7_HUMAN	MXRA7	Matrix-remodeling-associated protein 7
GCYA1_HUMAN	GUCY1A1	Guanylate cyclase soluble subunit alpha-1*
NOSTN_HUMAN	NOSTRIN	Nostrin*
NOS3_HUMAN	NOS3	Nitric oxide synthase, endothelial*
MAST1_HUMAN	MAST1	Microtubule-associated serine/threonine-protein kinase 1*
NOX5_HUMAN	NOX5	NADPH oxidase 5

(continued)

Table 2 (continued)

UniProt name	Gene name	Common name
Module 4		
GSTM4_HUMAN	GSTM4	Glutathione S-transferase Mu 4*
GSTM3_HUMAN	GSTM3	Glutathione S-transferase Mu 3*
GSTM5_HUMAN	GSTM5	Glutathione S-transferase Mu 5*
GSTM1_HUMAN	GSTM1	Glutathione S-transferase Mu 1*
GSTM2_HUMAN	GSTM2	Glutathione S-transferase Mu 2*
ECT2L_HUMAN	ECT2L	Epithelial cell-transforming sequence 2 oncogene-like*
OXDA_HUMAN	DAO	D-amino-acid oxidase*
GNB1L_HUMAN	GNB1L	Guanine nucleotide-binding protein subunit beta-like protein 1*
Module 5		
NF2L3_HUMAN	NFE2L3	Nuclear factor erythroid 2-related factor 3*
MAFK_HUMAN	MAFK	Transcription factor MafK*
NQO1_HUMAN	NQO1	NAD(P)H dehydrogenase [quinone] 1*
NF2L2_HUMAN	NFE2L2	Nuclear factor erythroid 2-related factor 2*
NFE2_HUMAN	NFE2	Transcription factor NF-E2 45 kDa subunit*
PAQR3_HUMAN	PAQR3	Progestin and adipoQ receptor family member*
ENC1_HUMAN	ENC1	Ectoderm-neural cortex protein 1*
Module 6		
GPX4_HUMAN	GPX4	Phospholipid hydroperoxide glutathione peroxidase *
DMRTD_HUMAN	DMRTC2	Doublesex- and mab-3-related transcription factor C2
TRXR2_HUMAN	TXNRD2	Thioredoxin reductase 2*
GLRX2_HUMAN	GLRX2	Glutaredoxin-2, mitochondrial
GSTO2_HUMAN	GSTO2	Glutathione S-transferase omega-2*
TRXR1_HUMAN	TXNRD1	Thioredoxin reductase 1*
PWP3A_HUMAN	PWWP3A	PWWP domain-containing DNA repair factor 3A*
Module 7		
GSTK1_HUMAN	GSTK1	Glutathione S-transferase kappa 1*
RBGPR_HUMAN	RAB3GAP2	Rab3 GTPase-activating protein non-catalytic subunit*
DHB11_HUMAN	HSD17B11	Estradiol 17-beta-dehydrogenase 11*
DCXR_HUMAN	DCXR	L-xylulose reductase*
MIC10_HUMAN	MICOS10	MICOS complex subunit MIC10 *
Module 8		
KPCD1_HUMAN	PRKD1	Serine/threonine-protein kinase D1*
ZDH23_HUMAN	ZDHHC23	Palmitoyltransferase ZDHHC23
NOS1_HUMAN	NOS1	Nitric oxide synthase, brain*
DLGP2_HUMAN	DLGAP2	Disks large-associated protein 2*
Module 9		
XDH_HUMAN	XDH	Xanthine dehydrogenase/oxidase
BT1A1_HUMAN	BTN1A1	Butyrophilin subfamily 1 member A1, BT
CC110_HUMAN	CCDC110	Coiled-coil domain-containing protein 110*
Module 10		
GST2_HUMAN	GSTT2	Glutathione S-transferase theta-2*

(continued)

Table 2 (continued)

UniProt name	Gene name	Common name
GSTT2_HUMAN	GSTT2B	Glutathione S-transferase theta-2B*
GSTP1_HUMAN	GSTP1	Glutathione S-transferase P*
Module 11		
AOFA_HUMAN	MAOA	Monoamine oxidase type A*
AOFB_HUMAN	MAOB	Monoamine oxidase type B*
Module 12		
PERM_HUMAN	MPO	Myeloperoxidase
SFTA1_HUMAN	SFTPA1	Pulmonary surfactant-associated protein A1*
Isolated Proteins		
TRXR3_HUMAN	TXNRD3	Thioredoxin reductase 3*
MGST1_HUMAN	MGST1	Microsomal glutathione S-transferase 1*
MAAI_HUMAN	GSTZ1	Maleylacetoacetate isomerase*
GCYB2_HUMAN	GUCY1B2	Guanylate cyclase soluble subunit beta-2*
NOX4_HUMAN	NOX4	NADPH oxidase 4*
NOX3_HUMAN	NOX3	NADPH oxidase 3*
MGST2_HUMAN	MGST2	Microsomal glutathione S-transferase 2*
MGST3_HUMAN	MGST3	Microsomal glutathione S-transferase 3 *

Clinically validated seed proteins selected to start building the network have been highlighted in grey. The table includes the UniProt name (shown in the network), Gene name and most common name of each node or protein in the network. Proteins highlighted (*) have more than one name in the literature but only one was kept for simplicity

5 Protein-Metabolite Interactions and Metabolic Databases

PPIs are a powerful tool when investigating therapeutic approaches to ROSopathies, but they cannot be used in isolation, and certainly not with respect to ROS, as some effects of ROS are not due to PPI but indirectly by protein-metabolite-protein interaction. Thus, when scrutinizing ROS signalling, protein-metabolite interactions should also be considered. However, many databases with information on the interactivity of proteins and metabolites are somewhat patchy and lack important ROS-protein interactions. In Table 3, important substrates and ROS or toxic products of some essential ROS-generating and ROS-toxifying enzymes are shown. The connections between the enzymes and the listed substrates and products should be a minimum in databases when applied for ROS disease module identification.

When analysing the connections of pathways in highly curated databases like Kyoto Encyclopedia of Genes and Genomes (KEGG) (Ogata et al. 1999) and WikiPathways (Slenter et al. 2018), it becomes evident that essential information for systems medicine approaches to ROSopathies is missing. We located pathways from KEGG and WikiPathways that included the enzymes listed in the first column of Table 3. To do this, several search queries were made for each protein as an

Table 3 The substrates and products of ROS-generating and ROS-toxifying enzymes

Enzyme	Important substrates	ROS or toxic products	Literature
NOS1,2,3	L-Arginine, NADPH, O_2	L-Citrulline, NO, $NADP^+$	Knowles and Moncada (1994)
NOX1,2,3,5	NADPH, O_2, NADPH	O_2^- (mainly), H_2O_2, $NADP^+$	Casas et al. (2015)
NOX4	NADPH, O_2, NADPH	O_2^-, H_2O_2 (mainly), $NADP^+$	Casas et al. (2015), Nisimoto et al. (2014)
MPO	H_2O_2, Cl^-, Br^-	HOBr, HOCl, HOSCN	Casas et al. (2015); Lane et al. (2010)
XO/XDH	(Hypo)Xanthine, H_2O, O_2	O_2^-, H_2O_2	Casas et al. (2015)
MAO A,B	amines*, O_2	H_2O_2, aldehydes*	Casas et al. (2015), Edmondson (2014)

NOS nitric oxide synthase, *NOX* NADPH oxidase, *MPO* myeloperoxidase, *XO/xanthine* oxidase/xanthine dehydrogenase, *MAO* monoamine oxidase

Amines, based on tyrosine or tryptophan; aldehydes, corresponding aldehyde of the given amine

attempt to cover the possible names the proteins go by. The used search terms and resulting pathways can be found in Table 4.

Some pathways of WikiPathways have captured all the listed substrates and products of the NOS proteins of Table 3 [n6, n8, n13]. Pathways in KEGG also show important substrates and products of NOS proteins in their pathways; however, NADPH and $NADP^+$ are missing (see Table 4).

For both, KEGG and WikiPathways, the connectivity data of NOX proteins is even more lacking. NOX1,2 in KEGG's pathways are connected to superoxide and hydrogen peroxide [hsa04380, hsa05418, hsa04216, hsa04066], but NOX3 and NOX5 are not in any pathways and therefore connected to nothing. NOX4 is only in a single pathway where it is simply linked to "ROS" (see Table 4). WikiPathways includes all NOX isoforms in their pathways and has, in some of their pathways maps, incorporated all important substrates and products except for hydrogen peroxide, which is never connected to any NOX protein [n3, n8, n18].

In quite a few pathways, KEGG and WikiPathways do not distinguish between the type of ROS produced by the different NOX'es [hsa04933, hsa04621, hsa04933, n2, n40]. Also, in some pathways from both databases, NOX and NOS isoforms are treated as functionally identical units, respectively, having close to or same role in the signalling map [hsa04933, hsa00220, hsa00330, hsa04371, n3, n6, n8, n13]. While this might be true in certain scenarios, evidence suggests that NOX isoforms are distributed differently with respect to subcellular localizations and expression in various tissue types (Hilenski et al. 2004). The same goes for the NOS proteins (Villanueva and Giulivi 2010). Thus, it is unlikely that the NOX and NOS isoforms, respectively, simultaneously participate in the same signalling events as identical functional entities. The isoforms are different and will most likely bring about their own unique touch to inter- and intracellular signalling.

Table 4 Literature-based ROS metabolic signalling pathways in KEGG and WikiPathways

Search queries	KEGG	WikiPathways
Nitric oxide synthase 1, NOS1, bNOS, nNOS	hsa00220, hsa00330, hsa01100, hsa04020, hsa04145, hsa04371, hsa04713, hsa04730, hsa04926, hsa04970, hsa05010, hsa05014	n6, n7, n8, n9, n10, n11, n12, n13, n27, n28, n29, n30, n31, n32, n33, n52, n58
Nitric oxide synthase 2, NOS2, iNOS	hsa04066, hsa04146, hsa05132, hsa05133, hsa05140, hsa05142, hsa05145, hsa05146, hsa05152, hsa05200, hsa05222	n6, n8, n13, n29, n31, n32, n33, n34, n35, n36, n37, n38, n39, n40, n41, n42, n52, n56, n57
Nitric oxide synthase 3, NOS3, eNOS	hsa04022, hsa04066, hsa04151, hsa04370, hsa04611, hsa04915, hsa04921, hsa04931, hsa04933, hsa05418	n6, n8, n11, n13, n16, n31, n32, n40, n43, n15, n44, n45, n46, n47, n48, n49, n50, n51, n52, n53, n54, n55
NADPH oxidase 1, NOX1	hsa04380, hsa04933, hsa05418	n2, n3, n14, n25, n26, n40
NADPH oxidase 2, NOX2, CYBB	hsa04066, hsa04145, hsa04216, hsa04217, hsa04621, hsa04670, hsa05140	n8, n2, n3, n17, n18, n19, n15, n40, n51, n59, n60, n61, n62
NADPH oxidase 3, NOX3	–	n2, n3, n40
NADPH oxidase 4, NOX4	hsa04933	n2, n15, n18, n33, n40
NADPH oxidase 5, NOX5	–	n2, n18
Myeloperoxidase, MPO	hsa00983, hsa04145, hsa05202, hsa05221	n20, n21, n22, n62, n63
Xanthine dehydrogenase, xanthine oxidase, XO, XDH	hsa00230, hsa00232, hsa00983, hsa04146	n2, n23, n24, n6, n64, n65
Monoamine oxidase A, MAOA	hsa00260, hsa00330, hsa00340, hsa00350, hsa00360, hsa00380, hsa00982, hsa04726, hsa04728, hsa05030, hsa05031, hsa05034	n1, n2, n3, n4, n5, n42, n58, n66, n67, n68, n69, n70
Monoamine oxidase B, MAOB	(same as MAOA)	n3, n5, n71

Different ROS-forming enzymes and isoforms were considered, i.e. (1) NOS1-2-3, (2) NOX1-2-3-4-5, (3) MPO, (4) XO and (5) MAOA-B. Several search queries were used for each enzyme in order to include all the possible names that these enzymes are known as. Output resulting pathways in KEGG and WikiPathways have been collected

KEGG and WikiPathways show that MPO catalyses formation of various compounds [hsa00983, hsa04145, n22, n63] including hypochlorous acid and the highly oxidizing compound hydrogen peroxide. Only in WikiPathways is the enzyme interacting with a chlorine ion. In none of the databases, it is linked to hypobromous acid or hypothiocyanous acid nor the bromide ion. Though the latter acids are not types of ROS, these connections might still be important for systems medicine approaches as they still are powerful oxidizers (Lane et al. 2010; Casas et al. 2015). XO interacts with xanthine and hypoxanthine in both KEGG and WikiPathways. But it is only in WikiPathways that XO is involved in ROS signalling as a producer of hydrogen peroxide and ROS [n24, n2]. MAO A,B catalyse formation of hydrogen peroxide in both KEGG and WikiPathways [hsa05030, hsa05034, n3, n5]. But they are treated as identical enzymes in KEGG with respect to function and site of action. WikiPathways distinguishes between the two isoforms as they are represented individually in certain pathways [n2, n71]. However, when looking at KEGG's chemical reactions of proteins, you find a very extensive list of substrates and products. These connections cannot be found in the pathways where they would add invaluable information to signalling pathways and, consequently, to systems medicine approaches.

Essential information on ROS signalling is also missing in the curated Human Metabolome Database (HMDB) (Wishart et al. 2007), which is considered the standard metabolomic resource for human metabolic studies (Wishart et al. 2018). Here, there are no connections between superoxide and the human NOX1,2,3,4,5 proteins (version 4.0).

6 Non-canonical ROS and Hybrid Signalling Modules

In systems medicine, diseases are viewed as perturbations in biological networks (Langhauser et al. 2018; Menche et al. 2015). Drug targeting the mechanism that creates such perturbations restores the physiological status of signalling pathways, thereby treating the disease (Casas et al. 2019a, b). Reliable biological networks are crucial for this type of approach for medicine and treatment development. In Fig. 2, we isolated 12 potential disease-relevant modules for ROSopathies using PPI experimental data from IID. Consequently, the classical ROS signalling pathway shown in Fig. 1 has been redrawn to show, in a more physiologically relevant manner, how these ROS disease signalling modules might actually behave (Fig. 3).

Different proteins appeared to be isolated and without any connections to the other modules in the network. This does not mean that they do not interact with anything but rather that they might not form signalling modules with other ROS enzymes. Not only that, for instance, IID has prediction interaction data for NOX3 and TRXR3 but neither of them nor GCYB2 have any experimental interaction data, hence why they appear isolated in the network. Here we show a clear example of study biases previously discussed. Further research needs to be done on these targets to link these enzymes to the extracted disease modules or perhaps to raise new ones.

On the other hand, PPI methods do not consider possible interactions between the proteins and enzymes through metabolites. Casas et al. have linked and validated NOX4 connection to NOS enzymes through metabolites using a guilt-by-association analysis (Casas et al. 2019a, b). Consequently, NOX4 could be linked to NOS enzymes emerging a clinically relevant ROS-cGMP (ROCG) disease module for network pharmacology and synergistic drug treatment. This emerged ROCG disease signalling cluster has previously been described, linked and validated to cerebro-cardiovascular metabolic disease phenotypes such as ischemic stroke (Kleinschnitz

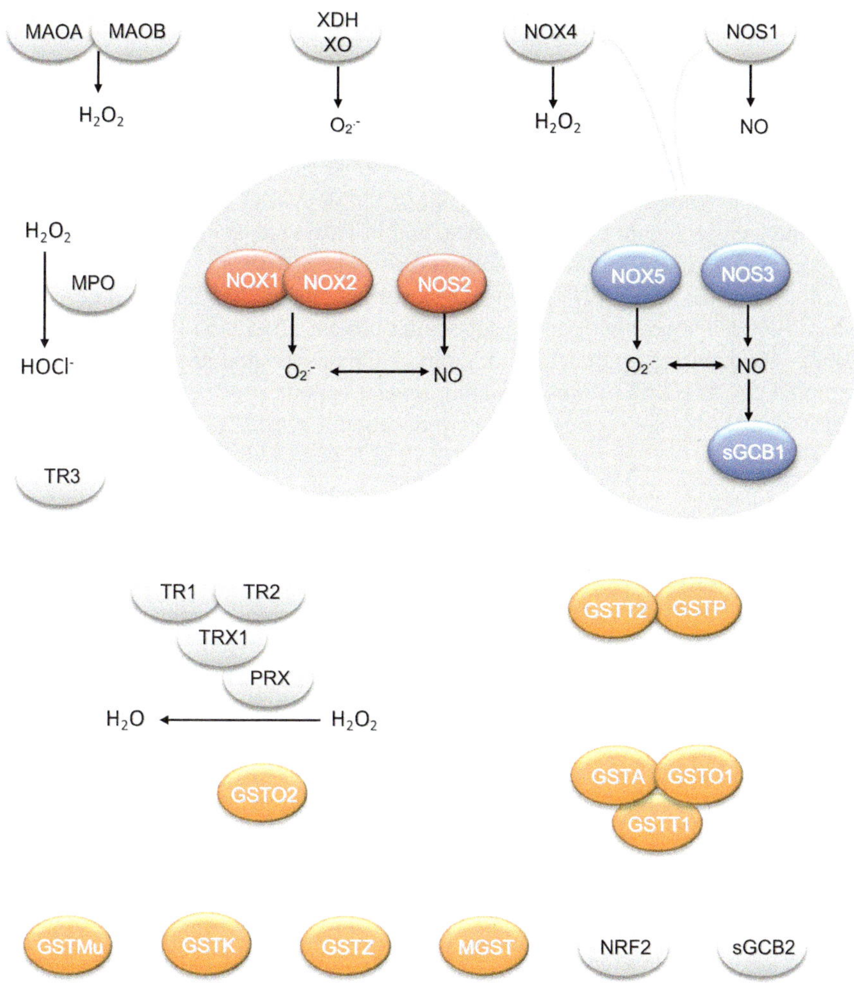

Fig. 3 ROS signalling pathway revisited. Classical ROS signalling pathways have been rearranged according to the potential ROS disease-relevant modules. Further research and metabolic interactions need to be included to further refine the signalling modules, e.g. NOX4 and NOS have already been linked to the NOX5/sGC module (Casas et al. 2019a)

et al. 2016; Casas et al. 2017, 2019b), diabetes (Jha et al. 2014), atherosclerosis (Gray et al. 2013, 2016) and heart failure (Paulus and Tschöpe 2013; Takimoto et al. 2005) among others.

The hybrid NOX1,2/NOS2 signalling module could potentially be playing a joint role in the immune defence response in the gastrointestinal (GI) track. NOX1 loss-of-function has been observed in patients with inflammatory bowel disease, and it is expressed in the membrane of the epithelial cells of the intestines where NOX1-dependent superoxide production appears to be essential (Schwerd et al. 2018).

Together with NOS2, NOX1 function in the GI track appears to be linked to maintaining the homeostasis of the gut microbiome through the production of physiologically relevant ROS (Matziouridou et al. 2018; Aviello and Knaus 2018). Here, NOS-derived NO may toxify superoxide through intermediate reactive nitrogen species such as peroxynitrite.

In a similar manner, MPO/SP-A interaction in the MPO disease module has already been studied. Surfactant proteins (SP) SP-A and SP-D are involved in apoptotic cell recognition and clearance, and in vitro studies have shown that they interact with neutrophils through MPO (Jäkel et al. 2010).

Moreover, the NRF2/ENC-1 (Fig. 2) signalling module has also been described, where ENC1 interacts with NRF2 supressing protein translation (Wang and Zhang 2009). Mechanistic approaches to target NRF2-associated diseases have been explored due to NRF2 role in regulating several cellular processes such as maintenance of redox balance and inflammation (Cuadrado et al. 2018, 2019; Pajares et al. 2017).

7 Conclusion

In conclusion, we extracted 12 disease-relevant signalling modules for ROSopathies and showed how some of them have already been explored therapeutically. Others need to be further studied and associated to diseases. Nevertheless, not all of them will be relevant for disease. Disease-relevant signalling modules represent the beginning of a journey to map disease phenotypes to clinically relevant mechanisms for network pharmacology, moving away from symptomatic treatment of diseases, i.e. from phenotypic to mechanotypic therapy. Once a mechanism is understood and validated, it will be possible to stratify and map patients to specific ROSopathies leading to precise disease interventions. Moreover, not all ROS (related proteins) are equal. Some will be associated to some diseases; others, despite forming the same product (e.g. superoxide), will be associated to different diseases. Fruitless hypothesis-driven discovery projects can be prevented, drug repurposing enabled, and translational speed of basic science into the clinic enhanced. ROSopathies can be one of the first disease mechanisms providing proof of concept for a new era of precision medicine.

References

Alanis-Lobato G, Andrade-Navarro MA (2017) A reliable and unbiased human protein network with the disparity filter. BioRxiv. https://doi.org/10.1101/207761

Altenhöfer S, Kleikers PWM, Radermacher KA, Peter S, Rob Hermans JJ, Schiffers P, Ho H, Wingler K, Schmidt HHHW (2012) The NOX toolbox: validating the role of NADPH oxidases in physiology and disease. Cell Mol Life Sci 69:2327. https://doi.org/10.1007/s00018-012-1010-9

Alvarez-Ponce D, Feyertag F, Chakraborty S (2017) Position matters: network centrality considerably impacts rates of protein evolution in the human protein-protein interaction network. Genome Biol Evol 9:1742. https://doi.org/10.1093/gbe/evx117

Aviello G, Knaus UG (2018) NADPH oxidases and ROS signaling in the gastrointestinal tract review-article. Mucosal Immunol 11:1011. https://doi.org/10.1038/s41385-018-0021-8

Bickers DR, Athar M (2006) Oxidative stress in the pathogenesis of skin disease. J Investig Dermatol 126:2565. https://doi.org/10.1038/sj.jid.5700340

Bigarella CL, Liang R, Ghaffari S (2014) Stem cells and the impact of ROS signaling. Development (Cambridge) 141:4206. https://doi.org/10.1242/dev.107086

Björklund ÅK, Light S, Hedin L, Elofsson A (2008) Quantitative assessment of the structural Bias in protein-protein interaction assays. Proteomics 8:4657. https://doi.org/10.1002/pmic.200800150

Casas AI, Dao VTV, Daiber A, Maghzal GJ, Di Lisa F, Kaludercic N, Leach S et al (2015) Reactive oxygen-related diseases: therapeutic targets and emerging clinical indications. Antioxid Redox Signal 23:1171. https://doi.org/10.1089/ars.2015.6433

Casas AI, Geuss E, Kleikers PWM, Mencl S, Herrmann AM, Buendia I, Egea J et al (2017) NOX4-dependent neuronal autotoxicity and BBB breakdown explain the superior sensitivity of the brain to ischemic damage. Proc Natl Acad Sci U S A 114:12315. https://doi.org/10.1073/pnas.1705034114

Casas AI, Hassan AA, Larsen SJ, Gomez-Rangel V, Elbatreek M, Kleikers PWM, Guney E et al (2019a) From single drug targets to synergistic network pharmacology in ischemic stroke. Proc Natl Acad Sci U S A 116:7129. https://doi.org/10.1073/pnas.1820799116

Casas AI, Kleikers PWM, Geuss E, Langhauser F, Adler T, Busch DH, Gailus-Durner V et al (2019b) Calcium-dependent blood-brain barrier breakdown by NOX5 limits postreperfusion benefit in stroke. J Clin Investig 129:1772. https://doi.org/10.1172/JCI124283

Cuadrado A, Manda G, Hassan A, Alcaraz MJ, Barbas C, Daiber A, Ghezzi P et al (2018) Transcription factor NRF2 as a therapeutic target for chronic diseases: a systems medicine approach. Pharmacol Rev 70:348. https://doi.org/10.1124/pr.117.014753

Cuadrado, Antonio, Ana I. Rojo, Geoffrey Wells, John D. Hayes, Sharon P. Cousin, William L. Rumsey, Otis C. Attucks, et al. 2019. "Therapeutic targeting of the NRF2 and KEAP1 partnership in chronic diseases." Nat Rev Drug Discov Doi: https://doi.org/10.1038/s41573-018-0008-x, 18, 295

Dröge W (2002) Free radicals in the physiological control of cell function. Physiol Rev 82:47. https://doi.org/10.1152/physrev.00018.2001

Edmondson D (2014) Hydrogen peroxide produced by mitochondrial monoamine oxidase catalysis: biological implications. Curr Pharm Des 20:155. https://doi.org/10.2174/13816128113190990406

Frijhoff J, Winyard PG, Zarkovic N, Davies SS, Stocker R, Cheng D, Knight AR et al (2015) Clinical relevance of biomarkers of oxidative stress. Antioxid Redox Signal 23(14):1144–1170. https://doi.org/10.1089/ars.2015.6317

Gracia C, Karla DL-C, Husi H (2017) CVD and oxidative stress. J Clin Med 6. https://doi.org/10.3390/jcm6020022

Gray SP, Di Marco E, Okabe J, Szyndralewiez C, Heitz F, Montezano AC, De Haan JB et al (2013) NADPH oxidase 1 plays a key role in diabetes mellitus-accelerated atherosclerosis. Circulation 127:1888. https://doi.org/10.1161/CIRCULATIONAHA.112.132159

Gray SP, Di Marco E, Kennedy K, Chew P, Okabe J, El-Osta A, Calkin AC et al (2016) Reactive oxygen species can provide atheroprotection via NOX4-dependent inhibition of inflammation and vascular remodeling. Arterioscler Thromb Vasc Biol 36:295. https://doi.org/10.1161/ATVBAHA.115.307012

He FJ, MacGregor GA (2007) Blood pressure is the most important cause of death and disability in the world. Eur Heart J Suppl 9(B):23–28. https://doi.org/10.1093/eurheartj/sum005

Hilenski LL, Clempus RE, Quinn MT, David Lambeth J, Griendling KK (2004) Distinct subcellular localizations of Nox1 and Nox4 in vascular smooth muscle cells. Arterioscler Thromb Vasc Biol 24:677. https://doi.org/10.1161/01.ATV.0000112024.13727.2c

Hochman JS, Alexander JH, Reynolds HR, Stebbins AL, Dzavik V, Harrington RA, Van De Werf F (2007) Effect of Tilarginine acetate in patients with acute myocardial infarction and cardiogenic shock: the TRIUMPH randomized controlled trial. J Am Med Assoc 297:1657. https://doi.org/10.1001/jama.297.15.joc70035

Ivanic J, Yu X, Wallqvist A, Reifman J (2009) Influence of protein abundance on high-throughput protein-protein interaction detection. PLoS One 4:e5815. https://doi.org/10.1371/journal.pone.0005815

Jäkel A, Clark H, Reid KBM, Sim RB (2010) Surface-bound myeloperoxidase is a ligand for recognition of late apoptotic neutrophils by human lung surfactant proteins A and D. Protein Cell 1:563. https://doi.org/10.1007/s13238-010-0076-0

Jensen LJ, Bork P (2008) Biochemistry: not comparable, but complementary. Science 322:56. https://doi.org/10.1126/science.1164801

Jha JC, Gray SP, Barit D, Okabe J, El-Osta A, Namikoshi T, Thallas-Bonke V et al (2014) Genetic targeting or pharmacologic inhibition of NADPH oxidase Nox4 provides renoprotection in long-term diabetic nephropathy. J Am Soc Nephrol 25:1237. https://doi.org/10.1681/asn.2013070810

Kleinschnitz C, Mencl S, Kleikers PWM, Schuhmann MK, López MG, Casas AI, Sürün B, Reif A, Schmidt HHHW (2016) NOS knockout or inhibition but not disrupting PSD-95-NOS interaction protect against ischemic brain damage. J Cereb Blood Flow Metab 36:1508. https://doi.org/10.1177/0271678X16657094

Knowles RG, Moncada S (1994) Nitric oxide synthases in mammals. Biochem J 298:249. https://doi.org/10.1042/bj2980249

Kotlyar M, Pastrello C, Malik Z, Jurisica I (2019) IID 2018 update: context-specific physical protein-protein interactions in human, model organisms and domesticated species. Nucleic Acids Res 47:D581. https://doi.org/10.1093/nar/gky1037

Lane AE, Tan JTM, Hawkins CL, Heather AK, Davies MJ (2010) The myeloperoxidase-derived oxidant HOSCN inhibits protein tyrosine phosphatases and modulates cell Signalling via the mitogen-activated protein kinase (MAPK) pathway in macrophages. Biochem J 430:161. https://doi.org/10.1042/BJ20100082

Langhauser F, Casas AI, Dao VTV, Guney E, Menche J, Geuss E, Kleikers PWM et al (2018) A diseasome cluster-based drug repurposing of soluble guanylate cyclase activators from smooth muscle relaxation to direct neuroprotection. NPJ Syst Biol Appl 4:8. https://doi.org/10.1038/s41540-017-0039-7

Lapchak PA (2010) A critical assessment of Edaravone acute ischemic stroke efficacy trials: is Edaravone an effective neuroprotective therapy? Expert Opin Pharmacother 11:1753. https://doi.org/10.1517/14656566.2010.493558

Matziouridou C, Rocha SDC, Haabeth OA, Rudi K, Carlsen H, Kielland A (2018) INOS- and NOX1-dependent ROS production maintains bacterial homeostasis in the ileum of mice. Mucosal Immunol 11:774. https://doi.org/10.1038/mi.2017.106

Menche J, Sharma A, Kitsak M, Ghiassian SD, Vidal M, Loscalzo J, Barabási AL (2015) Uncovering disease-disease relationships through the incomplete Interactome. Science 347 (6224):841. https://doi.org/10.1126/science.1257601

Moloney JN, Cotter TG (2018) ROS Signalling in the biology of cancer. Semin Cell Dev Biol 80:50. https://doi.org/10.1016/j.semcdb.2017.05.023

Nisimoto Y, Diebold BA, Constentino-Gomes D, David Lambeth J (2014) Nox4: a hydrogen peroxide-generating oxygen sensor. Biochemistry 53:5111. https://doi.org/10.1021/bi500331y

Ogata H, Goto S, Sato K, Fujibuchi W, Bono H, Kanehisa M (1999) KEGG: Kyoto encyclopedia of genes and genomes. Nucleic Acids Res 27:29. https://doi.org/10.1093/nar/27.1.29

Ogden LG, He J, Lydick E, Whelton PK (2000) Long-term absolute benefit of lowering blood pressure in hypertensive patients according to the JNC VI risk stratification. Hypertension 35 (2):539–543. https://doi.org/10.1161/01.HYP.35.2.539

Ogedegbe G, Shah NR, Phillips C, Goldfeld K, Roy J, Yu G, Gyamfi J, Torgersen C, Capponi L, Bangalore S (2015) Comparative effectiveness of angiotensin-converting enzyme inhibitor-based treatment on cardiovascular outcomes in hypertensive blacks versus whites. J Am Coll Cardiol 66(11):1224–1233. https://doi.org/10.1016/j.jacc.2015.07.021

Pajares M, Cuadrado A, Rojo AI (2017) Modulation of Proteostasis by transcription factor NRF2 and impact in neurodegenerative diseases. Redox Biol 11:543. https://doi.org/10.1016/j.redox.2017.01.006

Paulus WJ, Tschöpe C (2013) A novel paradigm for heart failure with preserved ejection fraction: comorbidities drive myocardial dysfunction and remodeling through coronary microvascular endothelial inflammation. J Am Coll Cardiol 62:263. https://doi.org/10.1016/j.jacc.2013.02.092

Schaefer MH, Serrano L, Andrade-Navarro MA (2015) Correcting for the study Bias associated with protein-protein interaction measurements reveals differences between protein degree distributions from different Cancer types. Front Genet 6. https://doi.org/10.3389/fgene.2015.00260

Schmidt HHHW, Stocker R, Vollbracht C, Paulsen G, Riley D, Daiber A, Cuadrado A (2015) Antioxidants in translational medicine. Antioxid Redox Signal 23:1130. https://doi.org/10.1089/ars.2015.6393

Schork NJ (2015) Personalized medicine: time for one-person trials. Nature 520:609. https://doi.org/10.1038/520609a

Schwerd T, Bryant RV, Pandey S, Capitani M, Meran L, Cazier JB, Jung J et al (2018) NOX1 loss-of-function genetic variants in patients with inflammatory bowel disease. Mucosal Immunol 11:562. https://doi.org/10.1038/mi.2017.74

Slenter DN, Kutmon M, Hanspers K, Riutta A, Windsor J, Nunes N, Mélius J et al (2018) WikiPathways: a multifaceted pathway database bridging metabolomics to other omics research. Nucleic Acids Res 46:D661. https://doi.org/10.1093/nar/gkx1064

Takimoto E, Champion HC, Li M, Belardi D, Shuxun R, Rene Rodriguez E, Bedja D, Gabrielson KL, Wang Y, Kass DA (2005) Chronic inhibition of cyclic GMP phosphodiesterase 5A prevents and reverses cardiac hypertrophy. Nat Med 11:214. https://doi.org/10.1038/nm1175

Thanan R, Oikawa S, Hiraku Y, Ohnishi S, Ma N, Pinlaor S, Yongvanit P, Kawanishi S, Murata M (2014) Oxidative stress and its significant roles in neurodegenerative diseases and Cancer. Int J Mol Sci 16:193. https://doi.org/10.3390/ijms16010193

Villanueva C, Giulivi C (2010) Subcellular and cellular locations of nitric oxide synthase isoforms as determinants of health and disease. Free Radic Biol Med 49:307. https://doi.org/10.1016/j.freeradbiomed.2010.04.004

von Mering C, Krause R, Snel B, Cornell M, Oliver SG, Fields S, Bork P (2002) Comparative assessment of large-scale data sets of protein-protein interactions. Nature 417:399. https://doi.org/10.1038/nature750

Wang XJ, Zhang DD (2009) Ectodermal-neural cortex 1 down-regulates Nrf2 at the translational level. PLoS One 4:e5492. https://doi.org/10.1371/journal.pone.0005492

Wilkinson-Berka JL, Deliyanti D, Rana I, Miller AG, Agrotis A, Armani R, Szyndralewiez C et al (2014) NADPH oxidase, NOX1, mediates vascular injury in ischemic retinopathy. Antioxid Redox Signal 20:2726. https://doi.org/10.1089/ars.2013.5357

Wishart DS, Tzur D, Knox C, Eisner R, Guo AC, Young N, Cheng D et al (2007) HMDB: the human metabolome database. Nucleic Acids Res 35:D521. https://doi.org/10.1093/nar/gkl923

Wishart DS, Feunang YD, Marcu A, Guo AC, Liang K, Vázquez-Fresno R, Sajed T et al (2018) HMDB 4.0: the human metabolome database for 2018. Nucleic Acids Res. https://doi.org/10.1093/nar/gkx1089

Yang JL, Mukda S, Der Chen S (2018) Diverse roles of mitochondria in ischemic stroke. Redox Biol 16:263. https://doi.org/10.1016/j.redox.2018.03.002

Zhao Y, Hu X, Liu Y, Dong S, Wen Z, He W, Zhang S, Huang Q, Shi M (2017) ROS signaling under metabolic stress: cross-talk between AMPK and AKT pathway. Mol Cancer 16:79. https://doi.org/10.1186/s12943-017-0648-1

Part II

Targeting Antioxidant Responses

Development of *Telintra* as an Inhibitor of Glutathione S-Transferase P

Jie Zhang, Zhi-Wei Ye, Yvonne Janssen-Heininger, Danyelle M. Townsend, and Kenneth D. Tew

Contents

Abstract

Glutathione S-transferase P (GSTP) is a component of a complex series of pathways that provide cellular redox homeostasis. It is an abundant protein in certain tumors and is over-expressed in cancer drug resistance. It has diverse cellular functions that include, thiolase activities with small electrophilic agents or susceptible cysteine residues on the protein to mediate S-glutathionylation, and chaperone binding with select protein kinases. Preclinical and clinical testing of a

J. Zhang · Z.-W. Ye · K. D. Tew (✉)
Department of Cell and Molecular Pharmacology and Experimental Therapeutics, Medical University of South Carolina, Charleston, SC, USA
e-mail: tewk@musc.edu

Y. Janssen-Heininger
Department of Pathology and Laboratory Medicine, The University of Vermont, Burlington, VT, USA

D. M. Townsend
Department of Pharmaceutical and Biomedical Sciences, Medical University of South Carolina, Charleston, SC, USA

H. H. H. W. Schmidt et al. (eds.), *Reactive Oxygen Species*,
Handbook of Experimental Pharmacology 264, https://doi.org/10.1007/164_2020_392

nanomolar inhibitor of GSTP, TLK199 (Telintra; Ezatiostat) has indicated a role for the enzyme in hematopoiesis and utility for the drug in the treatment of patients with myelodysplastic syndrome.

Graphical Abstract

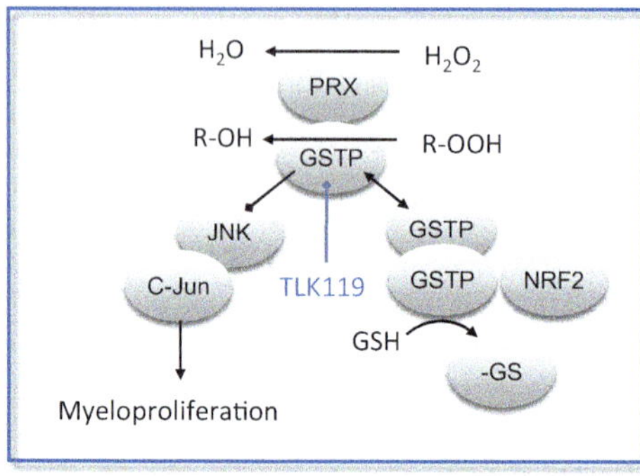

Keywords

c-Jun N-terminal kinases · Clinical trials · Cyclophosphamide · Drug resistance · Ezatiostat · Glutathione · Glutathione S-transferase · Glutathionylation · Hematopoiesis · Lung diseases · Myelodysplastic syndromes · Myeloproliferation · Oxidative stress · Telintra

1 Historical Perspectives

In the 1980s and 1990s, there was a great deal of emphasis on understanding those mechanistic aspects of cancer cell drug resistance that characteristically caused failure of chemotherapeutic response and patient relapse. Numerous conferences and books were dedicated to the subject, and while the term multidrug resistance (MDR) was primarily judged to be synonymous with overexpression of the P-glycoprotein (now called ABCB1), there were a few investigators with interests in glutathione pathways and their impact on cancer drug resistance. Since drug combination therapies were *de rigueur* for clinical treatments, much interest was focused on modulating existing cancer drugs with non-toxic agents that might serve to reverse resistance or, at worst, enhance a therapeutic index. Out of this discipline came the principle that resistance that resulted as a consequence of overexpression of various thiol-based enzyme pathways might be combatted through concomitant inhibition of certain redox-active enzymes. For example, increased levels of intracellular glutathione (GSH) were combatted by treatment with buthionine sulfoximine (BSO) (Griffith and Meister 1979), an inhibitor of γ-glutamylcysteine synthetase (now called glutamate-cysteine ligase), the rate-limiting enzyme in de novo GSH synthesis. Clinical trials with BSO proved to be of limited value (Bailey et al. 1994; O'Dwyer et al. 1992), but were a forerunner of efforts to modulate various GSH pathways linked with drug resistance. The early observation that GST isozymes were overexpressed in drug-resistant cells (Wang and Tew 1985) created an opportunity for a similar modulatory approach. While ethacrynic acid (EA) was first tried, the clinical toxicity associated with its diuretic properties limited its utility (O'Dwyer et al. 1991). Nevertheless, around this time, under the leadership of Dr. Larry Kauvar, Terrapin Technologies of South San Francisco was developing patented technologies that focused on the development of GSH analogs and paralog panels comprising GSH mimics. From this work, TLK199 was identified and became the lead agent that progressed through preclinical studies to eventual clinical testing and is the focus of the remainder of this chapter. Starting life as TER199 (reflecting the original name of Terrapin Technologies), preclinical studies are described in this chapter under the TLK199 moniker. As the drug moved through preclinical development, the name was changed to *Telintra* and finally *Ezatiostat*. In each case the essentially chemical entity of the drug is the same, although GLP/GMP manufacturing altered formulation.

2 Why Target GSTP?

When GST isozymes were first described (Boyland and Chasseaud 1969), most early publications detailed their catalytic functions in catalyzing the thioether conjugation of small molecule agents with GSH, usually accommodating a diverse range of electrophilic substrates. For GSTP, these reactions have proved to be quite restricted, and in subsequent decades, more and different biological functions have been ascribed to this GST isozyme family. These additional functions take on direct relevance to understanding why a GSTP inhibitor such as *Telintra* has translatable utility and has been tested in early stage clinical trials. The fact that GSTP is so highly expressed in certain solid tumors and in cancer cells that have acquired resistance to various anticancer drugs provided the initial rationale for inhibiting GSTP (Tew 1994). In principle, modulating GSTP in these setting could carry therapeutic benefit. Since these studies, the disparate roles of GSTP in cellular growth and stress response pathways have broadened the interest base for this GST isozyme. In particular, its role in regulating stress response pathways of various kinds and its catalytic functions in facilitating the forward reaction of the S-glutathionylation cycle (Townsend et al. 2009) have added to the potential importance of inhibiting such pathways. The following sections detail much of the material.

3 Stress Signaling Pathways

c-Jun N-terminal kinases (JNKs) are a family of stress kinases subject to transient activation in response to a variety of cellular stresses, including reactive oxygen or nitrogen species, heat shock, and perhaps of particular interest in terms of myeloproliferation, growth factors, or inflammatory cytokines (Davis 2000). JNK-mediated phosphorylation of the transcription factors c-Jun and activating transcription factor 2 (ATF2) can effectively facilitate the stress response. Because there would be few advantages in cells existing in a constant stressed condition, by necessity, basal JNK activity is maintained at a low level, and it is at this nexus that GSTP can act as an endogenous negative regulatory switch for the kinase. GSTP has ligand-binding properties that manifest as protein complexes where JNK activity can be regulated through a series of protein:protein interactions. In unstressed cells, low JNK activity is maintained through sequestration of the kinase in a multiprotein complex that includes GSTP-JNK. In this regard, treatment with *Telintra* has been shown to cause GSTP to dissociate from this complex, accumulating GSTP oligomers, with resultant activation of JNK impacting downstream events as divergent as proliferation or apoptosis (Yin et al. 2000). JNK-dependent stress-induced apoptosis may be suppressed during tumor development, and in this regard, the high levels of GSTP found in many solid tumors, or in drug-resistant cells, may act to sequester JNK in an inactive state, perhaps explaining why elevated GSTP levels can be found even when the selecting drug is not a substrate for GSH conjugation (Tew 1994). Recent studies confirm that for GSTP, binding to other protein partners is quite common. While these events might initially appear promiscuous, they are likely driven by the propensity for GSTP to act as a protein thiolase (see later discussion on S-glutathionylation). However, for the kinase regulatory effects, homology between GSTA and P family members may explain why GSTA1 by a similar mechanism can also suppress JNK signaling caused by inflammatory cytokines or ROS (Romero et al. 2006). Moreover, GSTP also regulates tumor necrosis factor alpha (TNFα) signaling through a protein ligand interaction with tumor necrosis factor receptor-associated factor 2 (TRAF2; Wu et al. 2006).

GSTP inhibits TRAF2-induced activation of both JNK and p38 (but not NFkB) attenuates TRAF2-enhanced apoptosis signal-regulating kinase 1 (ASK1) autophosphorylation and inhibits TRAF2-ASK1-induced apoptosis by suppressing the interaction of these two proteins. When GSTP interacts, its catalytic activity is unaffected, implicating sites distant to those involved in GSH or substrate binding. This would be in agreement with the principle that the interaction occurs in the first place as a conduit to the thiolase activity, i.e., that GSTP interacts as a prelude to S-glutathionylating the adjacent protein. A further example of functional redundancy within the GST family is afforded by the fact that GSTM1 binds to, and inhibits, the activity of ASK1 (Cho et al. 2001). Similar to GSTP:JNK, the interaction of the GSTM1:ASK1 complex is dissociated under stress conditions, leading to GSTM oligomerization and subsequent activation of ASK1 (Dorion et al. 2002). Because ASK1 can activate the JNK and p38 pathways, this disassociation could also serve to activate cytokine- and stress-induced apoptosis (Ichijo et al. 1997). A general conclusion from all these studies is that GST isozymes are not acting in a detoxification fashion, rather they serve to augment intermediary kinase regulation. In this regard, an inhibitor such as *Telintra* can potentially impact these stress kinase pathways through interference with this regulation.

4 GSTP and S-Glutathionylation

In previous reviews, we have drawn parallels between the processes of phosphorylation and S-glutathionylation (Ye et al. 2017). Each exact critical regulatory control functions on target proteins susceptible to these post-translational modifications. However, S-glutathionylation of either kinases or phosphatases (7) can be critical in maintaining the cyclical nature of phosphorylation/dephosphorylation and demonstrates the layered nature of how sulfur-based post-translational modifications may actually supersede those of phosphorus. S-glutathionylation generally occurs on cysteines in basic environments within the protein (e.g., vicinal to Arg, His, or Lys residues). GSTP can lower the pKa of the cysteine thiol of GSH, producing a nucleophilic thiolate anion (Graminski et al. 1989). Under the right conditions, cysteines on the surfaces of proteins may undergo spontaneous S-glutathionylation (Ghezzi 2005); nevertheless GSTP can influence the rate and extent of the process in a catalytic manner. In this regard, *Telintra* has been shown to interfere with S-glutathionylation, with subsequent influence on the structure and function of a variety of target proteins (McMillan et al. 2016; Jones et al. 2016).

Relative to the proteome, the number of S-glutathionylated proteins is not proportionally large, and those can be categorized into functional clusters. These include enzymes with catalytically important cysteines especially those involved with protein folding/stability, nitric oxide regulation, and redox homeostasis; cytoskeletal; signaling – particularly kinases and phosphatases; transcription factors; ras proteins; heat shock proteins; ion channels, calcium homeostasis; and energy metabolism and glycolysis. Under stress conditions, the half-life of S-glutathionylation approximates 4 h (Townsend et al. 2009), although this value is contingent upon both stress-induced conditions and cell type. However, relevant to the utility of *Telintra*, there are instances (as described in this chapter) where interference with S-glutathionylation can have a plausible therapeutic effect. In this regard, the enhanced myeloproliferative phenotype of the GSTP knockout mouse (Gate et al. 2004), together with the other indications where pharmacological inhibition of GSTP influences bone marrow proliferation and migration (Zhang et al. 2014), dictates that S-glutathionylation is an important factor in regulating myeloproliferation in the bone marrow. In this regard, the next two sections detail how preclinical studies with *Telintra* have created opportunities for its use in either myeloproliferative or lung diseases.

5 Modulation of Drug Resistance

Elevated levels of GSTs, especially GSTP1-1, are often associated with an increased resistance of tumors to a variety of anticancer drugs (Tew 1994; Tew et al. 1997; Townsend and Tew 2003). Potentiation of the cytotoxicity by GST inhibitors, e.g., ethacrynic acid, has been observed both in vitro and in vivo (O'Dwyer et al. 1991; Petrini et al. 1993; Tew et al. 1988). EA inhibits GSTs by binding to the H-site (substrate-binding site) of the isozyme, as well as by depleting its cofactor, GSH, via covalent binding (Michael addition), with K_i (μM) of 4.6–6.0, 0.3–1.9, and 3.3–4.8 for GSTA1-1, GSTM1-1, and GSTP1-1, respectively (Ploemen et al. 1993). EA was shown to potentiate the toxicity of chlorambucil in several cancer cell lines (Tew et al. 1988) and increase the sensitivity of melphalan in xenograft models in SCID mice. The therapeutic value of EA as a chemosensitizer has also been reported in patients (O'Dwyer et al. 1991; Petrini et al. 1993). However, EA is not GST isozyme specific (Ploemen et al. 1993) and causes extreme diuresis (O'Dwyer et al. 1991), a side effect that proved to be an important dose-limiting toxicity, making it less suitable for clinical modulation. Partly as a consequence, further attempts at selective inhibition of GST isozymes focused on synthesis of a number of GSH (γ-Glu-Cys-Gly) analogs (Flatgaard et al. 1993; Lyttle et al. 1994). The GSH analogs were designed and synthesized based on the observations that the γ-glutamyl residue of GSH was absolutely critical for binding (Adang et al. 1990), whereas substituting the C-terminal glycine of GSH and functionalizing the sulfur of the cysteine residue of GSH with different alkyl and aryl groups only affected the potency and selectivity of the GSH analogs as GST inhibitors (Adang et al. 1990; Askelof et al. 1975). Among those GSH analogs, TLK117 (γ-glutamyl-S-(benzyl)cysteinyl-R(−)-phenylglycine), which contains substituents at both the glycine α-carbon and cysteine thiol group, was produced as a specific GSTP1-1 inhibitor ($K_i = 0.4$ μM). Its binding affinity to the G-site of GSTP1-1 is greater than that of GSH, and its selectivity for GSTP1-1 is over 50-fold higher compared with GSTA1-1, GSTM1-1, and GSTM2-2 (Flatgaard et al. 1993; Lyttle et al. 1994). The high-resolution (2.0 Å) crystal structure of GSTP1-1 in complex with TLK117 provides an explanation as to why this compound inhibits the pi-class GST much better than the other GST classes. The phenyl moiety of TLK117 is stacked against the benzyl moiety and interacts with Phe8 and Trp38 in a lipophilic region of GSTP1-1. However, in the case of GSTA1-1, the phenyl substitution would clash with Phe220 and Phe222, while in the case of GSTM1-1, it would clash with Trp7, Met34, and Arg42 (Oakley et al. 1997). TLK117 was designed for efficient inhibition of the most abundant allelic variant *GSTP1*A* (Ile105, Ala114), but it also competitively inhibits *GSTP1*B* (Val105, Ala114) with similar potency (Johansson et al. 2000). The inhibitory effects of TLK117 on *GSTP1*C* (Val105, Val114) and *GSTP1*D* (Ile105, Ala114) have not been determined. Such considerations are quite relevant since there is evidence that of the four allelic variants of GST, the wild type GSTP*1A has the highest catalytic efficiency for the forward S-glutathionylation reaction (Manevich et al. 2013), and there is significant evidence that racial differences in expression of the polymorphic variants exist (Zhang et al. 2019a).

However, since TLK117 has two free carboxyl groups, the resulting charge was expected to inhibit cell uptake of the compound. Indeed, when tested in tumor cells which express GSTP1-1 as the predominant GST isozyme, TLK117 had neither toxicity nor potentiation. However, the diethyl ester form, TLK199 (γ-glutamyl-S-(benzyl)cysteinyl-R(−)-phenylglycine diethyl ester) did. The IC_{50} values were >200 μM for TLK117 compared with 22 μM for TLK199 in HT29 human colon adenocarcinoma cells line that express high levels of GSTP1-1 (Morgan et al. 1996). Similar IC_{50} values (26–28 μM) for TLK199 were obtained with other human colon

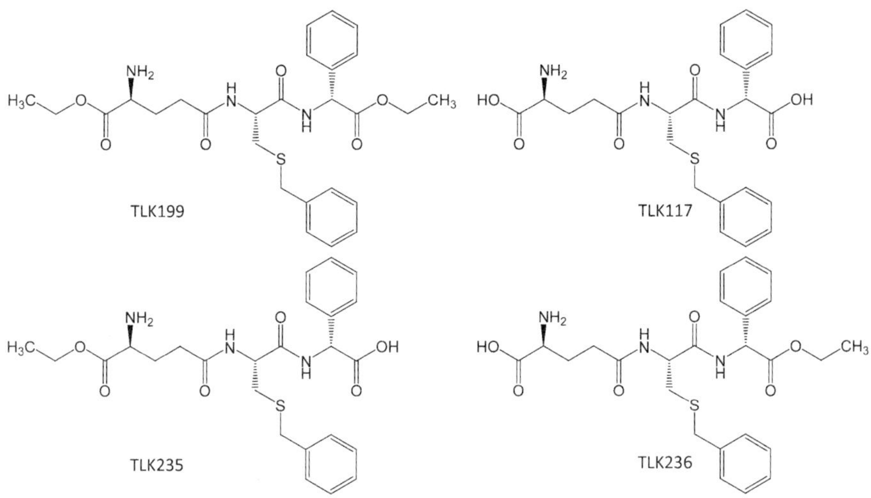

Fig. 1 Structures of Telintra (TLK199) and its metabolites

adenocarcinoma cell lines, e.g., SW620, LoVo, and Caco2 (Beaumont et al. 1998). TLK199 is easily taken up by the cells, rapidly converted to phenylglycine monoethyl ester TLK236, and then gradually converted to TLK117. TLK199 undergoes deesterification to glutamyl monoethyl ester TLK235 as well, but this metabolite is only produced in very limited quantities (Figs. 1 and 2) (Morgan et al. 1996). The absorption, distribution, metabolism, and elimination properties of TLK199 were characterized in rat and dog. The primary metabolites are TLK236 and TLK117. Unchanged TLK199 was not detected in the blood, although the metabolites TLK117 and TLK236 were, indicating that the systemic clearance of the parent compound was both rapid and extensive (Raza et al. 2009a). TLK199 has a half-life in rodents of <1 min and in monkeys of ~15 min (Kauvar et al. 1998).

Since GSTP1-1 is frequently overexpressed in tumors and correlates with the development of drug resistance, combinations of TLK199 with several chemotherapeutic agents has been tested to determine whether TLK199 would act as a chemosensitizer. In human cancer cell lines overexpressing GSTP1-1, e.g., HT29 colon adenocarcinoma, HT4-1 (HT29 subclone), SKOV3 ovarian carcinoma, and SK VLB (vinblastine-resistant variant of SKOV3), TLK199 (12.5 or 25 μM) was found to potentiate the toxicity of chlorambucil and doxorubicin by up to 2.5-fold (Morgan et al. 1996). In a separate study, the GSTP1-1 antisense cDNA construct was shown to sensitize the human colon cancer cells to doxorubicin about as well as TLK199 (Ban et al. 1996). Furthermore, increased sensitivity of melphalan (5 mg/kg), measured by human colon tumor growth in SCID mice, was achieved by TLK199 (60 mg/kg). No tumor growth inhibition was observed with TLK199 as a single agent (Morgan et al. 1996). As a chemosensitizer, TLK199's activity seemed easy to understand. TLK199 inhibits GSTP1-1, thus interferes with cellular phase II

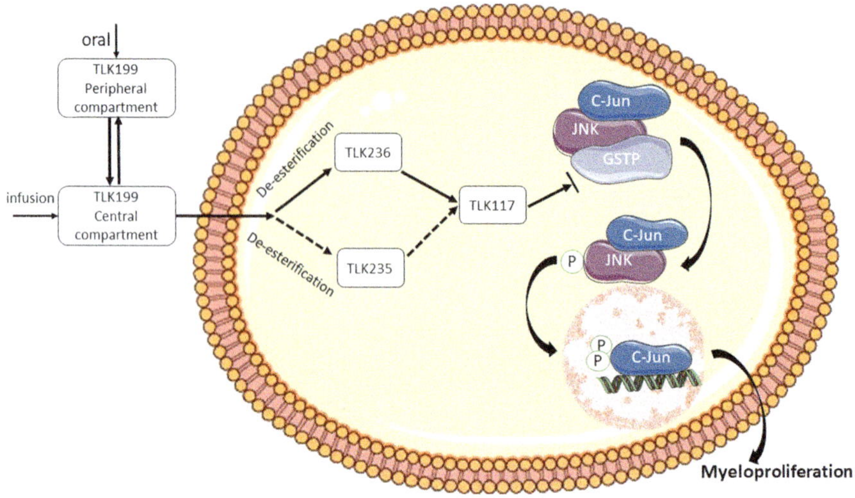

Fig. 2 Schematic of the influence of TLK199 on myeloproliferation

detoxification, leaving a cell susceptible to chemotherapeutic agents such as chlorambucil or melphalan. However, it was not immediately clear why TLK199 potentiated such a wide variety of drugs, including doxorubicin, which is not a specific GSTP1-1 substrate. We have posited at least one possible explanation in that TLK199 interfered with the GSTP1-1 and JNK interaction, leading to JNK activation (Bailey et al. 1994) and subsequent promotion of cancer cell apoptosis (Adler et al. 1999). In addition to GSTP1-1 inhibition, TLK199 has been implicated as an effective inhibitor (O'Brien et al. 1999) of multidrug resistance-associated protein 1 (MRP1 coded by the ABCC1 gene). MRP1 is an ATP-binding cassette transporter protein that plays an active role in multidrug resistance by its ability to efflux a vast array of anticancer drugs to sub-lethal levels (Cole et al. 1992). Using MRP1-transfected NIH3T3 mouse fibroblast cells with little detectable GSTP1-1 levels, TLK199 significantly inhibited the ATP-dependent transport, enhanced the accumulation, and subsequently reversed the resistance of numerous anticancer reagents, e.g., vincristine, etoposide, doxorubicin, daunorubicin, and mitoxantrone (O'Brien et al. 1999). Moreover, information now available indicates that a variety of genes, e.g., dihydrodiol dehydrogenase and γ-glutamylcysteine synthetase (rate-limiting enzyme in glutathione biosynthesis), are induced by TLK199 in cultured tumor cells (O'Dwyer et al. 1995). Modulation of expression of some of these genes might occur through the drug-induced perturbation of redox sensitive transcription factors such as Nrf2 and contribute further to the pharmacological actions of TLK199.

To elucidate how tumor cells may acquire resistance to TLK199, resistant cell lines were established. A human promyelocytic leukemia cell line (HL-60), which expresses GSTP1-1 as the predominant isozyme, was made tenfold resistant to TLK199. Both mRNA and protein levels of MRP1 were significantly increased by approximately 52- and 10-fold in the resistant cell line (HL60/TLK199). In addition, the HL60/TLK199 cells exhibited a drug resistance profile commensurate with a MRP1 overexpressing phenotype, with resistance to vinca alkaloids, epipodophyllotoxins, and anthracyclines (O'Brien et al. 1999). Further analysis of these cells revealed that TLK199 resistance was also associated with increased kinase activities of JNK1 (with threefold increase of basal expression levels) and ERK1/ERK2 (without modification of basal protein levels) (Ruscoe et al. 2001). The increased ERK1/ERK2 activities were suggested to protect the HL60/TLK199 cells against UV-induced apoptosis (Ruscoe et al. 2001) and PMA (phorbol 12-myristate 13-acetate)-induced cell growth arrest during monocyte/macrophage cytodifferentiation (Gate et al. 2003).

6 Hematopoiesis

Although TLK199 acted as a moderately effective chemosensitizer, the potentiation of therapeutic index was generally modest, a reflection of the state of the field at that time (Kauvar et al. 1998). Therefore, its effect on sensitizing normal cells to cytotoxins was examined, particularly in the case of the bone marrow where dose-limiting toxicity can sometimes predominate. Surprisingly, TLK199 showed a striking hematopoietic stimulatory effect.

In normal *Gstp1/p2*$^{+/+}$ mice, treatment with TLK199 (75 mg/kg i.p.) caused a twofold increase of circulating white blood cells, whereas no increase in white blood cell count was observed in *Gstp1/p2*$^{-/-}$ mice (Ruscoe et al. 2001). In addition, TLK199 administration caused significant increases in neutrophil levels in rodents and dogs (Hamilton and Batist 2005). Moreover, using a granulocyte/macrophage colony forming unit (CFU-GM) assay, direct effects of TLK199 on mouse bone marrow progenitor cell proliferation were found by in vivo (75 mg/kg i.p.) or in vitro (10 μM) treatments, and in each case, TLK199 induced a proliferative response, approximately twofold above vehicle control. Similar effects were also found in human bone marrow progenitor cells with TLK199 (1–10 μM) treatment. Furthermore, increased mobilization of the GM progenitors from mouse bone marrow to the spleen and peripheral blood was observed following treatment with TLK199. In contrast, the bone marrow from *Gstp1/p2*$^{-/-}$ mice did not respond to TLK199 (Kauvar et al. 1998; Ruscoe et al. 2001). These data are consistent with the results that *Gstp1/p2*$^{-/-}$ mice had higher basal levels of white blood cells compared with *Gstp1/p2*$^{+/+}$ mice, and cytokines (IL-3, GM-CSF, G-CSF, SCF, TPO, and Flt3L) were more effective at stimulating hematopoietic cell proliferation in *Gstp1/p2*$^{-/-}$ than in *Gstp1/p2*$^{+/+}$ mice (Gate et al. 2004; Zhang et al. 2014; Ruscoe et al. 2001). In addition, *Gstp1/p2*$^{-/-}$ mouse embryo fibroblast (MEF) cells doubled faster than *Gstp1/p2*$^{+/+}$ cells (26.2 versus 33.6 h) (Ruscoe et al. 2001).

Taken together, such evidence suggests that the presence, as well as the subsequent inhibition of GSTP1-1, is critical for the proliferative effects of TLK199. However, the mechanism underlying TLK199's hematopoietic stimulatory effects is not fully understood. These effects might be explainable by the ability of

TLK199 to disrupt the GSTP1-1 and JNK interaction (Adler et al. 1999), resulting in the activation of the JNK pathway that regulates proliferation, differentiation, and survival of hematopoietic cells (Geest and Coffer 2009) (Fig. 2). Indeed, treatment of TLK199 led to twofold increase in basal JNK activity in *Gstp1/p2*$^{+/+}$ cells (Adler et al. 1999), and the JNK inhibitor SP600125 completely inhibited the myelostimulant effects of TLK199 (Gate et al. 2004). Consistently, *Gstp1/p2*$^{-/-}$ cells exhibited higher basal levels of JNK activity (Adler et al. 1999), and SP600125 abrogated the differential myeloproliferation between *Gstp1/p2*$^{-/-}$ and *Gstp1/p2*$^{-/-}$ cells (Gate et al. 2004). Sustained activation of STAT proteins has been associated with increased proliferation of *Gstp1/p2*$^{-/-}$ bone marrow and mast cells (Gate et al. 2004). GSTP1-1 has been shown to be a negative regulator of STAT3. It binds to STAT3 and protects cells against EGF and angiotensin II-induced proliferation and migration through inhibition of STAT3 phosphorylation (Chen et al. 2014; Kou et al. 2013). However, whether TLK199 could interrupt the interaction between GSTP1-1 and STAT and in this way regulate myeloproliferation needs further investigation. Moreover, GSTP1-1 has the potential to mediate the S-glutathionylation of a number of proteins that may be involved in myeloproliferative events (Townsend et al. 2009), and this may provide a framework for explaining the myelostimulatory effects of TLK199.

The impact of TLK199 on normal animals and human bone marrow certainly provides opportunities for clinical application; however, the drug's effects on myelosuppressed subjects may also prove to be clinically relevant. Therefore, several preclinical studies have been performed in which TLK-199 was administered following chemotherapy. Data on the use of TLK199 in rodents demonstrated that TLK199 accelerated the recovery of circulating neutrophil levels following 5-fluorouracil treatment. Mice treated with TLK199 in addition to cisplatin or 5-fluorouracil demonstrated 60 or 100% of normal CFU-GM, respectively, compared with <10% observed when the cytotoxins were administered alone. In murine experiments, comparable results with TLK199 were also obtained following carboplatin and cyclophosphamide treatment. Overall, preclinical studies have demonstrated that (1) TLK199 reduces the severity of the cell count nadir in some animals; (2) cell count recovery to normal levels is accelerated by at least the same margin as that provided by G-CSF; and (3) the effects observed apply to both neutrophils and platelets, an advantage over G-CSF, which generally increases neutrophil numbers only (Kauvar et al. 1998; Hamilton and Batist 2005). TLK199 was non-toxic when parenterally administered daily for 7 days to both rats and dogs at doses up to 480 mg/m^2 and 800 mg/m^2, respectively (Hamilton and Batist 2005). No significant toxicities were observed in rats and dogs following daily oral administration of TLK199 for 14 days at doses up to 1,000 mg/kg and 20 mg/kg, respectively (Raza et al. 2009a). The collective preclinical results have been translated into phase I and phase II clinical trials in patients with myelodysplastic syndromes (MDS).

7 Use in Myelodysplastic Syndrome

Myelodysplastic syndromes represent a diverse group of bone marrow stem cell disorders predominantly affecting older individuals, with a median age at diagnosis of 65–70 years. The syndromes are characterized by ineffective hematopoiesis leading to cytopenia and in a third of patients, by progression to acute myeloid leukemia (AML) (Ades et al. 2014). The treatment options available are largely based on the patient's age and their prognosis as determined by the International Prognostic Scoring System (IPSS) (Greenberg et al. 1997, 2012). For patients in the IPSS low/intermediate [Int]-1 risk categories, the goal of the treatment is to improve infective hematopoiesis while providing the appropriate supportive care, including RBC and platelet transfusions, use of hematopoietic growth factors, antibiotics, and use of iron chelation therapy as appropriate. In higher-risk patients, the goal is to extend survival and delay transformation to AML. Currently, there are three FDA-approved drugs: the hypomethylating agents (HMAs) azacitidine and decitabine beneficial for higher-risk MDS patients and lenalidomide specific for lower-risk transfusion-dependent patients with del(5q) cytogenetic abnormalities. These agents, in addition to supportive care, immunosuppressive therapies, and allogeneic stem cell transplantation (allo-SCT), constitute the most commonly used therapeutic interventions (Zeidan et al. 2013). Overall, outside of a curative intent allo-SCT, the rest of the treatment modalities are palliative (Mahadevan and Sutton 2015). Even for those patients who proceed to allo-SCT, significant treatment-related mortality and morbidity and high relapse rates compromise long-term disease-free survival (Luger et al. 2012). There remains a clear need for new treatment options.

In this regard, TLK199, *Telintra* in its FDA-approved formulation, ezatiostat hydrochloride, was employed to treat MDS patients with low to intermediate risk. The drug company utilized the preclinical results and claimed that the promotion of proliferation and differentiation in normal hematopoietic cells and apoptosis of malignant cells was a sound rationale for phase I/II trials. Such a molecular mechanism was further supported by MDS patient pretreatment genomic data. Pathway analysis of the response profiles revealed that the genes comprising the JNK pathway, which is known to be activated by TLK199, are underexpressed in patients who were responders and overexpressed in patients who were non-responders to TLK199, suggesting that both the biology of the disease and the molecular mechanisms of action of the drug are positively correlated (Galili et al. 2012). There have been several clinical trials with MDS patients showing the safety and ezatiostat alone and in combination with lenalidomide (Table 1). In these trials, efficacy was based on the International Working Group (IWG) 2000 or 2006 criteria for hematologic improvement (HI) in the erythroid (HI-E), platelet (HI-P), or neutrophil (HI-N) lineages (Cheson et al. 2000, 2006). Adverse events (AEs) were graded by the National Cancer Institute Common Toxicity Criteria for Adverse Events (NCI-CTCAE) version 3.0.

The first-in-human phase I-IIa study of the intravenous (IV) formulation of ezatiostat was designed on the basis of safety demonstrated in multidose toxicity

Table 1 Summary of clinical trials with ezatiostat

Phase	Formulation	Dose	Patient	Response rate	Ref.
I	Tablet, orally	200, 400, 1,000, 1,400, 200, 2,400, 3,000, 4,000, 5,000, or 6,000 mg divided into two oral doses twice daily on days 1–7 of a 21-day treatment cycle	All WHO classification types of MDS with IPSS low to intermediate-2 risk	IWG 2000: HI-E: 6/29 (21%) HI-N: 4/19 (21%) HI-P: 7/21 (33%) Unilineage: 3/14 (21%) Bilineage: 3/25 (12%) IWG 2006: RBC transfusion reduction: 14/23 (61%) RBC transfusion independence: 8/23 (35%)	Raza et al. (2009a)
I	Tablet, orally	2000 mg divided into two oral doses twice daily in combination with lenalidomide 10 mg oral dose once daily on days 1–21 of a 28-day treatment cycle	All WHO classification types of MDS with IPSS low to intermediate-1 risk	IWG 2006: HI-E: 4/10 (40%) HI-N: 1/3 (33%) HI-P: 3/5 (60%) Bilineage: 5/11 (45%) Trilineage: 1/3 (33%) RBC transfusion independence: 3/7 (43%)	Raza et al. (2012b)
		2,500 mg divided into two oral doses twice daily in combination with lenalidomide 10 mg oral dose once daily on days 1–21 of a 28-day treatment cycle		IWG 2006: HI-E: 1/4 (25%) HI-N: 0/2 (0%) HI-P: 0/2 (0%) Bilineage: 0/5 (0%) Trilineage: 0/1 (0%) RBC transfusion independence: 0/2 (0%)	

(continued)

Table 1 (continued)

Phase	Formulation	Dose	Patient	Response rate	Ref.
I	Liposomes, intravenous injection	50, 100, 200, 400, and 600 mg/m2 daily at a constant rate infusion over 60 min on days 1–5 of a 14-day treatment cycle	All FAB classification types of MDS with IPSS low to high risk	IWG 2000: HI-E: 9/38 (24%) HI-N: 11/26 (42%) HI-P: 12/24 (50%) Unilineage: 7/14 (50%) Bilineage: 1/13 (8%) Trilineage: 4/16 (25%) RBC transfusion reduction: 5/31 (16%)	Raza et al. (2009b)
IIa		600 mg/m^2 daily on days 1–5 or days 1–3 of a 21-day treatment cycle			
II	Tablet, orally	3,000 mg divided into two oral doses twice daily on days 1–14 of a 21-day treatment cycle	All WHO classification types of MDS with IPSS low to intermediate-1 risk	IWG 2006: HI-E: 7/29 (24%) HI-N: 1/10 (10%) HI-P: 1/14 (7%) Bilineage: 1/9 (11%) Trilineage: 1/6 (17%) RBC transfusion reduction: 6/20 (30%) RBC transfusion independence: 1/20 (5%)	Raza et al. (2012a)
		2000 mg divided into two oral doses twice daily on days 1–21 of a 28-day treatment cycle		IWG 2006: HI-E: 6/32 (19%) HI-N: 3/11 (27%) HI-P: 0/0 (0%) Bilineage: 3/11 (27%) Trilineage: 0/0 (0%) RBC transfusion reduction: 5/18 (28%)	

(continued)

Table 1 (continued)

Phase	Formulation	Dose	Patient	Response rate	Ref.
				RBC transfusion independence: 3/18 (17%)	

FAB French-American-British, *IPSS* International Prognostic Scoring System, *IWG* International Working Group, *MDS* myelodysplastic syndromes, *RBC* red blood cells, *WHO* World Health Organization

studies and efficacy reported in animal models. Fifty-four MDS patients were enrolled. Phase I patients received liposomal ezatiostat at five dose levels (50, 100, 200, 400, and 600 mg/m^2) intravenously on days 1–5 of a 14-day cycle until MDS progression or unacceptable toxicity. In phase IIa, ezatiostat was administered on two dose schedules (DSs): 600 mg/m^2 IV on days 1–5 or days 1–3 of a 21-day treatment cycle. The most common AEs were grades 1 or 2 and non-hematologic, including chills, back pain, flushing, nausea, bone pain, fatigue, extremity pain, dyspnea, and diarrhea related to acute infusion hypersensitivity reactions. Pharmacokinetic parameters were estimated and derived for TLK199, TLK236, and TLK117. The ezatiostat elimination half-life was 0.20 h, an AUC/dose of 0.008 h/L, and a distribution half-life of 0.03 h. The active metabolite TLK236 had a half-life of 2.65 h, with an AUC/dose of 0.341 h/L, and TLK117 had a half-life of 0.24–0.60 h with an AUC/dose of 0.0116 h/L. Overall, trilineage responses were observed in 25% patients with trilineage cytopenia. HI-E, HI-N, and HI-P were observed in 24%, 42%, and 50% patients, respectively. These responses were accompanied by improvement in clinical symptoms and independence or reduction in red blood cell (RBC) and platelet transfusion requirements (Raza et al. 2009b).

Based on the promising clinical results obtained from the intravenous formulation of ezatiostat, a phase I study with an oral formulation (ezatiostat tablets) was initiated. Forty-five patients with low to Int-2 risk MDS were enrolled and received ten dose levels (200, 400, 1,000, 1,400, 2000, 2,400, 3,000, 4,000, 5,000, and 6,000 mg) of ezatiostat tablets divided into two oral doses twice daily on days 1–7 of a 21-day cycle. No dose-limiting toxicities were observed. The most common treatment-related AEs were non-hematologic and mild or moderate in grade (1 or 2), including nausea, diarrhea, vomiting, abdominal pain, constipation, anorexia, and dyspepsia. Levels of the major metabolite TLK236 increased proportionate to ezatiostat dosage. Eleven of the seventeen HI responses were observed at doses of 4,000–6,000 mg/day, reflecting a dose response. HI responses occurred in all lineages including three bilineage and one complete cytogenetic response. Decreased numbers of RBC and platelet transfusions and in some cases transfusion independence were attained (Raza et al. 2009a). These findings supported the further development of extended dose schedules of ezatiostat tablets in MDS.

Subsequently, a phase II study was conducted to evaluate two extended dose schedules of oral ezatiostat in 89 heavily pretreated patients with low to Int-1 risk MDS. In DS1, patients received 3,000 mg of ezatiostat tablets divided into two oral doses twice daily for 14 days of a 21-day cycle, and in DS2, patients received 2000 mg of ezatiostat tablets divided into two oral doses twice daily for 21 days of a 28-day cycle. Most common ezatiostat-related AEs were grade 1 and 2 gastrointestinal, including nausea, diarrhea, and vomiting. Overall, 29% of the RBC transfusion-dependent patients had transfusion reduction, with 11% achieving transfusion independence. The median duration of HI-E response was 34 weeks. Multilineage responses were observed. There was one cytogenetic complete response in a del (5q) MDS patient. An important trend was the effect of prior therapy on response. A 40% HI-E rate was observed in patients who had prior lenalidomide and were HMA naive, with 45% patients achieving significant RBC transfusion reduction and 27% achieving transfusion independence. In contrast, a 28% HI-E rate was observed in patients who were both lenalidomide and HMA naive, with 50% patients achieving clinically significant RBC transfusion reductions. The higher responses of ezatiostat in the subsets of patients previously treated with lenalidomide suggested a potential role for combining the two drugs. In addition, DS2 was selected for further ezatiostat studies due to its longer median duration of HI-E response (46 weeks), better tolerability of the lower daily dose and the greater convenience for patients of dosing with two tablets twice a day (Raza et al. 2012a).

Therefore, a phase I study was conducted to determine the safety and efficacy of ezatiostat in combination with lenalidomide. Nineteen patients with non-del(5q) MDS received one of two doses of ezatiostat (2000 mg or 2,500 mg/day) in combination with 10 mg of lenalidomide on days 1–21 of a 28-day cycle. No unexpected toxicities occurred, and the incidence and severity of AEs were consistent with those expected for each drug alone. All multilineage responses were observed in the 2000/10 mg doses, recommended for future studies. In the 2000/10 mg dose group, 4 of 10, 1 of 3, and 3 of 5 evaluable patients experienced an HI-E, HI-N, and HI-P response, respectively. Bilineage responses, HI-E/HI-P, HI-E/HI-N, and HI-N and HI-P, occurred in 3 of 5, 1 of 3, and 1 of 3 patients, respectively. One of three patients with pancytopenia experienced a complete trilineage response. In addition, three of seven RBC transfusion-dependent patients became RBC transfusion independent, including one patient for whom prior lenalidomide monotherapy was ineffective (Raza et al. 2012b).

Additionally, there are two case reports of MDS patients who responded unexpected well to ezatiostat (Quddus et al. 2010; Lyons et al. 2011). Both patients participated in the phase II study comparing two DSs of ezatiostat tablets for low to intermediate-1 risk MDS. The first patient was a 77-year-old male who relapsed after a short course of lenalidomide, with the disappearance of the del(5q) but the appearance of a new clonal abnormality t(2; 3) upon relapse. The patient discontinued lenalidomide and was randomized to receive ezatiostat tablets at 3000 mg/day for 14 days of a 21-day cycle. Five days into his second treatment cycle, he was withdrawn from the study due to intolerable side effects. However, striking improvement in all three blood counts had been observed since the initiation of the study and continued to remain high a year post-therapy, suggesting a role of ezatiostat in the treatment of patients who are resistant to lenalidomide (Quddus et al. 2010). Another patient was a 64-year-old female who suffered from longstanding idiopathic chronic neutropenia (ICN) with frequent episodes of sepsis, and had an inadequate response to G-CSF. She was randomized to receive ezatiostat tablets at 2000 mg/day for 21 days of a 28-day cycle. She responded by the end of the first cycle of treatment with stabilization of her absolute neutrophil count (ANC), clearing of fever and healing of areas of infection. Following eight cycles of treatment, she had continued to show remarkable improvement of ANC, suggesting a potential role of ezatiostat in the treatment of patients with ICN who are not responsive to G-CSF (Lyons et al. 2011).

Overall, the available clinical data have shown favorable tolerability and hematopoietic-promoting activity profiles for ezatiostat in MDS patients and indicated that the drug was worthy of further evaluation in randomized phase II and phase III trials. Missing from all of these clinical efforts was any type of "precision medicine" approach to patient selection. Given the time period that these drugs emerged and the subsequent trial design, there were no efforts to strategize patient selection on the basis of GST polymorphisms and no trial components to use possible biomarkers to assess drug efficacy. Given the variable catalytic efficiencies for GSTP variants (Manevich et al. 2013) and the recent indications of the utility of S-glutathionylated serum biomarkers in predicting response to electrophilic stress in patients (Zhang et al. 2019b), the absence of any pharmacogenetics approach to trial design may have restricted the chances for positive outcomes.

8 Lung Diseases

One of us (Y.J.) has specialized in studying lung pathologies such as idiopathic pulmonary fibrosis (IPF) and chronic obstructive pulmonary disease (COPD). The former is characterized by excessive collagen production and fibrogenesis and the latter by airway wall thickening and/or emphysema. Obviously, the lung is exposed to high oxygen concentrations, and aberrant GSH homeostasis has also been implicated in the presentation of each of these two disease states. As a consequence, various evidence have been generated to identify that GSTP-mediating S-glutathionylation is involved in these pathologies and that, by extension, *Telintra* may have relevance in their management.

As a first example, pro-inflammatory signaling cascades frequently begin with stimulation of the transcription factor nuclear factor kappa B (NfκB). To this end, S-glutathionylation has been shown to regulate the activity of inhibitory kappa B kinase beta (IKKβ), and GSTP also interacts with the adaptor protein TRAF2, a known regulator of NfκB (Jones et al. 2016). In mouse lung alveolar epithelial cells, a constitutive association between GSTP and IκBα was reported in unstimulated cells, rapidly lost when treated with LPS, but at a time that preceded IκBα degradation. In principle, this GSTP/IκBα interaction could prevent the phosphorylation and ubiquitination of IκBα, thereby preventing NFκB activation. LPS-induced nuclear contents of RelA and RelA phosphorylation were increased in cells following siRNA-mediated GSTP knockdown. In this regard, both GSTP knockdown and

TLK117 (active metabolite see Fig. 1) treatments mediated GSTP inhibition and enhanced LPS-induced NF-κB luciferase activity and cytokine production, suggesting a potential regulatory function of GSTP in preventing IκBα phosphorylation and/or degradation. Cysteine189 of IκBα is the site of S-glutathionylation causing a decrease in phosphorylation by IKK that can attenuate ubiquitination (Kil et al. 2008), limiting its degradation and subsequent activation of NF-κB (Seidel et al. 2011). There are reasons to believe that the protein interaction(s) between GSTP and IκBα may be stabilized in some manner by the process of S-glutathionylating the target cysteine residue. In context, these events may then control activation and/or assembly of the IKK signalsome. Since GSTP does not affect IKKβ S-glutathionylation until 6 h after LPS exposure (a time at which NF-κB transcriptional activity is beginning to subside (Jones et al. 2016)), S-glutathionylation of IKK proteins may represent a mechanism whereby GSTP can attenuate NF-κB. This possible model predicts that in the absence of stimulus, GSTP prevents degradation of endogenous IκBα and that GSTP-mediated S-glutathionylation shuts down IKK activity providing a versatile mechanism for regulation of NFκB by GSTP. Overall, in light of the reported relevance of GSTP polymorphisms in allergic asthma (McCunney 2005), pharmacological manipulation of GSTP by drugs like *Telintra* may prove, in the future, to be a useful therapeutic approach to regulate pro-inflammatory signaling in these types of lung diseases.

Lung tissue remodeling in chronic obstructive pulmonary disease (COPD) is characterized by airway wall thickening and/or emphysema. Surfactant protein C (SPC)-TNF-α mice showed remodeling in alveolar and airway walls similar to those observed in patients with COPD. Epithelial cells are able to undergo a phenotypic shift, gaining mesenchymal properties, a process in which JNK signaling is involved. Consequently, TNF-α induces JNK-dependent epithelial plasticity, contributing to lung matrix remodeling. A pharmacological inhibitor of JNK attenuated this phenotypic shift, indicating the role of JNK signaling in this process. Activation of JNK signaling was also present in the lungs of SPC-TNF-α mice and patients with COPD. Together, these studies provide evidence for the involvement of the TNF-alpha-JNK axis in extracellular matrix remodeling. In light of the known connections between JNK and GSTP, this may also indicate a role for *Telintra* in impacting JNK activity, particularly since the drug is known to interfere with the interactions between the two proteins.

IPF is a debilitating disease characterized by the development of excess fibrous tissue that causes thickening of alveolar walls and diminished lung function (Lomas et al. 2012). It is the most common subtype of interstitial lung disease, impacting >120,000 Americans with 40,000 deaths each year (Blackwell et al. 2014; Raghu et al. 2006). Apoptosis in lung epithelial cells is a critical determinant for the extent of disease progression, since increased loss of these cells promotes fibroblast activation and remodeling. Changes in glutathione and GST expression patterns have been reported in IPF patients (Anathy et al. 2012), and this provided an opportunity to consider a therapeutic intervention strategy with *Telintra* (McMillan et al. 2016). GSTP mediates lung fibrogenesis in part through FAS

S-glutathionylation, a critical event in epithelial cell apoptosis. GSTP expression (as well as the FAS-GSTP interaction) is increased in the lungs of IPF patients, mostly within type II epithelial cells. Bleomycin- and AdTGFβ-induced increases in collagen content, α-SMA, FAS S-glutathionylation, and total protein S-glutathionylation were strongly attenuated in GSTP knockout mice (McMillan et al. 2016). Oropharyngeal administration of TLK117, at a time when fibrosis was already apparent, attenuated bleomycin- and AdTGFβ-induced remodeling, α-SMA, caspase activation, FAS S-glutathionylation, and total protein S-glutathionylation. GSTP is an important driver of protein S-glutathionylation and lung fibrosis, and GSTP inhibition via inhalation of the *Telintra* active moiety may prove to be a novel therapeutic strategy for the future management of IPF.

9 Future Perspectives

In general terms the clinical success of drugs designed to target redox homeostasis have had limited success. There are many potential reasons for this, highlighted perhaps by the necessary redundancy inherent in maintaining cellular redox homeostasis (Zhang et al. 2018) and the evolutionary importance of oxidative regulation of a variety of transcription factors that control critical cell function (Hayes et al. 2020). Although clinical trials in MDS indicate that the drug has activity, the present absence of a corporate sponsor and supply of available GMP drug suggests that instigation of further clinical studies may be limited by these exigencies. When *Telik, Inc.* was reverse merged into privately held *MabVax Therapeutics Inc.* in May 2014, *Telintra* development was deemphasized, and this year, this company filed for bankruptcy. Patent coverage of the drug has expired, perhaps contributing to the reduction of corporate interest, but should remove limitations in further academic developments. In moving forward, the fact that the time-consuming components of formulation and initial IND application have already been accomplished does provide opportunities for more rapid clinical development. It should be noted though that positive preclinical results in lung disorders made use of a nasopharangeal administration route. Moreover, there are emerging examples of where S-glutathionylated proteins are critical in regulating important pathways. Since inhibition of GSTP limits this post-translational modification, there may prove to be a role for the drug in this area. We and others have discussed previously the importance of sulfur amino acid homeostasis to the bone marrow environment (Gate et al. 2004). Indeed, many leukemias, including CLL, rely upon cystine transporters from the surrounding marrow stromal environment to provide sufficient cystine as a precursor of cysteine (Zhang et al. 2012). Since both qualitative and quantitative aspects of protein S-glutathionylation will depend upon a balanced supply of GSH and GSTP, drugs like *Telintra* may hold promise in delineating the physiological importance of these pathways in both normal and malignant marrow tissues.

References

Adang AE et al (1990) The glutathione-binding site in glutathione S-transferases. Investigation of the cysteinyl, glycyl and gamma-glutamyl domains. Biochem J 269(1):47–54

Ades L, Itzykson R, Fenaux P (2014) Myelodysplastic syndromes. Lancet 383(9936):2239–2252

Adler V et al (1999) Regulation of JNK signaling by GSTp. EMBO J 18(5):1321–1334

Anathy V et al (2012) Redox-based regulation of apoptosis: S-glutathionylation as a regulatory mechanism to control cell death. Antioxid Redox Signal 16(6):496–505

Askelof P et al (1975) Purification and characterization of two glutathione S-aryltransferase activities from rat liver. Biochem J 147(3):513–522

Bailey HH et al (1994) Phase I clinical trial of intravenous L-buthionine sulfoximine and melphalan: an attempt at modulation of glutathione. J Clin Oncol 12(1):194–205

Ban N et al (1996) Transfection of glutathione S-transferase (GST)-pi antisense complementary DNA increases the sensitivity of a colon cancer cell line to adriamycin, cisplatin, melphalan, and etoposide. Cancer Res 56(15):3577–3582

Beaumont PO et al (1998) Role of glutathione S-transferases in the resistance of human colon cancer cell lines to doxorubicin. Cancer Res 58(5):947–955

Blackwell TS et al (2014) Future directions in idiopathic pulmonary fibrosis research. An NHLBI workshop report. Am J Respir Crit Care Med 189(2):214–222

Boyland E, Chasseaud LF (1969) Glutathione S-aralkyltransferase. Biochem J 115(5):985–991

Chen D et al (2014) GSTpi protects against angiotensin II-induced proliferation and migration of vascular smooth muscle cells by preventing signal transducer and activator of transcription 3 activation. Biochim Biophys Acta 1843(2):454–463

Cheson BD et al (2000) Report of an international working group to standardize response criteria for myelodysplastic syndromes. Blood 96(12):3671–3674

Cheson BD et al (2006) Clinical application and proposal for modification of the International Working Group (IWG) response criteria in myelodysplasia. Blood 108(2):419–425

Cho SG et al (2001) Glutathione S-transferase mu modulates the stress-activated signals by suppressing apoptosis signal-regulating kinase 1. J Biol Chem 276(16):12749–12755

Cole SP et al (1992) Overexpression of a transporter gene in a multidrug-resistant human lung cancer cell line. Science 258(5088):1650–1654

Davis RJ (2000) Signal transduction by the JNK group of MAP kinases. Cell 103(2):239–252

Dorion S, Lambert H, Landry J (2002) Activation of the p38 signaling pathway by heat shock involves the dissociation of glutathione S-transferase Mu from Ask1. J Biol Chem 277 (34):30792–30797

Flatgaard JE, Bauer KE, Kauvar LM (1993) Isozyme specificity of novel glutathione-S-transferase inhibitors. Cancer Chemother Pharmacol 33(1):63–70

Galili N et al (2012) Prediction of response to therapy with ezatiostat in lower risk myelodysplastic syndrome. J Hematol Oncol 5:20

Gate L, Lunk A, Tew KD (2003) Resistance to phorbol 12-myristate 13-acetate-induced cell growth arrest in an HL60 cell line chronically exposed to a glutathione S-transferase pi inhibitor. Biochem Pharmacol 65(10):1611–1622

Gate L et al (2004) Increased myeloproliferation in glutathione S-transferase pi-deficient mice is associated with a deregulation of JNK and Janus kinase/STAT pathways. J Biol Chem 279 (10):8608–8616

Geest CR, Coffer PJ (2009) MAPK signaling pathways in the regulation of hematopoiesis. J Leukoc Biol 86(2):237–250

Ghezzi P (2005) Regulation of protein function by glutathionylation. Free Radic Res 39 (6):573–580

Graminski GF, Kubo Y, Armstrong RN (1989) Spectroscopic and kinetic evidence for the thiolate anion of glutathione at the active site of glutathione S-transferase. Biochemistry 28 (8):3562–3568

Greenberg P et al (1997) International scoring system for evaluating prognosis in myelodysplastic syndromes. Blood 89(6):2079–2088

Greenberg PL et al (2012) Revised international prognostic scoring system for myelodysplastic syndromes. Blood 120(12):2454–2465

Griffith OW, Meister A (1979) Potent and specific inhibition of glutathione synthesis by buthionine sulfoximine (S-n-butyl homocysteine sulfoximine). J Biol Chem 254(16):7558–7560

Hamilton D, Batist G (2005) TLK-199 (Telik). IDrugs 8(8):662–669

Hayes JD et al (2020) Oxidative stress in cancer. Cancer Cell. https://doi.org/10.1016/j.ccell.2020.06.001

Ichijo H et al (1997) Induction of apoptosis by ASK1, a mammalian MAPKKK that activates SAPK/JNK and p38 signaling pathways. Science 275(5296):90–94

Johansson AS, Ridderstrom M, Mannervik B (2000) The human glutathione transferase P1-1 specific inhibitor TER 117 designed for overcoming cytostatic-drug resistance is also a strong inhibitor of glyoxalase I. Mol Pharmacol 57(3):619–624

Jones JT et al (2016) Glutathione S-transferase pi modulates NF-kappaB activation and pro-inflammatory responses in lung epithelial cells. Redox Biol 8:375–382

Kauvar LM et al (1998) Glutathione based approaches to improving cancer treatment. Chem Biol Interact 111-112:225–238

Kil IS, Kim SY, Park JW (2008) Glutathionylation regulates IkappaB. Biochem Biophys Res Commun 373(1):169–173

Kou X et al (2013) GSTP1 negatively regulates Stat3 activation in epidermal growth factor signaling. Oncol Lett 5(3):1053–1057

Lomas NJ et al (2012) Idiopathic pulmonary fibrosis: immunohistochemical analysis provides fresh insights into lung tissue remodelling with implications for novel prognostic markers. Int J Clin Exp Pathol 5(1):58–71

Luger SM et al (2012) Similar outcomes using myeloablative vs reduced-intensity allogeneic transplant preparative regimens for AML or MDS. Bone Marrow Transplant 47(2):203–211

Lyons RM et al (2011) Oral ezatiostat HCl (Telintra(R), TLK199) and idiopathic chronic neutropenia (ICN): a case report of complete response of a patient with G-CSF resistant ICN following treatment with ezatiostat, a glutathione S-transferase P1-1 (GSTP1-1) inhibitor. J Hematol Oncol 4:43

Lyttle MH et al (1994) Isozyme-specific glutathione-S-transferase inhibitors: design and synthesis. J Med Chem 37(1):189–194

Mahadevan D, Sutton GR (2015) Ezatiostat hydrochloride for the treatment of myelodysplastic syndromes. Expert Opin Investig Drugs 24(5):725–733

Manevich Y et al (2013) Allelic variants of glutathione S-transferase P1-1 differentially mediate the peroxidase function of peroxiredoxin VI and alter membrane lipid peroxidation. Free Radic Biol Med 54:62–70

McCunney RJ (2005) Asthma, genes, and air pollution. J Occup Environ Med 47(12):1285–1291

McMillan DH et al (2016) Attenuation of lung fibrosis in mice with a clinically relevant inhibitor of glutathione-S-transferase pi. JCI Insight 1(8):e85717

Morgan AS et al (1996) Isozyme-specific glutathione S-transferase inhibitors potentiate drug sensitivity in cultured human tumor cell lines. Cancer Chemother Pharmacol 37(4):363–370

O'Dwyer PJ et al (1992) Depletion of glutathione in normal and malignant human cells in vivo by buthionine sulfoximine: clinical and biochemical results. J Natl Cancer Inst 84(4):264–267

Oakley AJ et al (1997) The structures of human glutathione transferase P1-1 in complex with glutathione and various inhibitors at high resolution. J Mol Biol 274(1):84–100

O'Brien ML et al (1999) Glutathione peptidomimetic drug modulator of multidrug resistance-associated protein. J Pharmacol Exp Ther 291(3):1348–1355

O'Dwyer PJ et al (1991) Phase I study of thiotepa in combination with the glutathione transferase inhibitor ethacrynic acid. Cancer Res 51(22):6059–6065

O'Dwyer PJ et al (1995) Modulation of glutathione and related enzymes in reversal of resistance to anticancer drugs. Hematol Oncol Clin North Am 9(2):383–396

Petrini M et al (1993) Reversing of chlorambucil resistance by ethacrynic acid in a B-CLL patient. Br J Haematol 85(2):409–410

Ploemen JH et al (1993) Ethacrynic acid and its glutathione conjugate as inhibitors of glutathione S-transferases. Xenobiotica 23(8):913–923

Quddus F et al (2010) Oral Ezatiostat HCl (TLK199) and Myelodysplastic syndrome: a case report of sustained hematologic response following an abbreviated exposure. J Hematol Oncol 3:16
Raghu G et al (2006) Incidence and prevalence of idiopathic pulmonary fibrosis. Am J Respir Crit Care Med 174(7):810–816
Raza A et al (2009a) Phase 1 multicenter dose-escalation study of ezatiostat hydrochloride (TLK199 tablets), a novel glutathione analog prodrug, in patients with myelodysplastic syndrome. Blood 113(26):6533–6540
Raza A et al (2009b) Phase 1-2a multicenter dose-escalation study of ezatiostat hydrochloride liposomes for injection (Telintra, TLK199), a novel glutathione analog prodrug in patients with myelodysplastic syndrome. J Hematol Oncol 2:20
Raza A et al (2012a) A phase 2 randomized multicenter study of 2 extended dosing schedules of oral ezatiostat in low to intermediate-1 risk myelodysplastic syndrome. Cancer 118 (8):2138–2147
Raza A et al (2012b) Phase 1 dose-ranging study of ezatiostat hydrochloride in combination with lenalidomide in patients with non-deletion (5q) low to intermediate-1 risk myelodysplastic syndrome (MDS). J Hematol Oncol 5:18
Romero L et al (2006) Human GSTA1-1 reduces c-Jun N-terminal kinase signalling and apoptosis in Caco-2 cells. Biochem J 400(1):135–141
Ruscoe JE et al (2001) Pharmacologic or genetic manipulation of glutathione S-transferase P1-1 (GSTpi) influences cell proliferation pathways. J Pharmacol Exp Ther 298(1):339–345
Seidel P et al (2011) IkappaBalpha glutathionylation and reduced histone H3 phosphorylation inhibit eotaxin and RANTES. Eur Respir J 38(6):1444–1452
Tew KD (1994) Glutathione-associated enzymes in anticancer drug resistance. Cancer Res 54 (16):4313–4320
Tew KD, Bomber AM, Hoffman SJ (1988) Ethacrynic acid and piriprost as enhancers of cytotoxicity in drug resistant and sensitive cell lines. Cancer Res 48(13):3622–3625
Tew KD, Dutta S, Schultz M (1997) Inhibitors of glutathione S-transferases as therapeutic agents. Adv Drug Deliv Rev 26(2–3):91–104
Townsend DM, Tew KD (2003) The role of glutathione-S-transferase in anti-cancer drug resistance. Oncogene 22(47):7369–7375
Townsend DM et al (2009) Novel role for glutathione S-transferase pi. Regulator of protein S-glutathionylation following oxidative and nitrosative stress. J Biol Chem 284(1):436–445
Wang AL, Tew KD (1985) Increased glutathione-S-transferase activity in a cell line with acquired resistance to nitrogen mustards. Cancer Treat Rep 69(6):677–682
Wu Y et al (2006) Human glutathione S-transferase P1-1 interacts with TRAF2 and regulates TRAF2-ASK1 signals. Oncogene 25(42):5787–5800
Ye ZW et al (2017) Glutathione S-transferase P-mediated protein S-glutathionylation of resident endoplasmic reticulum proteins influences sensitivity to drug-induced unfolded protein response. Antioxid Redox Signal 26(6):247–261
Yin Z et al (2000) Glutathione S-transferase p elicits protection against H2O2-induced cell death via coordinated regulation of stress kinases. Cancer Res 60(15):4053–4057
Zeidan AM, Linhares Y, Gore SD (2013) Current therapy of myelodysplastic syndromes. Blood Rev 27(5):243–259
Zhang W et al (2012) Stromal control of cystine metabolism promotes cancer cell survival in chronic lymphocytic leukaemia. Nat Cell Biol 14(3):276–286
Zhang J et al (2014) Glutathione S-transferase P influences redox and migration pathways in bone marrow. PLoS One 9(9):e107478
Zhang J et al (2018) An evolving understanding of the S-glutathionylation cycle in pathways of redox regulation. Free Radic Biol Med 120:204–216
Zhang J et al (2019a) Racial disparities, cancer and response to oxidative stress. Adv Cancer Res 144:343–383
Zhang L et al (2019b) S-glutathionylated serine proteinase inhibitors as biomarkers for radiation exposure in prostate cancer patients. Sci Rep 9(1):13792

Perspectives on the Clinical Development of NRF2-Targeting Drugs

Diego Lastra, Raquel Fernández-Ginés, Gina Manda, and Antonio Cuadrado

Contents

Gina Manda and Antonio Cuadrado contributed equally to this work.

D. Lastra · R. Fernández-Ginés
Instituto de Investigaciones Biomédicas "Alberto Sols" UAM-CSIC, Instituto de Investigación Sanitaria La Paz (IdiPaz), Madrid, Spain

Department of Biochemistry, Faculty of Medicine, Autonomous University of Madrid, Madrid, Spain

Centro de Investigación Biomédica en Red sobre Enfermedades Neurodegenerativas (CIBERNED), ISCIII, Madrid, Spain

G. Manda
Victor Babes National Institute of Pathology, Bucharest, Romania

A. Cuadrado (✉)
Instituto de Investigaciones Biomédicas "Alberto Sols" UAM-CSIC, Instituto de Investigación Sanitaria La Paz (IdiPaz), Madrid, Spain

Department of Biochemistry, Faculty of Medicine, Autonomous University of Madrid, Madrid, Spain

Centro de Investigación Biomédica en Red sobre Enfermedades Neurodegenerativas (CIBERNED), ISCIII, Madrid, Spain

Victor Babes National Institute of Pathology, Bucharest, Romania
e-mail: antonio.cuadrado@uam.es

H. H. H. W. Schmidt et al. (eds.), *Reactive Oxygen Species*,
Handbook of Experimental Pharmacology 264, https://doi.org/10.1007/164_2020_381

Abstract

The transcription factor NRF2 (nuclear factor erythroid 2-related factor 2) triggers homeostatic responses against a plethora of environmental or endogenous deviations in redox metabolism, inflammation, proteostasis, etc. Therefore, pharmacological activation of NRF2 is a promising therapeutic strategy for several chronic diseases that are underlined by low-grade oxidative inflammation and dysregulation of redox metabolism, such as neurodegenerative, cardiovascular, and metabolic diseases. While NRF2 activation is useful in inhibiting carcinogenesis, its inhibition is needed in constituted tumors where NRF2 provides a survival advantage in the challenging tumor niche. This review describes the electrophilic and non-electrophilic NRF2 activators with clinical projection in various chronic diseases. We also analyze the status of NRF2 inhibitors, which are for the moment in a proof-of-concept stage. Advanced in silico screening and medicinal chemistry are expected to provide new or repurposing small molecules with increased potential for fostering the development of targeted NRF2 modulators.

Graphical Abstract

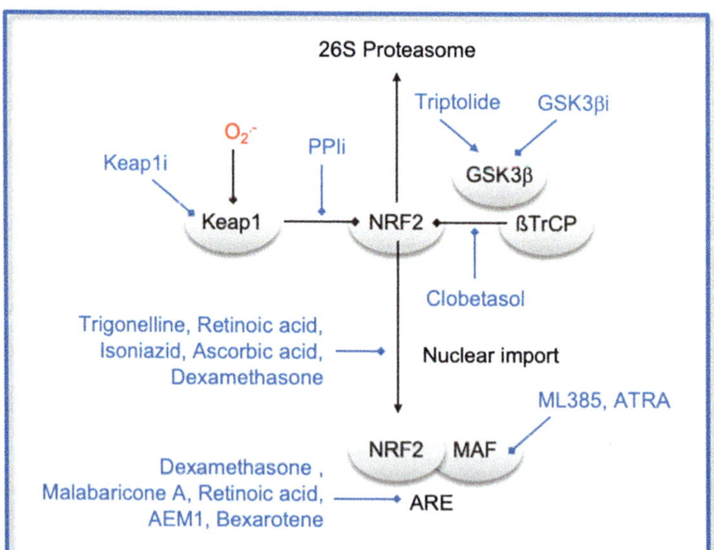

The nuclear factor erythroid 2 (NFE2)-related factor 2 (NRF2) is rapidly degraded by proteasomes under a basal condition in a Keap1-dependent manner. ROS oxidatively modifies Keap1 to release NRF2 and allow its nuclear translocation. Here it binds to the antioxidant response element to regulate gene transcription. An alternative mechanism controlling NRF2 stability is glycogen synthase kinase 3 (GSK-3)-induced phosphorylation. Indicated in blue are NRF2-activating and NRF2-inhibiting drugs.

Keywords

Chronic diseases · Cytoprotection · Inflammation

Abbreviations

AD	Alzheimer's disease
AHR	Aryl hydrocarbon receptor
BACH1	BTB domain and CNC homolog 1
BTB	Broad complex, tramtrack, bric-a-brac
CUL3	Cullin 3
DMF	Dimethyl fumarate
DRG	Double glycine repeat
GSH	Glutathione
GSK-3	Glycogen synthase kinase
IVR	Intervening region
KEAP1	Kelch-like ECH-associated protein 1
MMF	Monomethyl fumarate
MS	Multiple sclerosis
NFE2L2	Gene encoding NRF2
NRF2	Nuclear factor erythroid 2-related factor 2
PD	Parkinson's disease
PPI	Protein-protein interaction
RBX1	RING-box protein 1
ROS	Reactive oxygen species
SFN	Sulforaphane
SQSTM1	Sequestosome 1
XRE	Xenobiotic response element
β-TrCP	Beta-transducin repeat containing E3 ubiquitin protein ligase

1 Introduction

Chronic low-grade ROS formation and inflammation are underlining various chronic diseases (cardiovascular, neurodegenerative, and metabolic disorders) as well as cancer, being active long before disease-specific symptoms become clinically overt. Of note, aging and its abovementioned comorbidities share common pathological mechanisms that converge to persistent inflammation. In fact, the "inflammaging" theory proposes that aging is a consequence of the loss of homeostatic responses to inflammation and dysregulation of redox metabolism (Franceschi et al. 2018). One possible cause of uncontrolled and persistent inflammation is a chronically deregulated redox balance which exacerbates inflammatory reactions. In turn, inflammatory reactions accentuate redox alterations, hence generating a feed-forward loop (Biswas 2016). The origin of the persistent alterations in redox metabolism is still unknown (Schmidlin et al. 2019).

Chronic treatment with antioxidant supplements (beta carotene, vitamin A, vitamin C, vitamin E, and selenium) did not show long-term efficacy, excepting the case when a deficiency of micronutrients occurs due to inadequate intakes or malnutrition (Biesalski et al. 2010). The study of Myung et al. (2013) concluded that there is no evidence to support the use of vitamin and antioxidant supplements for prevention of chronic cardiovascular diseases (Myung et al. 2013). Additionally, the meta-analysis performed by Bjelakovic et al. (2006) on 17 randomized trials with 620 participants could not provide convincing evidence that antioxidant supplements have significant beneficial effects on primary or secondary prevention of colorectal adenoma. Of note, the meta-analysis performed by Bjelakovic et al. (2007) on 68 randomized trials with 232,606 participants (385 publications) has shown that beta carotene, vitamin A, and vitamin E may even increase all-cause mortality (Bjelakovic et al. 2007). Accordingly, pharmacological modulation of the endogenous antioxidant system, performed in a personalized way, could be a more targeted and safer therapeutic option for controlling redox homeostasis in chronic diseases and cancer.

One promising pharmacological option for addressing deregulated redox metabolism chronic diseases is to target the nuclear factor erythroid 2-related factor 2 (NRF2) transcription factor (Cuadrado et al. 2018, 2019). NRF2, a member of the cap′n′collar transcription factor family, regulates the transcription of more than 250 genes exhibiting antioxidant response elements (ARE; 5′-TGACNNNGC-3′) in their promoter region that are involved in a broad array of homeostatic mechanisms related to redox metabolism and signaling, inflammation, and proteostasis (Pajares et al. 2015, 2016, 2017; de la Vega et al. 2016; Raghunath et al. 2018). A wealth of NRF2 activators, with different mechanisms of action and outcomes in chronic diseases, are being described in the literature. This review will describe first the mechanism of NRF2 regulation with a pharmacological projection, then on the NRF2 activators that are under clinical development, and finally on the promising results obtained so far in preclinical cancer models with candidate NRF2 inhibitors.

2 Physiologic Regulation of NRF2

2.1 NRF2 in Non-stressed Conditions

The primary structure of NRF2 and its main regulator, KEAP1, are described in Fig. 1. NRF2 is continuously produced by the *NFE2L2* gene and is immediately degraded through the ubiquitin-proteasome system. This apparently futile

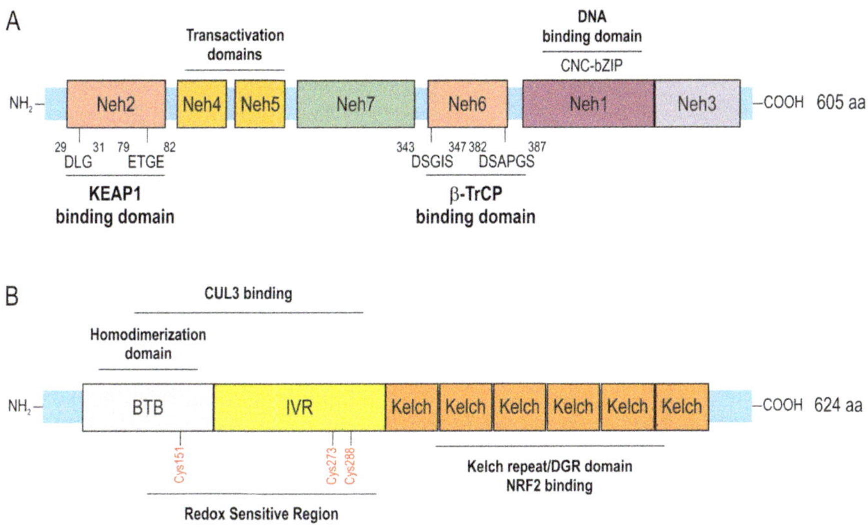

Fig. 1 Protein structure of the transcription factor NRF2 and its main modulator KEAP1. (**a**) NRF2 contains seven conserved domains called Neh1–Neh7. Mainly, Neh1 serves as the DNA binding and heterodimerization domain with sMAF proteins. Neh2 and Neh6 target NRF2 to degradation by the proteasome. Within the Neh2 domain, the ETGE (high-affinity) and DLG (low-affinity) motifs are zoomed in. Within the Neh6 domain, the positions of the DSGIS and DSAPGS motifs necessary for binding to β-TrCP are specified. The transactivation activity of NRF2 lies in Neh4 and Neh5. (**b**) Domain structure of a KEAP1 monomer. KEAP1 possesses four characteristic domains: the N-terminal broad complex, tramtrack, and bric-a-brac (BTB) domain that participates in homodimerization and binding to CUL3/RBX1; the C-terminal DGR (double glycine repeat) or Kelch repeat that binds NRF2-Neh2 domain; and the intervening region (IVR) that connects BTB and DGR domains and is particularly rich in redox-sensitive cysteine residues. Red cysteine residues in KEAP1 are the most relevant for electrophile reactivity

mechanism is extremely useful for allowing cells to respond rapidly to potentially harmful oxidative and electrophilic challenges.

Under non-stressed conditions, NRF2 is targeted for ubiquitin-proteasome degradation through the interaction with a KEAP1 homodimer. KEAP1 contains a broad complex, tramtrack, bric-a-brac (BTB) homodimerization domain, an intervening region (IVR), and a C terminal Kelch domain with a double glycine repeat (DGR). This Kelch domain binds to the Neh2 domain of NRF2 at two amino acid sequences, DLG and ETGE, the ETGE exhibiting about 100 times higher affinity for KEAP1 than the DLG motif (Tong et al. 2006). KEAP1 presents NRF2 for ubiquitination by the E3 ligase complex formed by Cullin3 and RBX1 proteins (CUL3/RBX1) (Tong et al. 2007), resulting in NRF2 degradation by the proteasome 26S (Hayes and Dinkova-Kostova 2014; Suzuki et al. 2013). An alternative mechanism controlling NRF2 stability is related to the glycogen synthase kinase 3 (GSK-3) and the E3 ligase adapter β-TrCP. The serine/threonine protein kinases GSK-3α and β are maintained in an inactive form under normal conditions by AKT-mediated phosphorylation at their N-terminal pseudosubstrate domain, as well as by sequestration in protein complexes. In absence of receptor signaling, active GSK-3 phosphorylates directly NRF2 at the Neh6 domain (DSGIS). GSK-mediated NRF2 phosphorylation then triggers the recruitment of β-TrCP and favors the interaction with pSGIpS and the CUL1/RBX1 complex for ubiquitin-proteasome degradation of NRF2 (Cuadrado 2015). β-TrCP also recognizes the DSAPGS motif in the Neh6 domain of NRF2 which is constitutively phosphorylated in a GSK-3-independent manner (Chowdhry et al. 2013).

Of note is that several proteins contain an (E/S)TGE motif that resembles the high-affinity ETGE motif of NRF2, such as dipeptidyl peptidase 3, partner and localizer of BRCA2, and SQSTM1/p62. Accordingly, these proteins have the ability to compete with NRF2 for KEAP1 binding, leading to a non-canonical mechanism of NRF2 stabilization (Hast et al. 2013).

Additional degradation mechanisms are proposed for NRF2, such as the inositol-requiring enzyme (IRE1)/E3 ubiquitin ligase synoviolin (HRD1) present in the endoplasmic reticulum (ER) (Wu et al. 2014a), whose expression is enhanced by the activation of the XBP1-HRD1 arm of the ER stress pathway, as demonstrated in cirrhosis to be a protective mechanism (Wu et al. 2014a). The interaction of HRD1 with the Neh4 and Neh5 domains of NRF2 mediates its degradation, independently of KEAP1, leading to enhanced NRF2 activity and consequent transcription of cytoprotective genes.

2.2 Activation of NRF2 Transcriptional Activity

KEAP1 is not only the main repressor of NRF2 but also a highly reactive redox sensor through its 27 cysteine residues (found in humans) (Sihvola and Levonen 2017). Excessive ROS oxidize thiols and induce glutathionylation and alkylation of macromolecules. Of utmost importance for the pharmacological activation of NRF2 are electrophile reactions with particular KEAP1 cysteines (Cys151, Cys273, and Cys288), leading to adduct formation and consequent inhibition of NRF2 ubiquitination (Taguchi et al. 2011). Altogether, KEAP1 cysteines are modified under redox challenging conditions (Holland et al. 2008), and the resulting conformational alteration compromises the interaction between NRF2 and KEAP1, leading to NRF2 stabilization and activation of its transcriptional activity (Cuadrado et al. 2018, 2019), as we will describe below.

NRF2 molecules escaping proteasomal degradation translocate specifically to the nucleus through a nuclear localization signal contained in NRF2 and/or phosphorylation mediated by several kinases such as protein kinase C (Huang et al. 2000) and mitogen-activated protein kinases (Yu et al. 2000). In the nucleus, the dimerization of NRF2 with the cognate bZip partners MAF G, K, and F and maybe JUN or ATF proteins favors the binding of NRF2 to the ARE sequence and recruitment of the transcriptional co-activator CBP/p300, leading to a complex array of transcriptional events. Resolution of NRF2 transcriptional activity occurs through the late induction of FOS proteins or oxidative modification of BACH proteins, leading to their nuclear accumulation and consequent competition with NRF2 for binding to ARE (Reddy 2008).

NRF2 itself can be regulated at transcriptional level, which is of utmost importance for maintaining a critical NRF2 pool. For instance, NRF2 can regulate its own expression through an ARE-like element located in the proximal region of its promoter (Kwak et al. 2002). The *NFE2L2* gene promoter presents several other regulatory sequences: (a) the xenobiotic response element (XRE) and two XRE-like sequences that are recognized by the transcription factor aryl hydrocarbon receptor (AHR) (Miao et al. 2005); (b) the 12-O-tetradecanoylphorbol-13-acetate-response element (TRE) (TGCGTCA) that is activated by the oncogenic KRAS (Tao et al. 2014), BRAF, and MYC (DeNicola et al. 2011) which are critically involved in carcinogenesis; and (c) NF-κB binding sites that respond to various inflammatory stimuli (Rushworth et al. 2012). Moreover, epigenetic changes encompassing promoter methylation, microRNAs (e.g., miR-144 (Sangokoya et al. 2010), miR-28 (Yang et al. 2011), miR93 (Wang et al. 2016), and miR-98-5p (Sun et al. 2018)), as well as long noncoding RNA deregulation (Fabrizio et al. 2018) contribute to expression changes in the *NFE2L2* gene.

Besides complex mechanisms of gene regulation, the analysis of the NRF2 interactome evidenced biological functions in close correlation with NRF2 (Cuadrado et al. 2018), sustaining the key role of NRF2 in far many more chronic pathologies than initially expected. This perspective also fosters the development of novel therapeutic strategies that specifically target the KEAP1-NRF2 system and guide the biological consequences.

Of utmost importance for designing innovative therapeutic strategies in chronic diseases underlined by low-grade inflammation and dysregulation of redox metabolism is the crosstalk between the transcription factor NRF2 which controls important antioxidant responses and the transcription factor NF-kB which regulates inflammatory processes (Wardyn et al. 2015). Heme oxygenase 1 (HO-1) encoded by a NRF2 target gene (*HMOX1*) is partly responsible for the NRF2-mediated NF-κB inhibition. In turn, the canonical NF-κB subunit p65 can exert a negative effect on ARE-linked gene expression through competition with NRF2 for the transcriptional co-activator CBP/p300 protein complex (Liu et al. 2008). Moreover, it was found that NRF2 contains κB sites in its proximal promoter, which are targeted by p65 for transcription initiation (Smale 2011). Inflammatory responses, such as those triggered by lipopolysaccharide, can be resolved by the generation of the small GTPase RAC1 (Ras-related C3 botulinum toxin substrate 1) which activates NRF2 and dampens therefore NF-kB-mediated pro-inflammatory responses (Cuadrado et al. 2014). Moreover, it was demonstrated that NRF2 can act as an upstream inhibitor of pro-inflammatory cytokine production, such as IL-6 and IL-1β (Kobayashi et al. 2016), in addition to the redox control that it exerts on inflammatory networks. Of note is also the observation that the E3 ubiquitin ligase β-TrCP that is involved in proteasomal degradation of NRF2 mediates also the degradation of IκB, the main cytoplasmic repressor of the pro-inflammatory NF-KB transcription factor, evidencing the inverse regulation for these transcription factors (Kanarek and Ben-Neriah 2012).

3 Pharmacologic Activators of NRF2

NRF2 activators address mainly KEAP1 (Magesh et al. 2012), the main cytoplasmic NRF2 repressor which impedes on its transcriptional activity by targeting NRF2 for proteasomal degradation. According to their specific interaction mechanisms with KEAP1, NRF2 activators can be classified as electrophiles and protein-protein interaction (PPI) inhibitors, as well as multi-target compounds (Fig. 2).

3.1 Electrophilic Compounds

Electrophilic molecules modify covalently, by oxidation or alkylation, critical cysteine residues (Cys) within the thiol-rich KEAP1 repressor, such as Cys-151, Cys-273, and Cys-288 (Levonen et al. 2004; Wakabayashi et al. 2004) that are highly susceptible to electrophilic reactions (Yamamoto et al. 2008; Saito et al. 2015), but also Cys-226, Cys-434, and Cys-613 in particular cases. Changes of the "Cys code" in NRF2 lead to a dysfunctional state of KEAP1 that is no longer able of targeting NRF2 for ubiquitination, despite its interaction with both of the critical Neh2 motifs of NRF2 (DLG and ETGE). Accordingly, there is a loss of free KEAP1 molecules for further interaction with the newly synthesized NRF2 which avoids in this way KEAP1-mediated proteolytic degradation (Baird et al. 2013). An alternative mechanism for pharmacological NRF2 activation addresses the interaction of KEAP1 with the CUL3/RBX1 complex that is required for NRF2 ubiquitination. It was found that adduct formation of electrophiles with Cys-151 in the BTB domain

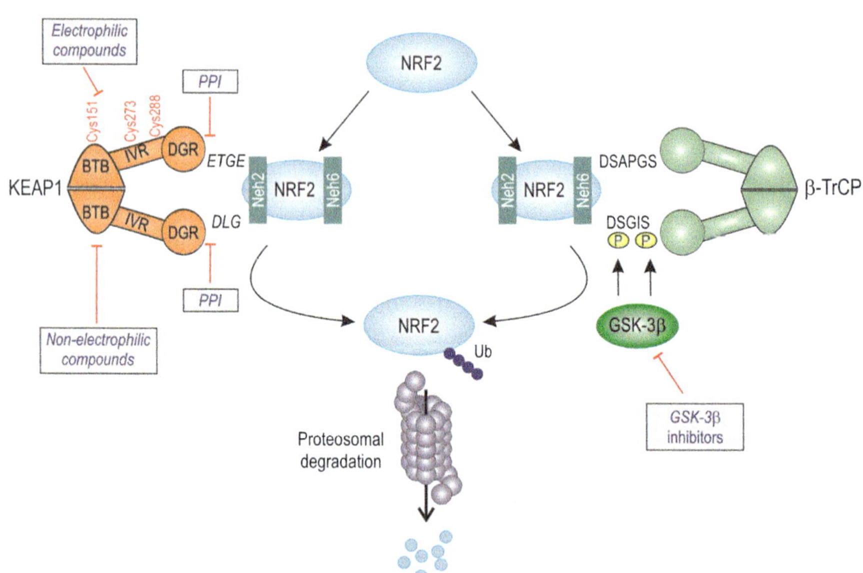

Fig. 2 Summary of the main pharmacological strategies to increase NRF2 activity

of KEAP1 most likely disrupts the interaction with CUL3 (Cleasby et al. 2014; Iso et al. 2016; Dayalan Naidu et al. 2018). A NRF2-bound conformation of KEAP1 is induced, that is, unable to drive NRF2 ubiquitination, and the newly formed NRF2 molecules consequently escape proteasomal degradation. Selected electrophilic activators of NRF2 that are in various stages of clinical development are presented in Table 1.

The most clinically developed NRF2 activator is dimethyl fumarate (DMF), a fumaric acid ester that has been authorized for the treatment of psoriasis (Hoxtermann et al. 1998). Under the trade name of Tecfidera (Biogen), DMF has been approved later by the Food and Drug Administration (FDA) and the European Medicines Agency (EMA) also for relapsing-remitting multiple sclerosis (MS), an autoimmune inflammatory demyelinating disease that may result in sustained neurologic damage (Xu et al. 2015; Schimrigk et al. 2006; Gold et al. 2012; Fox et al. 2012). Upon ingestion, DMF is rapidly metabolized by intestinal esterases into the biologically active form of the agent, mono methyl fumarate (MMF). Of note is that the direct biological actions of DMF and MMF in vitro are not identical (Booth et al. 2016).

DMF is an alkylating agent which non-specifically and covalently modifies nucleophilic groups in proteins, including cysteine thiols (Albrecht et al. 2012; Li et al. 2007; Lin et al. 2011) in KEAP1. DMF was also shown to induce in MS a prominent reduction of inflammatory events (Havrdova et al. 2013), such as the number of peripheral T cells, especially $CD8^+$ T cells (Mills et al. 2018a; Ghadiri et al. 2017); total B lymphocytes counts, especially due to the decrease of memory B cells; and a decrease in the levels of pro-inflammatory factors such as IL-6, TNFα, and granulocyte-macrophage colony-stimulating factor (Li et al. 2017; Smith et al. 2017). It is noteworthy that these effects may be exerted both through NRF2-dependent and NRF2-independent mechanisms, as we will explain later in this review (Schulze-Topphoff et al. 2016). Recently, DMF was shown to induce mitochondrial biogenesis in a NRF2-dependent manner (increase of mtDNA copy number and mitochondrial complex mRNA expression) in preclinical models and humans, without having mitochondrial complex inhibition activity. This newly evidenced mechanism of action may explain, at least partly, the efficacy of DMF in MS and opens new avenues for the treatment of mitochondriopathies (Hayashi et al. 2017). Using the MS mice model of experimental allergic encephalomyelitis (EAE), DMF was shown to induce the activation of NRF2 in the central nervous system (Linker et al. 2011), which correlated well with the observed improvement in the clinical course of MS, favored axon preservation, and increased astrocyte activation. The fact that these beneficial DMF effects did not occur in NRF2-null mice clearly indicated that DMF is acting mainly by targeting the NRF2 pathway which further modulates local inflammatory reactions (Mills et al. 2018a; Ghadiri et al. 2017; Li et al. 2017; Smith et al. 2017).

DMF is mostly converted to MMF by intestinal esterases, but only a small fraction is found in blood conjugated with glutathione (Dibbert et al. 2013). Therefore, an oral formulation of a MMF derivative, diroximel fumarate (2-(2,5-dioxo-1-pyrrolidinyl)ethyl ester; ALKS-8700; Alkermes), was designed for improving

Table 1 Selected modulators of NRF2 that are under clinical development

Compound	Type	Mechanism of action	Disease	Clinical trial	ClinicalTrials.gov identifier
Electrophilic activators					
a. Synthetic compounds					
Dimethyl fumarate 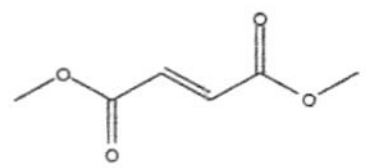	Fumaric acid ester	Electrophilic modification of KEAP1-Cys-151	MS	Approved	
			Psoriasis	Approved	
			Rheumatoid arthritis	Phase II	NCT00810836
			Adult brain glioblastoma	Phase I	NCT02337426
			Cutaneous T cell lymphoma	Phase II	NCT02546440
			Obstructive sleep apnea	Phase II	NCT02438137
			Chronic lymphocytic leukemia Small lymphocytic lymphoma	Phase I	NCT02784834
			Glioblastoma	Phase I	NCT02337426
ALKS-8700 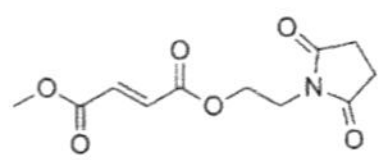	Fumaric acid ester (MMF-derivative)	Electrophilic modification of KEAP1-Cys-151	MS	Phase III	NCT02634307
Bardoxolone-methyl (CDDO-Me) 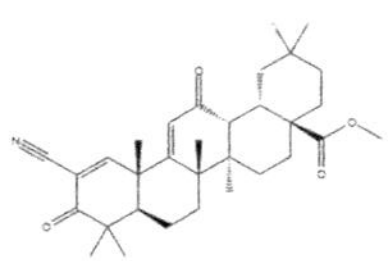	Synthetic triterpenoids	Electrophilic modification of KEAP1-Cys-151	Diabetic nephropathy	Phase II	NCT00811889
			IgA nephropathy CKD associated with type 1 DM Focal segmental glomerulosclerosis Autosomal dominant polycystic kidney	Phase II	NCT03366337

			CKD type 2 DM nephropathy	Phase III	NCT01351675
			Liver disease	Phase I/II	NCT00550849
			Hepatic impairment Healthy	Phase I	NCT01563562
			Advanced solid tumors lymphoid malignancies	Phase I	NCT00529438 NCT00508807
			Alport syndrome	Phase II/III cardinal	NCT03019185
			Pulmonary hypertension	Phase III RANGER	NCT03068130
			Pulmonary arterial hypertension	Phase III	NCT02657356
			Renal insufficiency, type 2 DM	Phase II	NCT01053936
RTA-408 (omaveloxolone) 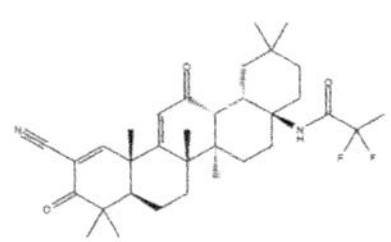	Synthetic triterpenoids	Electrophilic modification of KEAP1-Cys-151	Mitochondrial myopathy	Phase II	NCT02255422
			Friedreich's ataxia	Phase II	NCT02255435
			Inflammation and pain following ocular surgery	Phase II	NCT02065375
			Corneal endothelial cell loss Ocular pain Ocular inflammation Cataract surgery	Phase II	NCT02128113
			Melanoma	Phase I/II	NCT02259231
			Breast cancer	Phase II	NCT02142959

(continued)

Table 1 (continued)

Compound	Type	Mechanism of action	Disease	Clinical trial	ClinicalTrials.gov identifier
Oltipraz 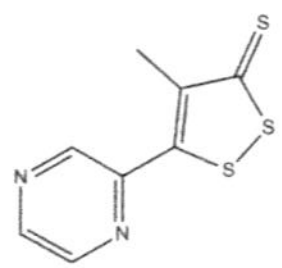	Organosulfur compound	Electrophilic modification of KEAP1-Cys-151	Non-alcoholic steatohepatitis	Phase III	NCT02068339
			Schistosomiasis	Approved	
			Lung cancer	Phase I	NCT00006457
Ursodiol 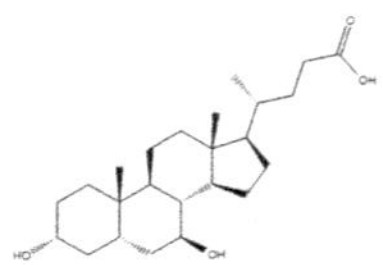		Electrophilic modification of KEAP1-Cys-151	Cholestasis	Phase II/III	NCT00846963
			Diarrhea	Phase IV	NCT02748616
			Cholelithiasis	Phase III	NCT02721862
			Primary biliary cirrhosis	Phase IV	NCT01510860
			Barrett esophagus Low-grade dysplasia	Phase II	NCT01097304
			Chronic hepatitis C	Phase III	NCT00200343
			Type 2 DM	Phase II	NCT02033876
b. *Natural compounds*					
Sulforaphane 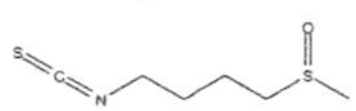	Isothiocyanate	Electrophilic modification of KEAP1-Cys-151	Schizophrenia	Phase II	NCT02880462
				Phase II	NCT02810964
				Phase II	NCT01716858
			COPD	Phase II	NCT01335971
			Atopic asthma	Phase I	NCT01845493
			Autism spectrum disorder	Phase II	NCT01474993
				Phase II	NCT02909959
				Phase II	NCT02677051
				Phase III	NCT02654743
				Phase I/II	NCT02561481

			Healthy	Phase I	NCT01008826
				Phase I	NCT02023931
			Melanoma	Phase I	NCT01568996
			Asthma	Phase I	NCT01845493
				N/A	NCT01183923
			Prostate cancer	Phase II	NCT01228084
			Breast cancer	Phase II	NCT00843167
			Lung cancer	Phase II	NCT03232138
			Environmental carcinogenesis	Phase II	NCT01437501
			Alcohol sensitivity	Phase II	NCT01845220
			Aging	Phase II	NCT03126539
			Rhinitis Allergy	Phase II	NCT02885025
			Helicobacter pylori infection	Phase IV	NCT03220542
			DM, non-insulin-dependent	Phase II	NCT02801448
Sulforadex (SFX-01) 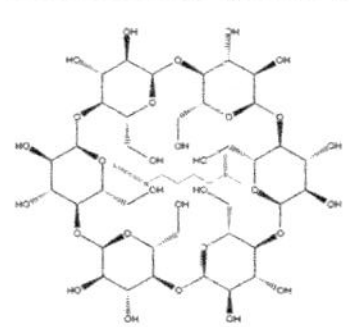	Sulforaphane/alpha-cyclodextrin complex	Electrophilic modification of KEAP1-Cys-151	Subarachnoid hemorrhage	Phase II	NCT02614742
			Breast neoplasm	Phase II	NCT02970682
			Prostate cancer	Phase I	NCT02055716 NCT01948362

(continued)

Table 1 (continued)

Compound	Type	Mechanism of action	Disease	Clinical trial	ClinicalTrials.gov identifier
ITH12674 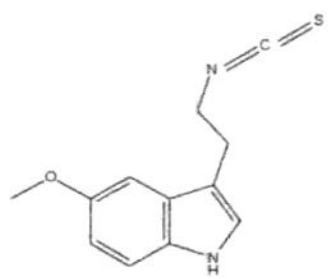	Melatonin-sulforaphane hybrid	Electrophilic modification of KEAP1-Cys-151	Brain ischemia	Preclinical PK	Not clinical trials available
Curcumin 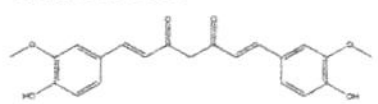	Stilbene	Electrophilic modification of KEAP1-Cys-151	Type 2 DM Pre-diabetes Insulin resistance Cardiovascular risk	Phase IV	NCT01052025
			Schizophrenia Cognition Psychosis	Phase I/II	NCT02104752
			Acute kidney injury Abdominal aortic aneurysm	Phase II/III	NCT01225094
			CKD Type 2 DM	Phase II/III	NCT03262363
			Alzheimer's disease	Phase I/II	NCT00164749
			Neoplasms	Phase II	NCT02944578
			Crohn's disease	Phase III	NCT02255370
			Chronic schizophrenia	Phase IV	NCT02298985
			Mild cognitive impairment	Phase II	NCT01811381
			Prostate cancer	Phase III	NCT02064673
			Major depression	Phase IV	NCT01750359

Resveratrol 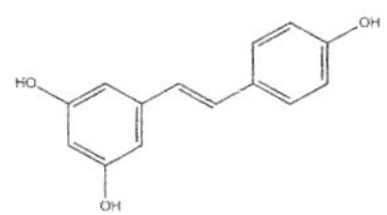	(*E*)-Stilbene derivate	Electrophilic modification of KEAP1-Cys-151	Type 2 DM	Phase I	NCT01677611
			Colon cancer	Phase I	NCT00256334
			COPD	N/A	NCT02245932
			Friedreich ataxia	Phase I/II	NCT01339884
			Non-alcoholic fatty liver	Phase II/III	NCT02030977
			Non-ischemic cardiomyopathy	Phase III	NCT01914081
			Endometriosis	Phase IV	NCT02475564
			Chronic renal insufficiency	Phase III	NCT02433925
			Metabolic syndrome X	Phase II	NCT02114892
			Chronic subclinic inflammation Redox status	Phase III	NCT01492114
			Alzheimer's disease	Phase II	NCT01504854
				Phase III	NCT00743743
			Huntington disease	N/A	NCT02336633
CXA-10 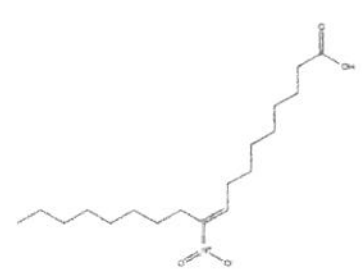	Nitro-fatty acid (NFA)	Electrophilic modification of KEAP1-Cys-273 and Cys-288	Acute kidney injury	Phase I	NCT02248051
			Pulmonary arterial hypertension	Phase II	NCT03449524
			Primary focal segmental glomerulosclerosis	Phase II	NCT03422510

(continued)

Table 1 (continued)

Compound	Type	Mechanism of action	Disease	Clinical trial	ClinicalTrials.gov identifier
KEAP1-independent NRF2 activators					
Tideglusib 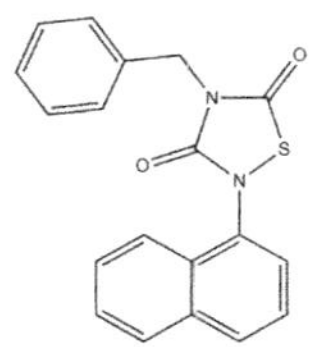	GSK-3 inhibition		Autism spectrum disorders	Phase II	NCT02586935
			Myotonic dystrophy 1	Phase II	NCT02858908
			Alzheimer's disease	Phase II	NCT01350362
Nordihydroguaiaretic acid (NDGA) 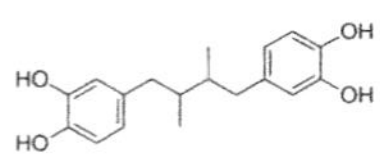	GSK-3 inhibition		Prostate cancer	Phase II	NCT00678015
				Phase I	NCT00313534
			Brain and central nervous system tumors	Phase I/II	NCT00404248
Terameprocol (NDGA derivative) 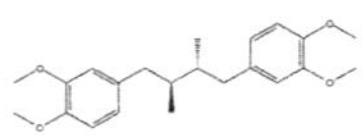	GSK-3 inhibition		High-grade glioma	Phase I	NCT02575794
			Leukemias Acute myeloid leukemia Acute lymphocytic leukemia	Phase I	NCT00664677
			Refractory solid tumors Lymphoma	Phase I	NCT00664586
Enzastaurin 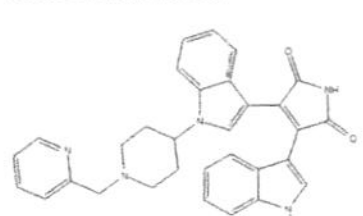	GSK-3 inhibition		Diffuse large B-cell lymphoma	Phase III	NCT03263026
			Solid tumor Lymphoma, malignant	Phase I	NCT01432951

LS-102 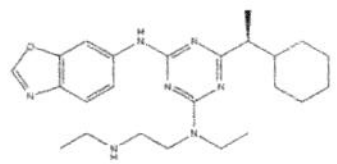	HRD1 inhibition	–	–	Not clinical trials available
Rapamycin 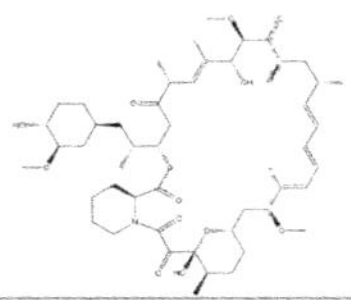	p62/SQSTM1 activation	Type 1 DM	Phase III	NCT01060605
		Systemic lupus erythematosus	Phase II	NCT00779194
		Autosomal dominant polycystic kidney disease	Phase II/III	NCT00920309

bioavailability and efficacy and is currently under phase III trial for MS (Sun et al. 2017; Zeidan et al. 2014). The therapeutic effects of fumaric acid esters are not fully characterized, and KEAP1/NRF2-independent effects were highlighted. For instance, the study of Schulze-Topphoff et al. (2016) showed that DMF can protect both "wild-type" and Nrf2$^{-/-}$ mice against the development of inflammation in acute EAE through the reduction of Th1 and Th17 cells as well as through the induction of anti-inflammatory M2 monocytes, indicating that Nrf2 may not be required for many of the observed beneficial effects of DMF (Schulze-Topphoff et al. 2016). Moreover, it was shown that DMF and MMF can activate the nicotinic receptor hydroxycarboxylic acid receptor 2 (HCAR2) which is expressed in immune cells and gut epithelial cells, resulting in NRF2-independent anti-inflammatory responses (von Glehn et al. 2018).

Synthetic triterpenoids were developed as electrophilic NRF2 activators. For instance, promising compounds have been derived from natural oleanolic acid by addition of enone and ciano groups to the A ring and another enone group to the C ring (Dinkova-Kostova et al. 2005; Liby and Sporn 2012), aiming to endow these derivatives with stronger Michael acceptor reactivity. Such a compound, bardoxolone methyl (CDDO-Me or RTA 402), reached clinical trials for the treatment of advanced chronic kidney disease (CKD), type 2 diabetes mellitus (DM), and cancer (Pergola et al. 2011). Clinical studies provided promising results in terms of efficacy; CDDO-Me was discontinued at phase III trial due to cardiovascular side effects (Zhang 2013). A new phase II clinical trial on patients with rare chronic kidney diseases has recently started for better defining the safety and efficacy profiles of CDDO-Me. Moreover, a phase II/III trial (NCT03019185) on the efficacy and safety of CDDO-Me in patients with Alport syndrome is ongoing. A second-generation difluoromethyl acetamide derivative of bardoxolone methyl (RTA-408, Omaveloxone) was designed for improving the safety profile and is now under phase II trial for Friedreich's ataxia, ocular inflammation, and pain after ocular surgery (Lynch et al. 2019). Recently, a preclinical study evaluating RTA-408 for diabetic wounds emphasized that upregulation of NRF2 is responsible for the reinforcement of regenerative processes (Rabbani et al. 2018).

Some other compounds target NRF2 at the level of other cysteine residues Cys-151, the cysteine residue generally targeted by most NRF2 activators (Kansanen et al. 2011). For instance, 9-nitro-octadec-9-enoic acid ($OA\text{-}NO_2$) is a nitro-fatty acid with anti-inflammatory properties that reacts with Cys-273 and Cys-288 of KEAP1, and its activity seems to be independent of CXA-10 (10-nitro-9(E)-octadec-9-enoic acid), an isomer of $OA\text{-}NO_2$, has proven efficacy in the uni-nephrectomized deoxycorticosterone acetate-high salt mouse model of CKD (Arbeeny et al. 2019), and is under several phase I clinical trial for the treatment of this disease (Batthyany and Lopez 2015). Moreover, CXA-10 is under phase II trials for the treatment of pulmonary arterial hypertension (NCT03449524) and primary focal segmental glomerulosclerosis (NCT03422510). Other NRF2 activators, such as 15-deoxy-Δ12,14-prostaglandin J_2, were also shown to interact with Cys-273 and Cys-288 of the KEAP1 homodimer (Yamamoto et al. 2008), resulting in increased NRF2 activity in models of ureteral obstruction (Nilsson et al. 2017), hepatic ischemia-reperfusion injury (Chen et al. 2017), and atherosclerosis (Lu et al. 2017). In a recent study (Mills et al. 2018b), the metabolite itaconate was described as a NRF2 activator that alkylates many cysteine residues in KEAP1 (Cys-151, Cys-257, Cys-288, Cys-273, and Cys-297). A cell-permeable itaconate derivate, 4-octyl itaconate, that triggers stronger NRF2 activation than DMF was shown to exert important anti-inflammatory effects that protect against lipopolysaccharide toxicity (Mills et al. 2018b).

Electrophilic drugs that are rendered active by the very same ROS that they have to counteract (pathologically activated therapeutics) were shown to exhibit increased specificity for the diseased tissues and less side effects due to the lack of reaction with other thiols that can result in lowering the glutathione pool and inducing ROS dysregulation in normal cells (Satoh et al. 2013). One example is carnosic acid found in *Rosmarinus officinalis* (with adjacent or "*ortho-*" position hydroxyl groups) which proved efficacy against AD and other neurologic conditions in rodent models (Lipton et al. 2016; Liu et al. 2016).

Some natural compounds have been identified as electrophilic NRF2 inducers, including sulforaphane, curcumin, resveratrol, quercetin, genistein, and more recently andrographolide (Wong et al. 2018). Extensive evidence was gathered regarding the beneficial effects of sulforaphane (SFN), an isothiocyanate found in cruciferous vegetables under the inactive form of glucoraphanin which is further processed enzymatically to the biologically active form (Abdull Razis and Noor 2013). SFN is rapidly entering into cells where it is conjugated with glutathione through a reaction mediated by glutathione-S-transferase, which greatly contributes to its accumulation (Kolm et al. 1995) SFN efficacy has been attributed, at least partly, to its capacity to activate the KEAP1-NRF2 system and trigger the expression of phase II detoxifying enzymes such as NQO1 (Boddupalli et al. 2012), hence being therapeutically active in preventing chemically induced carcinogenesis (Watanabe et al. 1997). Additional activities have been evidenced, such as the inhibition of histone deacetylase 6 (HDAC6) activity and cell cycle progression (Myzak and Dashwood 2006). SFN exerts anti-inflammatory effects by directly attenuating the production of IκB-α and by inhibiting NF-kB translocation, besides impairing its redox-sensitive DNA binding and transactivation (Heiss et al. 2001). SFN has been successfully used for the treatment of type II DM (Bahadoran et al. 2012; Axelsson et al. 2017). It has been also found that SFN can exert also beneficial effects in hypoxic-ischemic injury in rats (Ping et al. 2010; Zhao et al. 2006). Moreover, being able to cross the blood-brain barrier, SFN can provide protection in neurodegenerative disorders, as demonstrated in preclinical models of disease. Thus, SFN was shown to protect against the neurotoxicity of $A\beta_{1-42}$ peptides in neuronal cells (Park et al. 2009) and ameliorated cognitive impairment in an acute mouse model of Alzheimer's disease (AD) (Kim et al. 2013). In Parkinson's disease (PD), SFN protected dopaminergic cells against the cytotoxic effects of 6-hydroxydopamine (Han et al. 2007), counteracted astrogliosis and microgliosis, and significantly reduced dopaminergic neuron death in a mouse model of PD (Jazwa et al. 2011; Tarozzi et al. 2013; Houghton et al. 2016). Evgen Pharma has developed a cyclodextrin formulation, SFX-01, with improved stability, which is under phase II clinical trial (safety, tolerability, pharmacokinetic, and pharmacodynamic study) for the treatment of subarachnoid hemorrhage (NCT02614742). Moreover, a hybrid molecule of SFN and melatonin (ITH12674) was designed for a dual "drug-prodrug" mechanism of action in the treatment of brain ischemia (Egea et al. 2015). ITH12674 (Egea et al. 2015) proved to exert concentration-dependent neuroprotective effects in cortical neurons subjected to increasing ROS production, increasing GSH concentrations, and enhancing the NRF2-triggered antioxidant response. Moreover, it protected organotypic cultures of hippocampal slices subjected to oxygen and glucose deprivation and re-oxygenation from stress by increasing the expression of HO-1, one of the main targets of NRF2.

The list of electrophilic compounds that are able to interact with KEAP1 and modulate thereby the transcriptional activity of NRF2 is continuously growing, but most of these compounds have not evolved beyond preclinical experiments. Moreover, electrophiles, such as DMF, also produce considerable systemic side effects, in part due to non-specific S-alkylation of cysteine thiols, resulting in the depletion of the antioxidant glutathione pool. A long way needs to be covered to characterize the pharmacodynamic properties, the clinical safety profile, and the efficacy of candidate electrophilic drugs.

3.2 Protein-Protein Interaction Inhibitors of the KEAP1-NRF2 System

Protein-protein interaction (PPI) inhibitors can activate NRF2 by hindering its interaction with the Kelch propeller of KEAP1 through non-covalent interactions (Cuadrado et al. 2019). Therefore, they exhibit a higher selectivity for KEAP1 than electrophilic compounds, avoiding the formation of unwanted adducts with other cysteines than those contained in KEAP1 (Richardson et al. 2015) and consequently uncontrollable side effects like the deleterious depletion of the anti-inflammatory glutathione pool.

Based on the crystal structure of KEAP1 (Padmanabhan et al. 2006), small PPI inhibitors have been designed to hinder the binding to KEAP1 of either the high-affinity ETGE motif (Lo et al. 2006) or the low-affinity DLG motif (Tong et al. 2007) contained in NRF2. The first PPI inhibitors were designed from a series of truncated NRF2 peptides (Inoyama et al. 2012; Chen et al. 2011) and evidenced that the 9-mer sequence LDEETGEFL is the minimal binding sequence of NRF2 required for docking to KEAP1 (Inoyama et al. 2012; Chen et al. 2011; Hancock et al. 2012). The following peptides were found to act as PPI inhibitors (Robledinos-Anton et al. 2019): LDEETGEFL-NH2 (Inoyama et al. 2012) (Chen et al. 2011); DEETGE-CAL-Tat (NH_2-RKKRRQRRR-PLFAERLDEETGEFLPNH_2) (Tu et al. 2015); Ac-DPETGEL-OH, FITCβ-DEETGEF-OH, FITC-β-LDEETGEFL-OH, Ac-DEETGEF-OH, and Ac-DPETGEL-OH (Hancock et al. 2012); FITC-LDEETGEFL-NH_2 (Inoyama et al. 2012); FAM-LDEETGEFL-NH_2 (Jiang et al. 2014); LQLDEETGEFLPIQGK(MR121)-OH (Marcotte et al. 2013); Ac-LDEETGEFL-NH_2 (Inoyama et al. 2012) (Chen et al. 2011); Ac-DPETGEL-

NH_2, Ac-NPETGEL-OH, and St-DPETGEL-OH (Hancock et al. 2013); YGRKKRRQRRRLQLDEETGEFLPIQ (Steel et al. 2012); and c [GQLDPETGEFL] (Lu et al. 2018). A customized peptide was designed for increasing cellular uptake by adding the cleavage sequence of calpain and the Tat sequence of the human immunodeficiency virus (-Cal-Tat). It was shown to provide neuroprotection and cognition-preserving effects in a mouse model of cerebral ischemia (Tu et al. 2015). Moreover, hybrid peptides based on the interaction regions between KEAP1 and NRF2 at the level of the ETGE motif and between KEAP1 and p62/SQSTM1 exhibited superior binding activity (Hancock et al. 2013). A major drawback of peptides as therapeutic agents is their low oral bioavailability and cellular permeability. This is the reason why research has been lately focused on the development of small molecules as PPI inhibitors, comprising tetrahydroisoquinoline (Richardson et al. 2015; Jnoff et al. 2014), thiopyrimidine (Marcotte et al. 2013), naphthalene (Jiang et al. 2014), carbazone (Ranjan et al. 2014), and urea derivatives (Sato et al. 2013).

Several PPI inhibitors with improved selectivity over electrophiles have been identified in silico through screening of small molecule libraries, such as SRS-5, benzenesulfonyl-pyrimidone 2, N-phenyl-benzenesulfonamide, and a series of 1,4-diphenyl-1,2,3-triazole (Jnoff et al. 2014; Hu et al. 2013; Wen et al. 2015; Bertrand et al. 2015; Nasiri et al. 2016). The first KEAP1-NRF2 inhibitor that was identified using molecular binding determinant analysis and proved efficacy at nanomolar concentrations is CPUY192002 (Jiang et al. 2014). The compound was further developed by optimizing solubility using medicinal chemistry methods. The new PPI inhibitor, CPUY192018, showed potent NRF2 activation effects both in vitro and in vivo. The compound proved efficacy in the dextran sodium sulfate-induced experimental colitis model (Lu et al. 2016) and in LPS-induced chronic renal inflammation (Lu et al. 2019) by activating the NRF2-dependent antioxidative pathways and by inhibiting inflammatory responses mediated by the transcription factor NF-κB (Lu et al. 2019).

Huge promises in developing small molecules as PPI inhibitors may derive also from the complex biochemical protocol that lately was designed for better identifying reversible modifiers of the NRF2-KEAP1 interaction. This protocol puts together time-resolved fluorescence resonance energy transfer as primary screening tool, combined with surface plasmon resonance for evaluating the affinity of KEAP1 binders, and the ^{1}H-^{15}N heteronuclear single quantum coherence nuclear magnetic resonance assay for further analyzing the binding mode (Bresciani et al. 2017).

3.3 Other Mechanisms for Pharmacological NRF2 Activation

As described at point 2.1, NRF2 phosphorylation by GSK-3 leads to its ubiquitination by the E3 ligase β-TrCP and subsequent proteasomal degradation. Considering that an aberrant activity of GSK-3 is linked with several pathologies such as AD, cardiovascular diseases, or cancer (Hooper et al. 2008; Silva et al. 2014;

Luo 2009; Lal et al. 2015), several clinical trials are now being focused on GSK-3 inhibitors (Saraswati et al. 2018). For instance, the GSK-3-inhibitor tideglusib, a thiadiazolidinone compound, was studied in phase II trials for AD in the ARGO study (Lovestone et al. 2015). The study evidenced that short-term (26 weeks) tideglusib had an acceptable safety profile but produced no clinical benefit. However, given the non-linear dose response, especially in mildly affected patients, further dose finding studies in early disease stages and for longer duration are warranted to examine GSK-3 inhibition in AD patients. Another GSK-3 inhibitor, enzastaurin, intended for the treatment of various types of cancers, including lymphoma, showed promising results at preclinical level, but phase II and III trials involving more than 3,000 patients evidenced poor efficacy (Bourhill et al. 2017; Lombardi et al. 2017). According to the analysis performed by Bourhill et al. (2017), inappropriate end point analysis, limited standards in phase I clinical trials, insufficient use of biomarker analysis, and also patient stratification apparently contributed to the failure to achieve approval of enzastaurin as an anticancer therapeutic (Bourhill et al. 2017). GSK-3-dependent NRF2 phosphorylation was also shown to be inhibited by nordihydroguaiaretic acid (Rojo et al. 2012) and its derivative terameprocol (Kimura and Huang 2016; Chao et al. 2018) which is in phase I and II clinical trials for the treatment of several types of cancers, such as gliomas and leukemias (Table 1) (Palomo and Martinez 2017).

Taking advantage of the consistent knowledge gains in the mechanisms of NRF2 stabilization, new strategies may be developed to modulate therapeutically the activation of the NRF2 system. A new strategy for developing KEAP1-independent NRF2 activators is to develop small molecules for disrupting the docking of NRF2 to β-TrCP (Rada et al. 2012). Besides the E3 ubiquitin ligases KEAP1 and β-TrCP, HRD1 is a E3 ubiquitin ligase linked to KEAP1-independent NRF2 degradation. That crosstalk between NRF2 and HRD1 was investigated by Wu et al. (2014) in the context of cirrhotic liver (Wu et al. 2014a). HRD1 is a transcriptional target of XBP1 that is upregulated upon activation of IRE1 during endoplasmic reticulum (ER) stress related to cirrhotic conditions. The study evidenced that NRF2 is a bona fide substrate of HRD1 through the direct binding of the cytosolic C-terminal domain of HRD1 and the Neh4–5 domains of NRF2. This inverse correlation in the expression of NRF2 and HRD1 was observed in both human and mouse cirrhotic livers, and inhibitors of HRD1 and IRE1 were able to restore NRF2-mediated responses in liver cirrhosis (Wu et al. 2014a).

Proteins containing (E/S)TGE motifs similar to NRF2 were shown to compete with NRF2 for binding to KEAP1 and induce therefore NRF2 stabilization and its translocation to the nucleus (Jain et al. 2010; Lau et al. 2010; Komatsu et al. 2010). Such a NRF2 competitor is SQSTM1/p62, a protein that transports specific cargos to the autophagosome, including KEAP1. Compounds that increase SQSTM1/p62 levels, like rapamycin (Sarkar and Rubinsztein 2008) and trehalose (Mizunoe et al. 2018), are under investigation in several phase II and III trials in connection with DM, systemic lupus erythematosus, and autosomal dominant polycystic kidney disease.

The transcriptional activity of NRF2 can be pharmacologically manipulated by impeding its interaction with critical partners in the nucleus. The BTB domain and CNC homolog 1 (BACH1) is a transcriptional repressor of NRF2 which belongs to the cap′n′collar, b-Zip family. BACH1 competes with NRF2 in the nucleus to form heterodimers with small MAF proteins and blocks therefore the expression of ARE genes (Dhakshinamoorthy et al. 2005). A recent study characterized the HPP-4382 compound as an inhibitor of BACH1 repressive activity in vitro (Attucks et al. 2014) which favors the transcriptional activity of NRF2.

Another therapeutic approach for increasing NRF2 accumulation is to address more closely its proteasomal degradation by modulating the Cul3-based E3 ligase which selectively binds protein substrates intended to be degraded (Kobayashi et al. 2004). One option is to inhibit the process of cullin 3 neddylation through which NEDD8 (neural precursor cell expressed developmentally downregulated 8) tags cullin 3 and regulates its ubiquitin ligase activity (Duda et al. 2008), hence inhibiting NRF2 ubiquitylation and degradation of NRF2. This has been achieved with the small molecule MLN4924 (pevonedistat) which is a selective inhibitor of the NEDD8-activating enzyme (NAE) involved in the first step of neddylation (Soucy et al. 2009) that was shown to induce NRF2 accumulation (Zhao et al. 2014) with various biological consequences. For instance, it has been shown that MLN4924 exerts neuroprotective effects by increasing NRF2 accumulation in the cytoplasm and nucleus in primary cultures of cerebellar granule neurons exposed to hydrogen peroxide (Anderica-Romero et al. 2016). Interestingly, although cancer cells are "addicted" to NRF2 (Kitamura and Motohashi 2018) and inhibition of its proteasomal degradation results in NRF2 accumulation, consequently favoring tumorigenesis, MLN4924 was shown to inhibit in vitro and in mice xenografts the proliferation of head and neck squamous cell carcinoma cells, as well as to increase their radio-sensitivity (Vanderdys et al. 2018). Moreover, MLN4924 can induce apoptosis in acute myelogenous leukemia cell lines and clinical samples, but this was apparently independent of ROS production (Knorr et al. 2015). Various phase I clinical trials of MLN9424 (NCT00677170, NCT00722488, NCT00911066, NCT01011530, NCT01814826, NCT01862328) indicated that its safety profile was acceptable, with hepatotoxicity and sepsis as main adverse effects (Swords et al. 2017). The hepatotoxicity in patients with acute myeloid leukemia may be partly due to the therapy-induced decrease of the activation threshold for tumor necrosis factor-mediated cell death (Wolenski et al. 2015). In turn, this immunomodulatory activity of MLN4924 might be also responsible for its therapeutic anticancer activity, as MLN4924 was shown to exert NF-κB-mediated cytotoxicity against a subset of diffuse large B cell lymphoma and multiple myeloma cells (Milhollen et al. 2010). The clinical pharmacokinetic profile of MLN4924 was comparable in patients with solid tumors or hematological malignancies, and particular cytostatic drugs were shown to decrease its clearance (Faessel et al. 2019). A phase I study of MLN4924 suggests therapeutic activity of MLN4924 in lymphoma and evidenced a tolerable safety profile in patients with relapsed/refractory multiple myeloma or lymphoma (Shah et al. 2016). In 2019 was launched a clinical trial of the NAE inhibitor MLN4924 combined with an inhibitor of the chymotrypsin-like proteolytic (β5) site of the 20S proteasome (MLN9708, ixazomib) in treating patients with multiple myeloma (NCT03770260). Possibly, the anticancer effect of MLN4924 relies on the inhibition of the proteasomal degradation of other Cul3-based E3 ligase targets than NRF2, or other cullins might be concurrently inhibited by MLN4924 that stabilize proteins involved in DNA replication such as CDT1 in colon and lung cancer (Lin et al. 2010), c-myc-mediated apoptosis of acute myelogenous leukemia cells (Knorr et al. 2015), or Redd1-mediated suppression of AKT and mTOR pro-survival and proliferative signaling in multiple myeloma (Gu et al. 2014). Further clinical studies are needed for defining the therapeutic and safety profiles of MLN4924, as well as its action mechanism in various hematological and non-hematological malignancies. Through structure-based design and extensive medicinal chemistry optimization, DI-591 was found to be a high-affinity, cell-permeable, drug-like small molecule inhibitor of the DCN1-UBC12 interaction that is critical in specific neddylation of cullin 3. Treatment of cells with DI-591 selectively converts cellular cullin 3 into an un-neddylated inactive form that leads to NRF2 upregulation (Zhou et al. 2017).

All these alternative mechanisms and therapeutic tools for NRF2 stabilization and activation in various pathological conditions suggest that a combinatorial pharmaceutical approach will be the best way to activate the cytoprotective responses mediated by NRF2 while keeping off-target and side effects under control.

3.4 Dual Effects of NRF2 Activators

Knowing that cancer cells are NRF2 addicted, in-depth investigations have to be performed for clarifying the intriguing anticancer activity of some NRF2 activators at higher, still clinically acceptable concentrations, going beyond their chemopreventive action that impedes on the initial steps of carcinogenesis by antioxidant and anti-inflammatory mechanisms (Hu et al. 2010). There is extensive preclinical evidence indicating that SFN is a promising candidate as anticancer agent (Wu et al. 2020). For instance, it was shown that SFN can enhance cisplatin sensitivity of ovarian carcinoma cells (DNA damage and accumulation of intracellular cisplatin) through miR-30a-3p upregulation (Gong et al. 2020). Moreover, SFN can induce the suppression of human prostate cancer cell growth (ROS-induced G2/M phase cell cycle arrest and apoptosis) and slow tumor progression in a tumor-bearing mouse model (Singh et al. 2009). The study also showed that SFN can boost the anti-tumor immune response in vitro (enhanced cytotoxicity of natural killer cells and activation of dendritic cells against prostate cancer cells), which was translated in the investigated animal model by infiltration of T cells in the neoplastic lesions and increased levels of interleukin-12 production by dendritic cells. In turn, the pro-oxidative action of SFN was shown to induce the suppression of particular T cell populations that may negatively impact the efficacy of T cell-based immunotherapies in cancer (Liang et al. 2019). Recent studies also evidenced that DMF can be repurposed as an anticancer agent, either as monotherapy or in combination with other drugs (Booth et al. 2016). It is noteworthy that another anti-MS drug, fingolimod (Gilenya®), that acts through downregulation of sphingosine-1-phosphate signaling also exhibits anticancer properties and can synergistically combine with MMF to kill many tumor cells types, including multiple genetically diverse primary human glioblastoma cell isolates (Booth et al. 2016). There is a high need for specifically addressing the dual effects of NRF2 activators in various pathologic settings for better defining fit-to-purpose dosages and potential off-target effects.

3.5 NRF2 Activators and COVID-19

A promising therapeutic application of NRF2 activators, including SFN, MMF is related to their potential use in viral diseases, for counteracting the deleterious oxidative burst triggered by the virus in the host tissues (Lee 2018; Sharma et al. 2020). Moreover, viruses are endowed with specific mechanisms for sustaining the oxidative metabolism necessary for their replication, without killing the host cell, and for taking away from the host the NRF2-mediated control of redox homeostasis (Lee 2018). Pharmacologic NRF2 activation in viral diseases gains a higher importance in the context of the COVID-19 pandemics. PB125® (Pathways Bioscience), designed as dietary supplement that contains carnosol, withaferin A, and luteolin, proved to be a selective and potent NRF2 activator in vitro, as demonstrated by the Ingenuity Pathway Analysis using HepG2 cells (Hybertson et al. 2019). It was demonstrated recently that PB125 downregulates the mRNA expression of ACE2, of the surface receptor recognizing the SARS-CoV-2 virus, and also of TMPRSS2 which activates the spike protein for virus entry into host cells in human liver-derived HepG2 cells. Moreover, PB125 markedly downregulated 36 genes encoding inflammatory factors, out of which IL-1β, IL-6, and TNFα have been specifically identified in the "cytokine storm" observed in fatal cases of COVID-19 (McCord et al. 2020). Extensive studies have to be performed for assessing the role of NRF2 in viral replication and protection of the host.

4 Pharmacologic Inhibitors of NRF2

Inhibitors of NRF2 activity activators are a promising therapeutic strategy especially for cancer treatment, but their clinical development is being slow due to poor specificity of currently available molecules. Figure 3 shows some of the available compounds and pharmacological strategies to inhibit NRF2.

The role of NRF2 in tumor development is still controversial. On the one hand, some studies show how NRF2 is able to prevent the development of tumors in mice models of chemically induced carcinogenesis due to its cytoprotective and detoxifying effects (Ramos-Gomez et al. 2003; Xu et al. 2006). On the other hand, many tumors exhibit an exacerbated activation of NRF2 which is associated with poor disease prognosis, as NRF2 promotes a cell survival and growth advantage while it also renders cells resistant to chemo- and radiotherapy (Solis et al. 2010;

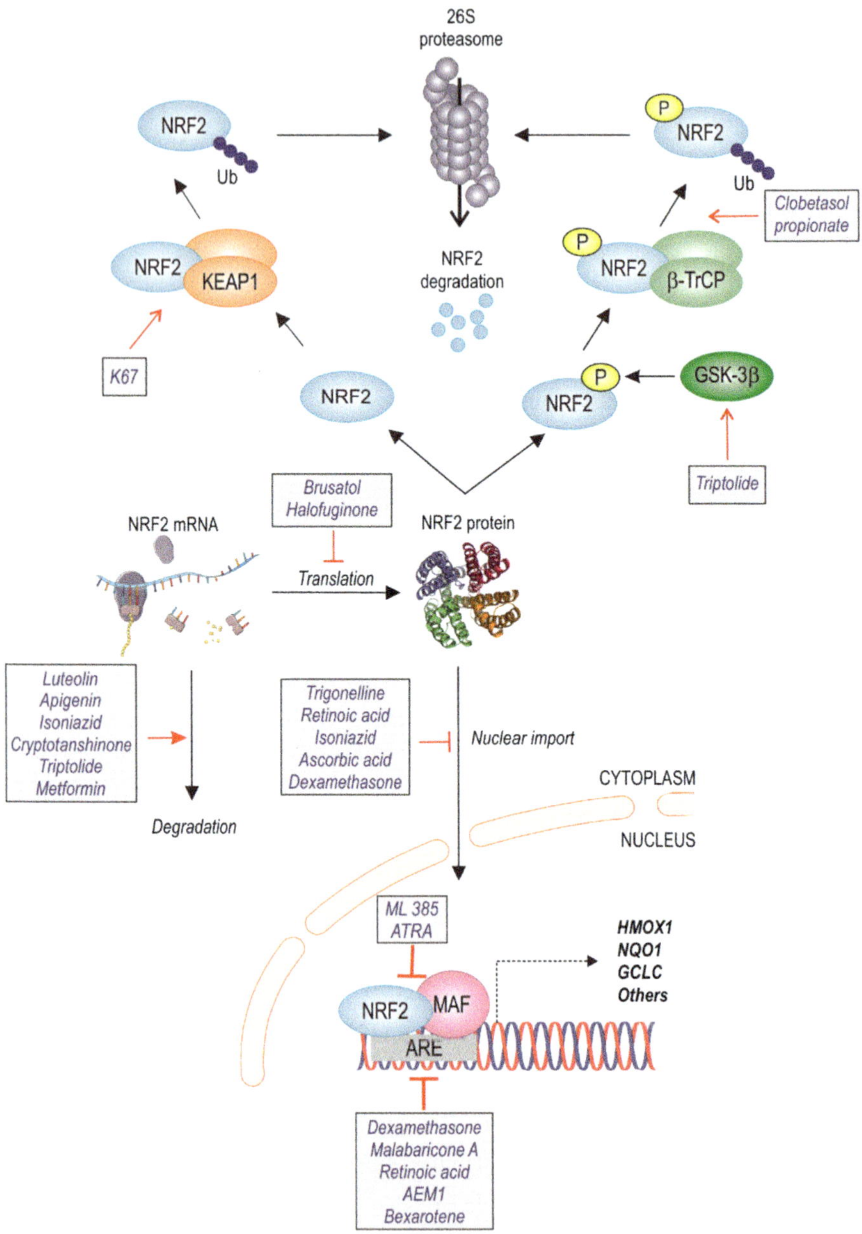

Fig. 3 Summary of available compounds and pharmacological strategies to inhibit NRF2 activity

Shibata et al. 2008; Ohta et al. 2008). Taken together, these studies show that NRF2 plays a dual role in cancer progression: it prevents cancer development by acting as a chemo-preventive agent, while, once disease is progressing, tumor cells become "addicted" to NRF2 which provides a selective advantage to cancer cells for surviving in the harsh tumor niche and for resisting to chemo- and radiotherapy (Kitamura and Motohashi 2018; Milkovic et al. 2017). Consequently, it is expected that molecules that inhibit NRF2 would sensitize tumor cells to conventional anti-tumor treatments, being therefore a promising novel co-therapy in cancer.

4.1 NRF2 Natural Inhibitors

Small molecules of natural origin (especially those found in plants) have a huge variety of functions and structures. Consequently, they are highly promising candidates for drug discovery, including NRF2 inhibition, due to the relatively uncomplicated isolation procedure and pharmaceutically relevant biological activity. Selected NRF2 inhibitors are presented in Table 2.

Brusatol is a natural quassinoid obtained from the seeds of *Brucea sumatrana*. Several studies have shown how brusatol treatment reduces NRF2 protein levels in a variety of cancer cell lines, such as A549 (non-small-cell lung carcinoma); MDA-MB 231 (human breast cancer); Ishikawa and SPEC-2 (human endometrial carcinoma); PATU-8988, PANC1, and BxPC-3 (human pancreatic cancer); and HCT116 and CT26 (human and murine colorectal cancer, respectively) (Zhu et al. 2016; Ren et al. 2011; Xiang et al. 2018; Evans et al. 2018). Brusatol also enhances both in vivo and in vitro the anti-tumor effect of different drugs, such as cisplatin in A549 cells; gemcitabine in PATU-8988, PANC1, and BxPC-3 cell lines; or irinotecan in the CT26 cell line (Ren et al. 2011; Xiang et al. 2018; Evans et al. 2018). However, it was evidenced that brusatol is not a specific NRF2 inhibitor as it inhibits cap-dependent and cap-independent protein translation, being therefore able to inhibit not only NRF2 but many other short-lived proteins (Harder et al. 2017). Due to the consequent off-target effects of brusatol, it has not been tested in clinical trials. Nevertheless, the lack of effect of brusatol at nanomolar concentrations on the levels of some proteins with short (cyclin A, HIF-1α, p53, and survivin) and long (Keap1, p62, and actin) half-lives indicates that the brusatol-induced NRF2 depletion might be in fact specific (Ren et al. 2011; Olayanju et al. 2015).

Other candidates for NRF2 inhibition are flavonoids, small molecules found in a variety of plants, vegetables, and fruits that are well known for their antioxidant effects and anti-tumor activity (Zhu et al. 2016; Ju et al. 2007). Among them, several inhibit NRF2 activity, as follows. Luteolin was shown to reduce mRNA and protein levels of NRF2 in non-small-cell lung carcinoma A549 cells or colorectal cancer cells (HCT116 and SW620) (Tang et al. 2011; Chian et al. 2014). It also sensitized A549 non-small-cell lung carcinoma cells to oxaliplatin, bleomycin, and doxorubicin in vitro and to cisplatin in vivo and the human colon tumor cells HCT116 and SW620 to oxaliplatin, cisplatin, and doxorubicin in vitro (Tang et al. 2011; Chian et al. 2014). It seems that luteolin is able to discriminate between normal and tumor cells. Thus, luteolin can arrest cells at the G1/S stage of the cell cycle, reduce mitochondrial membrane potential, and trigger apoptosis of liver carcinoma cells, while normal liver cells (HL-7702) were almost not affected (Ding et al. 2014).

Table 2 Selected NRF2 inhibitors

Compound	Type	Mechanism of action	Disease	Clinical trial	ClinicalTrials.gov identifier
Natural compounds					
Brusatol 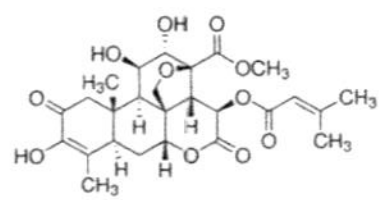	Natural product extracted from *Brucea javanica*	Inhibition de novo synthesis of NRF2	–	–	–
Luteolin 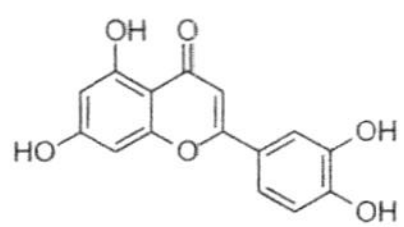	Natural flavonoid compound	Decrease mRNA and protein levels of NRF2	Tongue neoplasms Carcinoma	Early Phase 1	NCT03288298
Apigenin 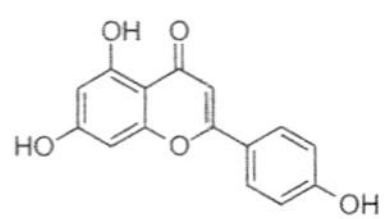	Natural flavonoid product	Decrease the mRNA and protein levels of NRF2	–	–	–
Malabaricone A 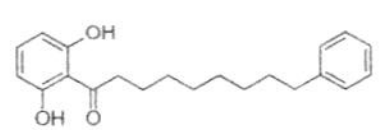	Natural product derived from *Myristica malabárica*	Inhibition NRF2 transcriptional activity	–	–	–
Cryptotanshinone 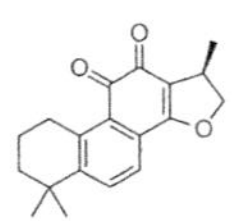	Natural product isolated from *Salvia miltiorrhiza*	Inhibition NRF2 protein expression	–	–	–

Triptolide 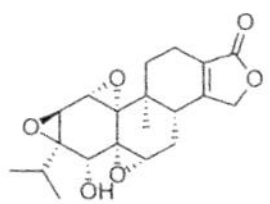	Natural product isolated from *Tripterygium wilfordii*	Inhibition mRNA and protein expression of NRF2	Pancreatic cancer	Phase II	NCT03117920
Trigonelline 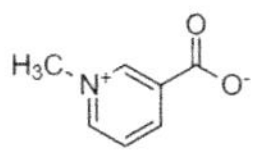	Alkaloid product constituent of coffee	Prevent nuclear translocation of NRF2	–	–	–
Synthetic compounds					
ML385 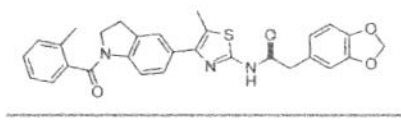	Synthetic compound	Decrease DNA binding activity of the NRF2-MAFG protein complex	–	–	–
AEM1 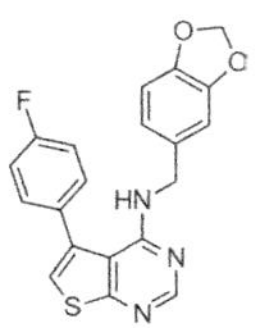	Synthetic compound	Inhibition transcriptional activity of NRF2	–	–	–
K67 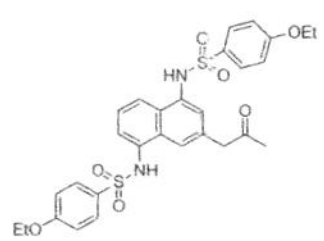	Synthetic compound	Inhibition of the phosphorylated p62-KEAP1 interaction	–	–	–
Halofuginone 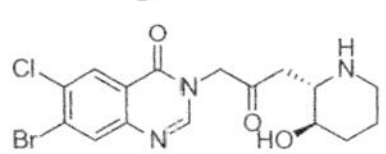	Synthetic derivate of febrifugine	Global inhibition of protein synthesis	Solid tumor	Phase I	NCT00027677
			Kaposi sarcoma	Phase II	NCT00064142

(continued)

Table 2 (continued)

Compound	Type	Mechanism of action	Disease	Clinical trial	ClinicalTrials.gov identifier
Repurposed drugs					
Dexamethasone 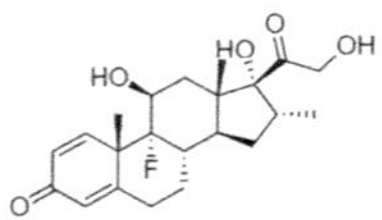	Synthetic glucocorticoid (agonist glucocorticoid receptor)	Inhibition transcriptional activity and nuclear translocation of NRF2	Prostate cancer	Phase II	NCT00006002
			Prostate cancer	Phase II	NCT00524589
Clobetasol propionate 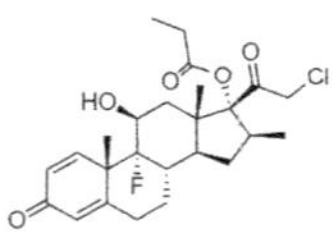	Synthetic glucocorticoid (agonist glucocorticoid receptor)	Promote β-TrCP-dependent degradation of NRF2	–	–	–
All-trans retinoic acid 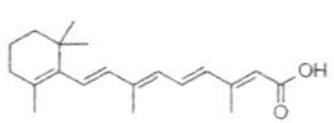	Synthetic retinoid acid (RXR agonist)	Inhibition nuclear import and transcriptional activity of NRF2	Lung cancer	Phase III	NCT01041833
			Multiple myeloma	Phase I/phase II	NCT02751255
Bexarotene 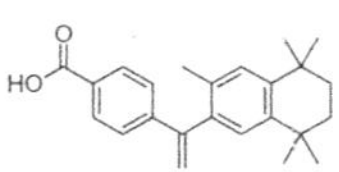	Synthetic retinoid acid (RXR agonist)	Inhibition transcriptional activity of NRF2	Breast cancer	Phase II	NCT00003752
			Lung cancer	Phase III	NCT00050973
Ascorbic acid (vitamin C) 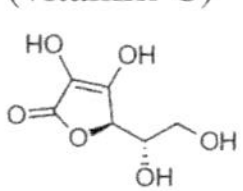	Natural product (vitamin)	Inhibition nuclear translocation and decrease the levels of the NRF2/ARE complex	Prostatic neoplasms	Phase II	NCT01080352
			Colorectal neoplasms	Phase III	NCT02969681

Isoniazid 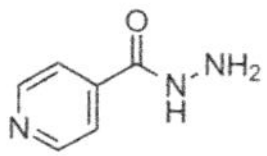	Antitubercular drug	Inhibition transcription activity and nuclear translocation of NRF2	–	–	–
Ethionamide 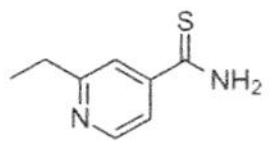	Antitubercular drug	Specific mechanism is not known yet	–	–	–
Metformin 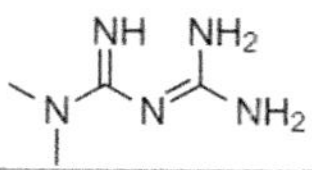	Synthetic compound	Inhibition NRF2 mRNA and protein expression	Prostate cancer	Phase II	NCT03137186
			Colon cancer	Phase II	NCT03359681

There are still gaps in understanding the mechanisms underlining the effects of luteolin on NRF2, as several studies have shown that luteolin might inhibit or activate NRF2 activity, depending on the concentration and context. Luteolin was able to epigenetically induce NRF2 in human colon carcinoma cells HCT116 and HT29 and to decrease cell proliferation, hence acting as an anti-tumor agent (Zuo et al. 2018). In turn, intragastric treatment with luteolin protected diabetic rats against ischemia-reperfusion injury due to NRF2 activation (Xiao et al. 2019). More interestingly, a recent study has shown that luteolin can prevent malignant transformation of human bronchial epithelial cells caused by the environmental carcinogen Cr(VI) through NRF2 activation while decreasing constitutive NRF2 activation in already transformed cells (Son et al. 2017). Although the dual effect of luteolin as chemo-preventive and anti-tumoral agent is quite promising, its precise modulatory effect and mechanism over NRF2 are still unclear, and it seems to be dependent of the transformation state of the cell.

Apigenin is another flavonoid with already known anti-tumoral effects, which inhibits the growth of cancer cells but does not significantly affect normal cells, being therefore a promising anti-tumoral drug (Gupta et al. 2001; Patel et al. 2007). The inhibitory effects of apigenin over NRF2 were highlighted. Apigenin can decrease the mRNA and protein levels of NRF2 in hepatocellular carcinoma BEL-7402/ADM cells, hence reducing the expression of NRF2 targets and sensitizing tumor cells to doxorubicin, both in vitro and in vivo, in a tumor xenograft mice model. This inhibitory effect seems to be dependent on the inhibition of the PI3K/AKT pathway (Gao et al. 2013). Meanwhile, a recent study suggests that apigenin can inhibit NRF2 through the expression of miR-101 which targets the 3'-UTR region of *NFE2L2* (Gao et al. 2017). Nevertheless, as also shown for luteolin, apigenin can activate NRF2 in human hepatoma HepG2 cells, through the PI3K/AKT pathway. Moreover, apigenin was shown to prevent non-alcoholic fatty liver disease by inducing NRF2 nuclear translocation and activation in a mouse model (Paredes-Gonzalez et al. 2015; Feng et al. 2017).

Although the effect of flavonoids as anti-tumor and chemo-sensitizing drugs is quite promising, their effects on NRF2 activity seem to be dependent on the cell type and the transformation state of the cells. Therefore, the anti-tumor effects of NRF2 inhibition by flavonoids should be systematically addressed using standard investigation protocols.

Malabaricone A is another plant natural product, obtained from *Myristica malabárica*, which has anti-tumor properties in leukemic cells (U937, MOLT-3, and CCRF CEM) by eliciting increased levels of ROS and consequent ROS-mediated apoptosis (Manna et al. 2012, 2015). This redox imbalance has been linked to the inhibitory effect of malabaricone A over NRF2, as it reduced NRF2 and HO-1 protein levels (Lo et al. 2006). However, in vivo studies should be done before considering its further development and potential clinical application.

Cryptotanshinone is a tanshinone obtained from *Salvia miltiorrhiza*, a traditional Chinese medicinal herb. Cryptotanshinone has promising anti-tumoral effects by reducing cell growth and apoptosis in several cancer cell lines such as colorectal cancer, breast cancer, squamous cell carcinoma, ovarian cancer, or leukemia cell lines (Li et al. 2015a, 2015b; Wu et al. 2016; Jiang et al. 2017; Wang et al. 2017). Cryptotanshinone has also been found to sensitize A549 cells to cisplatin by reducing NRF2 protein levels and also its transcriptional activity, probably through the MAPKs, Akt, and STAT3 pathway (Xia et al. 2015). Nevertheless, despite its promising anti-tumoral action, the effect of cryptotanshinone over NRF2 has become controversial recently, as some studies showed that cryptotanshinone can also activate NRF2. For instance, cryptotanshinone reduced the inflammatory response of microglia by activating NRF2 through the PI3K/Akt signaling pathway, while it attenuated inflammation in mice with unilateral ureteral obstruction through NRF2 and HO-1 activation (Wang et al. 2018; Zhou et al. 2019).

Triptolide is obtained from a Chinese traditional medicinal herb *Tripterygium wilfordii* and proved anti-tumor activities in a wide variety of cancer types, including acute myeloid leukemia, breast and ovarian cancer, osteosarcoma, lung cancer, prostate cancer, neuroblastoma, as well as several gastrointestinal cancers (Noel et al. 2019). However, its effect on NRF2 activation is ambiguous. There is increasing evidence concerning its inhibitory effect, as triptolide was shown to increase the sensitivity of AML leukemic stem-like KG1a cell line to idarubicin and of A549 cells to cisplatin, etoposide, and epirubicin, both in vitro and in vivo, by reducing the mRNA and protein levels of NRF2 (Liu et al. 2013; Zhu et al. 2018). However, another study indicated GSK3-β activation as the main mechanism of NRF2 inhibition (Pan et al. 2019). Meanwhile, some studies also evidenced a NRF2 activation effect caused by triptolide. Triptolide protected rats from myocardial ischemia-reperfusion injuries through the activation of NRF2 and HO-1 and induced NRF2 expression in the A549 cell line through increased ROS production (Kumar et al. 2016; Yu et al. 2016). These apparently controversial data require further research for precisely assessing the effect of triptolide over NRF2 activation, in relation with concentration and cell type. Triptolide is under investigation in a phase II clinical trial in refractory pancreatic cancer as Minnelide (University of Minnesota), a water-soluble pro-drug of triptolide that was designed to release triptolide in the bloodstream faster.

Trigonelline is a coffee alkaloid natural derivative that was first found to reduce NRF2 protein levels in HT29 colon carcinoma cells (Boettler et al. 2011). Since then, several studies have revealed promising data about its inhibitory effects over NRF2. It was found to reduce both basal and induced NRF2 activity in human pancreatic carcinoma (Panc1, Colo357, and MiaPaca2 cells) and duct cells (H6c7), by limiting NRF2 nuclear import as main mechanism. Moreover, trigonelline sensitized tumor cells to TRAIL-induced apoptosis and blocked the NRF2-dependent expression of proteasomal genes, both in vitro and in vivo in tumor-bearing mice (Arlt et al. 2013). Trigonelline has also been found to reduce the migratory activity of human hepatocarcinoma cells Hep3b through NRF2 downregulation (Liao et al. 2015). Moreover, it prevented artesunate resistance in some head and neck cancer cell lines (HN3, HN4, and HN9) and sensitized them to the inhibition of the antioxidant GSH and Trx systems, resulting in significant growth suppression and cell death both in vitro and in vivo (Roh et al. 2017a,

2017b). Even though the anti-tumoral effects of trigonelline are documented mostly in vitro, more studies in animal models are needed, especially for characterizing its bioavailability, dosage, and pharmacodynamics before starting clinical trials.

Taken altogether, in most of the cases, the mechanism of action of the currently available NRF2 inhibitors is either poorly understood or not specific, and more systematic work is needed for further development. Therefore, NRF2 natural inhibitors are still far from a clinical use.

4.2 Synthetic NRF2 Inhibitors

Although natural products are showing an important therapeutic potential for NRF2 inhibition, their lack of specificity, potential side effects, and their controversial role over NRF2 activity are still a major issue to overcome. Therefore, new approaches have been used to design and develop targeted NRF2-inhibiting drugs. This is the case of ML385, a first-in-class compound found through a quantitative high-throughput screening on a library of 400,000 compounds. Specifically, ML385 interferes with NRF2 heterodimerization with MAF proteins, hence preventing NRF2 binding to DNA that leads to an impairment of ARE-dependent gene transcription. ML385 sensitizes to carboplatin and other chemotherapeutics the KEAP1-deficient non-small cell lung cancer cell lines A549 and H460 both in vitro and in vivo, as demonstrated in an orthotopic lung tumor model (Singh et al. 2016). However, only few studies have tested ML385 activity in preclinical models, including its potential off-target effects, and therefore its clinical application is still difficult to plan.

Following a similar approach, the NRF2 inhibitor ARE expression modulator 1 (AEM1) was found through a quantitative high-throughput screening of a library of 30,000 heterocyclic and biologically active compounds. AEM1 (27634172) decreased NRF2-dependent gene transcription in cells with KEAP1 mutations that activate NRF2, such as the non-small-cell lung cancer cells H838, H460, or A549. Moreover, it increased the chemotherapeutic sensitivity and decreased the cell growth of A549 tumor cells, both in vivo and in vitro. However, AEM1 can also activate NRF2 in some cell types. As such, its action mechanism is still poorly understood, and further studies in preclinical models should be done before starting any clinical trial (Bollong et al. 2015).

Another NRF2 inhibitor found through high-throughput screening is K67. K67 inhibited phosphorylated p62-KEAP1 interaction and thus enabled KEAP1-dependent NRF2 degradation in hepatocellular carcinoma cell lines. Moreover, K67 reduced cell growth due to its sensitization effect against chemotherapeutic drugs (Saito et al. 2016). Considering the particular action mechanism of K67, this inhibitor could be of utmost importance in the treatment of tumors such as hepatocellular carcinoma, which accumulate phosphorylated p62 which is a KEAP1 inhibitor and therefore a NRF2 activator (Saito et al. 2016; Taguchi and Yamamoto 2017). Nevertheless, despite its promising effects, few preclinical studies have been carried out to ascertain its anti-tumoral effects.

Halofuginone is a less toxic synthetic derivative of febrifugine, a bioactive component of *Dichroa febrifuga* (Tsuchida et al. 2017). Using a high-throughput chemical library screening for NRF2 inhibitors, halofuginone was found to reduce NRF2 protein levels and to decrease its transcriptional activity in "NRF2-addicted" A549 and KYSE70 human esophageal cancer cells. Additionally, halofuginone enhanced the anti-tumoral effect of cisplatin and doxorubicin in these cell lines, while "non-NRF2-addicted" cancer cell lines were not responsive to halofuginone (Tsuchida et al. 2017). Despite this apparently NRF2-selective effect, the mechanism of action of halofuginone seems to be unspecific. Similar to the effect of febrifugine, halofuginone inhibits global protein synthesis through a cellular amino acid starvation response that derives from the inhibition of prolyl-tRNA synthetase (Tsuchida et al. 2017; Keller et al. 2012). Halofuginone has been tested in several clinical trials, including a phase I study for dose safety in solid tumor patients (NCT00064142) and a phase II study in HIV-related Kaposi's sarcoma patients NCT00064142).

4.3 Repurposed Drugs as NRF2 Inhibitors

Given the high cost and time needed to develop new drugs, repurposing already approved compounds to treat different diseases is an emerging approach. This is the case of the corticosteroids dexamethasone and clobetasol propionate that are actually used as anti-inflammatory drugs. Dexamethasone was first found to inhibit NRF2-dependent transcription through its binding to glucocorticoid receptor (GR), although the exact mechanism of action is not completely elucidated (Kratschmar et al. 2012; Jung et al. 2018). The generally accepted mechanism involves the recruitment of the co-repressor silencing mediator for retinoid and thyroid hormone receptors (SMRT) and NRF2 by GR, leading to reduced transcription of ARE genes. Another mechanism is related to an impairment of NRF2 nuclear translocation (Jung et al. 2018; Ki et al. 2005). Dexamethasone is already used in the treatment of hematopoietic cancers, and its effects in solid tumors, such as prostate cancer, are being tested in several clinical trials (Frankfurt and Rosen 2004). More recently, clobetasol propionate, another synthetic glucocorticoid, has also been found to inhibit NRF2, as demonstrated by a drug repositioning screening of 4,000 clinical compounds. Clobetasol propionate reduced NRF2 nuclear accumulation, while it increased β-TrCP-dependent NRF2 proteasomal degradation, hence inhibiting the growth of non-small-cell lung cancer cell lines A549 and H2228 both in vitro and in vivo (Choi et al. 2017). Despite these promising results, the effects of glucocorticoids for cancer treatment should be further investigated in disease-relevant preclinical models, before advancing to clinical trials.

All-trans-retinoic acid, also known as ATRA, is the physiologically active form of vitamin A, and it is mainly used to treat acute promyelocytic leukemia, with a complete remission in most cases (Schenk et al. 2014). Since then, the anti tumoral effect of ATRA and its analogues (such as bexarotene) has been tested in different tumor types. ATRA inhibits NRF2 by binding to the retinoic acid receptor alpha (RARα), which heterodimerizes with NRF2, hence inhibiting NRF2 binding to ARE elements (Wang et al. 2007). In addition, NRF2 inhibition by ATRA was described as a possible mechanism through which ATRA sensitizes acute promyelocytic leukemia cells to arsenic trioxide treatment (Valenzuela et al. 2014). Another ATRA nuclear receptor, retinoid X receptor alpha (RXRα), is also able to inhibit NRF2 by directly interacting with the Neh7 domain of NRF2 in absence of its ligands (Wang et al. 2013). However, this inhibitory effect is dose-dependently increased by the ATRA analogue bexarotene (Wu et al. 2014b). Finally, ATRA has also been found to inhibit the NRF2 transcriptional activity in non-small-cell lung cancer A549 cells by increasing ROS production and by reducing the expression of DNA repair proteins, consequently sensitizing cells to chemo-pharmaceuticals such as cisplatin (de Miranda Ramos et al. 2019). Due to its quite promising anti-tumoral effect which goes beyond the treatment of acute promyelocytic leukemia, ATRA has been tested as a sensitizing agent for other types of tumors in several clinical trials. These include a phase III study in non-small-cell lung cancer in combination with platinum-based chemotherapy (NCT01041833) and a phase I/II study in multiple myeloma in combination with daratumumab (NCT02751255).

Ascorbic acid (vitamin C) is a well-known antioxidant found in a wide variety of fruits and vegetables. Ascorbic acid was also shown to inhibit NRF2 transcriptional activity in HUH7 liver cells and leukemia KCL22/SR cells through inhibition of NRF2 nuclear translocation and through a reduction of the NRF2/ARE complex levels. Besides, it sensitized KCL22/SR cells to imatinib (Tarumoto et al. 2004; Wagner et al. 2011). However, ascorbic acid is also able to activate NRF2 in normal cells, such as RAW 264.7 mice macrophages, through an enhancement of its nuclear translocation. Further investigation is needed to determine the context-dependent effect on NRF2 ascorbic acid. Ascorbic acid has been already tested in clinical trials for cancer treatment, such as in prostatic and colorectal neoplasms, in order to determine its potential anti-tumoral activity due to its antioxidant properties. However, its therapeutic effect is ambiguous, considering that high doses of vitamine C may act as prooxidant rather than as antioxidant (Wilson et al. 2014). Consequently, clinical trials testing the dose-effect connection should be performed, and further investigation is needed to determine if the NRF2 inhibitory effect of vitamin C is its main anti-tumoral activity.

Other repurposed compounds that are able to inhibit NRF2 are the antitubercular drug isoniazid and its analogue ethionamide (Vilcheze and Jacobs Jr. 2014). In 3T3-L1 preadipocytes, isozianid was able to inhibit the transcriptional activity of NRF2, producing a reduction in adipogenesis and adipogenic differentiation (Chen et al. 2013). Moreover, isozianid prevented NRF2 nuclear translocation via ERK1 dephosphorylation in the hepatocellular carcinoma cell lines Hep3B and HepG2, sustaining ROS production and apoptosis (Verma et al. 2015). In addition, both isozianid and ethionamide reduced NRF2 transcriptional activity in acute monocytic leukemia THP-1 cells, increasing their sensitivity to arsenic trioxide (Peng et al. 2016). Despite the promising chemo-sensitizing effects of isoniazid and ethionamide, only few in vitro studies have been done for testing their anti-tumoral activity, and further progression toward in vivo studies is highly needed before clinical development.

Metformin is a first-line treatment for type 2 DM. Its antidiabetic effect is caused, at least in part, by AMPK activation. Interestingly, some studies have shown that metformin can exert also anti-tumoral effects. A systematic meta-analysis was carried out to determine cancer incidence and mortality in DM patients. Those diabetes patients treated with metformin showed a reduced risk of cancer development, compared to patients that were treated with other antidiabetic drugs (Decensi et al. 2010). Additionally, metformin's anti-tumoral effect was demonstrated both in vitro and in vivo in several cancer types such as breast, liver, or endometrial cancer (Morales and Morris 2015). Metformin-induced inhibitory effects on NRF2 were evidenced in several cancer cell lines (human hepatic carcinoma HepG2 cells, cervical HeLa cells, and non-small-cell lung cancer A549 cells) and comprised the decrease of both NRF2 mRNA and protein levels that probably occurred through Raf-ERK inhibition. Metformin also sensitized A549 tumor cells to epigallocatechin-3-gallate in vitro and in vivo (Do et al. 2013; Yu et al. 2017). However, another study suggested that metformin can inhibit NRF2 through microRNA-34a induction, hence increasing ROS production and apoptosis in the breast cancer cell line MCF-7 (Do et al. 2014). Taken together, all these data suggest that the anti-tumoral effect of metformin is partially dependent on NRF2. Interestingly, metformin has been tested in cancer treatment in several clinical trials, such as phase II studies in advanced prostate (NCT03137186) and colon cancer (NCT03359681) as well as in stage/II breast cancer (NCT00984490).

5 Conclusions

Comprehensive evidence, obtained mostly at preclinical level, argues for the therapeutic benefits of pharmacological modulation of the NRF2-KEAP1 system for controlling redox metabolism and hence a broad array of derived pathological events in various chronic diseases and cancer. While a plethora of NRF2 activators reached clinical trials, with DMF already put on the market, NRF2 inhibitors are far behind in the drug development pipeline. In-depth knowledge regarding NRF2 biology, complemented by sophisticated in silico methods, will for sure advance in the near future promising drug candidates, especially small molecules that would be able to specifically inhibit NRF2 in "addicted" tumor cells for inducing their death and for raising their responsivity to conventional anticancer therapies. Additionally, drug repurposing provides a valuable strategy for speeding up clinical translation and also for reducing the economic burden of drug development. Important issues regarding the pharmacologic modulation of NRF2 are also related to unwanted side effects that are apparently arising from insufficient specificity of the selected compounds and/or the complex signaling networks that were shown to be more or less affected by changes in NRF2 activity. To solve this problem, the existing drug candidates are modified using the tools of medicinal chemistry, and a system medicine approach started to be applied for smartly designing therapeutic approaches addressing the

NRF2-KEAP1 system. Another challenging issue is related that the effects of NRF2 modulators are dose- and context-dependent, encompassing both NRF2 activation and inhibition. Therefore, extensive pharmacokinetic and pharmacodynamic studies have to be performed for precisely establishing dosage in specific therapeutic applications. Finally, there is an urgent need to join forces for developing a systematic and comprehensive strategy aiming to boost preclinical and clinical research in the field of NRF2 medicine and therapeutics.

Acknowledgements This work was supported by grants SAF2016-76520-R of the Spanish Ministry of Economy and Competitiveness, B2017/BMD-3827 of the Autonomous Community of Madrid, and P_37_732/2016 REDBRAIN of the European Regional Development Fund, Competitiveness Operational Program 2014–2020. RFG and DL are recipient of FPI and FPU contracts, respectively, of the Spanish Ministry of Economy and Competitiveness.

Conflict of Interest The authors declare that there is no conflict of interest regarding the publication of this paper.

References

Abdull Razis AF, Noor NM (2013) Sulforaphane is superior to glucoraphanin in modulating carcinogen-metabolising enzymes in Hep G2 cells. Asian Pac J Cancer Prev 14(7):4235–4238

Albrecht P, Bouchachia I, Goebels N et al (2012) Effects of dimethyl fumarate on neuroprotection and immunomodulation. J Neuroinflammation 9:163

Anderica-Romero AC, Hernandez-Damian J, Vazquez-Cervantes GI et al (2016) The MLN4924 inhibitor exerts a neuroprotective effect against oxidative stress injury via Nrf2 protein accumulation. Redox Biol 8:341–347

Arbeeny C, Ling H, Smith MM et al (2019) CXA-10, a nitrated fatty acid, is renoprotective in deoxycorticosterone acetate-salt nephropathy. J Pharmacol Exp Ther 369:503–510

Arlt A, Sebens S, Krebs S et al (2013) Inhibition of the Nrf2 transcription factor by the alkaloid trigonelline renders pancreatic cancer cells more susceptible to apoptosis through decreased proteasomal gene expression and proteasome activity. Oncogene 32(40):4825–4835

Attucks OC, Jasmer KJ, Hannink M et al (2014) Induction of heme oxygenase I (HMOX1) by HPP-4382: a novel modulator of Bach1 activity. PLoS One 9(7):e101044

Axelsson AS, Tubbs E, Mecham B et al (2017) Sulforaphane reduces hepatic glucose production and improves glucose control in patients with type 2 diabetes. Sci Transl Med 9(394)

Bahadoran Z, Mirmiran P, Hosseinpanah F et al (2012) Broccoli sprouts powder could improve serum triglyceride and oxidized LDL/LDL-cholesterol ratio in type 2 diabetic patients: a randomized double-blind placebo-controlled clinical trial. Diabetes Res Clin Pract 96 (3):348–354

Baird L, Lleres D, Swift S et al (2013) Regulatory flexibility in the Nrf2-mediated stress response is conferred by conformational cycling of the Keap1-Nrf2 protein complex. Proc Natl Acad Sci U S A 110(38):15259–15264

Batthyany CI, Lopez GV (2015) Nitroalkene tocopherols and analogs thereof for use in the treatment and prevention of inflammation related conditions. Complexa Inc.

Bertrand HC, Schaap M, Baird L et al (2015) Design, synthesis, and evaluation of Triazole derivatives that induce Nrf2 dependent gene products and inhibit the Keap1-Nrf2 protein-protein interaction. J Med Chem 58(18):7186–7194

Biesalski HK, Grune T, Tinz J et al (2010) Reexamination of a meta-analysis of the effect of antioxidant supplementation on mortality and health in randomized trials. Nutrients 2 (9):929–949

Biswas SK (2016) Does the interdependence between oxidative stress and inflammation explain the antioxidant paradox? Oxidative Med Cell Longev 2016:5698931

Bjelakovic G, Nikolova D, Gluud LL et al (2007) Mortality in randomized trials of antioxidant supplements for primary and secondary prevention: systematic review and meta-analysis. JAMA 297(8):842–857

Boddupalli S, Mein JR, Lakkanna S et al (2012) Induction of phase 2 antioxidant enzymes by broccoli sulforaphane: perspectives in maintaining the antioxidant activity of vitamins A, C, and E. Front Genet 3:7

Boettler U, Sommerfeld K, Volz N et al (2011) Coffee constituents as modulators of Nrf2 nuclear translocation and ARE (EpRE)-dependent gene expression. J Nutr Biochem 22(5):426–440

Bollong MJ, Yun H, Sherwood L et al (2015) A small molecule inhibits deregulated NRF2 transcriptional activity in cancer. ACS Chem Biol 10(10):2193–2198

Booth L, Malkin M, Dent P (2016) Repurposing Tecfidera for cancer. Aging (Albany NY) 8 (7):1289–1290

Bourhill T, Narendran A, Johnston RN (2017) Enzastaurin: a lesson in drug development. Crit Rev Oncol Hematol 112:72–79

Bresciani A, Missineo A, Gallo M et al (2017) Nuclear factor (erythroid-derived 2)-like 2 (NRF2) drug discovery: biochemical toolbox to develop NRF2 activators by reversible binding of Kelch-like ECH-associated protein 1 (KEAP1). Arch Biochem Biophys 631:31–41

Chao A, Lin CY, Wu RC et al (2018) The combination of everolimus and terameprocol exerts synergistic antiproliferative effects in endometrial cancer: molecular role of insulin-like growth factor binding protein 2. J Mol Med (Berl) 96(11):1251–1266

Chen Y, Inoyama D, Kong AN et al (2011) Kinetic analyses of Keap1-Nrf2 interaction and determination of the minimal Nrf2 peptide sequence required for Keap1 binding using surface plasmon resonance. Chem Biol Drug Des 78(6):1014–1021

Chen Y, Xue P, Hou Y et al (2013) Isoniazid suppresses antioxidant response element activities and impairs adipogenesis in mouse and human preadipocytes. Toxicol Appl Pharmacol 273 (3):435–441

Chen K, Li JJ, Li SN et al (2017) 15-Deoxy-Delta(12,14)-prostaglandin J2 alleviates hepatic ischemia-reperfusion injury in mice via inducing antioxidant response and inhibiting apoptosis and autophagy. Acta Pharmacol Sin 38(5):672–687

Chian S, Li YY, Wang XJ et al (2014) Luteolin sensitizes two oxaliplatin-resistant colorectal cancer cell lines to chemotherapeutic drugs via inhibition of the Nrf2 pathway. Asian Pac J Cancer Prev 15(6):2911–2916

Choi EJ, Jung BJ, Lee SH et al (2017) A clinical drug library screen identifies clobetasol propionate as an NRF2 inhibitor with potential therapeutic efficacy in KEAP1 mutant lung cancer. Oncogene 36(37):5285–5295

Chowdhry S, Zhang Y, McMahon M et al (2013) Nrf2 is controlled by two distinct beta-TrCP recognition motifs in its Neh6 domain, one of which can be modulated by GSK-3 activity. Oncogene 32(32):3765–3781

Cleasby A, Yon J, Day PJ et al (2014) Structure of the BTB domain of Keap1 and its interaction with the triterpenoid antagonist CDDO. PLoS One 9(6):e98896

Cuadrado A (2015) Structural and functional characterization of Nrf2 degradation by glycogen synthase kinase 3/beta-TrCP. Free Radic Biol Med 88(Pt B):147–157

Cuadrado A, Martin-Moldes Z, Ye J et al (2014) Transcription factors NRF2 and NF-kappaB are coordinated effectors of the rho family, GTP-binding protein RAC1 during inflammation. J Biol Chem 289(22):15244–15258

Cuadrado A, Manda G, Hassan A et al (2018) Transcription factor NRF2 as a therapeutic target for chronic diseases: a systems medicine approach. Pharmacol Rev 70(2):348–383

Cuadrado A, Rojo AI, Wells G et al (2019) Therapeutic targeting of the NRF2 and KEAP1 partnership in chronic diseases. Nat Rev Drug Discov 18:295–317

Dayalan Naidu S, Muramatsu A, Saito R et al (2018) C151 in KEAP1 is the main cysteine sensor for the cyanoenone class of NRF2 activators, irrespective of molecular size or shape. Sci Rep 8 (1):8037

de la Vega MR, Dodson M, Gross C et al (2016) Role of Nrf2 and autophagy in acute lung injury. Curr Pharmacol Rep 2(2):91–101

de Miranda Ramos V, Gasparotto J, Figueiro F et al (2019) Retinoic acid downregulates thiol antioxidant defences and homologous recombination while promotes A549 cells sensitization to cisplatin. Cell Signal 62:109356

Decensi A, Puntoni M, Goodwin P et al (2010) Metformin and cancer risk in diabetic patients: a systematic review and meta-analysis. Cancer Prev Res (Phila) 3(11):1451–1461

DeNicola GM, Karreth FA, Humpton TJ et al (2011) Oncogene-induced Nrf2 transcription promotes ROS detoxification and tumorigenesis. Nature 475(7354):106–109

Dhakshinamoorthy S, Jain AK, Bloom DA et al (2005) Bach1 competes with Nrf2 leading to negative regulation of the antioxidant response element (ARE)-mediated NAD(P)H:quinone oxidoreductase 1 gene expression and induction in response to antioxidants. J Biol Chem 280 (17):16891–16900

Dibbert S, Clement B, Skak-Nielsen T et al (2013) Detection of fumarate-glutathione adducts in the portal vein blood of rats: evidence for rapid dimethylfumarate metabolism. Arch Dermatol Res 305(5):447–451

Ding S, Hu A, Hu Y et al (2014) Anti-hepatoma cells function of luteolin through inducing apoptosis and cell cycle arrest. Tumour Biol 35(4):3053–3060

Dinkova-Kostova AT, Liby KT, Stephenson KK et al (2005) Extremely potent triterpenoid inducers of the phase 2 response: correlations of protection against oxidant and inflammatory stress. Proc Natl Acad Sci U S A 102(12):4584–4589

Do MT, Kim HG, Khanal T et al (2013) Metformin inhibits heme oxygenase-1 expression in cancer cells through inactivation of Raf-ERK-Nrf2 signaling and AMPK-independent pathways. Toxicol Appl Pharmacol 271(2):229–238

Do MT, Kim HG, Choi JH et al (2014) Metformin induces microRNA-34a to downregulate the Sirt1/Pgc-1alpha/Nrf2 pathway, leading to increased susceptibility of wild-type p53 cancer cells to oxidative stress and therapeutic agents. Free Radic Biol Med 74:21–34

Duda DM, Borg LA, Scott DC et al (2008) Structural insights into NEDD8 activation of cullin-RING ligases: conformational control of conjugation. Cell 134(6):995–1006

Egea J, Buendia I, Parada E et al (2015) Melatonin-sulforaphane hybrid ITH12674 induces neuroprotection in oxidative stress conditions by a 'drug-prodrug' mechanism of action. Br J Pharmacol 172(7):1807–1821

Evans JP, Winiarski BK, Sutton PA et al (2018) The Nrf2 inhibitor brusatol is a potent antitumour agent in an orthotopic mouse model of colorectal cancer. Oncotarget 9(43):27104–27116

Fabrizio FP, Sparaneo A, Trombetta D et al (2018) Epigenetic versus genetic deregulation of the KEAP1/NRF2 Axis in solid tumors: focus on methylation and noncoding RNAs. Oxidative Med Cell Longev 2018:2492063

Faessel HM, Mould DR, Zhou X et al (2019) Population pharmacokinetics of pevonedistat alone or in combination with standard of care in patients with solid tumours or haematological malignancies. Br J Clin Pharmacol 85(11):2568–2579

Feng X, Yu W, Li X et al (2017) Apigenin, a modulator of PPARgamma, attenuates HFD-induced NAFLD by regulating hepatocyte lipid metabolism and oxidative stress via Nrf2 activation. Biochem Pharmacol 136:136–149

Fox RJ, Miller DH, Phillips JT et al (2012) Placebo-controlled phase 3 study of oral BG-12 or glatiramer in multiple sclerosis. N Engl J Med 367(12):1087–1097

Franceschi C, Garagnani P, Parini P et al (2018) Inflammaging: a new immune-metabolic viewpoint for age-related diseases. Nat Rev Endocrinol 14(10):576–590

Frankfurt O, Rosen ST (2004) Mechanisms of glucocorticoid-induced apoptosis in hematologic malignancies: updates. Curr Opin Oncol 16(6):553–563

Gao AM, Ke ZP, Wang JN et al (2013) Apigenin sensitizes doxorubicin-resistant hepatocellular carcinoma BEL-7402/ADM cells to doxorubicin via inhibiting PI3K/Akt/Nrf2 pathway. Carcinogenesis 34(8):1806–1814

Gao AM, Zhang XY, Ke ZP (2017) Apigenin sensitizes BEL-7402/ADM cells to doxorubicin through inhibiting miR-101/Nrf2 pathway. Oncotarget 8(47):82085–82091

Ghadiri M, Rezk A, Li R et al (2017) Dimethyl fumarate-induced lymphopenia in MS due to differential T-cell subset apoptosis. *Neurol Neuroimmunol Neuroinflamm* 4(3):e340

Gold R, Kappos L, Arnold DL et al (2012) Placebo-controlled phase 3 study of oral BG-12 for relapsing multiple sclerosis. N Engl J Med 367(12):1098–1107

Gong TT, Liu XD, Zhan ZP et al (2020) Sulforaphane enhances the cisplatin sensitivity through regulating DNA repair and accumulation of intracellular cisplatin in ovarian cancer cells. Exp Cell Res:112061

Gu Y, Kaufman JL, Bernal L et al (2014) MLN4924, an NAE inhibitor, suppresses AKT and mTOR signaling via upregulation of REDD1 in human myeloma cells. Blood 123 (21):3269–3276

Gupta S, Afaq F, Mukhtar H (2001) Selective growth-inhibitory, cell-cycle deregulatory and apoptotic response of apigenin in normal versus human prostate carcinoma cells. Biochem Biophys Res Commun 287(4):914–920

Han JM, Lee YJ, Lee SY et al (2007) Protective effect of sulforaphane against dopaminergic cell death. J Pharmacol Exp Ther 321(1):249–256

Hancock R, Bertrand HC, Tsujita T et al (2012) Peptide inhibitors of the Keap1-Nrf2 protein-protein interaction. Free Radic Biol Med 52(2):444–451

Hancock R, Schaap M, Pfister H et al (2013) Peptide inhibitors of the Keap1-Nrf2 protein-protein interaction with improved binding and cellular activity. Org Biomol Chem 11(21):3553–3557

Harder B, Tian W, La Clair JJ et al (2017) Brusatol overcomes chemoresistance through inhibition of protein translation. Mol Carcinog 56(5):1493–1500

Hast BE, Goldfarb D, Mulvaney KM et al (2013) Proteomic analysis of ubiquitin ligase KEAP1 reveals associated proteins that inhibit NRF2 ubiquitination. Cancer Res 73(7):2199–2210

Havrdova E, Hutchinson M, Kurukulasuriya NC et al (2013) Oral BG-12 (dimethyl fumarate) for relapsing-remitting multiple sclerosis: a review of DEFINE and CONFIRM. Evaluation of: Gold R, Kappos L, Arnold D, et al. Placebo-controlled phase 3 study of oral BG-12 for relapsing multiple sclerosis. N Engl J Med 2012;367:1098–107; and Fox RJ, Miller DH, Phillips JT, et al. Placebo-controlled phase 3 study of oral BG-12 or glatiramer in multiple sclerosis. N Engl J Med 2012;367:1087–97. Expert Opin Pharmacother 14(15):2145–2156

Hayashi G, Jasoliya M, Sahdeo S et al (2017) Dimethyl fumarate mediates Nrf2-dependent mitochondrial biogenesis in mice and humans. Hum Mol Genet 26(15):2864–2873

Hayes JD, Dinkova-Kostova AT (2014) The Nrf2 regulatory network provides an interface between redox and intermediary metabolism. Trends Biochem Sci 39(4):199–218

Heiss E, Herhaus C, Klimo K et al (2001) Nuclear factor kappa B is a molecular target for sulforaphane-mediated anti-inflammatory mechanisms. J Biol Chem 276(34):32008–32015

Holland R, Hawkins AE, Eggler AL et al (2008) Prospective type 1 and type 2 disulfides of Keap1 protein. Chem Res Toxicol 21(10):2051–2060

Hooper C, Killick R, Lovestone S (2008) The GSK3 hypothesis of Alzheimer's disease. J Neurochem 104(6):1433–1439

Houghton CA, Fassett RG, Coombes JS (2016) Sulforaphane and other nutrigenomic Nrf2 activators: can the clinician's expectation be matched by the reality? Oxidative Med Cell Longev 2016:7857186

Hoxtermann S, Nuchel C, Altmeyer P (1998) Fumaric acid esters suppress peripheral CD4- and CD8-positive lymphocytes in psoriasis. Dermatology 196(2):223–230

Hu R, Saw CL, Yu R et al (2010) Regulation of NF-E2-related factor 2 signaling for cancer chemoprevention: antioxidant coupled with antiinflammatory. Antioxid Redox Signal 13 (11):1679–1698

Hu L, Magesh S, Chen L et al (2013) Discovery of a small-molecule inhibitor and cellular probe of Keap1-Nrf2 protein-protein interaction. Bioorg Med Chem Lett 23(10):3039–3043

Huang HC, Nguyen T, Pickett CB (2000) Regulation of the antioxidant response element by protein kinase C-mediated phosphorylation of NF-E2-related factor 2. Proc Natl Acad Sci U S A 97 (23):12475–12480

Hybertson BM, Gao B, Bose S et al (2019) Phytochemical combination PB125 activates the Nrf2 pathway and induces cellular protection against oxidative injury. Antioxidants (Basel) 8(5)

Inoyama D, Chen Y, Huang X et al (2012) Optimization of fluorescently labeled Nrf2 peptide probes and the development of a fluorescence polarization assay for the discovery of inhibitors of Keap1-Nrf2 interaction. J Biomol Screen 17(4):435–447

Iso T, Suzuki T, Baird L et al (2016) Absolute amounts and status of the Nrf2-Keap1-Cul3 complex within cells. Mol Cell Biol 36(24):3100–3112

Jain A, Lamark T, Sjottem E et al (2010) p62/SQSTM1 is a target gene for transcription factor NRF2 and creates a positive feedback loop by inducing antioxidant response element-driven gene transcription. J Biol Chem 285(29):22576–22591

Jazwa A, Rojo AI, Innamorato NG et al (2011) Pharmacological targeting of the transcription factor Nrf2 at the basal ganglia provides disease modifying therapy for experimental parkinsonism. Antioxid Redox Signal 14(12):2347–2360

Jiang ZY, Lu MC, Xu LL et al (2014) Discovery of potent Keap1-Nrf2 protein-protein interaction inhibitor based on molecular binding determinants analysis. J Med Chem 57(6):2736–2745

Jiang G, Liu J, Ren B et al (2017) Anti-tumor and chemosensitization effects of Cryptotanshinone extracted from Salvia miltiorrhiza Bge. On ovarian cancer cells in vitro. J Ethnopharmacol 205:33–40

Jnoff E, Albrecht C, Barker JJ et al (2014) Binding mode and structure-activity relationships around direct inhibitors of the Nrf2-Keap1 complex. ChemMedChem 9(4):699–705

Ju W, Wang X, Shi H et al (2007) A critical role of luteolin-induced reactive oxygen species in blockage of tumor necrosis factor-activated nuclear factor-kappaB pathway and sensitization of apoptosis in lung cancer cells. Mol Pharmacol 71(5):1381–1388

Jung BJ, Yoo HS, Shin S et al (2018) Dysregulation of NRF2 in Cancer: from molecular mechanisms to therapeutic opportunities. Biomol Ther (Seoul) 26(1):57–68

Kanarek N, Ben-Neriah Y (2012) Regulation of NF-kappaB by ubiquitination and degradation of the IkappaBs. Immunol Rev 246(1):77–94

Kansanen E, Bonacci G, Schopfer FJ et al (2011) Electrophilic nitro-fatty acids activate NRF2 by a KEAP1 cysteine 151-independent mechanism. J Biol Chem 286(16):14019–14027

Keller TL, Zocco D, Sundrud MS et al (2012) Halofuginone and other febrifugine derivatives inhibit prolyl-tRNA synthetase. Nat Chem Biol 8(3):311–317

Ki SH, Cho IJ, Choi DW et al (2005) Glucocorticoid receptor (GR)-associated SMRT binding to C/EBPbeta TAD and Nrf2 Neh4/5: role of SMRT recruited to GR in GSTA2 gene repression. Mol Cell Biol 25(10):4150–4165

Kim HV, Kim HY, Ehrlich HY et al (2013) Amelioration of Alzheimer's disease by neuroprotective effect of sulforaphane in animal model. Amyloid 20(1):7–12

Kimura K, Huang RC (2016) Tetra-O-methyl nordihydroguaiaretic acid broadly suppresses cancer metabolism and synergistically induces strong anticancer activity in combination with etoposide, Rapamycin and UCN-01. PLoS One 11(2):e0148685

Kitamura H, Motohashi H (2018) NRF2 addiction in cancer cells. Cancer Sci 109(4):900–911

Knorr KL, Schneider PA, Meng XW et al (2015) MLN4924 induces Noxa upregulation in acute myelogenous leukemia and synergizes with Bcl-2 inhibitors. Cell Death Differ 22 (12):2133–2142

Kobayashi A, Kang MI, Okawa H et al (2004) Oxidative stress sensor Keap1 functions as an adaptor for Cul3-based E3 ligase to regulate proteasomal degradation of Nrf2. Mol Cell Biol 24 (16):7130–7139

Kobayashi EH, Suzuki T, Funayama R et al (2016) Nrf2 suppresses macrophage inflammatory response by blocking proinflammatory cytokine transcription. Nat Commun 7:11624

Kolm RH, Danielson UH, Zhang Y et al (1995) Isothiocyanates as substrates for human glutathione transferases: structure-activity studies. Biochem J 311(Pt 2):453–459

Komatsu M, Kurokawa H, Waguri S et al (2010) The selective autophagy substrate p62 activates the stress responsive transcription factor Nrf2 through inactivation of Keap1. Nat Cell Biol 12 (3):213–223

Kratschmar DV, Calabrese D, Walsh J et al (2012) Suppression of the Nrf2-dependent antioxidant response by glucocorticoids and 11beta-HSD1-mediated glucocorticoid activation in hepatic cells. PLoS One 7(5):e36774

Kumar A, Corey C, Scott I et al (2016) Minnelide/triptolide impairs mitochondrial function by regulating SIRT3 in P53-dependent manner in non-small cell lung cancer. PLoS One 11(8): e0160783

Kwak MK, Itoh K, Yamamoto M et al (2002) Enhanced expression of the transcription factor Nrf2 by cancer chemopreventive agents: role of antioxidant response element-like sequences in the nrf2 promoter. Mol Cell Biol 22(9):2883–2892

Lal H, Ahmad F, Woodgett J et al (2015) The GSK-3 family as therapeutic target for myocardial diseases. Circ Res 116(1):138–149

Lau A, Wang XJ, Zhao F et al (2010) A noncanonical mechanism of Nrf2 activation by autophagy deficiency: direct interaction between Keap1 and p62. Mol Cell Biol 30(13):3275–3285

Lee C (2018) Therapeutic modulation of virus-induced oxidative stress via the Nrf2-dependent Antioxidative pathway. Oxidative Med Cell Longev 2018:6208067

Levonen AL, Landar A, Ramachandran A et al (2004) Cellular mechanisms of redox cell signalling: role of cysteine modification in controlling antioxidant defences in response to electrophilic lipid oxidation products. Biochem J 378(Pt 2):373–382

Li J, Calkins MJ, Johnson DA et al (2007) Role of Nrf2-dependent ARE-driven antioxidant pathway in neuroprotection. Methods Mol Biol 399:67–78

Li W, Saud SM, Young MR et al (2015a) Cryptotanshinone, a Stat3 inhibitor, suppresses colorectal cancer proliferation and growth in vitro. Mol Cell Biochem 406(1–2):63–73

Li S, Wang H, Hong L et al (2015b) Cryptotanshinone inhibits breast cancer cell growth by suppressing estrogen receptor signaling. Cancer Biol Ther 16(1):176–184

Li R, Rezk A, Ghadiri M et al (2017) Dimethyl Fumarate treatment mediates an anti-inflammatory shift in B cell subsets of patients with multiple sclerosis. J Immunol 198(2):691–698

Liang J, Hansch GM, Hubner K et al (2019) Sulforaphane as anticancer agent: a double-edged sword? Tricky balance between effects on tumor cells and immune cells. Adv Biol Regul 71:79–87

Liao JC, Lee KT, You BJ et al (2015) Raf/ERK/Nrf2 signaling pathway and MMP-7 expression involvement in the trigonelline-mediated inhibition of hepatocarcinoma cell migration. Food Nutr Res 59:29884

Liby KT, Sporn MB (2012) Synthetic oleanane triterpenoids: multifunctional drugs with a broad range of applications for prevention and treatment of chronic disease. Pharmacol Rev 64 (4):972–1003

Lin JJ, Milhollen MA, Smith PG et al (2010) NEDD8-targeting drug MLN4924 elicits DNA rereplication by stabilizing Cdt1 in S phase, triggering checkpoint activation, apoptosis, and senescence in cancer cells. Cancer Res 70(24):10310–10320

Lin SX, Lisi L, Dello Russo C et al (2011) The anti-inflammatory effects of dimethyl fumarate in astrocytes involve glutathione and haem oxygenase-1. ASN Neuro 3(2):AN20100033

Linker RA, Lee DH, Ryan S et al (2011) Fumaric acid esters exert neuroprotective effects in neuroinflammation via activation of the Nrf2 antioxidant pathway. Brain 134(Pt 3):678–692

Lipton SA, Rezaie T, Nutter A et al (2016) Therapeutic advantage of pro-electrophilic drugs to activate the Nrf2/ARE pathway in Alzheimer's disease models. Cell Death Dis 7(12):e2499

Liu GH, Qu J, Shen X (2008) NF-kappaB/p65 antagonizes Nrf2-ARE pathway by depriving CBP from Nrf2 and facilitating recruitment of HDAC3 to MafK. Biochim Biophys Acta 1783 (5):713–727

Liu Y, Chen F, Wang S et al (2013) Low-dose triptolide in combination with idarubicin induces apoptosis in AML leukemic stem-like KG1a cell line by modulation of the intrinsic and extrinsic factors. Cell Death Dis 4:e948

Liu J, Su H, Qu QM (2016) Carnosic acid prevents beta-amyloid-induced injury in human neuroblastoma SH-SY5Y cells via the induction of autophagy. Neurochem Res 41 (9):2311–2323

Lo SC, Li X, Henzl MT et al (2006) Structure of the Keap1:Nrf2 interface provides mechanistic insight into Nrf2 signaling. EMBO J 25(15):3605–3617

Lombardi G, Pambuku A, Bellu L et al (2017) Effectiveness of antiangiogenic drugs in glioblastoma patients: a systematic review and meta-analysis of randomized clinical trials. Crit Rev Oncol Hematol 111:94–102

Lovestone S, Boada M, Dubois B et al (2015) A phase II trial of Tideglusib in Alzheimer's disease. J Alzheimers Dis 45(1):75–88

Lu MC, Ji JA, Jiang YL et al (2016) An inhibitor of the Keap1-Nrf2 protein-protein interaction protects NCM460 colonic cells and alleviates experimental colitis. Sci Rep 6:26585

Lu J, Guo S, Xue X et al (2017) Identification of a novel series of anti-inflammatory and anti-oxidative phospholipid oxidation products containing the cyclopentenone moiety in vitro and in vivo: implication in atherosclerosis. J Biol Chem 292(13):5378–5391

Lu MC, Jiao Q, Liu T et al (2018) Discovery of a head-to-tail cyclic peptide as the Keap1-Nrf2 protein-protein interaction inhibitor with high cell potency. Eur J Med Chem 143:1578–1589

Lu MC, Zhao J, Liu YT et al (2019) CPUY192018, a potent inhibitor of the Keap1-Nrf2 protein-protein interaction, alleviates renal inflammation in mice by restricting oxidative stress and NF-kappaB activation. Redox Biol 26:101266

Luo J (2009) Glycogen synthase kinase 3beta (GSK3beta) in tumorigenesis and cancer chemotherapy. Cancer Lett 273(2):194–200

Lynch DR, Farmer J, Hauser L et al (2019) Safety, pharmacodynamics, and potential benefit of omaveloxolone in Friedreich ataxia. Ann Clin Transl Neurol 6(1):15–26

Magesh S, Chen Y, Hu L (2012) Small molecule modulators of Keap1-Nrf2-ARE pathway as potential preventive and therapeutic agents. Med Res Rev 32(4):687–726

Manna A, Saha P, Sarkar A et al (2012) Malabaricone-A induces a redox imbalance that mediates apoptosis in U937 cell line. PLoS One 7(5):e36938

Manna A, Bauri AK, Chattopadhyay S et al (2015) Generation of redox imbalance mediates the cytotoxic effect of Malabaricone-a in a multidrug resistant cell line. Anti Cancer Agents Med Chem 15(9):1156–1163

Marcotte D, Zeng W, Hus JC et al (2013) Small molecules inhibit the interaction of Nrf2 and the Keap1 Kelch domain through a non-covalent mechanism. Bioorg Med Chem 21 (14):4011–4019

McCord JM, Hybertson BM, Cota-Gomez A, et al (2020) Nrf2 activator PB125(R) as a potential therapeutic agent against COVID-19. bioRxiv

Miao W, Hu L, Scrivens PJ et al (2005) Transcriptional regulation of NF-E2 p45-related factor (NRF2) expression by the aryl hydrocarbon receptor-xenobiotic response element signaling pathway: direct cross-talk between phase I and II drug-metabolizing enzymes. J Biol Chem 280 (21):20340–20348

Milhollen MA, Traore T, Adams-Duffy J et al (2010) MLN4924, a NEDD8-activating enzyme inhibitor, is active in diffuse large B-cell lymphoma models: rationale for treatment of NF-{kappa}B-dependent lymphoma. Blood 116(9):1515–1523

Milkovic L, Zarkovic N, Saso L (2017) Controversy about pharmacological modulation of Nrf2 for cancer therapy. Redox Biol 12:727–732

Mills EA, Ogrodnik MA, Plave A et al (2018a) Emerging understanding of the mechanism of action for dimethyl Fumarate in the treatment of multiple sclerosis. Front Neurol 9:5

Mills EL, Ryan DG, Prag HA et al (2018b) Itaconate is an anti-inflammatory metabolite that activates Nrf2 via alkylation of KEAP1. Nature 556(7699):113–117

Mizunoe Y, Kobayashi M, Sudo Y et al (2018) Trehalose protects against oxidative stress by regulating the Keap1-Nrf2 and autophagy pathways. Redox Biol 15:115–124

Morales DR, Morris AD (2015) Metformin in cancer treatment and prevention. Annu Rev Med 66:17–29

Myung SK, Ju W, Cho B et al (2013) Efficacy of vitamin and antioxidant supplements in prevention of cardiovascular disease: systematic review and meta-analysis of randomised controlled trials. BMJ 346:f10

Myzak MC, Dashwood RH (2006) Chemoprotection by sulforaphane: keep one eye beyond Keap1. Cancer Lett 233(2):208–218

Nasiri HR, Linge S, Ullmann D (2016) Thermodynamic profiling of inhibitors of Nrf2:Keap1 interactions. Bioorg Med Chem Lett 26(2):526–529

Nilsson L, Palm F, Norregaard R (2017) 15-Deoxy-Delta(12,14)-prostaglandin J2 exerts antioxidant effects while exacerbating inflammation in mice subjected to ureteral obstruction. Mediat Inflamm 2017:3924912

Noel P, von Hoff DD, Saluja AK et al (2019) Triptolide and its derivatives as cancer therapies. Trends Pharmacol Sci 40(5):327–341

Ohta T, Iijima K, Miyamoto M et al (2008) Loss of Keap1 function activates Nrf2 and provides advantages for lung cancer cell growth. Cancer Res 68(5):1303–1309

Olayanju A, Copple IM, Bryan HK et al (2015) Brusatol provokes a rapid and transient inhibition of Nrf2 signaling and sensitizes mammalian cells to chemical toxicity-implications for therapeutic targeting of Nrf2. Free Radic Biol Med 78:202–212

Padmanabhan B, Tong KI, Ohta T et al (2006) Structural basis for defects of Keap1 activity provoked by its point mutations in lung cancer. Mol Cell 21(5):689–700

Pajares M, Jimenez-Moreno N, Dias IH et al (2015) Redox control of protein degradation. Redox Biol 6:409–420

Pajares M, Jimenez-Moreno N, Garcia-Yague AJ et al (2016) Transcription factor NFE2L2/NRF2 is a regulator of macroautophagy genes. Autophagy 12(10):1902–1916

Pajares M, Cuadrado A, Rojo AI (2017) Modulation of proteostasis by transcription factor NRF2 and impact in neurodegenerative diseases. Redox Biol 11:543–553

Palomo V, Martinez A (2017) Glycogen synthase kinase 3 (GSK-3) inhibitors: a patent update (2014-2015). Expert Opin Ther Pat 27(6):657–666

Pan J, Shen F, Tian K et al (2019) Triptolide induces oxidative damage in NRK-52E cells through facilitating Nrf2 degradation by ubiquitination via the GSK-3beta/Fyn pathway. Toxicol In Vitro 58:187–194

Paredes-Gonzalez X, Fuentes F, Jeffery S et al (2015) Induction of NRF2-mediated gene expression by dietary phytochemical flavones apigenin and luteolin. Biopharm Drug Dispos 36(7):440–451

Park HM, Kim JA, Kwak MK (2009) Protection against amyloid beta cytotoxicity by sulforaphane: role of the proteasome. Arch Pharm Res 32(1):109–115

Patel D, Shukla S, Gupta S (2007) Apigenin and cancer chemoprevention: progress, potential and promise (review). Int J Oncol 30(1):233–245

Peng H, Wang H, Xue P et al (2016) Suppression of NRF2-ARE activity sensitizes chemotherapeutic agent-induced cytotoxicity in human acute monocytic leukemia cells. Toxicol Appl Pharmacol 292:1–7

Pergola PE, Raskin P, Toto RD et al (2011) Bardoxolone methyl and kidney function in CKD with type 2 diabetes. N Engl J Med 365(4):327–336

Ping Z, Liu W, Kang Z et al (2010) Sulforaphane protects brains against hypoxic-ischemic injury through induction of Nrf2-dependent phase 2 enzyme. Brain Res 1343:178–185

Rabbani PS, Ellison T, Waqas B et al (2018) Targeted Nrf2 activation therapy with RTA 408 enhances regenerative capacity of diabetic wounds. Diabetes Res Clin Pract 139:11–23

Rada P, Rojo AI, Evrard-Todeschi N et al (2012) Structural and functional characterization of Nrf2 degradation by the glycogen synthase kinase 3/beta-TrCP axis. Mol Cell Biol 32 (17):3486–3499

Raghunath A, Sundarraj K, Nagarajan R et al (2018) Antioxidant response elements: discovery, classes, regulation and potential applications. Redox Biol 17:297–314

Ramos-Gomez M, Dolan PM, Itoh K et al (2003) Interactive effects of nrf2 genotype and oltipraz on benzo[a]pyrene-DNA adducts and tumor yield in mice. Carcinogenesis 24(3):461–467

Ranjan N, Fulcrand G, King A et al (2014) Selective inhibition of bacterial topoisomerase I by alkynyl-bisbenzimidazoles. MedChemComm 5(6):816–825

Reddy SP (2008) The antioxidant response element and oxidative stress modifiers in airway diseases. Curr Mol Med 8(5):376–383

Ren D, Villeneuve NF, Jiang T et al (2011) Brusatol enhances the efficacy of chemotherapy by inhibiting the Nrf2-mediated defense mechanism. Proc Natl Acad Sci U S A 108(4):1433–1438

Richardson BG, Jain AD, Speltz TE et al (2015) Non-electrophilic modulators of the canonical Keap1/Nrf2 pathway. Bioorg Med Chem Lett 25(11):2261–2268

Robledinos-Anton N, Fernandez-Gines R, Manda G et al (2019) Activators and inhibitors of NRF2: a review of their potential for clinical development. Oxidative Med Cell Longev 2019:9372182

Roh JL, Kim EH, Jang H et al (2017a) Nrf2 inhibition reverses the resistance of cisplatin-resistant head and neck cancer cells to artesunate-induced ferroptosis. Redox Biol 11:254–262

Roh JL, Jang H, Kim EH et al (2017b) Targeting of the glutathione, thioredoxin, and Nrf2 antioxidant systems in head and neck cancer. Antioxid Redox Signal 27(2):106–114

Rojo AI, Medina-Campos ON, Rada P et al (2012) Signaling pathways activated by the phytochemical nordihydroguaiaretic acid contribute to a Keap1-independent regulation of Nrf2 stability: role of glycogen synthase kinase-3. Free Radic Biol Med 52(2):473–487

Rushworth SA, Zaitseva L, Murray MY et al (2012) The high Nrf2 expression in human acute myeloid leukemia is driven by NF-kappaB and underlies its chemo-resistance. Blood 120 (26):5188–5198

Saito R, Suzuki T, Hiramoto K et al (2015) Characterizations of three major cysteine sensors of Keap1 in stress response. Mol Cell Biol 36(2):271–284

Saito T, Ichimura Y, Taguchi K et al (2016) p62/Sqstm1 promotes malignancy of HCV-positive hepatocellular carcinoma through Nrf2-dependent metabolic reprogramming. Nat Commun 7:12030

Sangokoya C, Telen MJ, Chi JT (2010) microRNA miR-144 modulates oxidative stress tolerance and associates with anemia severity in sickle cell disease. Blood 116(20):4338–4348

Saraswati AP, Ali Hussaini SM, Krishna NH et al (2018) Glycogen synthase kinase-3 and its inhibitors: potential target for various therapeutic conditions. Eur J Med Chem 144:843–858

Sarkar S, Rubinsztein DC (2008) Small molecule enhancers of autophagy for neurodegenerative diseases. Mol BioSyst 4(9):895–901

Sato M, Aoki T, Inoue H et al (2013) Keap1 protein binding compound, Cristal of complex between the same and Keap1 protein, and method for producing the same. Toray Industries

Satoh T, McKercher SR, Lipton SA (2013) Nrf2/ARE-mediated antioxidant actions of pro-electrophilic drugs. Free Radic Biol Med 65:645–657

Schenk T, Stengel S, Zelent A (2014) Unlocking the potential of retinoic acid in anticancer therapy. Br J Cancer 111(11):2039–2045

Schimrigk S, Brune N, Hellwig K et al (2006) Oral fumaric acid esters for the treatment of active multiple sclerosis: an open-label, baseline-controlled pilot study. Eur J Neurol 13(6):604–610

Schmidlin CJ, Dodson MB, Madhavan L et al (2019) Redox regulation by NRF2 in aging and disease. Free Radic Biol Med 134:702–707

Schulze-Topphoff U, Varrin-Doyer M, Pekarek K et al (2016) Dimethyl fumarate treatment induces adaptive and innate immune modulation independent of Nrf2. Proc Natl Acad Sci U S A 113 (17):4777–4782

Shah JJ, Jakubowiak AJ, O'Connor OA et al (2016) Phase I study of the novel investigational NEDD8-activating enzyme inhibitor Pevonedistat (MLN4924) in patients with relapsed/refractory multiple myeloma or lymphoma. Clin Cancer Res 22(1):34–43

Sharma S, Ray A, Sadasivam B (2020) Metformin in COVID-19: a possible role beyond diabetes. Diabetes Res Clin Pract 164:108183

Shibata T, Ohta T, Tong KI et al (2008) Cancer related mutations in NRF2 impair its recognition by Keap1-Cul3 E3 ligase and promote malignancy. Proc Natl Acad Sci U S A 105 (36):13568–13573

Sihvola V, Levonen AL (2017) Keap1 as the redox sensor of the antioxidant response. Arch Biochem Biophys 617:94–100

Silva T, Reis J, Teixeira J et al (2014) Alzheimer's disease, enzyme targets and drug discovery struggles: from natural products to drug prototypes. Ageing Res Rev 15:116–145

Singh SV, Warin R, Xiao D et al (2009) Sulforaphane inhibits prostate carcinogenesis and pulmonary metastasis in TRAMP mice in association with increased cytotoxicity of natural killer cells. Cancer Res 69(5):2117–2125

Singh A, Venkannagari S, Oh KH et al (2016) Small molecule inhibitor of NRF2 selectively intervenes therapeutic resistance in KEAP1-deficient NSCLC tumors. ACS Chem Biol 11 (11):3214–3225

Smale ST (2011) Hierarchies of NF-kappaB target-gene regulation. Nat Immunol 12(8):689–694

Smith MD, Martin KA, Calabresi PA et al (2017) Dimethyl fumarate alters B-cell memory and cytokine production in MS patients. Ann Clin Transl Neurol 4(5):351–355

Solis LM, Behrens C, Dong W et al (2010) Nrf2 and Keap1 abnormalities in non-small cell lung carcinoma and association with clinicopathologic features. Clin Cancer Res 16(14):3743–3753

Son YO, Pratheeshkumar P, Wang Y et al (2017) Protection from Cr(VI)-induced malignant cell transformation and tumorigenesis of Cr(VI)-transformed cells by luteolin through Nrf2 signaling. Toxicol Appl Pharmacol 331:24–32

Soucy TA, Smith PG, Milhollen MA et al (2009) An inhibitor of NEDD8-activating enzyme as a new approach to treat cancer. Nature 458(7239):732–736

Steel R, Cowan J, Payerne E et al (2012) Anti-inflammatory effect of a cell-penetrating peptide targeting the Nrf2/Keap1 interaction. ACS Med Chem Lett 3(5):407–410

Sun H, Zhu J, Lin H et al (2017) Recent progress in the development of small molecule Nrf2 modulators: a patent review (2012-2016). Expert Opin Ther Pat 27(7):763–785

Sun X, Li X, Ma S et al (2018) MicroRNA-98-5p ameliorates oxygen-glucose deprivation/reoxygenation (OGD/R)-induced neuronal injury by inhibiting Bach1 and promoting Nrf2/ARE signaling. Biochem Biophys Res Commun 507(1–4):114–121

Suzuki T, Motohashi H, Yamamoto M (2013) Toward clinical application of the Keap1-Nrf2 pathway. Trends Pharmacol Sci 34(6):340–346

Swords RT, Watts J, Erba HP et al (2017) Expanded safety analysis of pevonedistat, a first-in-class NEDD8-activating enzyme inhibitor, in patients with acute myeloid leukemia and myelodysplastic syndromes. Blood Cancer J 7(2):e520

Taguchi K, Yamamoto M (2017) The KEAP1-NRF2 system in cancer. Front Oncol 7:85

Taguchi K, Motohashi H, Yamamoto M (2011) Molecular mechanisms of the Keap1-Nrf2 pathway in stress response and cancer evolution. Genes Cells 16(2):123–140

Tang X, Wang H, Fan L et al (2011) Luteolin inhibits Nrf2 leading to negative regulation of the Nrf2/ARE pathway and sensitization of human lung carcinoma A549 cells to therapeutic drugs. Free Radic Biol Med 50(11):1599–1609

Tao S, Wang S, Moghaddam SJ et al (2014) Oncogenic KRAS confers chemoresistance by upregulating NRF2. Cancer Res 74(24):7430–7441

Tarozzi A, Angeloni C, Malaguti M et al (2013) Sulforaphane as a potential protective phytochemical against neurodegenerative diseases. Oxidative Med Cell Longev 2013:415078

Tarumoto T, Nagai T, Ohmine K et al (2004) Ascorbic acid restores sensitivity to imatinib via suppression of Nrf2-dependent gene expression in the imatinib-resistant cell line. Exp Hematol 32(4):375–381

Tong KI, Katoh Y, Kusunoki H et al (2006) Keap1 recruits Neh2 through binding to ETGE and DLG motifs: characterization of the two-site molecular recognition model. Mol Cell Biol 26 (8):2887–2900

Tong KI, Padmanabhan B, Kobayashi A et al (2007) Different electrostatic potentials define ETGE and DLG motifs as hinge and latch in oxidative stress response. Mol Cell Biol 27 (21):7511–7521

Tsuchida K, Tsujita T, Hayashi M et al (2017) Halofuginone enhances the chemo-sensitivity of cancer cells by suppressing NRF2 accumulation. Free Radic Biol Med 103:236–247

Tu J, Zhang X, Zhu Y et al (2015) Cell-permeable peptide targeting the Nrf2-Keap1 interaction: a potential novel therapy for global cerebral ischemia. J Neurosci 35(44):14727–14739

Valenzuela M, Glorieux C, Stockis J et al (2014) Retinoic acid synergizes ATO-mediated cytotoxicity by precluding Nrf2 activity in AML cells. Br J Cancer 111(5):874–882

Vanderdys V, Allak A, Guessous F et al (2018) The Neddylation inhibitor Pevonedistat (MLN4924) suppresses and Radiosensitizes head and neck squamous carcinoma cells and tumors. Mol Cancer Ther 17(2):368–380

Verma AK, Yadav A, Dewangan J et al (2015) Isoniazid prevents Nrf2 translocation by inhibiting ERK1 phosphorylation and induces oxidative stress and apoptosis. Redox Biol 6:80–92

Vilcheze C, Jacobs WR Jr (2014) Resistance to isoniazid and ethionamide in Mycobacterium tuberculosis: genes, mutations, and causalities. Microbiol Spectr 2(4):MGM2-0014-2013

von Glehn F, Dias-Carneiro RPC, Moraes AS et al (2018) Dimethyl fumarate downregulates the immune response through the HCA2/GPR109A pathway: implications for the treatment of multiple sclerosis. Mult Scler Relat Disord 23:46–50

Wagner AE, Boesch-Saadatmandi C, Breckwoldt D et al (2011) Ascorbic acid partly antagonizes resveratrol mediated heme oxygenase-1 but not paraoxonase-1 induction in cultured hepatocytes - role of the redox-regulated transcription factor Nrf2. BMC Complement Altern Med 11:1

Wakabayashi N, Dinkova-Kostova AT, Holtzclaw WD et al (2004) Protection against electrophile and oxidant stress by induction of the phase 2 response: fate of cysteines of the Keap1 sensor modified by inducers. Proc Natl Acad Sci U S A 101(7):2040–2045

Wang XJ, Hayes JD, Henderson CJ et al (2007) Identification of retinoic acid as an inhibitor of transcription factor Nrf2 through activation of retinoic acid receptor alpha. Proc Natl Acad Sci U S A 104(49):19589–19594

Wang H, Liu K, Geng M et al (2013) RXRalpha inhibits the NRF2-ARE signaling pathway through a direct interaction with the Neh7 domain of NRF2. Cancer Res 73(10):3097–3108

Wang Y, Li F, Wang S (2016) MicroRNA93 is overexpressed and induces apoptosis in glaucoma trabecular meshwork cells. Mol Med Rep 14(6):5746–5750

Wang Y, Lu HL, Liu YD et al (2017) Cryptotanshinone sensitizes antitumor effect of paclitaxel on tongue squamous cell carcinoma growth by inhibiting the JAK/STAT3 signaling pathway. Biomed Pharmacother 95:1388–1396

Wang W, Wang X, Zhang XS et al (2018) Cryptotanshinone attenuates oxidative stress and inflammation through the regulation of Nrf-2 and NF-kappaB in mice with unilateral ureteral obstruction. Basic Clin Pharmacol Toxicol 123(6):714–720

Wardyn JD, Ponsford AH, Sanderson CM (2015) Dissecting molecular cross-talk between Nrf2 and NF-kappaB response pathways. Biochem Soc Trans 43(4):621–626

Watanabe G, Pena P, Albanese C et al (1997) Adrenocorticotropin induction of stress-activated protein kinase in the adrenal cortex in vivo. J Biol Chem 272(32):20063–20069

Wen X, Thorne G, Hu L et al (2015) Activation of NRF2 signaling in HEK293 cells by a first-in-class direct KEAP1-NRF2 inhibitor. J Biochem Mol Toxicol 29(6):261–266

Wilson MK, Baguley BC, Wall C et al (2014) Review of high-dose intravenous vitamin C as an anticancer agent. Asia Pac J Clin Oncol 10(1):22–37

Wolenski FS, Fisher CD, Sano T et al (2015) The NAE inhibitor pevonedistat (MLN4924) synergizes with TNF-alpha to activate apoptosis. Cell Death Discov 1:15034

Wong DPW, Ng MY, Leung JY et al (2018) Regulation of the NRF2 transcription factor by andrographolide and organic extracts from plant endophytes. PLoS One 13(10):e0204853

Wu T, Zhao F, Gao B et al (2014a) Hrd1 suppresses Nrf2-mediated cellular protection during liver cirrhosis. Genes Dev 28(7):708–722

Wu J, Wang H, Tang X (2014b) Rexinoid inhibits Nrf2-mediated transcription through retinoid X receptor alpha. Biochem Biophys Res Commun 452(3):554–559

Wu CF, Klauck SM, Efferth T (2016) Anticancer activity of cryptotanshinone on acute lymphoblastic leukemia cells. Arch Toxicol 90(9):2275–2286

Wu G, Yan Y, Zhou Y et al (2020) Sulforaphane: expected to become a novel anti-tumor compound. Oncol Res Featur Preclin Clin Cancer Therap

Xia C, Bai X, Hou X et al (2015) Cryptotanshinone reverses Cisplatin resistance of human lung carcinoma A549 cells through down-regulating Nrf2 pathway. Cell Physiol Biochem 37 (2):816–824

Xiang Y, Ye W, Huang C et al (2018) Brusatol enhances the chemotherapy efficacy of gemcitabine in pancreatic cancer via the Nrf2 Signalling pathway. Oxidative Med Cell Longev 2018:2360427

Xiao C, Xia ML, Wang J et al (2019) Luteolin attenuates cardiac ischemia/reperfusion injury in diabetic rats by modulating Nrf2 Antioxidative function. Oxidative Med Cell Longev 2019:2719252

Xu C, Huang MT, Shen G et al (2006) Inhibition of 7,12-dimethylbenz(a)anthracene-induced skin tumorigenesis in C57BL/6 mice by sulforaphane is mediated by nuclear factor E2-related factor 2. Cancer Res 66(16):8293–8296

Xu Z, Zhang F, Sun F et al (2015) Dimethyl fumarate for multiple sclerosis. Cochrane Database Syst Rev 4:CD011076

Yamamoto T, Suzuki T, Kobayashi A et al (2008) Physiological significance of reactive cysteine residues of Keap1 in determining Nrf2 activity. Mol Cell Biol 28(8):2758–2770

Yang M, Yao Y, Eades G et al (2011) MiR-28 regulates Nrf2 expression through a Keap1-independent mechanism. Breast Cancer Res Treat 129(3):983–991

Yu R, Chen C, Mo YY et al (2000) Activation of mitogen-activated protein kinase pathways induces antioxidant response element-mediated gene expression via a Nrf2-dependent mechanism. J Biol Chem 275(51):39907–39913

Yu H, Shi L, Zhao S et al (2016) Triptolide attenuates myocardial ischemia/reperfusion injuries in rats by inducing the activation of Nrf2/HO-1 defense pathway. Cardiovasc Toxicol 16 (4):325–335

Yu C, Jiao Y, Xue J et al (2017) Metformin sensitizes non-small cell lung Cancer cells to an Epigallocatechin-3-Gallate (EGCG) treatment by suppressing the Nrf2/HO-1 signaling pathway. Int J Biol Sci 13(12):1560–1569

Zeidan TA, Duncan S, Hencken CP et al (2014) Prodrugs of fumarates and their use in treating various diseases. Alkermes Pharma Ireland Limited

Zhang DD (2013) Bardoxolone brings Nrf2-based therapies to light. Antioxid Redox Signal 19 (5):517–518

Zhao J, Kobori N, Aronowski J et al (2006) Sulforaphane reduces infarct volume following focal cerebral ischemia in rodents. Neurosci Lett 393(2–3):108–112

Zhao Y, Morgan MA, Sun Y (2014) Targeting Neddylation pathways to inactivate cullin-RING ligases for anticancer therapy. Antioxid Redox Signal 21(17):2383–2400

Zhou H, Lu J, Liu L et al (2017) A potent small-molecule inhibitor of the DCN1-UBC12 interaction that selectively blocks cullin 3 neddylation. Nat Commun 8(1):1150

Zhou Y, Wang X, Ying W et al (2019) Cryptotanshinone attenuates inflammatory response of microglial cells via the Nrf2/HO-1 pathway. Front Neurosci 13:852

Zhu J, Wang H, Chen F et al (2016) An overview of chemical inhibitors of the Nrf2-ARE signaling pathway and their potential applications in cancer therapy. Free Radic Biol Med 99:544–556

Zhu J, Wang H, Chen F et al (2018) Triptolide enhances chemotherapeutic efficacy of antitumor drugs in non-small-cell lung cancer cells by inhibiting Nrf2-ARE activity. Toxicol Appl Pharmacol 358:1–9

Zuo Q, Wu R, Xiao X et al (2018) The dietary flavone luteolin epigenetically activates the Nrf2 pathway and blocks cell transformation in human colorectal cancer HCT116 cells. J Cell Biochem 119(11):9573–9582

Part III

Inhibiting ROS Formation and Toxification

NOX Inhibitors: From Bench to Naxibs to Bedside

Mahmoud H. Elbatreek, Hermann Mucke, and Harald H. H. W. Schmidt

Contents

M. H. Elbatreek (✉)
Department of Pharmacology and Personalised Medicine, School of MeHNS, Faculty of Health, Medicine and Life Sciences, Maastricht University, Maastricht, The Netherlands

Department of Pharmacology and Toxicology, Faculty of Pharmacy, Zagazig University, Zagazig, Egypt
e-mail: melbatreek@ppmlab.net

H. Mucke
H. M. Pharma Consultancy, Vienna, Austria

H. H. H. W. Schmidt
Department of Pharmacology and Personalised Medicine, School of MeHNS, Faculty of Health, Medicine and Life Sciences, Maastricht University, Maastricht, The Netherlands

H. H. H. W. Schmidt et al. (eds.), *Reactive Oxygen Species*,
Handbook of Experimental Pharmacology 264, https://doi.org/10.1007/164_2020_387

Abstract

Reactive oxygen species (ROS) are ubiquitous metabolic products and important cellular signaling molecules that contribute to several biological functions. Pathophysiology arises when ROS are generated either in excess or in cell types or subcellular locations that normally do not produce ROS or when non-physiological types of ROS (e.g., superoxide instead of hydrogen peroxide) are formed. In the latter scenario, antioxidants were considered as the apparent remedy but, clinically, have consistently failed and even sometimes induced harm. The obvious reason for that is the non-selective ROS scavenging effects of antioxidants which interfere with both qualities of ROS, physiological and pathological. Therefore, it is essential to overcome this "antidote or neutralizer" strategy. We here review the most promising alternative approach by identifying the disease-relevant enzymatic sources of ROS, target these selectively, but leave physiological ROS signaling through other sources intact. Among all ROS sources, NADPH oxidases (NOX1-5 and DUOX1-2) stand out as their sole function is to produce ROS, whereas most other enzymatic sources only produce ROS as a by-product or upon biochemical uncoupling or damage. This qualifies NOXs as the main potential drug-target candidates in diseases associated with dysfunction in ROS signaling. As a reflection of this, the development of several NOX inhibitors has taken place. Recently, the WHO approved a new stem, "naxib," which refers to *NA*DPH o*x*idase inh*ib*itors, and thereby recognized NOX inhibitors as a new therapeutic class. This has been announced while clinical trials with the first-in-class compound, setanaxib (initially known as GKT137831) had been initiated. We also review the differences between the seven NOX family members in terms of structure and function in health and disease and then focus on the most advanced NOX inhibitors with an exclusive focus on clinically relevant validations and applications.

Graphical Abstract

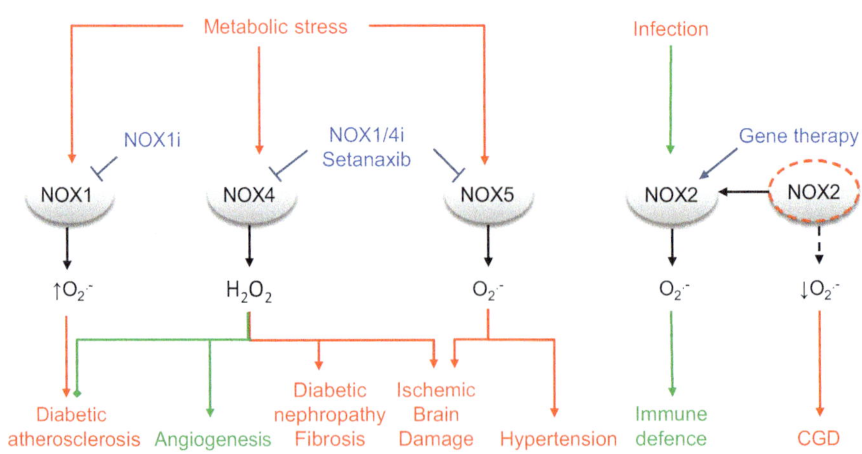

Therapeutically relevant NADPH oxidase isoforms type 1, 2, 4, and 5 (NOX1, NOX2, NOX4, NOX5). Of note, NOX5 is not present in mice and rats and thus pre-clinically less studied. NOX2, formerly termed gp91^{phox}, has been correlated with many, too many, diseases and is rather relevant as genetic deficiency in chronic granulomatous disease (CGD), treated by gene therapy. Overproduction of ROS through NOX1, NOX4, and NOX5 leads to the indicated diseases states including atherosclerosis (red), a condition where NOX4 is surprisingly protective.

Keywords

Mechanism-based redox therapeutics · NADPH oxidases · NOX inhibitors · Setanaxib · Reactive oxygen species

1 The NADPH Oxidase Family of Enzymes

NADPH oxidases (NOXs) are transmembrane enzymes that transfer electrons from NADPH in the cytoplasm across the cell membrane resulting in the formation of reactive oxygen species (ROS) (Cross and Segal 2004; Panday et al. 2015). The NOX family consists of seven members, NOX1–5 and the dual oxidases DUOX1–2. These enzymes are different in terms of enzyme complex composition, tissue and cellular distributions, subcellular localizations, mechanisms of activation, and the ROS type they produce. Thus, they are implicated in diverse physiological functions and disease conditions (Altenhofer et al. 2015; Casas et al. 2015; Elbatreek et al. 2019).

NOX2 (formerly known as gp91^{phox}) was the first NOX family member to be discovered (Rossi and Zatti 1964; Segal and Jones 1978). Other NOXs were discovered later and share certain sequence homology with this isoform, 56% for NOX1(aka Mox1), 58% for NOX3 (aka MOX2), 39% for NOX4 (aka Renox), 27% for NOX5, 57% for DUOX1 (aka ThOX1), and 43% for DUOX2 (aka ThOX2) (Cheng et al. 2001; De Deken et al. 2000; Suh et al. 1999). With respect to structure, all the family members possess a catalytic subunit, NOX, which is formed of six- or seven-transmembrane helices in NOX1–5 and DUOX1–2, respectively. NOX subunit binds two heme cofactors and allows for NADPH oxidation through a FAD/NADPH-binding domain in the cytosolic C-terminus (Cheng et al. 2001; Meitzler and Ortiz de Montellano 2009, 2011). In the case of NOX5 and DUOX1–2, NOX also binds to an intracellular Ca^{2+}-binding EF-hand region (Banfi et al. 2001, 2004b; Rigutto et al. 2009). Besides, NOXs differ in their membrane or cytosolic binding partners that are required for the enzymatic activity.

In NOX1–4, a membrane-bound subunit, p22^{phox}, is required for stabilization, whereas DUOX1 and 2 associate with the membrane maturation factors DUOXA1 and 2, respectively (Ambasta et al. 2004; Grasberger and Refetoff 2006; Parkos et al. 1987; Ueno et al. 2005). NOXs are also associated with cytosolic activator proteins (NOXA1 for NOX1 and NOX3, p67^{phox} and p40^{phox} for NOX2) which increase enzymatic ROS-forming activity and organizer proteins (NOXO1 for NOX1 and NOX3, p47^{phox} for NOX2) that help tether the activators with the NOX subunit (Banfi et al. 2003, 2004a; Volpp et al. 1988; Wientjes et al. 1993). In addition, other binding proteins help regulate NOX activity such as the small GTPase, RAC1, for NOX1–3 (Cheng et al. 2006; Diebold and Bokoch 2001; Ueyama et al. 2006), polymerase δ-interacting protein 2 (POLDIP2) for NOX4 (Lyle et al. 2009), and heat shock protein 90 (HSP90) for NOX1–3 and NOX5 (Chen et al. 2011).

Regarding their tissue and cellular distribution, the seven NOXs are widely expressed throughout different tissues (Fig. 1). NOX1 is predominantly expressed in colon epithelium (Szanto et al. 2005) and also in the uterus (Banfi et al. 2000; Suh et al. 1999), placenta (Cui et al. 2006), prostate (Banfi et al. 2000; Suh et al. 1999), pancreas (Xia et al. 2019), retina (Manea et al. 2005), keratinocytes (Chamulitrat et al. 2003), endothelium (Gray et al. 2013), and vascular smooth muscle cells (Lassegue et al. 2001). NOX2 is expressed in phagocytes which are present in numerous tissues and is often called "the phagocyte NADPH oxidase" (Bedard and Krause 2007); however, it can also be detected in other cell types including cardiomyocytes (Krijnen et al. 2003), skeletal muscle (Henriquez-Olguin et al. 2019), endothelial cells (Gorlach et al. 2000), hepatocytes (Reinehr et al. 2005), and neurons (Fan et al. 2019). NOX3 is highly abundant in the inner ear (Banfi et al. 2004a) in addition to other fetal tissues (Banfi et al. 2004a; Cheng et al. 2001), while NOX4 is highly expressed in kidney cells (Geiszt et al. 2000; Gorin et al. 2003; Jha et al. 2016), endothelium (Van Buul et al. 2005), vascular smooth muscle cells (Hoidal et al. 2003), cardiomyocytes (Brewer et al. 2011), fibroblasts (Cucoranu et al. 2005), adipocytes (Den Hartigh et al. 2017), and neurons (Casas et al. 2017). NOX5, which is absent in rodents, shows substantial expression in the testis, spleen, and lymph nodes (Banfi et al. 2001) and is also detected in the endothelial cells (BelAiba et al. 2007), vascular smooth muscle cells (Jay et al. 2008), kidney (Holterman et al. 2014; Jha et al. 2017a), and white blood cells (Manea et al. 2015). DUOX1–2 are predominantly found in the thyroid gland (De Deken et al. 2000) in addition to the lung epithelia (Fischer 2009) and prostate (D. Wang et al. 2005). DUOX1 is also expressed in epidermal keratinocytes (Ko et al. 2014) and DUOX2 in salivary ducts and the gastrointestinal tract (El Hassani et al. 2005; Geiszt et al. 2003b).

The subcellular localization/compartmentalization varies between NOXs in different cell types; however, the data are limited by the lack of high-quality antibodies against these enzymes (Zhang et al. 2019). NOX1 is localized in the endoplasmic reticulum, caveolae, and nucleus (Chamulitrat et al. 2003; Hilenski et al. 2004; Janiszewski et al. 2005), while NOX2 is present at the plasma membrane, perinuclear cytoskeleton, and endoplasmic reticulum (Huang et al. 1995; Krijnen et al. 2003; Segal and Jones 1978; Van Buul et al. 2005). There is barely any information about the subcellular localization of NOX3; however, one study showed the co-localization of NOX3 and p22^{phox} in the plasma membrane of transfected HEK-293 cells (Nakano et al. 2007). NOX4 and NOX5 are localized at the cell membrane, nucleus, endoplasmic reticulum, and mitochondria (Ago et al. 2010; BelAiba et al. 2007; Case et al. 2013; Hilenski et al. 2004; Marzaioli et al. 2017;

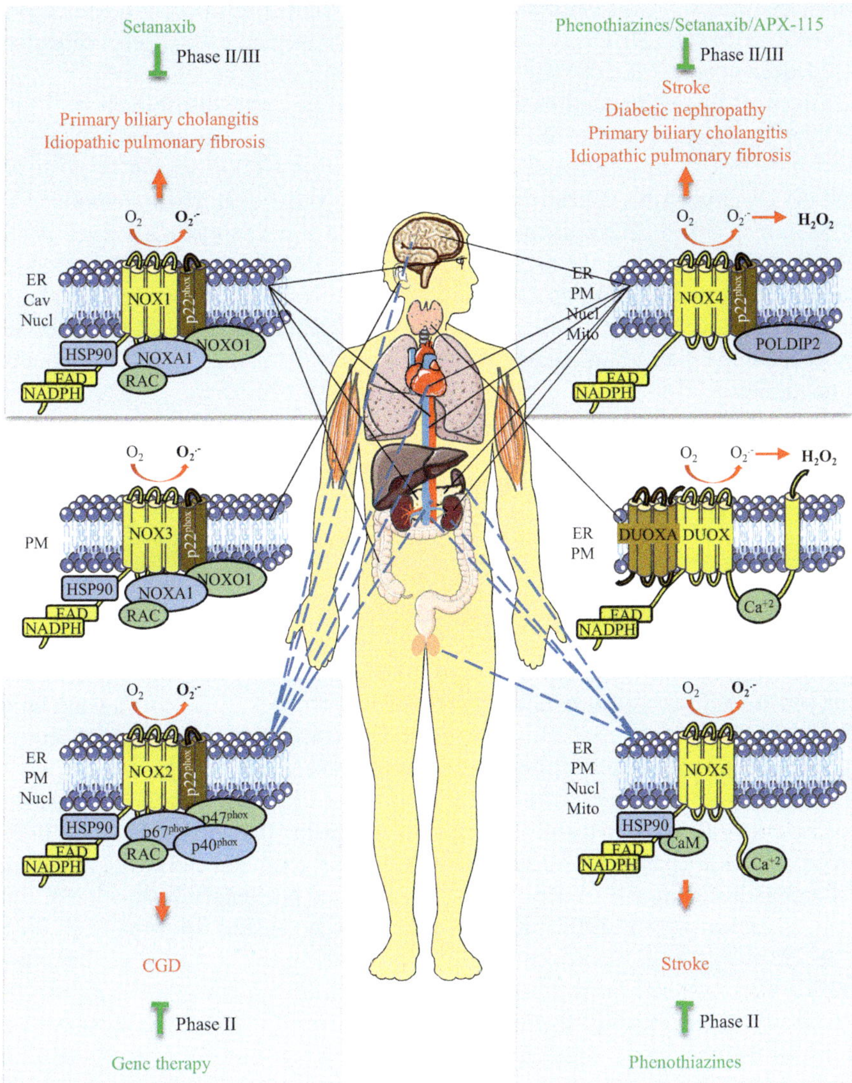

Fig. 1 NADPH oxidases and their clinical relevance. Seven NADPH oxidases, NOX1–5 and DUOX1–2, exist and possess a catalytic transmembrane subunit, NOX, which allows for NADPH oxidation through a FAD/NADPH-binding domain. NOXs have different membrane or cytosolic binding partners (p22^{phox}, DUOXA, NOXA1, NOXO1, RAC, HSP90, p67^{phox}, p40^{phox}, p47^{phox}, POLIDIP2) that are required for the enzymatic activity. NOXs are expressed in many organs: NOX1 in the colon, blood vessels, retina, and pancreas; NOX2 in the blood vessels, neurons, skeletal muscles, liver, and heart; NOX3 in the inner ear; NOX4 in the neurons, heart, blood vessels, kidney, and pancreas; NOX5 in the spleen, testis, kidney, and blood vessels; and DUOXs in the thyroid gland. NOXs are also expressed in different subcellular locations including endoplasmic reticulum (ER), caveolae (Cav), nucleus (Nucl), plasma membrane (PM), and mitochondria (Mito). The therapeutically relevant NOXs (in light blue boxes) include NOX1 and NOX4 being tested in

Matsushima et al. 2013; Perrotta et al. 2011; Van Buul et al. 2005; Wu et al. 2010; Yu et al. 2014) and DUOX1–2 at the apical membrane and endoplasmic reticulum (De Deken et al. 2000, 2002; Schwarzer et al. 2004).

Given all these structural characteristics, dissimilar tissue distribution, and subcellular localization of the NOX family members, they show distinct modes of activation (reviewed in (Brandes et al. 2014)) except NOX4 (Zhang et al. 2019) and the DUOXs (Aliasgharzadeh et al. 2019; Azmoonfar et al. 2018; Farhood et al. 2019) which are constitutively active and regulated at the expressional level. Additionally, NOXs differ in their ROS product, i.e., NOX1–3 and NOX5 produce superoxide, while NOX4 and DUOX1–2 produce hydrogen peroxide (Altenhofer et al. 2015). Overall, full characterization of the NOX enzymes is still deficient, yet by generating reliable high-quality antibodies and isoform-specific NOX inhibitors, it might be achievable.

2 NADPH Oxidases in Physiology

Being the sole and primary function of the NOX enzymes, ROS production should not be viewed mainly as disease trigger and metabolic waste. Indeed, ROS from NOXs among others contribute to several physiological functions such as host defense, angiogenesis, cell survival, tissue regeneration, hearing, hormone synthesis and sensitivity, vasodilation, and cell signaling (Elbatreek et al. 2019; Jiang et al. 2011). These functions need to be taken also into consideration as potential sources of side effects when NOX inhibitors are used therapeutically.

With respect to individual NOXs, NOX1-derived ROS slow down apoptosis of gastric mucosal cells and thereby regulate their growth (Teshima et al. 2000). In the colon, NOX1 is a part of the innate immune response, promotes cell proliferation and differentiation, stimulates mucosal wound repair, and prevents inflammation (Coant et al. 2010; Geiszt et al. 2003a; Kajla et al. 2012; Kato et al. 2016; Moll et al. 2018; Rokutan et al. 2006). Moreover, NOX1 plays a role in cell signaling via inhibiting protein tyrosine phosphatases and thus inactivation of peroxiredoxin 1, an enzyme that metabolizes/detoxifies hydrogen peroxide, thereby allowing the localized and transient accumulation of hydrogen peroxide for cell signaling (Woo et al. 2010). In the brain, NOX1 is suggested to suppress neuronal differentiation via inhibiting excessive neurite outgrowth (Ibi et al. 2006).

NOX2 is a key player in the innate host defense against infection. Mutations in genes encoding components of the NOX2 enzyme complex lead to chronic

Fig. 1 (continued) primary biliary cholangitis and idiopathic pulmonary fibrosis, NOX4 also in diabetic nephropathy and stroke, NOX2 in chronic granulomatous disease (CGD), and NOX5 in stroke. NOX inhibitors including setanaxib, APX-115, and phenothiazines and NOX2 gene therapy are being tested in Phase II/III clinical trials for these indications. Abbreviations: CaM, calmodulin; HSP90, heat shock protein 90; NOXA1, NADPH oxidase activator 1; NOXO1, NADPH oxidase organizer 1; POLDIP2, polymerase δ-interacting protein 2

granulomatous disease (CGD) which is characterized by immunodeficiency and recurrent and life-threatening infections (Panday et al. 2015). ROS from NOX2 can kill the attacking microorganisms directly by oxidative damage of proteins, lipids, and DNA and indirectly by activation of downstream signaling (Iles and Forman 2002). In addition to host defense, NOX2 might be involved in learning and memory as CGD patients show cognitive deficits and NOX2 mutant mice have mild memory impairment (Kishida et al. 2006; Pao et al. 2004). NOX2 also might have a protective function against colon inflammation as CGD patients also exhibit non-infective colitis (Pao et al. 2004). Moreover, NOX2 mediates the renal vasoconstriction effect of angiotensin and thus regulates the normal renal blood flow (Haque and Majid 2004) and enhances skeletal muscle metabolism and insulin sensitivity (Henriquez-Olguin et al. 2019). Apart from NOX2, the key physiological roles of NOX3 are mainly known in the inner ear. Mutation in the *NOX3* gene results in a lack of otoconia formation and vestibular dysfunction as shown in "head-tilt" mutant mice (Paffenholz et al. 2004). Also, recently, NOX3, together with NOX5, has been suggested to induce differentiation of human oligodendrocytes (Accetta et al. 2016).

NOX4 has a plethora of physiological and protective roles. This is probably explained by its constitutive activity, wide distribution, and production of hydrogen peroxide which is an omnipresent signaling molecule (Elbatreek et al. 2019; Guo and Chen 2015; Veal and Day 2011; Zhang et al. 2019). However, knocking out the *NOX4* gene in mice and rats does not result in an obvious phenotype or affect the life span of the animals (Kleinschnitz et al. 2010; Rezende et al. 2017). NOX4 enhances hormone-stimulated sodium and water transport in the kidney (Feraille et al. 2014; Lu et al. 2016), adipocytes differentiation (Schroder et al. 2009), insulin sensitivity in the liver and adipose tissue (Mahadev et al. 2004; Taniguchi et al. 2006), glucose-stimulated insulin secretion (Plecita-Hlavata et al. 2020), autophagy in cardiomyocytes (Kouroku et al. 2007), hippocampal neurogenesis, memory formation (Choi et al. 2019; Yoshikawa et al. 2019), angiogenesis, and vasodilation (Burgoyne et al. 2007; Drummond et al. 2000). NOX4 also activates downstream redox-sensitive proteins that play important roles in cell proliferation, migration, and apoptosis (Guo and Chen 2015). Further, NOX4 protects the vasculature from ischemic and inflammatory stress such as in atherosclerosis (Gray et al. 2016; Schroder et al. 2012).

NOX5 is the least studied NOX, and its physiological roles are not fully understood due to its absence in rodents. However, it has been suggested to regulate cell signaling and function (Fulton 2009) and contribute to sperm motility and viability (Ghanbari et al. 2018), as well as vascular smooth muscle cells contraction (Montezano et al. 2018). DUOX enzymes appear to be important for thyroid hormone synthesis. Mutations in the *DUOX2* lead to disruption of thyroid hormone synthesis and hypothyroidism (Moreno et al. 2002). DUOXs also play a role in host defense in the gastrointestinal tract and lung epithelia (van der Vliet et al. 2018).

Most of the abovementioned functions of NOXs derive from preclinical data and the physiological roles of NOXs in humans remain poorly understood. While the biological effects of NOXs are important for health, dysfunctions in these enzymes may lead to pathology.

3 NADPH Oxidases in Pathology

Several pathophysiological roles have been validated for NOX enzymes, and thus several diseases are largely based on NOX dysregulation (Casas et al. 2015; Dao et al. 2015) (Fig. 1).

3.1 NOX1

NOX1 is involved in fibrotic diseases in many organs (Kato and Hecker 2020). Current clinical studies to target NOX1, together with NOX4, are focused on idiopathic pulmonary fibrosis and primary biliary cholangitis (a fibrotic orphan disease) (Table 1). Moreover, NOX1 seems a clinically relevant target in GI disorders. On the one hand, defects in *NOX1* are found in patients with very-early-onset inflammatory bowel diseases (Hayes et al. 2015; Scherz-Shouval and Elazar 2011). Indeed, some variants in *NOX1* are associated with complete loss of function of the gene product and with loss of ROS production in IBD patients (Schwerd et al. 2018). On the other hand, NOX1 overexpression is associated with colon and gall bladder cancers (Wang et al. 2019; Juhasz et al. 2017; Laurent et al. 2008). Besides GI-related disorders, diabetic vascular complications including diabetes-accelerated atherosclerosis (Gray et al. 2013) and retinopathy (Wilkinson-Berka et al. 2014) are potential conditions for clinical testing of NOX1 inhibitors.

3.2 NOX2

NOX2 genetic defects or inhibition are associated with immune deficiency and increased risk of infection, particularly in diabetes (Gray et al. 2013). Mutations in *CYBB*, *CYBA*, *NCF1*, *NCF2*, and *NCF4* genes, encoding NOX2, p22phox, p47phox, p67phox, and p40phox, respectively, cause CGD. Around 70% of CGD cases are due to mutations in *CYBB* (called X-linked CGD) resulting in decreased NOX2 expression, activity, or both. Therefore, CGD leads to immunodeficiency and increases susceptibility to recurrent and life-threatening infections due to fungal or bacterial pathogens (Giardino et al. 2017; O'Neill et al. 2015). As a key enzyme of the innate and inflammatory response, NOX2 has been suggested to be involved in an excessive and unlikely number of disease models (Casas et al. 2015; Elbatreek et al. 2019) which might indicate a possible positive publication bias, as shown by a meta-analysis of NOX2 studies in stroke (Kleikers et al. 2015), or an epiphenomenon without therapeutic relevance.

Table 1 NOX inhibitors and gene therapy for CGD and their clinical status

	Isoform	Indication	Clinical trial	Results/status
NOX inhibitors				
GKT137831 (GKT-831 or setanaxib)	1,4	Type 2 diabetes mellitus nephropathy	Phase I (NCT03740217)	Safe
			A double-blind, placebo-controlled, randomized, multicenter, parallel group, Phase II (NCT02010242)	Reduced several secondary efficacy endpoints. However, improvements in albuminuria, the study's primary efficacy endpoint, was not achieved after 12 weeks of treatment
		Type 1 diabetes mellitus nephropathy	A double-blind, placebo-controlled, randomized, multicenter, with two parallel arms Phase II (U1111-1187-2609)	Ongoing in Australia and expanded to Europe and New Zealand
		Primary biliary cholangitis/ cirrhosis (PBC)	A double-blind, placebo-controlled, randomized, multicenter, parallel group, Phase II (NCT03226067)	Achieved rapid, dose- and time-dependent reductions in markers of cholestatic bile duct and liver injury. These reductions in disease activity were highly significant for both ALP and GGT
		Idiopathic pulmonary fibrosis (IPF)	A double-blind, placebo-controlled, randomized, multicenter, parallel group, Phase II (NCT03865927)	Not yet recruiting
APX-115	1,2,4	Type 1 diabetes mellitus nephropathy	Phase II (10.4062/ biomolther.2019.188)	Not yet recruiting
Perphenazine	4,5	Stroke	Phase II, repo-stroke (repo-trial.Eu) (EduraCT no. 2019–000474-31)	Not yet recruiting
Gene therapy				
Lentiviral gene therapy	2	X-linked chronic granulomatous disease (X-CGD)	Phase I/II, non-randomized, multicenter, open-label (NCT02234934 and NCT01855685)	The primary objective (to assess the safety and evaluate the efficacy and stability of biochemical and functional reconstitution in the

(continued)

Table 1 (continued)

	Isoform	Indication	Clinical trial	Results/status
				progeny of engrafted cells at 12 months) was met in six of the nine patients
			Phase I/II, non-randomized, open-label (NCT03645486)	Recruiting
			Phase I/II, non-randomized, monocentric open-label (NCT02757911)	Recruiting
Retroviral gene therapy			Phase I/II, non-randomized, single center, uncontrolled, open-label (NCT00778882)	In progress
			Phase I/II, non-randomized, open-label (NCT01906541)	Recruiting

3.3 NOX3

Being expressed in the inner ear, NOX3 appears to be a key target for hearing loss. Preclinical data show that noise exposure causes overexpression of NOX3 that results in cochlear inflammation, apoptosis, and eventually hearing loss (Dhukhwa et al. 2019). NOX3 also was shown to be associated with drug-induced hearing loss (Rybak et al. 2012). Genetic clinical studies show that NOX3 is associated with noise-induced hearing loss (Zhao et al. 2020) and pulmonary hypertension as shown in a recent GWAS (Yin et al. 2018). Further clinical and therapeutic validation of NOX3 in these conditions needs to be investigated.

3.4 NOX4

Preclinical data suggest that NOX4 is involved in many diseases including diabetic kidney disease (Jha et al. 2014, 2016), cancer (Lin et al. 2017), fibrosis of the liver (Lan et al. 2015) and lung (Carnesecchi et al. 2011), and ischemic stroke (Casas et al. 2017, 2019a; Kleinschnitz et al. 2010). Genetic clinical data shows that NOX4 is associated with an increased risk of stroke (He et al. 2018). Diversely, the role of NOX4 in cardiovascular disorders such as hypertension and atherosclerosis is likely limited. Indeed NOX4 seems protective in diabetes-accelerated atherosclerosis (Gray et al. 2016; Schurmann et al. 2015) and myocardial infarction-induced cardiac remodeling (Mongue-Din et al. 2017). Clinical studies on NOX4 are focused on stroke for acute indications and diabetic kidney disease and fibrotic conditions for chronic indications. Yet, due to the dual effects of NOX4 and its plentiful biological functions in many organs, chronic NOX4 inhibition seems a less attractive approach. In cancer, however, targeting NOX4 needs to be examined given its metabolic, anti-apoptotic, and pro-angiogenic properties.

3.5 NOX5

NOX5 appears as a promising target in cardiovascular diseases, i.e., hypertension and atherosclerosis (Guzik et al. 2008; Touyz et al. 2019). Our recent findings show that NOX5 levels in endothelial microparticles are increased in a subgroup of hypertensive patients leading to eNOS uncoupling and endothelial dysfunction. NOX5 might also be a potential target in stroke (Casas et al. 2019b), myocardial infarction (Hahn et al. 2012), cancer (Antony et al. 2017), diabetic nephropathy (Jha et al. 2017b), aortic aneurysm (Guzik et al. 2013), and hemorrhagic transformation (Won et al. 2011).

3.6 DUOXs

The clinical relevance of targeting DUOX isoforms is not yet clear. Preclinical evidence suggests that DUOXs might contribute to immune and allergic disorders (van der Vliet et al. 2018) and can be targeted for radiation-induced thyroid cancer (Ameziane-El-Hassani et al. 2015). Similar to *NOX1*, mutations in *DUOX2* are found in patients with very-early-onset inflammatory bowel diseases (Hayes et al. 2015).

Taken together, given the diverse effects of the NOX enzymes both in physiology and disease, benefit-risk assessments should be considered as exemplified by NOX2 inhibition which is associated with immunodeficiency and infection (Gray et al. 2013; Panday et al. 2015). Also, inhibition of DUOX2 can result in hypothyroidism and bowel inflammation (Hayes et al. 2015). Similarly, inhibiting NOX1 might enhance gut inflammation (Schwerd et al. 2018), and inhibiting NOX4 might promote atherosclerosis (Gray et al. 2016) and enhance the risk of kidney fibrosis (Nlandu Khodo et al. 2012) and liver cancer (Crosas-Molist et al. 2017). Acute indications such as ischemic stroke are likely to have, however, a low risk-benefit profile.

4 NADPH Oxidases Inhibitors

Despite the fact that NOX inhibitors are already in the clinic, the field has still to be considered relatively immature. There are no compounds available that deserve the term NOS isoform specific. Most compounds are pan-NOX inhibitors. Two recent analyses identified compounds with some isoform preference (Augsburger et al.

2019; Dao et al. 2019), and it has been shown that by using a panel of marginally selective inhibitors, specific isoforms, such as NOX4, could be validated pharmacologically (Dao et al. 2019). Considering the critical roles of NOXs in the pathogenesis of many diseases, they have been suggested as promising therapeutic targets. Several small molecules have been thought to inhibit NOX activity; however, majority were unspecific due to off-target effects. These molecules include, for example, diphenyleneiodonium (DPI) and apocynin. The former is a flavoprotein inhibitor and thus inhibits many other enzymes besides NOXs, while the latter has non-specific ROS scavenging properties (Altenhofer et al. 2015). Likewise, some other recently developed NOX inhibitors are unspecific such as VAS2870, ML-171, and GKT136901 (Augsburger et al. 2019; Dao et al. 2019). Only a few compounds are claimed to be specific NOX inhibitor in preclinical studies including GSK2795039 which selectively inhibits NOX2 (Hirano et al. 2015), GLX7013114 for NOX4 (Wang et al. 2018), and Ewha-18278 (APX-115) for NOX1, NOX2, and NOX4 (Cha et al. 2017).

NOX inhibitors currently being tested in the clinical phase are focused on fibrotic and neurovascular disease indications with NOX1, 4, and 5 as the main isoforms to be targeted. GKT137831 (setanaxib or GKT-831) claimed as a NOX1/4 dual inhibitor and a partial Nox5 inhibitor is the first-in-class NOX inhibitor to reach the clinical trial stage (Table 1). Setanaxib was safe and showed encouraging pharmacokinetic properties during Phase I study. Subsequently, it was tested in Phase II clinical trial for nephropathy in type 2 diabetes patients, yet the primary efficacy endpoint, i.e., albuminuria reduction, was not achieved. However, several other secondary efficacy endpoints were reached such as maximal inhibition of the renin-angiotensin-aldosterone system. In another Phase II trial focused on primary biliary cholangitis, setanaxib has succeeded and met its primary and secondary efficacy endpoints. Two additional Phase II clinical trials using setanaxib are ongoing, for idiopathic pulmonary fibrosis and kidney disease in type 1 diabetes. The second NOX inhibitor to reach the clinical trial stage is APX-115 which is moving from Phase I to II for diabetic kidney disease (Lee et al. 2020). Phenothiazines, already marketed for some indications, i.e., antipsychotic, show pan NOX inhibition activity in some preclinical studies (Seredenina et al. 2015, 2016). One clinical trial is planned to repurpose phenothiazines into stroke (Repo-Stroke).

Based on preclinical data, further indications for NOX inhibitors might also have potential toward the clinical application. For example, in ischemic retina disease and diabetic retinopathy, setanaxib and its analogue GKT136901 showed favorable effects (Appukuttan et al. 2018; Jiao et al. 2019; J L Wilkinson-Berka et al. 2013). Also, in cardiovascular disorders, including diabetes-associated atherosclerosis and hypertensive cardiac remodeling and hypertrophy, and liver fibrosis, setanaxib attenuated inflammatory and fibrotic markers (Gray et al. 2013; Sun et al. 2017; Zeng et al. 2019; Zhao et al. 2015) even when the treatment was delayed (Gray et al. 2017). VAS2870 which is a pan NOX inhibitor showed vascular protective effects in pulmonary hypertension (Li et al. 2019) and Alzheimer's disease (Abubaker et al. 2019). ML090 which has preferential activity toward NOX5 was beneficial in stroke

(Casas et al. 2019b; Dao et al. 2019). Collectively, NOX inhibition seems a promising therapeutic strategy with a broad range of clinical applications and warrants further investigations.

5 Advanced Therapies

Currently, the only known cure for CGD is allogeneic hematopoietic stem cell transplantation which is a high-risk procedure and associated with severe disability or death (Kang et al. 2011b). Only one drug is approved to treat/manage CGD, interferon gamma-1b that reduces the frequency and severity of serious infections associated with the disease (Miller et al. 2009). Current clinical research suggests that gene therapy holds great promise in curing CGD obviating the need for a transplantation donor and eliminating the risks associated with stem cell transplantation (Keller et al. 2018). What also makes gene therapy an attractive treatment for CGD is that restoration of normal NOX activity in only 10–20% of circulating neutrophils is sufficient to achieve significant clinical benefit (Keller et al. 2018). Early clinical trials on gene therapy for CGD were mainly based on γ-retroviral vectors that can only infect mitotically active cell types (Escors and Breckpot 2010). These studies failed to show efficacy and were associated with insertional mutagenesis, due to upregulation of proto-oncogene expression (Kang et al. 2011a; Keller et al. 2018). To overcome the latter issue, self-inactivating (SIN) retroviral vectors have been developed (Thornhill et al. 2008) and are being tested in clinical trials (NCT01906541). More recent clinical trials are using SIN lentiviral vectors (complex retroviruses), which unlike γ-retroviral vectors are capable of transducing quiescent cells and devoid of insertional toxicities (Escors and Breckpot 2010). The preliminary results from these lentiviral gene therapy trials (NCT02234934 and NCT01855685) are encouraging (Kohn et al. 2020) (Table 1).

6 Conclusions

NOX enzymes are primary sources of ROS, and their activation results in the activation of secondary ROS sources, i.e., ROS-dependent ROS production or the kindling-bonfire sequence. These secondary ROS sources include uncoupled nitric oxide synthase (NOS), xanthine oxidase, and dysfunctional mitochondria (Zhang et al. 2019). Therefore, NOX inhibition might represent an intelligent therapeutic strategy in ROS-related diseases as it targets the origin. However, none of the ROS sources act on their own, and different ROS forming enzymes will affect different targets. Thus, combinations are most likely more effective that single target strategies, which may lead to better efficacy and reduced side effects. As NOX inhibitors have entered clinical trials, two main aspects should be considered, specificity and isoform selectivity. Most of the NOX inhibitors in development are non-specific even the most advanced ones, setanaxib and GKT136901, have ROS scavenging activities (Augsburger et al. 2019; Dao et al. 2019). Isoform selectivity of the NOX inhibitors is also important given the physiological tissue- and cell-specific effects of NOXs. Applying a NOX inhibitor panel approach could be an option for NOX target validation (Dao et al. 2019). Further lead optimization of the current NOX inhibitors might help find isoform-selective compounds. Finally, ROS have important beneficial signaling functions. Thus, acute interventions such as in stroke (NOX4 and NOX5) appear safer than chronic therapies suppressing NOX1 or NOX4. Clinical trials in both directions are under way (NCT03865927, EudraCT No. 2019-000474-31) and will answer this by the early 2020s.

Acknowledgments This review project has received funding from the European Union's Horizon 2020 research and innovation program under grant agreement No 777111 (REPO-TRIAL). This reflects only the author's view, and the European Commission is not responsible for any use that may be made of the information it contains.

Conflict of Interest None

References

Abubaker AA, Vara D, Visconte C, Eggleston I, Torti M, Canobbio I, Pula G (2019) Amyloid peptide beta1-42 induces integrin alphaiibbeta3 activation, platelet adhesion, and thrombus formation in a NADPH oxidase-dependent manner. Oxidative Med Cell Longev 2019:1050476. https://doi.org/10.1155/2019/1050476

Accetta R, Damiano S, Morano A, Mondola P, Paterno R, Avvedimento EV, Santillo M (2016) Reactive oxygen species derived from NOX3 and NOX5 drive differentiation of human oligodendrocytes. Front Cell Neurosci 10:146. https://doi.org/10.3389/fncel.2016.00146

Ago T, Kuroda J, Pain J, Fu C, Li H, Sadoshima J (2010) Upregulation of Nox4 by hypertrophic stimuli promotes apoptosis and mitochondrial dysfunction in cardiac myocytes. Circ Res 106 (7):1253–1264. https://doi.org/10.1161/CIRCRESAHA.109.213116

Aliasgharzadeh A, Farhood B, Amini P, Saffar H, Motevaseli E, Rezapoor S et al (2019) Melatonin attenuates upregulation of Duox1 and Duox2 and protects against lung injury following chest irradiation in rats. Cell J 21(3):236–242. https://doi.org/10.22074/cellj.2019.6207

Altenhofer S, Radermacher KA, Kleikers PW, Wingler K, Schmidt HH (2015) Evolution of NADPH oxidase inhibitors: selectivity and mechanisms for target engagement. Antioxid Redox Signal 23(5):406–427. https://doi.org/10.1089/ars.2013.5814

Ambasta RK, Kumar P, Griendling KK, Schmidt HH, Busse R, Brandes RP (2004) Direct interaction of the novel Nox proteins with p22phox is required for the formation of a functionally active NADPH oxidase. J Biol Chem 279(44):45935–45941. https://doi.org/10.1074/jbc.M406486200

Ameziane-El-Hassani R, Talbot M, de Souza Dos Santos MC, Al Ghuzlan A, Hartl D, Bidart JM et al (2015) NADPH oxidase DUOX1 promotes long-term persistence of oxidative stress after an exposure to irradiation. Proc Natl Acad Sci U S A 112(16):5051–5056. https://doi.org/10.1073/pnas.1420707112

Antony S, Jiang G, Wu Y, Meitzler JL, Makhlouf HR, Haines DC et al (2017) NADPH oxidase 5 (NOX5)-induced reactive oxygen signaling modulates normoxic HIF-1alpha and p27(Kip1) expression in malignant melanoma and other human tumors. Mol Carcinog 56(12):2643–2662. https://doi.org/10.1002/mc.22708

Appukuttan B, Ma Y, Stempel A, Ashander LM, Deliyanti D, Wilkinson-Berka JL, Smith JR (2018) Effect of NADPH oxidase 1 and 4 blockade in activated human retinal endothelial cells. Clin Exp Ophthalmol 46(6):652–660. https://doi.org/10.1111/ceo.13155

Augsburger F, Filippova A, Rasti D, Seredenina T, Lam M, Maghzal G et al (2019) Pharmacological characterization of the seven human NOX isoforms and their inhibitors. Redox Biol 26:101272. https://doi.org/10.1016/j.redox.2019.101272

Azmoonfar R, Amini P, Saffar H, Rezapoor S, Motevaseli E, Cheki M et al (2018) Metformin protects against radiation-induced pneumonitis and fibrosis and attenuates upregulation of dual oxidase genes expression. Adv Pharm Bull 8(4):697–704. https://doi.org/10.15171/apb.2018.078

Banfi B, Maturana A, Jaconi S, Arnaudeau S, Laforge T, Sinha B et al (2000) A mammalian H+ channel generated through alternative splicing of the NADPH oxidase homolog NOH-1. Science 287(5450):138–142. https://doi.org/10.1126/science.287.5450.138

Banfi B, Molnar G, Maturana A, Steger K, Hegedus B, Demaurex N, Krause KH (2001) A Ca(2+)-activated NADPH oxidase in testis, spleen, and lymph nodes. J Biol Chem 276 (40):37594–37601. https://doi.org/10.1074/jbc.M103034200

Banfi B, Clark RA, Steger K, Krause KH (2003) Two novel proteins activate superoxide generation by the NADPH oxidase NOX1. J Biol Chem 278(6):3510–3513. https://doi.org/10.1074/jbc.C200613200

Banfi B, Malgrange B, Knisz J, Steger K, Dubois-Dauphin M, Krause KH (2004a) NOX3, a superoxide-generating NADPH oxidase of the inner ear. J Biol Chem 279(44):46065–46072. https://doi.org/10.1074/jbc.M403046200

Banfi B, Tirone F, Durussel I, Knisz J, Moskwa P, Molnar GZ et al (2004b) Mechanism of Ca2+ activation of the NADPH oxidase 5 (NOX5). J Biol Chem 279(18):18583–18591. https://doi.org/10.1074/jbc.M310268200

Bedard K, Krause KH (2007) The NOX family of ROS-generating NADPH oxidases: physiology and pathophysiology. Physiol Rev 87(1):245–313. https://doi.org/10.1152/physrev.00044.2005

BelAiba RS, Djordjevic T, Petry A, Diemer K, Bonello S, Banfi B et al (2007) NOX5 variants are functionally active in endothelial cells. Free Radic Biol Med 42(4):446–459. https://doi.org/10.1016/j.freeradbiomed.2006.10.054

Brandes RP, Weissmann N, Schroder K (2014) Nox family NADPH oxidases: molecular mechanisms of activation. Free Radic Biol Med 76:208–226. https://doi.org/10.1016/j.freeradbiomed.2014.07.046

Brewer AC, Murray TV, Arno M, Zhang M, Anilkumar NP, Mann GE, Shah AM (2011) Nox4 regulates Nrf2 and glutathione redox in cardiomyocytes in vivo. Free Radic Biol Med 51 (1):205–215. https://doi.org/10.1016/j.freeradbiomed.2011.04.022

Burgoyne JR, Madhani M, Cuello F, Charles RL, Brennan JP, Schroder E et al (2007) Cysteine redox sensor in PKGIa enables oxidant-induced activation. Science 317(5843):1393–1397. https://doi.org/10.1126/science.1144318

Carnesecchi S, Deffert C, Donati Y, Basset O, Hinz B, Preynat-Seauve O et al (2011) A key role for NOX4 in epithelial cell death during development of lung fibrosis. Antioxid Redox Signal 15 (3):607–619. https://doi.org/10.1089/ars.2010.3829

Casas AI, Dao VT, Daiber A, Maghzal GJ, Di Lisa F, Kaludercic N et al (2015) Reactive oxygen-related diseases: therapeutic targets and emerging clinical indications. Antioxid Redox Signal 23 (14):1171–1185. https://doi.org/10.1089/ars.2015.6433

Casas AI, Geuss E, Kleikers PWM, Mencl S, Herrmann AM, Buendia I et al (2017) NOX4-dependent neuronal autotoxicity and BBB breakdown explain the superior sensitivity of the brain to ischemic damage. Proc Natl Acad Sci U S A 114(46):12315–12320. https://doi.org/10.1073/pnas.1705034114

Casas AI, Hassan AA, Larsen SJ, Gomez-Rangel V, Elbatreek M, Kleikers PWM et al (2019a) From single drug targets to synergistic network pharmacology in ischemic stroke. Proc Natl Acad Sci U S A 116(14):7129–7136. https://doi.org/10.1073/pnas.1820799116

Casas AI, Kleikers PW, Geuss E, Langhauser F, Adler T, Busch DH et al (2019b) Calcium-dependent blood-brain barrier breakdown by NOX5 limits postreperfusion benefit in stroke. J Clin Invest 130:1772–1778. https://doi.org/10.1172/JCI124283

Case AJ, Li S, Basu U, Tian J, Zimmerman MC (2013) Mitochondrial-localized NADPH oxidase 4 is a source of superoxide in angiotensin II-stimulated neurons. Am J Physiol Heart Circ Physiol 305(1):H19–H28. https://doi.org/10.1152/ajpheart.00974.2012

Cha JJ, Min HS, Kim KT, Kim JE, Ghee JY, Kim HW et al (2017) APX-115, a first-in-class pan-NADPH oxidase (Nox) inhibitor, protects db/db mice from renal injury. Lab Investig 97 (4):419–431. https://doi.org/10.1038/labinvest.2017.2

Chamulitrat W, Schmidt R, Tomakidi P, Stremmel W, Chunglok W, Kawahara T, Rokutan K (2003) Association of gp91phox homolog Nox1 with anchorage-independent growth and MAP kinase-activation of transformed human keratinocytes. Oncogene 22(38):6045–6053. https://doi.org/10.1038/sj.onc.1206654

Chen F, Pandey D, Chadli A, Catravas JD, Chen T, Fulton DJ (2011) Hsp90 regulates NADPH oxidase activity and is necessary for superoxide but not hydrogen peroxide production. Antioxid Redox Signal 14(11):2107–2119. https://doi.org/10.1089/ars.2010.3669

Cheng G, Cao Z, Xu X, van Meir EG, Lambeth JD (2001) Homologs of gp91phox: cloning and tissue expression of Nox3, Nox4, and Nox5. Gene 269(1–2):131–140. https://doi.org/10.1016/s0378-1119(01)00449-8

Cheng G, Diebold BA, Hughes Y, Lambeth JD (2006) Nox1-dependent reactive oxygen generation is regulated by Rac1. J Biol Chem 281(26):17718–17726. https://doi.org/10.1074/jbc.M512751200

Choi SA, Kim YH, Park YH, Yang HJ, Jeong PS, Cha JJ et al (2019) Novel crosstalk between Vps26a and Nox4 signaling during neurogenesis. Cell Death Differ 26(9):1582–1599. https://doi.org/10.1038/s41418-018-0226-0

Coant N, Ben Mkaddem S, Pedruzzi E, Guichard C, Treton X, Ducroc R et al (2010) NADPH oxidase 1 modulates WNT and NOTCH1 signaling to control the fate of proliferative progenitor cells in the colon. Mol Cell Biol 30(11):2636–2650. https://doi.org/10.1128/MCB.01194-09

Crosas-Molist E, Bertran E, Rodriguez-Hernandez I, Herraiz C, Cantelli G, Fabra A et al (2017) The NADPH oxidase NOX4 represses epithelial to amoeboid transition and efficient tumour dissemination. Oncogene 36(21):3002–3014. https://doi.org/10.1038/onc.2016.454

Cross AR, Segal AW (2004) The NADPH oxidase of professional phagocytes--prototype of the NOX electron transport chain systems. Biochim Biophys Acta 1657(1):1–22. https://doi.org/10.1016/j.bbabio.2004.03.008

Cucoranu I, Clempus R, Dikalova A, Phelan PJ, Ariyan S, Dikalov S, Sorescu D (2005) NAD(P)H oxidase 4 mediates transforming growth factor-beta1-induced differentiation of cardiac fibroblasts into myofibroblasts. Circ Res 97(9):900–907. https://doi.org/10.1161/01.RES.0000187457.24338.3D

Cui XL, Brockman D, Campos B, Myatt L (2006) Expression of NADPH oxidase isoform 1 (Nox1) in human placenta: involvement in preeclampsia. Placenta 27(4–5):422–431. https://doi.org/10.1016/j.placenta.2005.04.004

Dao VT, Casas AI, Maghzal GJ, Seredenina T, Kaludercic N, Robledinos-Anton N et al (2015) Pharmacology and clinical drug candidates in redox medicine. Antioxid Redox Signal 23 (14):1113–1129. https://doi.org/10.1089/ars.2015.6430

Dao VT, Elbatreek MH, Altenhofer S, Casas AI, Pachado MP, Neullens CT et al (2019) Isoform-selective NADPH oxidase inhibitor panel for pharmacological target validation. Free Radic Biol Med 148:60–69. https://doi.org/10.1016/j.freeradbiomed.2019.12.038

De Deken X, Wang D, Many MC, Costagliola S, Libert F, Vassart G et al (2000) Cloning of two human thyroid cDNAs encoding new members of the NADPH oxidase family. J Biol Chem 275 (30):23227–23233. https://doi.org/10.1074/jbc.M000916200

De Deken X, Wang D, Dumont JE, Miot F (2002) Characterization of ThOX proteins as components of the thyroid H(2)O(2)-generating system. Exp Cell Res 273(2):187–196. https://doi.org/10.1006/excr.2001.5444

Den Hartigh LJ, Omer M, Goodspeed L, Wang S, Wietecha T, O'Brien KD, Han CY (2017) Adipocyte-specific deficiency of NADPH oxidase 4 delays the onset of insulin resistance and attenuates adipose tissue inflammation in obesity. Arterioscler Thromb Vasc Biol 37 (3):466–475. https://doi.org/10.1161/ATVBAHA.116.308749

Dhukhwa A, Bhatta P, Sheth S, Korrapati K, Tieu C, Mamillapalli C et al (2019) Targeting inflammatory processes mediated by TRPVI and TNF-alpha for treating noise-induced hearing loss. Front Cell Neurosci 13:444. https://doi.org/10.3389/fncel.2019.00444

Diebold BA, Bokoch GM (2001) Molecular basis for Rac2 regulation of phagocyte NADPH oxidase. Nat Immunol 2(3):211–215. https://doi.org/10.1038/85259

Drummond GR, Cai H, Davis ME, Ramasamy S, Harrison DG (2000) Transcriptional and posttranscriptional regulation of endothelial nitric oxide synthase expression by hydrogen peroxide. Circ Res 86(3):347–354. https://doi.org/10.1161/01.res.86.3.347

El Hassani RA, Benfares N, Caillou B, Talbot M, Sabourin JC, Belotte V et al (2005) Dual oxidase2 is expressed all along the digestive tract. Am J Physiol Gastrointest Liver Physiol 288(5):G933–G942. https://doi.org/10.1152/ajpgi.00198.2004

Elbatreek MH, Pachado MP, Cuadrado A, Jandeleit-Dahm K, Schmidt H (2019) Reactive oxygen comes of age: mechanism-based therapy of diabetic end-organ damage. Trends Endocrinol Metab 30(5):312–327. https://doi.org/10.1016/j.tem.2019.02.006

Escors D, Breckpot K (2010) Lentiviral vectors in gene therapy: their current status and future potential. Arch Immunol Ther Exp 58(2):107–119. https://doi.org/10.1007/s00005-010-0063-4

Fan LM, Geng L, Cahill-Smith S, Liu F, Douglas G, McKenzie CA et al (2019) Nox2 contributes to age-related oxidative damage to neurons and the cerebral vasculature. J Clin Invest 129 (8):3374–3386. https://doi.org/10.1172/JCI125173

Farhood B, Aliasgharzadeh A, Amini P, Saffar H, Motevaseli E, Rezapoor S et al (2019) Radiation-induced dual oxidase upregulation in rat heart tissues: protective effect of melatonin. Medicina 55(7):317. https://doi.org/10.3390/medicina55070317

Feraille E, Dizin E, Roth I, Derouette JP, Szanto I, Martin PY et al (2014) NADPH oxidase 4 deficiency reduces aquaporin-2 mRNA expression in cultured renal collecting duct principal cells via increased PDE3 and PDE4 activity. PLoS One 9(1):e87239. https://doi.org/10.1371/journal.pone.0087239

Fischer H (2009) Mechanisms and function of DUOX in epithelia of the lung. Antioxid Redox Signal 11(10):2453–2465. https://doi.org/10.1089/ARS.2009.2558

Fulton DJ (2009) Nox5 and the regulation of cellular function. Antioxid Redox Signal 11 (10):2443–2452. https://doi.org/10.1089/ars.2009.2587

Geiszt M, Kopp JB, Varnai P, Leto TL (2000) Identification of renox, an NAD(P)H oxidase in kidney. Proc Natl Acad Sci U S A 97(14):8010–8014. https://doi.org/10.1073/pnas.130135897

Geiszt M, Lekstrom K, Brenner S, Hewitt SM, Dana R, Malech HL, Leto TL (2003a) NAD(P)H oxidase 1, a product of differentiated colon epithelial cells, can partially replace glycoprotein 91phox in the regulated production of superoxide by phagocytes. J Immunol 171(1):299–306. https://doi.org/10.4049/jimmunol.171.1.299

Geiszt M, Witta J, Baffi J, Lekstrom K, Leto TL (2003b) Dual oxidases represent novel hydrogen peroxide sources supporting mucosal surface host defense. FASEB J 17(11):1502–1504. https://doi.org/10.1096/fj.02-1104fje

Ghanbari H, Keshtgar S, Kazeroni M (2018) Inhibition of the CatSper channel and NOX5 enzyme activity affects the functions of the progesterone-stimulated human sperm. Iran J Med Sci 43 (1):18–25

Giardino G, Cicalese MP, Delmonte O, Migliavacca M, Palterer B, Loffredo L et al (2017) NADPH oxidase deficiency: a multisystem approach. Oxidative Med Cell Longev 2017:4590127. https://doi.org/10.1155/2017/4590127

Gorin Y, Ricono JM, Kim NH, Bhandari B, Choudhury GG, Abboud HE (2003) Nox4 mediates angiotensin II-induced activation of Akt/protein kinase B in mesangial cells. Am J Physiol Renal Physiol 285(2):F219–F229. https://doi.org/10.1152/ajprenal.00414.2002

Gorlach A, Brandes RP, Nguyen K, Amidi M, Dehghani F, Busse R (2000) A gp91phox containing NADPH oxidase selectively expressed in endothelial cells is a major source of oxygen radical generation in the arterial wall. Circ Res 87(1):26–32. https://doi.org/10.1161/01.res.87.1.26

Grasberger H, Refetoff S (2006) Identification of the maturation factor for dual oxidase. Evolution of an eukaryotic operon equivalent. J Biol Chem 281(27):18269–18272. https://doi.org/10.1074/jbc.C600095200

Gray SP, Di Marco E, Okabe J, Szyndralewiez C, Heitz F, Montezano AC et al (2013) NADPH oxidase 1 plays a key role in diabetes mellitus-accelerated atherosclerosis. Circulation 127 (18):1888–1902. https://doi.org/10.1161/CIRCULATIONAHA.112.132159

Gray SP, Di Marco E, Kennedy K, Chew P, Okabe J, El-Osta A et al (2016) Reactive oxygen species can provide atheroprotection via NOX4-dependent inhibition of inflammation and vascular remodeling. Arterioscler Thromb Vasc Biol 36(2):295–307. https://doi.org/10.1161/ATVBAHA.115.307012

Gray SP, Jha JC, Kennedy K, van Bommel E, Chew P, Szyndralewiez C et al (2017) Combined NOX1/4 inhibition with GKT137831 in mice provides dose-dependent reno- and atheroprotection even in established micro- and macrovascular disease. Diabetologia 60 (5):927–937. https://doi.org/10.1007/s00125-017-4215-5

Guo S, Chen X (2015) The human Nox4: gene, structure, physiological function and pathological significance. J Drug Target 23(10):888–896. https://doi.org/10.3109/1061186X.2015.1036276

Guzik TJ, Chen W, Gongora MC, Guzik B, Lob HE, Mangalat D et al (2008) Calcium-dependent NOX5 nicotinamide adenine dinucleotide phosphate oxidase contributes to vascular oxidative stress in human coronary artery disease. J Am Coll Cardiol 52(22):1803–1809. https://doi.org/10.1016/j.jacc.2008.07.063

Guzik B, Sagan A, Ludew D, Mrowiecki W, Chwala M, Bujak-Gizycka B et al (2013) Mechanisms of oxidative stress in human aortic aneurysms--association with clinical risk factors for atherosclerosis and disease severity. Int J Cardiol 168(3):2389–2396. https://doi.org/10.1016/j.ijcard.2013.01.278

Hahn NE, Meischl C, Kawahara T, Musters RJ, Verhoef VM, van der Velden J et al (2012) NOX5 expression is increased in intramyocardial blood vessels and cardiomyocytes after acute myocardial infarction in humans. Am J Pathol 180(6):2222–2229. https://doi.org/10.1016/j.ajpath.2012.02.018

Haque MZ, Majid DS (2004) Assessment of renal functional phenotype in mice lacking gp91PHOX subunit of NAD(P)H oxidase. Hypertension 43(2):335–340. https://doi.org/10.1161/01.HYP.0000111137.15873.4a

Hayes P, Dhillon S, O'Neill K, Thoeni C, Hui KY, Elkadri A et al (2015) Defects in NADPH oxidase genes NOX1 and DUOX2 in very early onset inflammatory bowel disease. Cell Mol Gastroenterol Hepatol 1(5):489–502. https://doi.org/10.1016/j.jcmgh.2015.06.005

He W, Wang Q, Gu L, Zhong L, Liu D (2018) NOX4 rs11018628 polymorphism associates with a decreased risk and better short-term recovery of ischemic stroke. Exp Ther Med 16 (6):5258–5264. https://doi.org/10.3892/etm.2018.6874

Henriquez-Olguin C, Boronat S, Cabello-Verrugio C, Jaimovich E, Hidalgo E, Jensen TE (2019) The emerging roles of nicotinamide adenine dinucleotide phosphate oxidase 2 in skeletal muscle redox signaling and metabolism. Antioxid Redox Signal 31(18):1371–1410. https://doi.org/10.1089/ars.2018.7678

Hilenski LL, Clempus RE, Quinn MT, Lambeth JD, Griendling KK (2004) Distinct subcellular localizations of Nox1 and Nox4 in vascular smooth muscle cells. Arterioscler Thromb Vasc Biol 24(4):677–683. https://doi.org/10.1161/01.ATV.0000112024.13727.2c

Hirano K, Chen WS, Chueng AL, Dunne AA, Seredenina T, Filippova A et al (2015) Discovery of GSK2795039, a novel small molecule NADPH oxidase 2 inhibitor. Antioxid Redox Signal 23 (5):358–374. https://doi.org/10.1089/ars.2014.6202

Hoidal JR, Brar SS, Sturrock AB, Sanders KA, Dinger B, Fidone S, Kennedy TP (2003) The role of endogenous NADPH oxidases in airway and pulmonary vascular smooth muscle function. Antioxid Redox Signal 5(6):751–758. https://doi.org/10.1089/152308603770380052

Holterman CE, Thibodeau JF, Towaij C, Gutsol A, Montezano AC, Parks RJ et al (2014) Nephropathy and elevated BP in mice with podocyte-specific NADPH oxidase 5 expression. J Am Soc Nephrol 25(4):784–797. https://doi.org/10.1681/ASN.2013040371

Huang J, Hitt ND, Kleinberg ME (1995) Stoichiometry of p22-phox and gp91-phox in phagocyte cytochrome b558. Biochemistry 34(51):16753–16757. https://doi.org/10.1021/bi00051a024

Ibi M, Katsuyama M, Fan C, Iwata K, Nishinaka T, Yokoyama T, Yabe-Nishimura C (2006) NOX1/NADPH oxidase negatively regulates nerve growth factor-induced neurite outgrowth. Free Radic Biol Med 40(10):1785–1795. https://doi.org/10.1016/j.freeradbiomed.2006.01.009

Iles KE, Forman HJ (2002) Macrophage signaling and respiratory burst. Immunol Res 26 (1–3):95–105. https://doi.org/10.1385/IR:26:1-3:095

Janiszewski M, Lopes LR, Carmo AO, Pedro MA, Brandes RP, Santos CX, Laurindo FR (2005) Regulation of NAD(P)H oxidase by associated protein disulfide isomerase in vascular smooth muscle cells. J Biol Chem 280(49):40813–40819. https://doi.org/10.1074/jbc.M509255200

Jay DB, Papaharalambus CA, Seidel-Rogol B, Dikalova AE, Lassegue B, Griendling KK (2008) Nox5 mediates PDGF-induced proliferation in human aortic smooth muscle cells. Free Radic Biol Med 45(3):329–335. https://doi.org/10.1016/j.freeradbiomed.2008.04.024

Jha JC, Gray SP, Barit D, Okabe J, El-Osta A, Namikoshi T et al (2014) Genetic targeting or pharmacologic inhibition of NADPH oxidase nox4 provides renoprotection in long-term diabetic nephropathy. J Am Soc Nephrol 25(6):1237–1254. https://doi.org/10.1681/ASN.2013070810

Jha JC, Thallas-Bonke V, Banal C, Gray SP, Chow BS, Ramm G et al (2016) Podocyte-specific Nox4 deletion affords renoprotection in a mouse model of diabetic nephropathy. Diabetologia 59(2):379–389. https://doi.org/10.1007/s00125-015-3796-0

Jha JC, Banal C, Okabe J, Gray SP, Hettige T, Chow BSM et al (2017a) NADPH oxidase Nox5 accelerates renal injury in diabetic nephropathy. Diabetes 66(10):2691–2703. https://doi.org/10.2337/db16-1585

Jha JC, Banal C, Okabe J, Gray SP, Hettige T, Chow BSM et al (2017b) NADPH oxidase Nox5 accelerates renal injury in diabetic nephropathy. Diabetes 66(10):2691–2703. https://doi.org/10.2337/db16-1585

Jiang F, Zhang Y, Dusting GJ (2011) NADPH oxidase-mediated redox signaling: roles in cellular stress response, stress tolerance, and tissue repair. Pharmacol Rev 63(1):218–242. https://doi.org/10.1124/pr.110.002980

Jiao W, Ji J, Li F, Guo J, Zheng Y, Li S, Xu W (2019) Activation of the NotchNox4reactive oxygen species signaling pathway induces cell death in high glucose treated human retinal endothelial cells. Mol Med Rep 19(1):667–677. https://doi.org/10.3892/mmr.2018.9637

Juhasz A, Markel S, Gaur S, Liu H, Lu J, Jiang G et al (2017) NADPH oxidase 1 supports proliferation of colon cancer cells by modulating reactive oxygen species-dependent signal transduction. J Biol Chem 292(19):7866–7887. https://doi.org/10.1074/jbc.M116.768283

Kajla S, Mondol AS, Nagasawa A, Zhang Y, Kato M, Matsuno K et al (2012) A crucial role for Nox 1 in redox-dependent regulation of Wnt-beta-catenin signaling. FASEB J 26(5):2049–2059. https://doi.org/10.1096/fj.11-196360

Kang HJ, Bartholomae CC, Paruzynski A, Arens A, Kim S, Yu SS et al (2011a) Retroviral gene therapy for X-linked chronic granulomatous disease: results from phase I/II trial. Mol Ther 19 (11):2092–2101. https://doi.org/10.1038/mt.2011.166

Kang EM, Marciano BE, DeRavin S, Zarember KA, Holland SM, Malech HL (2011b) Chronic granulomatous disease: overview and hematopoietic stem cell transplantation. J Allergy Clin Immunol 127(6):1319–1326.; quiz 1327-1318. https://doi.org/10.1016/j.jaci.2011.03.028

Kato K, Hecker L (2020) NADPH oxidases: pathophysiology and therapeutic potential in age-associated pulmonary fibrosis. Redox Biol 33:101541. https://doi.org/10.1016/j.redox.2020.101541

Kato M, Marumo M, Nakayama J, Matsumoto M, Yabe-Nishimura C, Kamata T (2016) The ROS-generating oxidase Nox1 is required for epithelial restitution following colitis. Exp Anim 65(3):197–205. https://doi.org/10.1538/expanim.15-0127

Keller MD, Notarangelo LD, Malech HL (2018) Future of care for patients with chronic granulomatous disease: gene therapy and targeted molecular medicine. J Pediatric Infect Dis Soc 7 (suppl_1):S40–S44. https://doi.org/10.1093/jpids/piy011

Kishida KT, Hoeffer CA, Hu D, Pao M, Holland SM, Klann E (2006) Synaptic plasticity deficits and mild memory impairments in mouse models of chronic granulomatous disease. Mol Cell Biol 26(15):5908–5920. https://doi.org/10.1128/MCB.00269-06

Kleikers PW, Hooijmans C, Gob E, Langhauser F, Rewell SS, Radermacher K et al (2015) A combined pre-clinical meta-analysis and randomized confirmatory trial approach to improve data validity for therapeutic target validation. Sci Rep 5:13428. https://doi.org/10.1038/srep13428

Kleinschnitz C, Grund H, Wingler K, Armitage ME, Jones E, Mittal M et al (2010) Post-stroke inhibition of induced NADPH oxidase type 4 prevents oxidative stress and neurodegeneration. PLoS Biol 8(9):e1000479. https://doi.org/10.1371/journal.pbio.1000479

Ko E, Choi H, Kim B, Kim M, Park KN, Bae IH et al (2014) Testosterone stimulates Duox1 activity through GPRC6A in skin keratinocytes. J Biol Chem 289(42):28835–28845. https://doi.org/10.1074/jbc.M114.583450

Kohn DB, Booth C, Kang EM, Pai SY, Shaw KL, Santilli G et al (2020) Lentiviral gene therapy for X-linked chronic granulomatous disease. Nat Med 26(2):200–206. https://doi.org/10.1038/s41591-019-0735-5

Kouroku Y, Fujita E, Tanida I, Ueno T, Isoai A, Kumagai H et al (2007) ER stress (PERK/eIF2alpha phosphorylation) mediates the polyglutamine-induced LC3 conversion, an essential step for autophagy formation. Cell Death Differ 14(2):230–239. https://doi.org/10.1038/sj.cdd.4401984

Krijnen PA, Meischl C, Hack CE, Meijer CJ, Visser CA, Roos D, Niessen HW (2003) Increased Nox2 expression in human cardiomyocytes after acute myocardial infarction. J Clin Pathol 56(3):194–199. https://doi.org/10.1136/jcp.56.3.194

Lan T, Kisseleva T, Brenner DA (2015) Deficiency of NOX1 or NOX4 prevents liver inflammation and fibrosis in mice through inhibition of hepatic stellate cell activation. PLoS One 10(7):e0129743. https://doi.org/10.1371/journal.pone.0129743

Lassegue B, Sorescu D, Szocs K, Yin Q, Akers M, Zhang Y et al (2001) Novel gp91(phox) homologues in vascular smooth muscle cells: nox1 mediates angiotensin II-induced superoxide formation and redox-sensitive signaling pathways. Circ Res 88(9):888–894. https://doi.org/10.1161/hh0901.090299

Laurent E, McCoy JW 3rd, Macina RA, Liu W, Cheng G, Robine S et al (2008) Nox1 is over-expressed in human colon cancers and correlates with activating mutations in K-Ras. Int J Cancer 123(1):100–107. https://doi.org/10.1002/ijc.23423

Lee SR, An EJ, Kim J, Bae YS (2020) Function of NADPH oxidases in diabetic nephropathy and development of nox inhibitors. Biomol Ther 28(1):25–33. https://doi.org/10.4062/biomolther.2019.188

Li T, Luo XJ, Wang EL, Li NS, Zhang XJ, Song FL et al (2019) Magnesium lithospermate B prevents phenotypic transformation of pulmonary arteries in rats with hypoxic pulmonary hypertension through suppression of NADPH oxidase. Eur J Pharmacol 847:32–41. https://doi.org/10.1016/j.ejphar.2019.01.020

Lin XL, Yang L, Fu SW, Lin WF, Gao YJ, Chen HY, Ge ZZ (2017) Overexpression of NOX4 predicts poor prognosis and promotes tumor progression in human colorectal cancer. Oncotarget 8(20):33586–33600. https://doi.org/10.18632/oncotarget.16829

Lu X, Wang F, Liu M, Yang KT, Nau A, Kohan DE et al (2016) Activation of ENaC in collecting duct cells by prorenin and its receptor PRR: involvement of Nox4-derived hydrogen peroxide. Am J Physiol Renal Physiol 310(11):F1243–F1250. https://doi.org/10.1152/ajprenal.00492.2015

Lyle AN, Deshpande NN, Taniyama Y, Seidel-Rogol B, Pounkova L, Du P et al (2009) Poldip2, a novel regulator of Nox4 and cytoskeletal integrity in vascular smooth muscle cells. Circ Res 105(3):249–259. https://doi.org/10.1161/CIRCRESAHA.109.193722

Mahadev K, Motoshima H, Wu X, Ruddy JM, Arnold RS, Cheng G et al (2004) The NAD(P)H oxidase homolog Nox4 modulates insulin-stimulated generation of H2O2 and plays an integral role in insulin signal transduction. Mol Cell Biol 24(5):1844–1854. https://doi.org/10.1128/mcb.24.5.1844-1854.2004

Manea A, Raicu M, Simionescu M (2005) Expression of functionally phagocyte-type NAD(P)H oxidase in pericytes: effect of angiotensin II and high glucose. Biol Cell 97(9):723–734. https://doi.org/10.1042/BC20040107

Manea A, Manea SA, Gan AM, Constantin A, Fenyo IM, Raicu M et al (2015) Human monocytes and macrophages express NADPH oxidase 5; a potential source of reactive oxygen species in atherosclerosis. Biochem Biophys Res Commun 461(1):172–179. https://doi.org/10.1016/j.bbrc.2015.04.021

Marzaioli V, Hurtado-Nedelec M, Pintard C, Tlili A, Marie JC, Monteiro RC et al (2017) NOX5 and p22phox are 2 novel regulators of human monocytic differentiation into dendritic cells. Blood 130(15):1734–1745. https://doi.org/10.1182/blood-2016-10-746347

Matsushima S, Kuroda J, Ago T, Zhai P, Park JY, Xie LH et al (2013) Increased oxidative stress in the nucleus caused by Nox4 mediates oxidation of HDAC4 and cardiac hypertrophy. Circ Res 112(4):651–663. https://doi.org/10.1161/CIRCRESAHA.112.279760

Meitzler JL, Ortiz de Montellano PR (2009) Caenorhabditis elegans and human dual oxidase 1 (DUOX1) "peroxidase" domains: insights into heme binding and catalytic activity. J Biol Chem 284(28):18634–18643. https://doi.org/10.1074/jbc.M109.013581

Meitzler JL, Ortiz de Montellano PR (2011) Structural stability and heme binding potential of the truncated human dual oxidase 2 (DUOX2) peroxidase domain. Arch Biochem Biophys 512 (2):197–203. https://doi.org/10.1016/j.abb.2011.05.021

Miller CH, Maher SG, Young HA (2009) Clinical use of interferon-gamma. Ann N Y Acad Sci 1182:69–79. https://doi.org/10.1111/j.1749-6632.2009.05069.x

Moll F, Walter M, Rezende F, Helfinger V, Vasconez E, De Oliveira T et al (2018) NoxO1 controls proliferation of colon epithelial cells. Front Immunol 9:973. https://doi.org/10.3389/fimmu.2018.00973

Mongue-Din H, Patel AS, Looi YH, Grieve DJ, Anilkumar N, Sirker A et al (2017) NADPH oxidase-4 driven cardiac macrophage polarization protects against myocardial infarction-induced remodeling. JACC Basic Transl Sci 2(6):688–698. https://doi.org/10.1016/j.jacbts.2017.06.006

Montezano AC, De Lucca Camargo L, Persson P, Rios FJ, Harvey AP, Anagnostopoulou A et al (2018) NADPH oxidase 5 is a pro-contractile nox isoform and a point of cross-talk for calcium and redox signaling-implications in vascular function. J Am Heart Assoc 7(12):e009388. https://doi.org/10.1161/JAHA.118.009388

Moreno JC, Bikker H, Kempers MJ, van Trotsenburg AS, Baas F, de Vijlder JJ et al (2002) Inactivating mutations in the gene for thyroid oxidase 2 (THOX2) and congenital hypothyroidism. N Engl J Med 347(2):95–102. https://doi.org/10.1056/NEJMoa012752

Nakano Y, Banfi B, Jesaitis AJ, Dinauer MC, Allen LA, Nauseef WM (2007) Critical roles for p22phox in the structural maturation and subcellular targeting of Nox3. Biochem J 403 (1):97–108. https://doi.org/10.1042/BJ20060819

Nlandu Khodo S, Dizin E, Sossauer G, Szanto I, Martin PY, Feraille E et al (2012) NADPH-oxidase 4 protects against kidney fibrosis during chronic renal injury. J Am Soc Nephrol 23 (12):1967–1976. https://doi.org/10.1681/ASN.2012040373

O'Neill S, Brault J, Stasia MJ, Knaus UG (2015) Genetic disorders coupled to ROS deficiency. Redox Biol 6:135–156. https://doi.org/10.1016/j.redox.2015.07.009

Paffenholz R, Bergstrom RA, Pasutto F, Wabnitz P, Munroe RJ, Jagla W et al (2004) Vestibular defects in head-tilt mice result from mutations in Nox3, encoding an NADPH oxidase. Genes Dev 18(5):486–491. https://doi.org/10.1101/gad.1172504

Panday A, Sahoo MK, Osorio D, Batra S (2015) NADPH oxidases: an overview from structure to innate immunity-associated pathologies. Cell Mol Immunol 12(1):5–23. https://doi.org/10.1038/cmi.2014.89

Pao M, Wiggs EA, Anastacio MM, Hyun J, DeCarlo ES, Miller JT et al (2004) Cognitive function in patients with chronic granulomatous disease: a preliminary report. Psychosomatics 45 (3):230–234. https://doi.org/10.1176/appi.psy.45.3.230

Parkos CA, Allen RA, Cochrane CG, Jesaitis AJ (1987) Purified cytochrome b from human granulocyte plasma membrane is comprised of two polypeptides with relative molecular weights of 91,000 and 22,000. J Clin Invest 80(3):732–742. https://doi.org/10.1172/JCI113128

Perrotta I, Sciangula A, Perrotta E, Donato G, Cassese M (2011) Ultrastructural analysis and electron microscopic localization of Nox4 in healthy and atherosclerotic human aorta. Ultrastruct Pathol 35(1):1–6. https://doi.org/10.3109/01913123.2010.510261

Plecita-Hlavata L, Jaburek M, Holendova B, Tauber J, Pavluch V, Berkova Z et al (2020) Glucose-stimulated insulin secretion fundamentally requires H2O2 signaling by NADPH oxidase 4. Diabetes 7(12):e009388. https://doi.org/10.2337/db19-1130

Reinehr R, Becker S, Eberle A, Grether-Beck S, Haussinger D (2005) Involvement of NADPH oxidase isoforms and Src family kinases in CD95-dependent hepatocyte apoptosis. J Biol Chem 280(29):27179–27194. https://doi.org/10.1074/jbc.M414361200

Rezende F, Schurmann C, Schutz S, Harenkamp S, Herrmann E, Seimetz M et al (2017) Knock out of the NADPH oxidase Nox4 has no impact on life span in mice. Redox Biol 11:312–314. https://doi.org/10.1016/j.redox.2016.12.012

Rigutto S, Hoste C, Grasberger H, Milenkovic M, Communi D, Dumont JE et al (2009) Activation of dual oxidases Duox1 and Duox2: differential regulation mediated by camp-dependent protein kinase and protein kinase C-dependent phosphorylation. J Biol Chem 284(11):6725–6734. https://doi.org/10.1074/jbc.M806893200

Rokutan K, Kawahara T, Kuwano Y, Tominaga K, Sekiyama A, Teshima-Kondo S (2006) NADPH oxidases in the gastrointestinal tract: a potential role of Nox1 in innate immune response and carcinogenesis. Antioxid Redox Signal 8(9–10):1573–1582. https://doi.org/10.1089/ars.2006.8.1573

Rossi F, Zatti M (1964) Biochemical aspects of phagocytosis in polymorphonuclear leucocytes. NADH and NADPH oxidation by the granules of resting and phagocytizing cells. Experientia 20(1):21–23. https://doi.org/10.1007/bf02146019

Rybak LP, Mukherjea D, Jajoo S, Kaur T, Ramkumar V (2012) siRNA-mediated knock-down of NOX3: therapy for hearing loss? Cell Mol Life Sci 69(14):2429–2434. https://doi.org/10.1007/s00018-012-1016-3

Scherz-Shouval R, Elazar Z (2011) Regulation of autophagy by ROS: physiology and pathology. Trends Biochem Sci 36(1):30–38. https://doi.org/10.1016/j.tibs.2010.07.007

Schroder K, Wandzioch K, Helmcke I, Brandes RP (2009) Nox4 acts as a switch between differentiation and proliferation in preadipocytes. Arterioscler Thromb Vasc Biol 29 (2):239–245. https://doi.org/10.1161/ATVBAHA.108.174219

Schroder K, Zhang M, Benkhoff S, Mieth A, Pliquett R, Kosowski J et al (2012) Nox4 is a protective reactive oxygen species generating vascular NADPH oxidase. Circ Res 110 (9):1217–1225. https://doi.org/10.1161/circresaha.112.267054

Schurmann C, Rezende F, Kruse C, Yasar Y, Lowe O, Fork C et al (2015) The NADPH oxidase Nox4 has anti-atherosclerotic functions. Eur Heart J 36(48):3447–3456. https://doi.org/10.1093/eurheartj/ehv460

Schwarzer C, Machen TE, Illek B, Fischer H (2004) NADPH oxidase-dependent acid production in airway epithelial cells. J Biol Chem 279(35):36454–36461. https://doi.org/10.1074/jbc.M404983200

Schwerd T, Bryant RV, Pandey S, Capitani M, Meran L, Cazier JB et al (2018) NOX1 loss-of-function genetic variants in patients with inflammatory bowel disease. Mucosal Immunol 11 (2):562–574. https://doi.org/10.1038/mi.2017.74

Segal AW, Jones OT (1978) Novel cytochrome b system in phagocytic vacuoles of human granulocytes. Nature 276(5687):515–517. https://doi.org/10.1038/276515a0

Seredenina T, Chiriano G, Filippova A, Nayernia Z, Mahiout Z, Fioraso-Cartier L et al (2015) A subset of N-substituted phenothiazines inhibits NADPH oxidases. Free Radic Biol Med 86:239–249. https://doi.org/10.1016/j.freeradbiomed.2015.05.023

Seredenina T, Nayernia Z, Sorce S, Maghzal GJ, Filippova A, Ling SC et al (2016) Evaluation of NADPH oxidases as drug targets in a mouse model of familial amyotrophic lateral sclerosis. Free Radic Biol Med 97:95–108. https://doi.org/10.1016/j.freeradbiomed.2016.05.016

Suh YA, Arnold RS, Lassegue B, Shi J, Xu X, Sorescu D et al (1999) Cell transformation by the superoxide-generating oxidase Mox1. Nature 401(6748):79–82. https://doi.org/10.1038/43459

Sun Q, Zhang W, Zhong W, Sun X, Zhou Z (2017) Pharmacological inhibition of NOX4 ameliorates alcohol-induced liver injury in mice through improving oxidative stress and mitochondrial function. Biochim Biophys Acta Gen Subj 1861(1 Pt A):2912–2921. https://doi.org/10.1016/j.bbagen.2016.09.009

Szanto I, Rubbia-Brandt L, Kiss P, Steger K, Banfi B, Kovari E et al (2005) Expression of NOX1, a superoxide-generating NADPH oxidase, in colon cancer and inflammatory bowel disease. J Pathol 207(2):164–176. https://doi.org/10.1002/path.1824

Taniguchi CM, Emanuelli B, Kahn CR (2006) Critical nodes in signalling pathways: insights into insulin action. Nat Rev Mol Cell Biol 7(2):85–96. https://doi.org/10.1038/nrm1837

Teshima S, Kutsumi H, Kawahara T, Kishi K, Rokutan K (2000) Regulation of growth and apoptosis of cultured Guinea pig gastric mucosal cells by mitogenic oxidase 1. Am J Physiol Gastrointest Liver Physiol 279(6):G1169–G1176. https://doi.org/10.1152/ajpgi.2000.279.6.G1169

Thornhill SI, Schambach A, Howe SJ, Ulaganathan M, Grassman E, Williams D et al (2008) Self-inactivating gammaretroviral vectors for gene therapy of X-linked severe combined immunodeficiency. Mol Ther 16(3):590–598. https://doi.org/10.1038/sj.mt.6300393

Touyz RM, Anagnostopoulou A, Camargo LL, Rios FJ, Montezano AC (2019) Vascular biology of superoxide-generating NADPH oxidase 5-implications in hypertension and cardiovascular disease. Antioxid Redox Signal 30(7):1027–1040. https://doi.org/10.1089/ars.2018.7583

Ueno N, Takeya R, Miyano K, Kikuchi H, Sumimoto H (2005) The NADPH oxidase Nox3 constitutively produces superoxide in a p22phox-dependent manner: its regulation by oxidase organizers and activators. J Biol Chem 280(24):23328–23339. https://doi.org/10.1074/jbc.M414548200

Ueyama T, Geiszt M, Leto TL (2006) Involvement of Rac1 in activation of multicomponent Nox1- and Nox3-based NADPH oxidases. Mol Cell Biol 26(6):2160–2174. https://doi.org/10.1128/MCB.26.6.2160-2174.2006

Van Buul JD, Fernandez-Borja M, Anthony EC, Hordijk PL (2005) Expression and localization of NOX2 and NOX4 in primary human endothelial cells. Antioxid Redox Signal 7(3–4):308–317. https://doi.org/10.1089/ars.2005.7.308

van der Vliet A, Danyal K, Heppner DE (2018) Dual oxidase: a novel therapeutic target in allergic disease. Br J Pharmacol 175(9):1401–1418. https://doi.org/10.1111/bph.14158

Veal E, Day A (2011) Hydrogen peroxide as a signaling molecule. Antioxid Redox Signal 15(1):147–151. https://doi.org/10.1089/ars.2011.3968

Volpp BD, Nauseef WM, Clark RA (1988) Two cytosolic neutrophil oxidase components absent in autosomal chronic granulomatous disease. Science 242(4883):1295–1297. https://doi.org/10.1126/science.2848318

Wang D, De Deken X, Milenkovic M, Song Y, Pirson I, Dumont JE, Miot F (2005) Identification of a novel partner of duox: EFP1, a thioredoxin-related protein. J Biol Chem 280(4):3096–3103. https://doi.org/10.1074/jbc.M407709200

Wang X, Elksnis A, Wikstrom P, Walum E, Welsh N, Carlsson PO (2018) The novel NADPH oxidase 4 selective inhibitor GLX7013114 counteracts human islet cell death in vitro. PLoS One 13(9):e0204271. https://doi.org/10.1371/journal.pone.0204271

Wang FT, Hassan M, Ansari KH, Xu GL, Li XP, Fan YZ (2019) Upregulated NOX1 expression in gallbladder cancer-associated fibroblasts predicts a poor prognosis. Oncol Rep 42(4):1475–1486

Wientjes FB, Hsuan JJ, Totty NF, Segal AW (1993) p40phox, a third cytosolic component of the activation complex of the NADPH oxidase to contain src homology 3 domains. Biochem J 296 (Pt 3):557–561. https://doi.org/10.1042/bj2960557

Wilkinson-Berka JL, Deliyanti D, Rana I, Miller AG, Agrotis A, Armani R et al (2013) NADPH oxidase, NOX1, mediates vascular injury in ischemic retinopathy. Antioxid Redox Signal 20 (17):2726–2740. https://doi.org/10.1089/ars.2013.5357

Wilkinson-Berka JL, Deliyanti D, Rana I, Miller AG, Agrotis A, Armani R et al (2014) NADPH oxidase, NOX1, mediates vascular injury in ischemic retinopathy. Antioxid Redox Signal 20 (17):2726–2740. https://doi.org/10.1089/ars.2013.5357

Won SJ, Tang XN, Suh SW, Yenari MA, Swanson RA (2011) Hyperglycemia promotes tissue plasminogen activator-induced hemorrhage by increasing superoxide production. Ann Neurol 70(4):583–590. https://doi.org/10.1002/ana.22538

Woo HA, Yim SH, Shin DH, Kang D, Yu DY, Rhee SG (2010) Inactivation of peroxiredoxin I by phosphorylation allows localized H(2)O(2) accumulation for cell signaling. Cell 140 (4):517–528. https://doi.org/10.1016/j.cell.2010.01.009

Wu RF, Ma Z, Liu Z, Terada LS (2010) Nox4-derived H2O2 mediates endoplasmic reticulum signaling through local Ras activation. Mol Cell Biol 30(14):3553–3568. https://doi.org/10.1128/MCB.01445-09

Xia D, Halder B, Godoy C, Chakraborty A, Singla B, Thomas E et al (2019) NADPH oxidase 1 mediates caerulein-induced pancreatic fibrosis in chronic pancreatitis. Free Radic Biol Med 147:139–149. https://doi.org/10.1016/j.freeradbiomed.2019.11.034

Yin C, Li K, Yu Y, Huang H, Yu Y, Wang Z et al (2018) Genome-wide association study identifies loci and candidate genes for non-idiopathic pulmonary hypertension in Eastern Chinese Han population. BMC Pulm Med 18(1):158. https://doi.org/10.1186/s12890-018-0719-0

Yoshikawa Y, Ago T, Kuroda J, Wakisaka Y, Tachibana M, Komori M et al (2019) Nox4 promotes neural stem/precursor cell proliferation and neurogenesis in the hippocampus and restores memory function following trimethyltin-induced injury. Neuroscience 398:193–205. https://doi.org/10.1016/j.neuroscience.2018.11.046

Yu P, Han W, Villar VA, Yang Y, Lu Q, Lee H et al (2014) Unique role of NADPH oxidase 5 in oxidative stress in human renal proximal tubule cells. Redox Biol 2:570–579. https://doi.org/10.1016/j.redox.2014.01.020

Zeng SY, Yang L, Yan QJ, Gao L, Lu HQ, Yan PK (2019) Nox1/4 dual inhibitor GKT137831 attenuates hypertensive cardiac remodelling associating with the inhibition of ADAM17-dependent proinflammatory cytokines-induced signalling pathways in the rats with abdominal artery constriction. Biomed Pharmacother 109:1907–1914. https://doi.org/10.1016/j.biopha.2018.11.077

Zhang Y, Murugesan P, Huang K, Cai H (2019) NADPH oxidases and oxidase crosstalk in cardiovascular diseases: novel therapeutic targets. Nat Rev Cardiol 17(3):170–194. https://doi.org/10.1038/s41569-019-0260-8

Zhao QD, Viswanadhapalli S, Williams P, Shi Q, Tan C, Yi X et al (2015) NADPH oxidase 4 induces cardiac fibrosis and hypertrophy through activating Akt/mTOR and NFkappaB signaling pathways. Circulation 131(7):643–655. https://doi.org/10.1161/CIRCULATIONAHA.114.011079

Zhao T, Wang Y, Li Z, Xu X, Lei S, Huang L et al (2020) Associations of noise kurtosis, genetic variations in NOX3 and lifestyle factors with noise-induced hearing loss. Environ Health 19 (1):13. https://doi.org/10.1186/s12940-020-0566-3

Nitric Oxide Synthase Inhibitors into the Clinic at Last

Vu Thao-Vi Dao, Mahmoud H. Elbatreek, Thomas Fuchß, Ulrich Grädler, Harald H. H. W. Schmidt, Ajay M. Shah, Alan Wallace, and Richard Knowles

Contents

Vu Thao-Vi Dao and Mahmoud H. Elbatreek contributed equally to this work.

V. T.-V. Dao
Department of Psychiatry, Psychosomatic Medicine and Psychotherapy, University Hospital Frankfurt, Goethe University, Frankfurt am Main, Germany

M. H. Elbatreek
Department of Pharmacology and Personalised Medicine, MeHNS, FHML, Maastricht, The Netherlands

Department of Pharmacology and Toxicology, Faculty of Pharmacy, Zagazig University, Zagazig, Egypt

T. Fuchß · U. Grädler
Takeda GmbH (former Nycomed/Altana Pharma), Konstanz, Germany

H. H. H. W. Schmidt
Department of Pharmacology and Personalised Medicine, MeHNS, FHML, Maastricht, The Netherlands

A. M. Shah
King's College London British Heart Foundation Centre, School of Cardiovascular Medicine and Sciences, London, UK

A. Wallace
Health and Life Sciences, Coventry University, Coventry, UK

R. Knowles (✉)
Knowles Consulting Ltd., The Stevenage Bioscience Catalyst, Stevenage, UK

H. H. H. W. Schmidt et al. (eds.), *Reactive Oxygen Species*, Handbook of Experimental Pharmacology 264, https://doi.org/10.1007/164_2020_382

Abstract

The 1998 Nobel Prize in Medicine and Physiology for the discovery of nitric oxide, a nitrogen containing reactive oxygen species (also termed reactive nitrogen or reactive nitrogen/oxygen species) stirred great hopes. Clinical applications, however, have so far pertained exclusively to the downstream signaling of cGMP enhancing drugs such as phosphodiesterase inhibitors and soluble guanylate cyclase stimulators. All clinical attempts, so far, to inhibit NOS have failed even though preclinical models were strikingly positive and clinical biomarkers correlated perfectly. This rather casts doubt on our current way of target identification in drug discovery in general and our way of patient stratification based on correlating but not causal biomarkers or symptoms. The opposite, NO donors, nitrite and enhancing NO synthesis by *e*NOS/NOS3 recoupling in situations of NO deficiency, are rapidly declining in clinical relevance or hold promise but need yet to enter formal therapeutic guidelines, respectively. Nevertheless, NOS inhibition in situations of NO overproduction often jointly with enhanced superoxide (or hydrogen peroxide production) still holds promise, but most likely only in acute conditions such as neurotrauma (Stover et al., J Neurotrauma 31(19):1599–1606, 2014) and stroke (Kleinschnitz et al., J Cereb Blood Flow Metab 1508–1512, 2016; Casas et al., Proc Natl Acad Sci U S A 116 (14):7129–7136, 2019). Conversely, in chronic conditions, long-term inhibition of NOS might be too risky because of off-target effects on *e*NOS/NOS3 in particular for patients with cardiovascular risks or metabolic and renal diseases.

Graphical Abstract

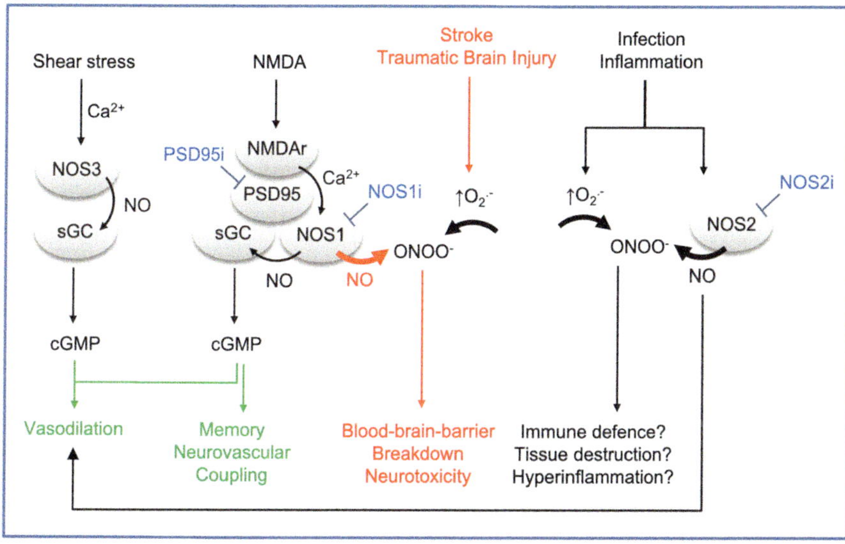

Nitric oxide synthases (NOS) and their role in health (green) and disease (red). Only neuronal/type 1 NOS (NOS1) has a high degree of clinical validation and is in late stage development for traumatic brain injury, followed by a phase II safety/efficacy trial in ischemic stroke. The pathophysiology of NOS1 (Kleinschnitz et al., J Cereb Blood Flow Metab 1508–1512, 2016) is likely to be related to parallel superoxide or hydrogen peroxide formation (Kleinschnitz et al., J Cereb Blood Flow Metab 1508–1512, 2016; Casas et al., Proc Natl Acad Sci U S A 114(46):12315–12320, 2017; Casas et al., Proc Natl Acad Sci U S A 116(14):7129–7136, 2019) leading to peroxynitrite and protein nitration, etc. Endothelial/type 3 NOS (NOS3) is considered protective only and its inhibition should be avoided. The preclinical evidence for a role of high-output inducible/type 2 NOS (NOS2) isoform in sepsis, asthma, rheumatic arthritis, etc. was high, but all clinical development trials in these indications were neutral despite target engagement being validated. This casts doubt on the role of NOS2 in humans in health and disease (hence the neutral, black coloring).

Keywords

Nitric oxide · Nitric oxide synthase · NOS · NOS inhibitor · NOS isoforms

Abbreviations

ADMA	Asymmetric dimethyl arginine
ADME	Absorption, distribution, metabolism, and excretion
CaM	Calmodulin
cHL	Classical Hodgkin lymphoma
CLL	Chronic lymphocytic leukemia
DAMP	Damage-associated molecular pattern
*e*NOS/NOS3	Endothelial nitric oxide synthase
GLP-2	Glucagon-like peptide-2
H4Bip	Tetrahydrobiopterin
HNSCC	Neck squamous cell carcinoma
Hsp	Heat shock protein
IDH	Intradialytic hypotension
*i*NOS/NOS2	Inducible nitric oxide synthase
JSN	Joint space narrowing
L-NIL	L-*N* iminoethyl lysine
L-NMMA	N^G-monomethyl-L-arginine
NF-κB	Nuclear factor kappa B
NMDA	*N*-methyl-D-aspartate
*n*NOS/NOS1	Neuronal nitric oxide synthase
PAMP	Pathogen-associated molecular pattern
PD-1	Programmed death-1
PSD95	Post synaptic domain
sGC	Soluble guanylate cyclase
SMTC	*S*-methyl-L-thiocitrulline

1 NOS Isoforms, Regulation and Dysregulation

Nitric oxide (NO) synthases (NOS) are homodimeric NADPH binding flavo-heme proteins additionally regulated by the redox-sensitive cofactor tetrahydrobiopterin (H4Bip), calmodulin and several other modulatory interactions (Nedvetsky et al. 2002) to convert L-arginine (Schmidt et al. 1988; Nedvetsky et al. 2002) to NO. Three isoforms exist, originally named according to their first observed cellular/tissue localization or expressional regulation, i.e., neuronal, inducible, and endothelial (i.e., NOS1, NOS2, and NOS3) (Schmidt et al. 1991; Förstermann et al. 1992; Chakrabarti et al. 2012; Liu et al. 2012; Caviedes et al. 2017).

All three NOS isoforms generate NO by conversion of L-arginine to L-citrulline by a stoichiometric five electron oxidation utilizing molecular oxygen (O_2) and reduced nicotinamide adenine dinucleotide phosphate (NADPH) as co-substrates and several cofactors including 6R-tetrahydrobiopetrin (BH4), flavin adenine dinucleotide (FAD), and flavin mononucleotide (FMN) (Bredt et al. 1991; Knowles and Moncada 1994).

The NOS monomer structure has two domains, the NOS heme containing amino-terminal and the carboxy-terminal domain harboring binding sites for FAD, FMN, and NADPH. Both domains are connected by linking to an calcium-binding regulatory protein, i.e., calmodulin (CaM) (Smith et al. 2013) enhancing electron flux within the reductase domain and thus is essential for O_2 to bind to NOS heme to start the NO synthesis. Ca^{2+}-dependent CaM binding occurs in NOS1 and NOS3 following an increase in intracellular Ca^{2+}concentrations, while NOS2 binds CaM at low intracellular Ca^{2+}level thus independent of elevation of intracellular Ca^{2+}level (Cho et al. 1992).

A key step of NOS maturation which has been shown for NOS1 and 2 involves the insertion of heme by associating to different heat shock proteins (Hsp) such as Hsp 90 and Hsp 70 (Ghosh et al. 2011; Peng et al. 2012). Then, in the presence of NOS heme, the monomers can form dimers by coupling their ferric hemes and thus become fully active to generate NO. This takes place in the N-terminal oxygenase site that catalyzes first hydroxylation then oxidation of L-arginine to L-citrulline and NO, initiated by O2 binding to reduced ferric heme that utilized electrons provided by the reductase C-terminal domain of the opposite monomer via NADPH to FAD to FMN (Stuehr et al. 2001; Forstermann and Sessa 2012; Ramasamy et al. 2014). Intermediate formed heme-dioxy species are then reduced by NOS coupled BH4 resulting in reactive heme-oxy species that react either with L-arginine or *N*-hydroxy-L-arginine (Masters et al. 1996; Stuehr and Haque 2019). Hence, oxidation of BH4 results in BH3 radical or radical cation that can be reversed by NOS itself transferring an electron (Crabtree and Channon 2011) or by ascorbic acid (Kuzkaya et al. 2003). In addition, all NOS enzymes contain a zinc-tetrathiolate binding two cysteine residues provided by each monomer Masters et al. 1996; Raman et al. 1998; Li et al. 1999; Hemmens et al. 2000), which is catalytically inactive at the dimer interface (Forstermann and Sessa 2012) but stabilizing the active homodimer and appear to promote BH4 binding (Chreifi et al. 2014). In addition, the zinc-tetrathiolate cluster can be targeted by selective *S*-nitrosation (Wynia-Smith and Smith 2017).

1.1 Neuronal NOS/Type 1 NOS (nNOS/NOS1)

NOS1 is abundantly expressed in neurons of the central and peripheral nervous systems (Dudzinski et al. 2006), e.g., in hypothalamic supraoptic nucleus and the paraventricular nucleus, in some parts of rat glial cells (Korzhevskiĭ et al. 2007), in rat astrocytes and the adventitia of rat brain blood vessels (Yuan et al. 2004), but is also found in skeletal muscle, pulmonary epithelium, the gastrointestinal system, and the genitourinary system (Zhou and Zhu 2009).

Amongst the three NOS isoforms only NOS1 encodes a unique regulatory protein–protein interaction domain, i.e., PSD/Disc-Large/ZO-1 (PDZ) at the N-terminus, relevant for changes of subcellular localization as it can interact directly with other proteins containing PDZ domains or adapter proteins (Courtney et al. 2014; Candemir et al. 2016). Further, regulatory aspects leading to activation of constitutively expressed NOS1 are besides CaM-binding, the AKT phosphorylation site at the C-terminus, that helps CaM to bind (Gantner et al. 2020). Another allosteric activator that is shown to activate NOS producing NO is Hsp90 which associates to *n*NOS, thus enhancing calmodulin binding to *n*NOS (Bender et al. 1999; Song et al. 2001). In addition, there is an autoinhibition site close to the CaM binding domain controlling NOS activity (Salerno et al. 1997).

NOS1-derived NO and its signaling has been implicated in not only antegrade but also retrograde signaling from the postsynaptic neuron, thus contributing to long-term potentiation (Böhme et al. 1993; O'Dell et al. 1994), fear conditioning (Ota et al. 2010), and neurogenesis (Chong et al. 2018). Furthermore, NO signaling is involved in mediating excitotoxicity in neurons driven by excessive glutamate-dependent overstimulation of NMDA receptors following Ca^{2+}overload in the cell, and consequently activation of Ca^{2+}-sensitive enzymes such as NOS1 (Sattler and Tymianski 2000; Chong et al. 2018) leading to cell death.

Mechanistically, neuronal NOS1 signaling requires adapter proteins harboring a PDZ motif such as syntrophin (Aquilano et al. 2014), PSD95 or PSD93 (Brenman et al. 1996) to anchor NOS1 and thereby target NOS1 to the proximity of the NMDA receptors at the postsynaptic membranes. Another adapter protein is NOS1AP (former Capon) that binds to the PDZ motif of NOS1 via its C-terminal domain (Jaffrey et al. 2002).

Besides the brain, NOS1 is involved in other physiological regulations such as skeletal muscle metabolism in response to exercise training (Percival 2011). This is regulated by different splice variants of the highly conserved NOS1 consisting of 29 exons and about 240 kb. These are an almost full length *n*NOSα, *n*NOSμ with 34 additional amino acid insertions, the PDZ lacking *n*NOSβ and *n*NOSγ and finally *n*NOS2 (Gantner et al. 2020).

The enzymatic activity of the splice variants may differ depending on their subcellular localizations but their activation also requires calcium and phosphorylation by PI3K/Akt (Gantner et al. 2020). Of those *n*NOSμ in skeletal muscle is bound to the dystrophin glycoprotein complex, at the sarcolemma and has been shown to contribute to better muscle blood flow, resisting fatigue by endurance training (Percival 2011). Furthermore, NOS1 has been implicated in inflammatory response (Baig et al. 2015) cardiac and smooth muscle physiology involving to cardiac protection and vascular tone (Seddon et al. 2008, 2009; Shabeeh et al. 2017).

1.2 Inducible NOS/Type 2 NOS (iNOS/NOS2)

Mammalian inducible NOS2 is a 131 kDa protein composed of 1,153 amino acids that lacks the PDZ domain and in contrast to NOS1&3 also the autoinhibition site and is not constantly expressed but only by induction of the cell. Depending on cell type and species strong stimulants of transcription of NOS2 expression are tumor necrosis factor (TNF), interleukin (IL-1β), interferon (IFN-γ), and lipopolysaccharide (LPS) exerting synergistic effects when combined (Cinelli et al. 2020). However, constitutively expressed NOS2 has been found tissue specific in the human colonic and lung epithelium as well as in primate lungs and could be a response to the local microbiota (Mattila and Thomas 2014).

Accumulated NOS2-derived NO plays important roles in innate and adaptive immunity such as regulating T-cells, B-cells, and myeloid-derived suppressor cells (Bogdan 2015) and helping macrophages to defend against pathogens. The latter has been best established so far (Weinberg et al. 1995; MacMicking et al. 1997; Fang 2004; Nathan 2006).

Once NO is generated, it rapidly reacts with superoxide to form radical peroxynitrite (ONOO−) that can cause, damage to DNA (Pacher et al. 2007), modifications of proteins (Casas et al. 2015; Dao et al. 2015; Bartesaghi and Radi 2018) and reactions with unsaturated lipids (Jones 2012). Thus, also host tissue can be targeted. Therefore, regulation of NOS2 gene expression is strictly bound to transcriptional processes (Scheschowitsch et al. 2015).

Briefly, pathogen (PAMP) and damage (DAMP) associated molecular patterns bind to pattern recognition receptors (Amarante-Mendes et al. 2018), and proinflammatory cytokines such as TNF-α and IL-1 bind to the cell surface, e.g., one popular PAMP, LPS, binds to toll like receptor 4 in macrophages (Hume et al. 2001), starting the signaling cascade by activation of transcription factors, including nuclear factor κB (NF-κB) and (STAT-1a) while IFN-γ activates the JAK/STAT-1α pathway to induce NOS2 mRNA expression (Dell'Albani et al. 2001; Ganster et al. 2001).

Control of NO output is also regulated by autoregulation of its own expression in a feedback manner (Ganster et al. 2001). For example, posttranslational regulations involving *S*-nitrosylation of NF-κB binding partners lead to a stop of mRNA transcription (Kelleher et al. 2007). Further, *S*-nitrosation at the Zn^{2+} tetrathiolate cluster discards the Zn^{2+} leading to destabilization of the active homodimer resulting in dissociation to the inactive monomeric form (Wynia-Smith and Smith 2017). In contrast, inhibition of phosphorylation on phosphotyrosine residues leads to increased NOS2 activity (Pan et al. 1996).

1.3 Endothelial NOS/Type 3 NOS (eNOS/NOS3)

NOS3 is constitutively expressed mostly in endothelial cells but also in cardiac myocytes, platelets (Forstermann and Sessa 2012), and macrophages (Mattila and Thomas 2014). NOS3-derived NO exerts various physiological functions such as vasodilation through its receptor soluble guanylyl cyclase leading to increasing cyclic GMP in smooth muscle cells (Förstermann et al. 1986), inhibition of platelet aggregation (Förstermann et al. 1986; Alheid et al. 1987), platelet and leukocytes adhesion to the vascular wall (Kubes et al. 1991), vascular remodeling, anti-inflammatory effects (Ahluwalia et al. 2004), and angiogenesis (Wei et al. 2020).

Its structure has important features, including (1) an autoinhibitory loop within the FMN binding domain where CaM can be removed in the absence of Ca^{2+} to stop catalytic reaction, (2) the loss of PDZ domain, (3) an AKT phosphorylation site at the C-terminal, and (4) an additional acylation site (palmitoylation and myristoylation) in the oxygenase domain.

Another regulatory element that differs from the other NOS isoform is a shorter and less active hinge, which is responsible for binding FMN to the reductase domain. Notably, mammalian NOS3 has the weakest activity amongst the NOS family (Haque et al. 2007, 2012).

NOS3 can be found in sarcolemmal caveolae where it is bound by posttranslational myristoylation and palmitoylation to caveolin-1 which tonically inhibits NOS3 activity. A rise of intracellular Ca^{2+} level induces CaM binding and interaction of NOS3 with heat shock protein 90 results in disruption of the NOS3-caveolin-1 heterodimer complex leading to NOS3 activation (Averna et al. 2008).

However, NOS3 activation can also be regulated by mechanical changes, i.e., fluid shear stress, leading to NOS3 up-regulation in endothelial cell and rodents (Nishida et al. 1992; Sessa et al. 1994; Awolesi et al. 1995; Fukai et al. 2000; Dao et al. 2016). The mode of shear stress has different effects on NOS3 regulation. While acute changes affect immediate vascular tone, chronic shear stress induces gene expression and remodeling of blood vessels (Garcia and Sessa 2019). The mechanism underlying shear stress involves not only intracellular calcium rise but depends directly on phosphorylation by serine/threonine (Ser/Thr) protein kinase Akt/PKA (Dimmeler et al. 1999). In this regard, studies in mutant *AKT1* mice have shown that AKT1 is an important NOS3 kinase in vivo that phosphorylates Ser1176 (human Ser1177) (Schleicher et al. 2009).

Besides fluid shear stress, other stimulants can regulate NOS3 activation such as vascular endothelial growth factor induced phosphorylation by AKT1 at Ser/Thr site, bradykinin-induced phosphorylation at Ser1177 by Ca^{2+}/calmodulin dependent protein kinase II, insulin-mediated Akt1 and AMP-activated protein kinase activation (Forstermann and Sessa 2012) or hydrogen peroxide (Drummond et al. 2000; Thomas et al. 2002; Searles 2006; Dao et al. 2011).

Several phosphorylation sites including serine, threonine, and tyrosine residues of NOS3 have been discovered such as Y81, S615, S633, and S1177 (equivalent to Y83, S617, S635, and S1179 of bovine NOS3) responsible for stimulation (Fulton et al. 2005; Fulton 2016), while S114, T495, and Y657 leads to inhibition of NOS3 activity (Loot et al. 2009; Fulton 2016). Briefly, phosphorylation of the Ser1177 increases, while constitutively phosphorylated Thr495 in endothelial cells appears to interfere with CaM binding (Heiss and Dirsch 2014). This plays a role in *e*NOS uncoupling (Lin et al. 2003), and decreases enzyme activity.

Other posttranslational modifications are also described to change NOS3 regulation, including *S*-nitrosylation (at C94 and C98) (Erwin et al. 2005) leading to reduced activity, while acetylation (K609, S765, and S771) increases its activity (Jung et al. 2010). Glutathionylation in the C-terminal reductase domain (C689 and C908) uncouples NOS3 thus forming superoxide anion (Chen et al. 2010).

At transcriptional level regulation of NOS3 mRNA expression is decreased by DNA methylation of the promoter thus reducing Sp1, Sp3, and Ets1 transcription factor binding (Chan et al. 2004) and controlled by histone modification at NOS3 promoter (Fish et al. 2005, 2010). In addition, NOS3 mRNA expression can be up-regulated by long noncoding RNAs (lncRNAs) in endothelial cells induced via transcription factor KLF2 (Man et al. 2018).

2 Dysregulation of NOS Isoforms

Qualitatively, NOS can exist in three different functional states: (1) normal state that produces physiological levels of NO which signals mainly via its receptor, soluble guanylate cyclase (sGC) (Schmidt et al. 1994), (2) uncoupled state that produces superoxide rather than NO resulting in endothelial dysfunction and cardiovascular diseases (Li et al. 2015), and (3) hyperactive state that produces excessive NO leading to cellular and tissue injury (Kleinschnitz et al. 2016), e.g., in stroke and myocardial infarction (Fig. 1).

The uncoupled state is induced upon depletion of the NOS substrate, L-arginine, or oxidation of its cofactor, tetrahydrobiopterin (H4Bip) (Schmidt et al. 1992;

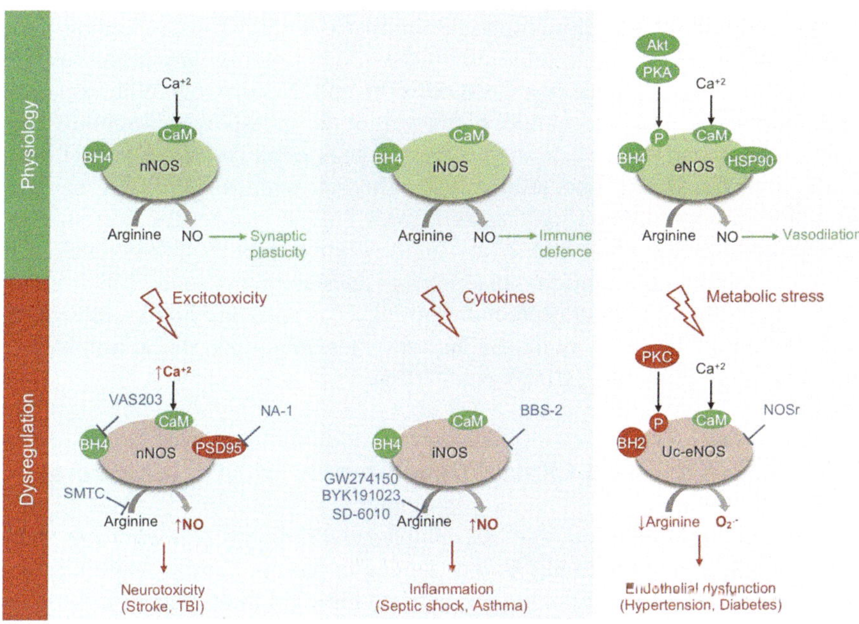

Fig. 1 NOS isoforms and their regulation and dysregulation. In physiology, the three NOS isoforms, *n*NOS, *i*NOS, and *e*NOS, bind to the substrate, arginine, the cofactor, BH4, and calmodulin (CaM) to produce NO. Ca^{+2} is required for CaM binding in *n*NOS and *e*NOS. The latter is phosphorylated by PKA/Akt. The NO produced by *n*NOS, *i*NOS, and *e*NOS performs several biological functions including synaptic plasticity, immune defense, and vasodilation, respectively. Under disease conditions, dysregulation of NOS enzymes takes place. In excitotoxicity, e.g., in stroke and traumatic brain injury (TBI), Ca^{+2} is increased leading to binding of *n*NOS to postsynaptic density protein 95 (PSD95) and increased NO production resulting in neurotoxicity. Under inflammatory conditions, e.g., in asthma and septic shock, cytokines activate *i*NOS to produce high levels of NO that further aggravates inflammation and induces cytotoxicity. Metabolic stress causes uncoupling of *e*NOS by reducing arginine, oxidation of BH4 to BH2 and PKC-induced phosphorylation resulting in formation of superoxide ($O_2^{\cdot}$-) rather than NO and endothelial dysfunction. *n*NOS and *i*NOS inhibitors and *e*NOS recoupling agents (NOSr) are written in blue

Bömmel et al. 1998), leading to NOS monomerization (Reif et al. 1999) and inhibition of the enzyme activity (Kotsonis et al. 1999; Reif et al. 1999). However, recently NOS monomerization has been shown to be irrelevant for uncoupling (Gebhart et al. 2019). NOS uncoupling can also be induced by the accumulation of methylated arginine analogs, in particular, asymmetric dimethyl arginine (ADMA), which is a competitive inhibitor of NOS (Antoniades et al. 2009).

The hyperactive state of NOS is induced in many conditions that involve inflammation and hypoxia leading to formation of high and non-physiological levels of NO that cause harmful effects, most likely in a cGMP-independent manner. Several mechanisms have been proposed that explain the high NO-induced cytotoxicity. These include protein *S*-nitrosylation (Shahani and Sawa 2012; Dao et al. 2020), formation of peroxynitrite via interaction with superoxide (Mendes et al. 2002; Pacher et al. 2007), activation of inflammatory signaling pathways, e.g., NF-κB (Mendes et al. 2002), decreased expression and activity of sGC, and formation of the NO-insensitive *apo*-sGC (Dao et al. 2020). Two NOS isoforms have been shown to be activated in disease conditions, *n*NOS/NOS1 and *i*NOS/NOS2. With respect to the former, it is activated to produce high levels of NO in response to excitotoxicity, which is an important event in the pathophysiology of stroke and traumatic brain injury (Ito et al. 2010; Luo et al. 2019). The activation *n*NOS/NOS1 in these conditions is dependent on its interaction with PSD-95 and results in neurotoxic effects (Zhou et al. 2010; Luo et al. 2019). With respect to *i*NOS/NOS2, it is activated mainly in inflammatory conditions in response to cytokines. The inappropriately high NO concentration produced by *i*NOS can result in cytotoxic effects and is associated with a variety of human diseases, including septic shock, asthma, and cancer (Zamora et al. 2000; Cinelli et al. 2020).

3 Discovery and Clinical Development of NOS Inhibitors

A large number of nitric oxide synthase inhibitors with various degrees of selectivity for NOS isoenzymes are available and claimed in patents Table 1. However, few have reached the clinical trial stage Table 2, and not all of these meet the criteria for pharmaceutical developability.

3.1 Which Site to Target?

NOS are proteins with a high number of binding and regulatory sites (Nedvetsky et al. 2002). The earliest attempts to develop NOS inhibitors were based on the physiological substrate, L-arginine (Alderton et al. 2001). Still to date, most compounds target that site. Later, cofactor and regulator sites were added, of which only those targeting BH4 (Bömmel et al. 1998; Fröhlich et al. 1999; Reif et al. 1999; Kotsonis et al. 2001; Pantke et al. 2001; Matter et al. 2002, 2005) and PSD-95 interaction (Cui et al. 2007) made it into clinical testing. Other sites such as calmodulin (ketoconazole) were never tested for NOS inhibition in humans.

Table 1 Overview of clinically applied and developed NOS inhibitors

Compound	Binding site	Reversibility
Unspecific		
L-NMMA	Arg	Reversible
L-NAME	Arg	Reversible
VAS203	H_4Bip	Reversible
2-Iminobiotin	Arg	Reversible
MTR 104	Arg	Reversible
nNOS/NOS1 specific		
S-methyl-L-thiocitrulline	Arg	Reversible
Tat-NR2B9c (NA-1)	PSD95	Reversible
ARL17477	Arg	Reversible
NXN-462 and NXN-188	Arg	Reversible
*i*NOS/NOS2 specific		
GW274150 and GW273629	Arg	NADPH dependent inactivation
Cindunistat (SD-6010)	Arg	Irreversible
BYK191023	Arg	Reversible/irreversible
L-NILTA (prodrug of L-NIL)	Arg	Reversible
BBS-2	Heme-containing *i*NOS monomer	Reversible

3.2 Nonspecific Inhibitors

L-NMMA (N^G-monomethyl-L-arginine) is approximately equipotent on all three isoforms of NOS (Alderton et al. 2001). The acetate salt is also known as tilarginine; the less hygroscopic HCl salt is also known as 546C88. L-NAME is an inactive prodrug of N^G-nitro-L-arginine, L-NOARG (Pfeiffer et al. 1996). This is modestly selective for *n*NOS and *e*NOS versus *i*NOS (Alderton et al. 2001). Other-nitroarginine esters include L-N^G-benzylarginine, L-N^G-aminoarginine, and iminoethylornithine. Moreover, specific inducible NOS inhibitors, i.e., acetamidine derivatives, have been synthesized and evaluated showing high isoform-specificity with submicromolar concentrations (Fantacuzzi et al. 2016). The currently most advanced NOS inhibitor under development is Vasopharm's ronopterin (VAS203; 4-amino-tetrahydrobiopterine) (Ott et al. 2019; Tegtmeier et al. 2020a, b). Ronopterin is a potent pterin-based inhibitor that competes with exogenous BH4 (Werner and Schmidt 2000; Schinzel and Tegtmeier 2017). This drug has shown promising results in patients with traumatic brain injury (see below). Vasopharm holds several patent applications on this compound and related pteridines (WO/2005/037286 and WO/2004/084906). Another nonspecific NOS1/NOS2 inhibitor, 2-iminobiotin, originally patented by the Amsterdam University Medical Center for the treatment and prevention of perinatal asphyxia (WO/2001/074351), is being developed by the Dutch company Neurophyxia B.V. The company has published an international patent application claiming the compound for cerebral

Table 2 Phase II and higher clinical trials involving NOS inhibitors

Drug name(s)	Originator company, *active companies	Therapy area, *active indications	Target-based actions	Highest status
L-NMMA.HCl (546C88)	Glaxo Group Ltd.	Infection; Neurology/psychiatric *Stroke; septic shock	Nonspecific NOS inhibitor	Discontinued (phase 3)
L-NMMA	Toronto University Health Network	GLP-2 mediated intestinal lipoprotein release	Nonspecific NOS inhibitor	Ongoing (phase 3)
Gingivex (GED; Inotek; Guanidinoethyl disulfide)	Rocket Pharmaceuticals Inc.	Gastrointestinal; inflammatory; ocular; endocrine/metabolic	*i*NOS/NOS2	Discontinued (phase 2)
GW274150	Glaxo Group Ltd.	Neurology/psychiatric; inflammatory; immune; gastrointestinal; respiratory	*i*NOS/NOS2	Discontinued (phase 2)
ONO-1714	Ono Pharmaceutical Co Ltd.	Cardiovascular; infection *Sepsis; hypotension	*i*NOS *e*NOS	Discontinued (phase 2)
Cindunistat hydrochloride maleate (PHA-728669F; SD-6010)	Pfizer Inc.	Inflammatory	*i*NOS/NOS2	Discontinued (phase 3)
Pimagedine (aminoguanidine)	Rockefeller University	Dermatologic; endocrine/metabolic; gastrointestinal; genitourinary/sexual function	Nitric oxide synthesis inhibitor	Discontinued (phase 3)
S-Ethylisothiourea diethylphosphate (MTR 104)	Meditor Pharmaceuticals Ltd., *TrioxBio Inc.	Neurology/psychiatric, *migraine	NOS inhibitor	Launched
		Cardiovascular, *hypotension	NOS inhibitor	Launched
		Neurology/psychiatric, *cluster headache	NOS inhibitor	Phase 2
		Cardiovascular, *hypotension	NOS inhibitor	Phase 2

(continued)

Table 2 (continued)

Drug name(s)	Originator company, *active companies	Therapy area, *active indications	Target-based actions	Highest status
XQ-1H (Ginkgo biloba lactone B mesylate)	Jiangsu Carephar pharmaceutical Co Ltd.	Neurology/ psychiatric, *brain ischemia	*i*NOS/NOS2; platelet activating factor receptor agonist	Phase 1
OsteoDex (ODX; dextran-guanidine-bisphosphonate conjugate)	Dextech medical AB	Cancer; musculoskeletal, *bone metastases; hormone refractory prostate cancer	NOS inhibitor	Phase 2
NXN-462	NeurAxon Inc.	Neurology/ psychiatric, *movement disorder; Parkinson's disease	*n*NOS/NOS1	Phase 2
2-iminobiotin	Universiteit Utrecht	Other/ miscellaneous, *asphyxia	*i*NOS/NOS2; *n*NOS/NOS1	Phase 2
Ronopterin (VAS203)	Vasopharm GmbH	Neurology/ psychiatric, *traumatic brain injury	NOS inhibitor	Phase 3
Nerinetide (NA-1, tat-NR2B9c)	NoNO Inc.	Neurology/ psychiatric, *brain ischemia; stroke; traumatic brain injury	Discs large homolog 4 inhibitor; NMDA receptor antagonist; *n*NOS/NOS1	Phase 3

hypoxia-ischemia and reperfusion injury (WO/2017/105237), and another one claiming pharmaceutical formulations (WO/2011/149349). Finally, MTR 104, a low molecular weight isothiourea derivative (*S*-ethylisothiouronium diethylphosphate) and nonspecific NOS inhibitor (Garvey et al. 1994), has received orphan drug designation from the FDA. Meditor Pharmaceuticals has begun development through TrioxBio.

3.3 *n*NOS/NOS1 Inhibitors

Although much of the focus on creating selective NO synthase inhibitors has focused on *i*NOS (NOS2) and certainly steered well clear of NOS3 (*e*NOS), a few *n*NOS selective inhibitor programs have emerged and one of the first of these came from Fisons' research laboratories in Rochester, New York. The best known of these compounds is ARL17477 (alternatively known as AR-R17477 or initially FPL17477). This compound represents a series of heterocyclic substituted amidines, the full synthesis of which has only been covered in the patent literature (WO 95/05363). Of relevance in this respect are some notable differences in the amino acid sequence between NOS isoforms around the substrate or inhibitor binding site. Some of these are highlighted in a paper by Fedorov and colleagues (Fedorov et al. 2004) who present crystal structures of AR-R17477 in the different NOS isoforms to explain why the compound selectively inhibits *n*NOS. A potentially important observation in this respect is a key difference in sequence that is observed within the active site. The aspartate residue (rat *n*NOS; N597) becomes an asparagine in *e*NOS (bovine *e*NOS; D368). Li et al. demonstrated the importance of this residue in providing selectivity between isoforms for a series of L-nitroarginine based dipeptide inhibitors. However, these papers used bovine *n*NOS—human would be the same; but in mouse and rat (the species most commonly used in pharmacological studies), the aspartate residue is conserved (Li et al. 2005). Therefore, selectivity may be seriously compromised in these experiments. Site directed mutagenesis showed that switching these residues did indeed alter selectivity profiles (Li et al. 2005). There is a distinct possibility that failing to recognize this structural feature may have compromised some in vivo pharmacological or safety studies, blocking further progression of otherwise interesting, selective NOS inhibitors.

The AR-R series of compounds failed to produce clinical candidates for AstraZeneca, but the series was adopted elsewhere, most notable by Neuraxon in Canada. Although initially (Annedi et al. 2011, 2012) Neuraxon presented analogs of the thiophene amidines such as AR-R17477 and particularly AR-R17338 (Reif et al. 2000) which led to novel inhibitors with good selectivity toward *n*NOS, one of these, NXN-323, showed some efficacy in animal models of allodynia (Felice et al. 2010). A further compound, NXN-462, is reported to have entered trial for post-herpetic pain; to date, no results from the study have been released. Subsequently, the group moved on to produce dual functionality ligands including molecules that combine NOS inhibition with μ-opiate agonism or noradrenaline re-uptake inhibition, but notably NXN-188 a mixed *n*NOS and 5-HT$_{1B/D}$ agonist. Whenever one sees a mixed function drug it is fair to question whether one or other pharmacological activity dominates. NXN-188 is said to be a selective *n*NOS inhibitor with potency similar to L-NMMA and to have similar 5-HT potency to Sumatriptan. This potential drug prevented CGRP-release from preparations of several migraine-relevant brain areas (dura mater, trigeminal nucleus caudalis, and trigeminal ganglion (Bhatt et al. 2013). After Phase I trials of the drug showed suitable pharmacokinetics and that it was well tolerated in both single and multiple dose protocols (Vaughan et al. 2010), the compound entered a single center, double-blind, randomized cross-over study in patients suffering migraine with aura (Hougaard et al. 2013). While the results of this study were seen as encouraging, with a reduction in patients reporting headache, the study's primary endpoints were not achieved as the trial suffered a high drop-out rate and only a small sample of patients completed the placebo-controlled cross-over study. Consequently, the data failed to achieve statistical significance. In 2015, this drug was licensed to Knight Therapeutics (Canada) and a further Phase II trial is reported to be ongoing (as of June 2015: source adisinsight.springer.com).

S-Methyl-L-thiocitrulline (SMTC) is an amino acid derivative that is selective for *n*NOS/NOS1. Lack of patentability has so far prevented further development of this compound. However, it has been employed in a number of investigational human studies that studied the effects of either local infusion in different vascular beds or systemic infusion in healthy volunteers. However, stability issues and (more importantly) also lack of patentability prevented its further development.

In excitotoxicity, the postsynaptic density protein PSD95 recruits the calcium-dependent *n*NOS to NMDA receptor channels leading to neurotoxic effects (Cui et al. 2007; Kleinschnitz et al. 2016). Therefore, inhibition of PSD-95/NMDA interactions has been suggested as a potential therapeutic approach for neurotoxicity, e.g., in stroke and traumatic brain injury (Cui et al. 2007). Tat-NR2B9c (NA-1 or nerinetide), a synthetic peptide and a PSD-95 inhibitor, has been tested clinically for stroke and preliminary results seem promising (Bruder 2012; Hill et al. 2012; Matsumoto 2013) (see below). ZL006 is also a small molecule drug that blocks the PSD95/*n*NOS interaction and has been tested only preclinically (Zhou et al. 2010). However, its binding to the PSD95/*n*NOS has been doubted (Bach et al. 2015).

3.4 iNOS/NOS2 Inhibitors

There is considerable evidence from animal models for a potential pathological role of excessive NO production in numerous chronic inflammatory diseases and for the beneficial effects of treatment with *i*NOS inhibitors (Cheshire 2001; Tinker and Wallace 2006). However, much of this data has been derived using compounds that are far from optimal with regard to potency and specificity. Considerable effort has been directed at discovering truly selective inhibitors of this isoform that will prevent over-production of NO while maintaining the basal formation of NO from constitutive NOS that is required for normal physiological function. The analogs of arginine have been widely used as inhibitors of NO synthases, with considerable success although they lack some of the "drug-like" properties sought by pharmaceutical research programs. The simplest analogs of arginine are highly hydrophilic, and so incapable of readily diffusing across biological membranes; hence, they may also rely upon cationic amino acid transporters to enter cells (Baydoun et al. 2006). Consequently, many companies have tried to design inhibitors that move away from these pharmacophores.

Compounds unrelated to arginine can also inhibit NO synthase. Often these are fairly simple, small compounds, such as 2-aminopyridines; featuring an aromatized amidine that can mimic the binding of the basic guanidino side chain in arginine to an active site glutamyl residue. Aminopyridines offer scope to design and synthesize novel inhibitors, with potential improvements in "drug-like" properties. Connolly et al. at AstraZeneca reported on a series of analogs of 2-amino-4-methyl pyridine leading to a potent (IC_{50} = 71 nM), selective *i*NOS inhibitor, AR-C133057 (Connolly et al. 2004). In crystal structures of the ligand bound to the active site of *i*NOS the compound was observed to have adopted a flip in binding of the pyridine ring in order to accommodate an *N*-(1-acyl-4-piperidinyl) group that could be further elaborated to derive the series of *i*NOS inhibitors. Unfortunately, these aminopyridine analogs often exhibited poor pharmacokinetics with low volumes of distribution and weak in vivo activity.

Probably the best known *i*NOS inhibitor to emerge from AstraZeneca was AR-C102222. This compound is a potent inhibitor of human *i*NOS (IC_{50} 35 nM) and also exquisitely selective (>1,000-fold against *e*NOS). The inhibitor arose through studies focusing on aminopyridine analogs intended to exhibit improved drug-like properties. Two studies (Beaton et al. 2001a, b) identified a pair of series of bicyclic amidines: 3,4-dihydro-1-isoquinolinamines and closely related thienopyridines. Some of these molecules exhibited reasonable potency against *i*NOS, with a range of selectivity. The breakthrough came with the introduction of a nitrogen atom to create quinazolinamine inhibitors followed by a limited parallel synthesis approach that identified spirocyclic dihydroquinazoline molecules as *i*NOS inhibitors. This modification removed a stereochemical center, making the molecules rather simpler to work with whilst simultaneously opening scope for further chemical elaboration. Further, parallel synthetic studies extended this substitution and identified some highly selective compounds, including AR-C102222 (Tinker et al. 2003). Surprisingly, it achieved much of its selectivity through a cascade of interactions involving conserved residues close to the active site. Crystal structures of the *i*NOS oxygenase domain with AR-C102222, or other similar ligands, bound in the active site show displacement of a glutamine residue (Gln257 in mouse *i*NOS). This conformation can only be achieved by further movements of residues beyond the immediate region of the active site (Garcin et al. 2008) and such movements are impaired in *n*NOS and largely blocked in *e*NOS by more bulky amino acid side chains in these more distant positions. The lead spirocyclic compound maintained reasonable potency against *i*NOS in cell-based assays and also offered good oral bioavailability.

There are very few NO synthase inhibitors that bind in the enzyme's active site, but do not include an isostere of the guanidinium present in arginine to bind to the crucial glutamic acid residue. However, a few examples of inhibitors that lack this functionality and therefore do not interact directly with this acidic side chain do exist. 7-nitroindazole and chlorzoxazone actually displace the glutamate carboxylic acid in order to bind, locating in the active site through a π–stacking interaction with the heme-porphyrin ring (Rosenfeld et al. 2002a). Indeed, the induced fit afforded by this movement was shown to compromise the binding affinity of these compounds.

Building upon this observation Cheshire et al. presented a new series of selective, non-amidine *i*NOS inhibitors that made a similar interaction with the heme, but could also access the region close to Gln257 that conferred selectivity to compounds such as AR-C102222 (Cheshire et al. 2011).

NOS inhibition by competition with L-arginine comes with potential difficulties. The L-arginine dependence of NO synthase activity has been reported in several studies. Typical values for K_m range between 1 and 10 μM (Bredt and Snyder 1990; Sherman et al. 1993) with similar potency reported in direct measurements of arginine binding (Berka et al. 1996). Inhibitors must compete with the relatively high concentrations of this substrate present in cells. Intracellular levels of arginine tend to be similar to plasma levels, around 100–200 μM (Armstrong and Stave 1973), some tenfold or more higher than K_m. This implies that an inhibitor acting by a purely substrate-competitive mechanism will need to have high affinity for the enzyme in order to show significant activity in cells and in vivo at a reasonable dose. Indeed, it is not uncommon to see potency losses of 30-fold to 100-fold when comparing *i*NOS inhibition in cells with potency in enzyme assays. Experimentally, this is further complicated by variations in the arginine content in different culture media. For example, RPMI-1640 is often used to culture human cells and typically contains around 1 mM arginine.

Circumventing this issue, one type of *i*NOS inhibitors has been described which have no effect on either the active NOS enzymes themselves or the stimulated production of NOS protein. Instead, these compounds appear to act by preventing the assembly of the initially synthesized monomeric NOS protein into the functional homodimer. The initial evidence for compounds acting by this mechanism came in a study of the antifungal imidazoles clotrimazole and miconazole from Stuehr's laboratory in Ohio (Sennequier et al. 1999). Although these compounds are fairly weak inhibitors, further work by from Berlex (later part of Schering AG) (Blasko et al. 2002) identified compounds with nanomolar activity toward *i*NOS. The Berlex group reported X-ray diffraction data for a complex between one of these compounds, BBS-1, and the *i*NOS monomer, which shows that the imidazole unit is acting as a ligand for the heme iron whilst the rest of the molecule binds to more remote parts of the NOS protein. This binding appears to cause conformational changes in the monomer which preclude dimer formation. Other companies followed suit with compounds from Fujisawa (FR-260330; see (Chida et al. 2005)) apparently using a pyridine to ligate the heme, SSP Co (PPA250; see (Ohtsuka et al. 2002)) and Adolor (Chu et al. 2009). Compounds from both the Berlex and Fujisawa series have shown beneficial effects in in vivo models of transplant rejection (Szabolcs et al. 2002; Ouyang et al. 2005) with the former also effective in a sepsis model in mice (Ichinose et al. 2003) and in lung injury caused by burns and smoke inhalation in sheep (Enkhbaatar et al. 2003). As these compounds prevent assembly and dimerization of the *i*NOS protein upon induction typical assays involve LPS treatment of mouse macrophage cell lines, or cytokine stimulation of human, DLD-1, cells followed by analysis of nitrite production or enzyme activity. Mallinder and colleagues reported technical information on using a proprietary cell-based, β-galactosidase enzyme complementation method to screen for *i*NOS inhibitors acting via this mode-of-action in a more convenient HTS format (Mallinder et al. 2009). The assay system is known as InteraX™ and employs fusion proteins of *i*NOS oxygenase domains and β-galactosidase mutants as reporter enzymes. The individual mutants are inactive, but dimerization of the *i*NOS domains, fused in suitable orientation to the β-galactosidase reconstituted the galactosidase activity, which could then be used as a functional readout. The assay technique was shown to identify dimerization inhibitors, but more traditional *i*NOS inhibitors that bind in the active site were found to enhance the signal suggesting that they can promote dimerization as has been shown with the natural substrate, arginine. This methodology was reported to have been used in a high-throughput screen of around 800,000 compounds, but no data on the output from the screen has been published.

Wellcome/GlaxoWellcome/GlaxoSmithKline (GSK) ran a large *i*NOS/NOS2 program. Whilst 1400W, one of the first selective *i*NOS/NOS2 inhibitors (Garvey et al. 1997; Thomsen et al. 1997; Kankuri et al. 2001; Vuolteenaho et al. 2001; Pérez-Asensio et al. 2005; Järvinen et al. 2008), a non-amino acid compound, never made it into the clinic because of preclinical toxicity, GW274150 and GW273629, GSK's lead selective *i*NOS/NOS2 inhibitors, both amino acids (Alderton et al. 2005), were taken into clinical studies. These were both highly selective for *i*NOS inhibition, with slow or no reversal of inhibition and both orally bioavailable. GW274150 is transported by amino acid transporters such as y+-LAT (Baydoun et al. 2006). The imidazo[4,5-*b*]pyridine derivative BYK191023 has been identified by Altana Pharma/Nycomed as an *i*NOS/NOS2 selective inhibitor, which binds to the L-arginine site (Grädler et al. 2011). In the absence of NADPH, BYK191023 acts as a reversible L-arginine competitive inhibitor, whereas an NADPH and time-dependent irreversible inactivation mechanism with heme depletion is observed at low L-arginine levels and in intact cells (Tiso et al. 2008). AstraZeneca and Berlex also developed a line of NOS inhibitors (Rosenfeld et al. 2002b, c, d, e; Cheshire et al. 2011). A selective irreversible *i*NOS inhibitor developed by Pfizer/Pharmacia, cindunistat (SD-6010), is close to GSK's GW274150 in structure. Its entry into cells or tissue may be impaired by the (α)-methyl which is unlikely to be a substrate for amino acid transporters. This compound was investigated by Pfizer in osteoarthritis (50 or 200 mg/day) and failed to slow the disease progression (Hellio le Graverand et al. 2013). One of the well-known arginine analogs that have been tested in humans is L-NIL (L-*N* iminoethyl lysine), in the form of a pro-drug known as L-NILTA (and also as SC51 or SD3651). This was a product of the research teams at Pharmacia at a time when many acquisitions and mergers were occurring in the pharmaceutical industry, the work originated within G.D. Searle & Company and is now part of Pfizer. L-NIL is widely recognized as a selective *i*NOS inhibitor, but the compound is hygroscopic. While this can be managed in a research laboratory it can pose a problem in clinical trials if it becomes difficult to confidently and consistently prepare exactly the same concentration solutions for dosing. Unlike most pro-drugs this substance was not designed to circumvent issues with ADME or PK; rather L-NILTA, the tetrazolinium amide of L-NIL, is a stable, non-hygroscopic solid. Upon dosing the amide is rapidly removed, in effect dosing the parent drug L-NIL.

4 Clinical Applications of NOS Inhibitors

Clinical applications have focused mainly on the use of NOS inhibitors to vasoconstrict, inhibit inflammation, and neuroprotect. The role of the involved NOS isoform has not always been entirely clarified. In conditions such as inflammation and traumatic and ischemic damage additional interaction with reactive oxygen species is likely.

4.1 Vasoconstriction in Sepsis and for Blood Flow Disruption

The earliest clinical translational attempt for a NOS inhibitor has been in sepsis. Increased production of nitric oxide has been demonstrated in both experimental and clinical sepsis; the increased production of nitric oxide has subsequently been associated with hypotension, decreased responsiveness to vasoconstrictors, and development of multiple organ dysfunction (Petros et al. 1994; Grover et al. 1999; López et al. 2004). Reducing the overproduction of nitric oxide by partial inhibition of NOS could be postulated as a beneficial intervention in the treatment of septic shock. Previous experimental studies have produced conflicting results from the use of NOS inhibitors in models of septic shock provoked by either endotoxin or bacterial challenge. Clinical studies have shown that the administration of NOS inhibitors (L-NMMA; nitroarginine, L-NNA) to patients with septic shock can restore hemodynamics and the vascular responsiveness to vasoconstrictor therapy without significant acute adverse effects. However, a phase 3 study of infusion of L-NMMA (as 546C88) in septic shock showed that mortality was increased overall (López et al. 2004). Although post hoc analysis of the mortality by dose suggested that low doses (546C88, 5 $mg\cdot kg^{1}\cdot hr^{1}$ or below) provided an overall survival benefit, this was not regarded as strong enough to progress and the project was discontinued.

Preclinical shock models provide some support for the hypothesis that selective *i*NOS inhibition would be a better therapeutic approach to septic shock, but given the large phase 3 trial that would be needed to test this on the required mortality endpoint it doesn't seem likely to be tried.

Recently, non-specific L-arginine derived NOS inhibitors such as tilarginine (L-NMMA) have been employed as blood flow disruptors; in the gut to prevent GLP-2 from releasing gut lipid stores (ongoing study sponsored by the Toronto University Health Network, NCT03534661); in cancer, to overcome cancer-related immunosuppression. The nature of the involved isoform is unclear, although in cancer *i*NOS/NOS2 has been suggested to be involved.

The gut is able to retain some fat for many hours after a fatty meal. The gut hormone glucagon-like peptide-2 (GLP-2) is known to release these fat stores in the gut, but it is not known how GLP-2 achieves this. One possibility is that GLP-2 increases blood flow in the gut. NG-monomethyl-L-arginine (L-NMMA) is a substance that inhibits nitric oxide synthase (an enzyme that helps make nitric oxide which increases blood flow). This protocol examines whether blocking gut blood flow with L-NMMA is able to prevent GLP-2 from releasing gut lipid stores. Healthy participants were treated with a combination of Teduglutide (a resistant form of GLP-2) and L-NMMA and their respective controls.

With a focus on elucidating blood pressure physiology and the role of NOS1 therein, the NOS1-specific inhibitor, *S*-methyl-thiocitrulline (SMTC), has been tested in experimental medicine investigational studies (Seddon et al. 2008, 2009; Melikian et al. 2009; Shabeeh et al. 2013; Khan et al. 2015). These studies demonstrated that *n*NOS and *e*NOS appear to have distinct roles in the regulation of vascular tone and blood flow, at least in healthy humans. Local intra-arterial infusion of SMTC in the forearm (Seddon et al. 2008) or intracoronary circulation (Seddon et al. 2008; Ammar et al. 2020), suggesting that *n*NOS contributes to the maintenance of basal blood flow in healthy humans. Locally infused SMTC also inhibits mental stress-induced increases in blood flow in the forearm and coronary circulations (Seddon et al. 2008; Khan et al. 2017). Systemic infusion of SMTC in healthy volunteers resulted in a significant increase in systemic vascular resistance and blood pressure without inhibiting *e*NOS-dependent flow-mediated dilatation (Shabeeh et al. 2017). These effects were of a similar magnitude to those previously observed with the infusion of non-selective L-NMMA, suggesting that the major NOS isoform involved in the regulation of blood pressure in healthy humans may be *n*NOS. This study does not establish the site of action on SMTC, i.e., central or peripheral, but ongoing work is examining the effects of SMTC in the human brain (unpublished data).

In persistent cardiogenic shock, systemic inflammation, including expression of inducible nitric oxide synthase (NOS) and generation of excess nitric oxide, is believed to contribute to pathogenesis and inappropriate vasodilatation. Preliminary, single-center studies had indeed suggested a beneficial effect of NOS inhibition on hemodynamics, renal function, and survival in these patients (TRIUMPH Investigators 2007). However, when tilarginine was tested in acute myocardial infarction complicated by refractory cardiogenic shock (the TRIUMPH trial, NCT00112281), it failed to reduce mortality (Bailey et al. 2007; Kielstein et al. 2007; Salem and Mebazaa 2007; Teerlink 2007; TRIUMPH Investigators 2007).

MTR 104 (Garvey et al. 1994) has received orphan drug designation from the FDA and addresses a variety of acute and chronic therapeutic indications associated with hypotension. Meditor Pharmaceuticals has begun through TrioxBio as developer a phase II clinical trial for MTR 104 in intradialytic hypotension (IDH). The double-blind clinical study will involve chronic renal failure patients who experience IDH.

4.2 Inflammatory Diseases and Conditions

A key target in NOS drugs discovery has been *i*NOS/NOS2 and its possible role in inflammation, although *n*NOS/NOS1 may play a role herein as well (Baig et al. 2015). Elevated exhaled NO is a characteristic feature of human atopic asthma and correlates with the degree of inflammation and can be further increased by exposure to allergens (Kharitonov et al. 1995). Exhaled NO is recognized as a suitable biomarker to guide asthma treatment (Smith et al. 2005). The majority of this is believed to come from *i*NOS, and consistent with that, selective *i*NOS inhibitor GW274150 was shown to decrease exhaled NO in asthma patients in a dose-dependent manner to a maximum inhibition of >90% and persisting over 24 h when dosed once daily. Acute animal model studies with GW274150 showed inhibition of the late asthmatic response to allergen in guinea pigs similar to that of prednisolone, along with inhibition of exhaled NO, and inhibition of airway hyper-responsiveness in sensitized and challenged mice. Furthermore, GW274150 was active in some other models of lung inflammation in mice and rats. On the basis of these results, GW274150 was taken into a clinical trial in atopic asthma, looking at the responses to an allergen challenge. Although the expected inhibition by GW274150 of exhaled NO was observed, this was not accompanied by any benefit on the endpoints of early or late airway response, airway responsiveness to methacholine or AMP, or airway inflammation (Singh et al. 2007). This puts into question the acute animal models of asthma that were current at the time, and indeed studies in more chronic, complex allergen models showed a lack of beneficial effects consistent with the clinical findings (Evans et al. 2012; Mercer et al. 2015). Another compound, the prodrug L-NILTA, showed efficacy against allergen challenge in rats (Eynott et al. 2002) and moved on to a double-blind, placebo-controlled study monitoring exhaled NO in asthmatics. This compound reduced exhaled nitric oxide in asthma patients and also resulted in some depression of basal levels in healthy volunteer controls (Hansel et al. 2003). Subsequently, the drug appears to have been discontinued amid rumors of animal toxicity observed with the parent compound. Overall it seems that selective inhibition of *i*NOS is unlikely to provide benefit in asthma.

A preclinical case was also made for selective inhibition of *i*NOS in migraine, including efficacy in models of pain, but clinical studies with GW274150 (for migraine prophylaxis) and GW273629 (for acute migraine treatment) (Van der Schueren et al. 2009; Høivik et al. 2010; Hoffmann and Goadsby 2012; Barbanti et al. 2014) were convincingly negative.

The experience of testing selective *i*NOS inhibition in rheumatoid arthritis is somewhat similar; again, a preclinical case was made for this, resulting in progression into a clinical study with GW274150 (Cuzzocrea et al. 2002; Seymour et al. 2012). Although there were some beneficial trends in the GW274150 arm after 28 days dosing, they did not achieve statistical significance, in contrast to the positive control arm on prednisolone.

Similarly, a substantial body of work with *i*NOS inhibitors in vitro and in vivo supported the hypothesis that *i*NOS inhibition could be therapeutic in osteoarthritis (OA). Cindunistat (SD6010) was tested in a 2-year, multinational, double-blind, placebo-controlled trial, which enrolled 1,457 patients with symptomatic knee OA randomly assigned to cindunistat (50 or 200 mg/day) or placebo (Hellio le Graverand et al. 2013). Cindunistat did not slow the rate of Joint Space Narrowing (JSN) versus placebo overall. After 48 weeks, a subset of patients showed less JSN; however, the improvement was not sustained at 96 weeks. Thus, the loss of efficacy over time and lack of effect in more advanced OA patients suggest that alternative biochemical catabolic pathways overcame the effects of NO inhibition and/or that the consequences of the increased intra-articular stress may not have been amenable to *i*NOS inhibition alone. It was not reported as to what degree of *i*NOS inhibition was achieved with 50 or 200 mg/day, either initially or on long-term dosing, so it could be that this was not sufficient or long-lasting enough to achieve efficacy. No further development of cindunistat has been reported.

AR-C102222, exhibited excellent efficacy in animal models of inflammation and arthritis following oral administration. AR-C102222 reduced plasma nitrate in LPS-treated rats and in adjuvant induced arthritis it was shown to be effective at reducing the onset and severity of symptoms and prevented the development of structural changes in the joints of these animals (Tinker et al. 2003). This is not an ideal model of human joint disease, not least because indomethacin is effective whereas it has little benefit in human arthritic disease. However, the results were very encouraging that *i*NOS inhibitors would be of great potential for therapeutic benefit in rheumatoid or osteoarthritis. Particularly when, in addition, AR-C102222 was found to abrogate a cytokine induced decrease in aggrecan production by human chondrocytes (Johnston et al. 2004). Apart from being effective in rodent models of joint disease this *i*NOS selective inhibitor also alleviated neuropathic and inflammatory pain in independent studies conducted by Adolor Corporation (LaBuda et al. 2006) In other work AR-C102222 has also been shown to ameliorate experimental pancreatitis and modulate gallbladder sphincter function (Sandstrom et al. 2004, 2005; Woods et al. 2007). In the studies from Adolor (LaBuda et al. 2006) it is noted that the inhibitor led to some reduction in motor activity in the experimental animals, and ultimately further concerns about reactivity with glutathione (Cheshire et al. 2011) prevented inhibitors from this series proceeding into clinical studies.

4.3 Cancer

The role of NO in tumor biology is complex (Vamvakas and Schmidt 1997). The immune system is normally the body's first defense against threats like cancer. However, sometimes cancer cells produce signals like programmed death-1 (PD-1) that prevent the immune system from detecting and killing them. Pembrolizumab blocks PD-1 so the immune system can detect and attack cancer cells. Most patients do not, or only incompletely, respond to PD-1 inhibitors due to cancer-related immunosuppression. The presence of nitric oxide synthase in the area around the cancer cells blocks the cancer-fighting ability of the immune system. In cancer, *i*NOS product, nitric oxide, is associated with the establishment of an immunosuppressive environment and poor survival due to increased tumor aggressiveness (Davila-Gonzalez et al. 2018). To help further boost the cancer-fighting ability of the immune system, L-NMMA is tested along with pembrolizumab. Thus, the use of L-NMMA and Pembrolizumab together may augment the immune response against cancer cells. Recently, L-arginine-derived NOS inhibitors, such as tilarginine, have been employed as potentially synergistic adjuncts to other candidate compounds to treat cancers when combined with the anti-PD-1 monoclonal antibody pembrolizumab (Merck & Co.'s Keytruda), again as a blood flow disruptor (NCT03236935; Phase Ib). The purpose of this Phase Ib study is to test the safety of L-NMMA and pembrolizumab when used together in participants with melanoma, non-small cell lung cancer (NSCLC), head and neck squamous cell carcinoma (HNSCC), classical Hodgkin lymphoma (cHL), urothelial carcinoma, or microsatellite instability-high (MSI-H)/mismatch repair deficient (dMMR) cancer (Tables 1 and 2).

In an interesting work from the group of Weinberg at Duke University, following up a previous observation that non-selective NOS inhibitors induced apoptosis in cultured CLL (chronic lymphocytic leukemia) cells a positive correlation was observed between potency of *n*NOS inhibition (but not NOS2 inhibition) and the ability to induce cell death and apoptosis in these cells (Levesque et al. 2008). AR-R 17477 was identified as the most potent *n*NOS inhibitor in the study and the most effective at inducing cell death. The compound increased Caspase 3 expression and Annexin V binding suggesting the inhibitor was, indeed, inducing apoptosis. At the time of this study the incidence of CLL in the USA was not considered high enough to pursue the effect as a treatment solely for a sub-set of leukemias. However, in more recent times the incidence (though not the morbidity) of CLL has risen considerably. In light of this and the move toward more personalized and targeted therapy this unique application of *n*NOS inhibitors could, perhaps, be re-addressed.

4.4 Neuroprotection

In traumatic brain injury, ronopterin is the only and first NOS inhibitor to be clinically investigated. In a phase IIa study (NOSTRA, NCT02012582), ronopterin showed a significant improvement in clinical outcomes, however, induced acute renal failure that was dose-related (Stover et al. 2014). The effect of this compound on renal function was then examined in healthy volunteers (NCT02992236), showing a pharmacodynamic inhibitory effect on renal perfusion that was reversible (Ott et al. 2019). Now, a Phase III clinical trial to test the efficacy of this drug in patients with moderate and severe traumatic brain injury (NOSTRA-III, NCT02794168) (Tegtmeier et al. 2020b) is ongoing.

In ischemic stroke, inhibition of *n*NOS seems a potential neuroprotective therapy (Casas et al. 2019), however, has not been tested clinically so far. Preclinical studies with *n*NOS knockout animals or selective *n*NOS inhibitors show promising results in stroke (Huang et al. 1994; Willmot et al. 2005; Kleinschnitz et al. 2016). Examples of these compounds include 7-NI (Nanri et al. 1998), TRIM (Haga et al. 2003), BN 80933 (Chabrier et al. 1999), and ARL17477 (O'Neill et al. 2000). Also, Reif et al. provide a general overview of this class of inhibitors showing some of their potency as *n*NOS/NOS1 inhibitors, selectivity toward *e*NOS/NOS3 and *i*NOS/NOS2 and basic pharmacology and pharmacokinetics (Reif et al. 2000). The heterocyclic, thiene-substituted amidine compound, ARL17477, demonstrated neuroprotection when reducing infarct volume in a transient ischemia model in rats (Zhang et al. 1996) and toward hypothermic circulatory arrest in dogs (Tseng et al. 1999). The ischemic model showed an inverted dose curve that Reif et al. suggest may be due to inhibition of *e*NOS (NOS3) at the higher doses.

The indirect inhibition of *n*NOS-induced NO production by using PSD-95 inhibitors has reached clinical trial stage. However, it is not known yet whether the efficacy of PSD-95 inhibition is superior to direct *n*NOS inhibition (Kleinschnitz et al. 2016). Indeed, infusion of Tat-NR2B9c (NA-1 or nerinetide), a PSD-95 inhibitor, resulted in fewer ischemic infarcts in patients with iatrogenic stroke compared to placebo in a small trial (NCT00728182) (Bruder 2012; Hill et al. 2012; Matsumoto 2013). Another trial for nerinetide (NCT02930018) has been completed with no better clinical outcomes in acute ischemic stroke patients. However, a subgroup analysis of patients not treated with alteplase showed that nerinetide was associated with improved outcomes compared to placebo (Hill et al. 2020). A third trial for nerinetide is ongoing (NCT02315443) and a fourth one has been withdrawn (NCT02056574).

The *n*NOS and *i*NOS inhibitor, 2-iminobiotin, has also reached the clinical trial stage for hypoxic brain injury. Neurophyxia B.V. has been recruiting patients for a Phase II study to prevent hypoxic brain injury in patients with out-of-hospital cardiac arrest (NCT02836340) using an intravenous administration within 6 h after the event. The current status of the trial is unknown (Zitta et al. 2017; van Hoogdalem et al. 2019; Biselele et al. 2020; Favié et al. 2020).

5 Conclusions

One major conclusion from studying *i*NOS/NOS2 inhibition preclinically and clinically is that the animal models have performed poorly in predicting outcomes in human clinical trials despite diligent attempts to match the preclinical and clinical characteristics and endpoints. One reason for this might be due to the difference in inducibility and the relevance of NOS2 between rodents and humans (Rico et al. 2007). In some cases, this is being addressed by evaluating more chronic, complex models; in other instances, it may be feasible to develop human/animal hybrid models, or to make greater use of more sophisticated in vitro human models with multiple cell types and/or matrices (Mercer et al. 2015). This makes the question of which NOS to target in which clinical indication particularly challenging. Four persistent questions remain after this massive international industry effort: (1) Were the previously chosen indications the best opportunities for *i*NOS/NOS2 inhibition? (2) Was the degree of inhibition sufficient to test the hypothesis fully? (3) Should *i*NOS/NOS2 inhibition be combined with other therapeutics/targets, i.e. for network pharmacology (Casas et al. 2019)? (4) Are there ways of selecting sub-populations of "responders to *i*NOS inhibition," i.e., mechanistic endophenotyping? With respect to *n*NOS/NOS1 inhibition the situation is probably more optimistic. Here acute indications were chosen, such as traumatic brain injury and stroke, with a dramatic phenotype and presumably uniform pathomechanism preserved in different species (Casas et al. 2017, 2019) so that preclinical animal models are likely to be more predictive. With respect to *e*NOS inhibition, the advice is clear: leave it alone or else stimulate it!

Acknowledgments This review project has received funding from the European Union's Horizon 2020 research and innovation program under grant agreement No 777111 (REPO-TRIAL). This reflects only the author's view and the European Commission is not responsible for any use that may be made of the information it contains. AMS is supported in part by the British Heart Foundation (CH/1999001/11735; RE/18/2/34213); the National Institute for Health Research Biomedical Research Centre at Guy's & St Thomas' NHS Foundation Trust and King's College London (IS-BRC-1215-20006); and the Foundation Leducq. MHE is supported by the PhD research grant from the Egyptian Ministry of Higher Education and Scientific Research.

Conflict of Interest HHHWS is a minor shareholder in Vasopharm GmbH.
RGK is a former employee of GSK.
AW is a former employee of AstraZeneca.

References

Ahluwalia A et al (2004) Antiinflammatory activity of soluble guanylate cyclase: cGMP-dependent down-regulation of P-selectin expression and leukocyte recruitment. Proc Natl Acad Sci U S A 101(5):1386–1391

Alderton WK, Cooper CE, Knowles RG (2001) Nitric oxide synthases: structure, function and inhibition. Biochem J 357(3):593–615

Alderton WK et al (2005) GW274150 and GW273629 are potent and highly selective inhibitors of inducible nitric oxide synthase in vitro and in vivo. Br J Pharmacol 145(3):301–312

Alheid U, Frölich JC, Förstermann U (1987) Endothelium-derived relaxing factor from cultured human endothelial cells inhibits aggregation of human platelets. Thromb Res 47(5):561–571

Amarante-Mendes GP et al (2018) Pattern recognition receptors and the host cell death molecular machinery. Front Immunol 9:2379

Ammar W et al (2020) Assessment of vascular stiffness using different modalities in patients with systemic lupus erythematosus: a case control study. Egypt Heart J (EHJ) 72(1):24

Annedi SC et al (2011) Discovery ofN-(3-(1-Methyl-1,2,3,6-tetrahydropyridin-4-yl)-1H-indol-6-yl) thiophene-2-carboximidamide as a selective inhibitor of human neuronal nitric oxide synthase (nNOS) for the treatment of pain. J Med Chem:7408–7416. https://doi.org/10.1021/jm201063u

Annedi SC et al (2012) Novel, druglike 1,7-disubstituted 2,3,4,5-tetrahydro-1H-benzo[b]azepine-based selective inhibitors of human neuronal nitric oxide synthase (nNOS). Bioorg Med Chem Lett 22(7):2510–2513

Antoniades C et al (2009) Association of plasma asymmetrical dimethylarginine (ADMA) with elevated vascular superoxide production and endothelial nitric oxide synthase uncoupling: implications for endothelial function in human atherosclerosis. Eur Heart J 30(9):1142–1150

Aquilano K, Baldelli S, Ciriolo MR (2014) Nuclear recruitment of neuronal nitric-oxide synthase by α-syntrophin is crucial for the induction of mitochondrial biogenesis. J Biol Chem:365–378. https://doi.org/10.1074/jbc.m113.506733

Armstrong MD, Stave U (1973) A study of plasma free amino acid levels. II. Normal values for children and adults. Metab Clin Exp 22(4):561–569

Averna M et al (2008) Functional role of HSP90 complexes with endothelial nitric-oxide synthase (eNOS) and calpain on nitric oxide generation in endothelial cells. J Biol Chem:29069–29076. https://doi.org/10.1074/jbc.m803638200

Awolesi MA, Sessa WC, Sumpio BE (1995) Cyclic strain upregulates nitric oxide synthase in cultured bovine aortic endothelial cells. J Clin Invest 96(3):1449–1454

Bach A et al (2015) Biochemical investigations of the mechanism of action of small molecules ZL006 and IC87201 as potential inhibitors of the nNOS-PDZ/PSD-95-PDZ interactions. Sci Rep 5:12157

Baig MS et al (2015) NOS1-derived nitric oxide promotes NF-κB transcriptional activity through inhibition of suppressor of cytokine signaling-1. J Exp Med:1725–1738. https://doi.org/10.1084/jem.20140654

Bailey A et al (2007) The tragedy of TRIUMPH for nitric oxide synthesis inhibition in cardiogenic shock: where do we go from here? Am J Cardiovasc Drugs 7(5):337–345

Barbanti P et al (2014) Drugs targeting nitric oxide synthase for migraine treatment. Expert Opin Investig Drugs 23(8):1141–1148

Bartesaghi S, Radi R (2018) Fundamentals on the biochemistry of peroxynitrite and protein tyrosine nitration. Redox Biol 14:618–625

Baydoun AR et al (2006) Y LAT-1 mediates transport of the potent and selective iNOS inhibitor, GW274150, in control J774 macrophages. Amino Acids:101–109. https://doi.org/10.1007/s00726-005-0311-9

Beaton H et al (2001a) 3,4-Dihydro-1-isoquinolinamines: a novel class of nitric oxide synthase inhibitors with a range of isoform selectivity and potency. Bioorg Med Chem Lett:1023–1026. https://doi.org/10.1016/s0960-894x(01)00119-6

Beaton H et al (2001b) Thienopyridines: nitric oxide synthase inhibitors with potent in vivo activity. Bioorg Med Chem Lett 11(8):1027–1030

Bender AT et al (1999) Neuronal nitric-oxide synthase is regulated by the hsp90-based chaperone systemin vivo. J Biol Chem:1472–1478. https://doi.org/10.1074/jbc.274.3.1472

Berka V, Chen PF, Tsai AL (1996) Spatial relationship between L-arginine and heme binding sites of endothelial nitric-oxide synthase. J Biol Chem 271(52):33293–33300

Bhatt DK et al (2013) NXN-188, a selective nNOS inhibitor and a 5-HT1B/1D receptor agonist, inhibits CGRP release in preclinical migraine models. Cephalalgia:87–100. https://doi.org/10.1177/0333102412466967

Biselele T et al (2020) A phase iia clinical trial of 2-Iminobiotin for the treatment of Birth Asphyxia in DR Congo, a low-income country. Paediatr Drugs 22(1):95–104

Blasko E et al (2002) Mechanistic studies with potent and selective inducible nitric-oxide synthase dimerization inhibitors. J Biol Chem:295–302. https://doi.org/10.1074/jbc.m105691200

Bogdan C (2015) Nitric oxide synthase in innate and adaptive immunity: an update. Trends Immunol 36(3):161–178

Böhme GA et al (1993) Altered synaptic plasticity and memory formation in nitric oxide synthase inhibitor-treated rats. Proc Natl Acad Sci U S A 90(19):9191–9194

Bömmel HM et al (1998) Anti-pterins as tools to characterize the function of tetrahydrobiopterin in NO synthase. J Biol Chem 273(50):33142–33149

Bredt DS, Snyder SH (1990) Isolation of nitric oxide synthetase, a calmodulin-requiring enzyme. Proc Natl Acad Sci U S A 87(2):682–685

Bredt DS et al (1991) Cloned and expressed nitric oxide synthase structurally resembles cytochrome P-450 reductase. Nature 351(6329):714–718

Brenman JE et al (1996) Interaction of nitric oxide synthase with the postsynaptic density protein PSD-95 and α1-syntrophin mediated by PDZ domains. Cell:757–767. https://doi.org/10.1016/s0092-8674(00)81053-3

Bruder N (2012) Faculty opinions recommendation of safety and efficacy of NA-1 in patients with iatrogenic stroke after endovascular aneurysm repair (ENACT): a phase 2, randomised, double-blind, placebo-controlled trial. Faculty Opin Post Publ Peer Rev Biomed Lit. https://doi.org/10.3410/f.717961804.793464512

Candemir E et al (2016) Interaction of NOS1AP with the NOS-I PDZ domain: implications for schizophrenia-related alterations in dendritic morphology. Eur Neuropsychopharmacol 26 (4):741–755

Casas AI et al (2015) Reactive oxygen-related diseases: therapeutic targets and emerging clinical indications. Antioxid Redox Signal 23(14):1171–1185

Casas AI et al (2017) NOX4-dependent neuronal autotoxicity and BBB breakdown explain the superior sensitivity of the brain to ischemic damage. Proc Natl Acad Sci U S A 114 (46):12315–12320

Casas AI et al (2019) From single drug targets to synergistic network pharmacology in ischemic stroke. Proc Natl Acad Sci U S A 116(14):7129–7136

Caviedes A et al (2017) Endothelial nitric oxide synthase is present in dendritic spines of neurons in primary cultures. Front Cell Neurosci 11:180

Chabrier PE et al (1999) BN 80933, a dual inhibitor of neuronal nitric oxide synthase and lipid peroxidation: a promising neuroprotective strategy. Proc Natl Acad Sci U S A 96 (19):10824–10829

Chakrabarti S et al (2012) Neuronal nitric oxide synthase regulates endothelial inflammation. J Leukoc Biol 91(6):947–956

Chan Y et al (2004) The cell-specific expression of endothelial nitric-oxide synthase. J Biol Chem:35087–35100. https://doi.org/10.1074/jbc.m405063200

Chen C-A et al (2010) S-glutathionylation uncouples eNOS and regulates its cellular and vascular function. Nature:1115–1118. https://doi.org/10.1038/nature09599

Cheshire DR (2001) Use of nitric oxide synthase inhibitors for the treatment of inflammatory disease and pain. IDrugs Invest Drugs J 4(7):795–803

Cheshire DR et al (2011) The discovery of novel, potent and highly selective inhibitors of inducible nitric oxide synthase (iNOS). Bioorg Med Chem Lett 21(8):2468–2471

Chida N et al (2005) Pharmacological profile of FR260330, a novel orally active inducible nitric oxide synthase inhibitor. Eur J Pharmacol:71–76. https://doi.org/10.1016/j.ejphar.2004.12.028

Cho HJ et al (1992) Calmodulin is a subunit of nitric oxide synthase from macrophages. J Exp Med 176(2):599–604

Chong C-M et al (2018) Roles of nitric oxide synthase isoforms in neurogenesis. Mol Neurobiol 55 (3):2645–2652

Chreifi G et al (2014) Communication between the zinc and tetrahydrobiopterin binding sites in nitric oxide synthase. Biochemistry:4216–4223. https://doi.org/10.1021/bi5003986

Chu G-H et al (2009) Design and synthesis of imidazopyrimidine derivatives as potent iNOS dimerization inhibitors. Open Med Chem J 3:8–13

Cinelli MA et al (2020) Inducible nitric oxide synthase: regulation, structure, and inhibition. Med Res Rev 40(1):158–189

Connolly S et al (2004) 2-aminopyridines as highly selective inducible nitric oxide synthase inhibitors. Differential binding modes dependent on nitrogen substitution. J Med Chem 47 (12):3320–3323

Courtney MJ, Li L-L, Lai YY (2014) Mechanisms of NOS1AP action on NMDA receptor-nNOS signaling. Front Cell Neurosci 8:252

Crabtree MJ, Channon KM (2011) Synthesis and recycling of tetrahydrobiopterin in endothelial function and vascular disease. Nitric Oxide Biol Chem 25(2):81–88

Cui H et al (2007) PDZ protein interactions underlying NMDA receptor-mediated excitotoxicity and neuroprotection by PSD-95 inhibitors. J Neurosci Off J Soc Neurosci 27(37):9901–9915

Cuzzocrea S et al (2002) Beneficial effects of GW274150, a novel, potent and selective inhibitor of iNOS activity, in a rodent model of collagen-induced arthritis. Eur J Pharmacol:119–129. https://doi.org/10.1016/s0014-2999(02)02338-5

Dao VT-V et al (2011) Catalase activity prevents exercise-induced up-regulation of vasoprotective proteins in venous tissue. J Cell Mol Med 15(11):2326–2334

Dao VT-V et al (2015) Pharmacology and clinical drug candidates in redox medicine. Antioxid Redox Signal 23(14):1113–1129

Dao VT-V et al (2016) Nitric oxide up-regulates endothelial expression of angiotensin II type 2 receptors. Biochem Pharmacol 112:24–36

Dao VT-V et al (2020) Non-canonical chemical feedback self-limits nitric oxide-cyclic GMP signaling in health and disease. Sci Rep. https://doi.org/10.1038/s41598-020-66639-w

Davila-Gonzalez D. et al. (2018) Abstract A202: evaluating the combination of anti-PD-1 and nitric oxide synthase inhibition therapy in 12 triple-negative breast cancer patient-derived xenografts using a human-derived immune system model. In: Immune checkpoints. abstracts: AACR-NCI-EORTC international conference: molecular targets and cancer therapeutics; October 26–30, 2017. American Association for Cancer Research, Philadelphia, pp A202–A202.

Dell'Albani P et al (2001) JAK/STAT signaling pathway mediates cytokine-induced iNOS expression in primary astroglial cell cultures. J Neurosci Res:417–424. https://doi.org/10.1002/jnr.1169

Dimmeler S et al (1999) Activation of nitric oxide synthase in endothelial cells by Akt-dependent phosphorylation. Nature:601–605. https://doi.org/10.1038/21224

Drummond GR et al (2000) Transcriptional and posttranscriptional regulation of endothelial nitric oxide synthase expression by hydrogen peroxide. Circ Res:347–354. https://doi.org/10.1161/01.res.86.3.347

Dudzinski DM et al (2006) The regulation and pharmacology of endothelial nitric oxide synthase. Annu Rev Pharmacol Toxicol:235–276. https://doi.org/10.1146/annurev.pharmtox.44.101802.121844

Enkhbaatar P et al (2003) The inducible nitric oxide synthase inhibitor BBS-2 prevents acute lung injury in sheep after burn and smoke inhalation injury. Am J Respir Crit Care Med 167 (7):1021–1026

Erwin PA et al (2005) Receptor-regulated dynamic S-nitrosylation of endothelial nitric-oxide synthase in vascular endothelial cells. J Biol Chem 280(20):19888–19894

Evans RL et al (2012) A comparison of antiasthma drugs between acute and chronic ovalbumin-challenged Guinea-pig models of asthma. Pulm Pharmacol Ther 25(6):453–464

Eynott PR et al (2002) Role of nitric oxide in allergic inflammation and bronchial hyperresponsiveness. Eur J Pharmacol:123–133. https://doi.org/10.1016/s0014-2999(02)02237-9

Fang FC (2004) Antimicrobial reactive oxygen and nitrogen species: concepts and controversies. Nat Rev Microbiol 2(10):820–832

Fantacuzzi M et al (2016) Screening of NOS activity and selectivity of newly synthesized acetamidines using RP-HPLC. J Pharm Biomed Anal 120:419–424

Favié LMA et al (2020) Pharmacokinetics and short-term safety of the selective NOS inhibitor 2-iminobiotin in asphyxiated neonates treated with therapeutic hypothermia. Pediatr Res 87 (4):689–696

Fedorov R et al (2004) Structures of nitric oxide synthase isoforms complexed with the inhibitor AR-R17477 suggest a rational basis for specificity and inhibitor design. Proc Natl Acad Sci:5892–5897. https://doi.org/10.1073/pnas.0306588101

Felice MD et al (2010) Triptan-induced enhancement of neuronal nitric oxide synthase in trigeminal ganglion dural afferents underlies increased responsiveness to potential migraine triggers. Brain:2475–2488. https://doi.org/10.1093/brain/awq159

Fish JE et al (2005) The expression of endothelial nitric-oxide synthase is controlled by a cell-specific histone code. J Biol Chem:24824–24838. https://doi.org/10.1074/jbc.m502115200

Fish JE et al (2010) Hypoxic repression of endothelial nitric-oxide synthase transcription is coupled with eviction of promoter histones. J Biol Chem 285(2):810–826

Forstermann U, Sessa WC (2012) Nitric oxide synthases: regulation and function. Eur Heart J:829–837. https://doi.org/10.1093/eurheartj/ehr304

Förstermann U et al (1986) Stimulation of soluble guanylate cyclase by an acetylcholine-induced endothelium-derived factor from rabbit and canine arteries. Circ Res:531–538. https://doi.org/10.1161/01.res.58.4.531

Förstermann U et al (1992) Induced RAW 264.7 macrophages express soluble and particulate nitric oxide synthase: inhibition by transforming growth factor-ß. Eur J Pharmacol Mol Pharmacol:161–165. https://doi.org/10.1016/0922-4106(92)90096-e

Fröhlich LG et al (1999) Inhibition of neuronal nitric oxide synthase by 4-amino pteridine derivatives: structure-activity relationship of antagonists of (6R)-5,6,7,8-tetrahydrobiopterin cofactor. J Med Chem 42(20):4108–4121

Fukai T et al (2000) Regulation of the vascular extracellular superoxide dismutase by nitric oxide and exercise training. J Clin Invest 105(11):1631–1639

Fulton DJR (2016) Transcriptional and posttranslational regulation of eNOS in the endothelium. Adv Pharmacol:29–64. https://doi.org/10.1016/bs.apha.2016.04.001

Fulton D et al (2005) Src kinase activates endothelial nitric-oxide synthase by phosphorylating Tyr-83. J Biol Chem 280(43):35943–35952

Ganster RW et al (2001) Complex regulation of human inducible nitric oxide synthase gene transcription by Stat 1 and NF- B. Proc Natl Acad Sci 98(15):8638–8643. https://doi.org/10.1073/pnas.151239498

Gantner BN, LaFond KM, Bonini MG (2020) Nitric oxide in cellular adaptation and disease. Redox Biol 34:101550. https://doi.org/10.1016/j.redox.2020.101550

Garcia V, Sessa WC (2019) Endothelial NOS: perspective and recent developments. Br J Pharmacol 176(2):189–196

Garcin ED et al (2008) Anchored plasticity opens doors for selective inhibitor design in nitric oxide synthase. Nat Chem Biol 4(11):700–707

Garvey EP et al (1994) Potent and selective inhibition of human nitric oxide synthases. Inhibition by non-amino acid isothioureas. J Biol Chem 269(43):26669–26676

Garvey EP et al (1997) 1400W is a slow, tight binding, and highly selective inhibitor of inducible nitric-oxide synthase in vitro and in vivo. J Biol Chem 272(8):4959–4963

Gebhart V et al (2019) Site and mechanism of uncoupling of nitric-oxide synthase: uncoupling by monomerization and other misconceptions. Nitric Oxide Biol Chem 89:14–21

Ghosh A, Chawla-Sarkar M, Stuehr DJ (2011) Hsp90 interacts with inducible NO synthase client protein in its heme-free state and then drives heme insertion by an ATP-dependent process. FASEB J:2049–2060. https://doi.org/10.1096/fj.10-180554

Grädler U et al (2011) Novel nanomolar imidazo[4,5-b]pyridines as selective nitric oxide synthase (iNOS) inhibitors: SAR and structural insights. Bioorg Med Chem Lett 21(14):4228–4232

Grover R et al (1999) An open-label dose escalation study of the nitric oxide synthase inhibitor, NG-methyl-L-arginine hydrochloride (546C88), in patients with septic shock. Crit Care Med:913–922. https://doi.org/10.1097/00003246-199905000-00025

Haga KK et al (2003) The neuronal nitric oxide synthase inhibitor, TRIM, as a neuroprotective agent: effects in models of cerebral ischaemia using histological and magnetic resonance imaging techniques. Brain Res 993(1–2):42–53

Hansel TT et al (2003) A selective inhibitor of inducible nitric oxide synthase inhibits exhaled breath nitric oxide in healthy volunteers and asthmatics. FASEB J 17(10):1298–1300

Haque MM et al (2007) A connecting hinge represses the activity of endothelial nitric oxide synthase. Proc Natl Acad Sci U S A 104(22):9254–9259

Haque MM et al (2012) Control of electron transfer and catalysis in neuronal nitric-oxide synthase (nNOS) by a hinge connecting its FMN and FAD-NADPH domains. J Biol Chem 287 (36):30105–30116

Heiss E, Dirsch V (2014) Regulation of eNOS enzyme activity by posttranslational modification. Curr Pharm Des:3503–3513. https://doi.org/10.2174/13816128113196660745

Hellio le Graverand M-P et al (2013) A 2-year randomised, double-blind, placebo-controlled, multicentre study of oral selective iNOS inhibitor, cindunistat (SD-6010), in patients with symptomatic osteoarthritis of the knee. Ann Rheum Dis 72(2):187–195

Hemmens B et al (2000) Role of bound zinc in dimer stabilization but not enzyme activity of neuronal nitric-oxide synthase. J Biol Chem 275(46):35786–35791

Hill MD et al (2012) Safety and efficacy of NA-1 in patients with iatrogenic stroke after endovascular aneurysm repair (ENACT): a phase 2, randomised, double-blind, placebo-controlled trial. Lancet Neurol 11(11):942–950

Hill MD et al (2020) Efficacy and safety of nerinetide for the treatment of acute ischaemic stroke (ESCAPE-NA1): a multicentre, double-blind, randomised controlled trial. Lancet 395 (10227):878–887

Hoffmann J, Goadsby PJ (2012) New agents for acute treatment of migraine: CGRP receptor antagonists, iNOS inhibitors. Curr Treat Options Neurol 14(1):50–59

Høivik HO et al (2010) Lack of efficacy of the selective iNOS inhibitor GW274150 in prophylaxis of migraine headache. Cephalalgia Int J headache 30(12):1458–1467

Hougaard A et al (2013) The nitric oxide synthase inhibitor and serotonin-receptor agonist NXN-188 during the aura phase of migraine with aura: a randomized, double-blind, placebo-controlled cross-over study. Scand J Pain 4(1):48–52

Huang Z et al (1994) Effects of cerebral ischemia in mice deficient in neuronal nitric oxide synthase. Science 265(5180):1883–1885

Hume DA et al (2001) Macrophages exposed continuously to lipopolysaccharide and other agonists that act via toll-like receptors exhibit a sustained and additive activation state. BMC Immunol 2:11

Ichinose F et al (2003) A selective inducible NOS dimerization inhibitor prevents systemic, cardiac, and pulmonary hemodynamic dysfunction in endotoxemic mice. Am J Phys Heart Circ Phys 285(6):H2524–H2530

Ito Y et al (2010) Nitric oxide production during cerebral ischemia and reperfusion in eNOS- and nNOS-knockout mice. Curr Neurovasc Res 7(1):23–31

Jaffrey SR et al (2002) Neuronal nitric-oxide synthase localization mediated by a ternary complex with synapsin and CAPON. Proc Natl Acad Sci:3199–3204. https://doi.org/10.1073/pnas.261705799

Järvinen K et al (2008) Selective iNOS inhibitor 1400W enhances anti-catabolic IL-10 and reduces destructive MMP-10 in OA cartilage. Survey of the effects of 1400W on inflammatory mediators produced by OA cartilage as detected by protein antibody array. Clin Exp Rheumatol 26(2):275–282

Johnston JE et al (2004) A novel inducible nitric oxide synthase (iNOS) inhibitor influences aggrecan mRNA levels in osteoarthritic (OA) chondrocytes. Rheumatology 43:ii50

Jones LH (2012) Chemistry and biology of biomolecule nitration. Chem Biol:1086–1092. https://doi.org/10.1016/j.chembiol.2012.07.019

Jung S-B et al (2010) Histone deacetylase 3 antagonizes aspirin-stimulated endothelial nitric oxide production by reversing aspirin-induced lysine acetylation of endothelial nitric oxide synthase. Circ Res 107(7):877–887

Kankuri E et al (2001) Suppression of acute experimental colitis by a highly selective inducible nitric-oxide synthase inhibitor, N-[3-(aminomethyl)benzyl]acetamidine. J Pharmacol Exp Ther 298(3):1128–1132

Kelleher ZT et al (2007) NOS2 regulation of NF-κB byS-Nitrosylation of p65. J Biol Chem:30667–30672. https://doi.org/10.1074/jbc.m705929200

Khan SG et al (2015) Impaired neuronal nitric oxide synthase-mediated vasodilator responses to mental stress in essential hypertension. Hypertension 65(4):903–909

Khan SG et al (2017) The human coronary vasodilatory response to acute mental stress is mediated by neuronal nitric oxide synthase. Am J Phys Heart Circ Phys:H578–H583. https://doi.org/10.1152/ajpheart.00745.2016

Kharitonov SA et al (1995) Allergen-induced late asthmatic reactions are associated with elevation of exhaled nitric oxide. Am J Respir Crit Care Med 151(6):1894–1899

Kielstein JT, Sydow K, Thum T (2007) Tilarginine in patients with acute myocardial infarction and cardiogenic shock. JAMA:969. https://doi.org/10.1001/jama.298.9.971-a

Kleinschnitz C et al (2016) NOS knockout or inhibition but not disrupting PSD-95-NOS interaction protect against ischemic brain damage. J Cereb Blood Flow Metab:1508–1512. https://doi.org/10.1177/0271678x16657094

Knowles RG, Moncada S (1994) Nitric oxide synthases in mammals. Biochem J:249–258. https://doi.org/10.1042/bj2980249

Korzhevskiĭ DE et al (2007) Immunocytochemical demonstration of neuronal NO-synthase in rat brain cells. Morfologiia 132(4):77–80

Kotsonis P et al (1999) Autoinhibition of neuronal nitric oxide synthase: distinct effects of reactive nitrogen and oxygen species on enzyme activity. Biochem J 340(Pt 3):745–752

Kotsonis P et al (2001) Structural basis for pterin antagonism in nitric-oxide synthase. Development of novel 4-oxo-pteridine antagonists of (6R)-5,6,7,8-tetrahydrobiopterin. J Biol Chem 276 (52):49133–49141

Kubes P, Suzuki M, Granger DN (1991) Nitric oxide: an endogenous modulator of leukocyte adhesion. Proc Natl Acad Sci:4651–4655. https://doi.org/10.1073/pnas.88.11.4651

Kuzkaya N et al (2003) Interactions of peroxynitrite, tetrahydrobiopterin, ascorbic acid, and thiols: implications for uncoupling endothelial nitric-oxide synthase. J Biol Chem 278 (25):22546–22554

LaBuda CJ et al (2006) Antinociceptive activity of the selective iNOS inhibitor AR-C102222 in rodent models of inflammatory, neuropathic and post-operative pain. Eur J Pain 10(6):505–512

Levesque MC et al (2008) CLL cell apoptosis induced by nitric oxide synthase inhibitors: correlation with lipid solubility and NOS1 dissociation constant. Leuk Res 32(7):1061–1070

Li H et al (1999) Crystal structures of zinc-free and -bound heme domain of human inducible nitric-oxide synthase. J Biol Chem:21276–21284. https://doi.org/10.1074/jbc.274.30.21276

Li H et al (2005) Exploring the binding conformations of bulkier dipeptide amide inhibitors in constitutive nitric oxide synthases. Biochemistry 44(46):15222–15229

Li Q, Youn J-Y, Cai H (2015) Mechanisms and consequences of endothelial nitric oxide synthase dysfunction in hypertension. J Hypertens 33(6):1128–1136

Lin MI et al (2003) Phosphorylation of threonine 497 in endothelial nitric-oxide synthase coordinates the coupling of l-arginine metabolism to efficient nitric oxide production. J Biol Chem:44719–44726. https://doi.org/10.1074/jbc.m302836200

Liu AC et al (2012) Induction of endothelial nitric oxide synthase expression by IL-17 in human vascular endothelial cells: implications for vascular remodeling in transplant vasculopathy. J Immunol 188(3):1544–1550

Loot AE et al (2009) Angiotensin II impairs endothelial function via tyrosine phosphorylation of the endothelial nitric oxide synthase. J Exp Med 206(13):2889–2896

López A et al (2004) Multiple-center, randomized, placebo-controlled, double-blind study of the nitric oxide synthase inhibitor 546C88: effect on survival in patients with septic shock. Crit Care Med 32(1):21–30

Luo P et al (2019) Preso regulates NMDA receptor-mediated excitotoxicity via modulating nitric oxide and calcium responses after traumatic brain injury. Cell Death Dis 10(7):496

MacMicking J, Xie QW, Nathan C (1997) Nitric oxide and macrophage function. Annu Rev Immunol 15:323–350

Mallinder PR, Wallace AV, Allenby G (2009) Identification of iNOS inhibitors using InteraX™. J Biomol Screen:263–272. https://doi.org/10.1177/1087057109331476

Man HSJ et al (2018) Angiogenic patterning by STEEL, an endothelial-enriched long noncoding RNA. Proc Natl Acad Sci:2401–2406. https://doi.org/10.1073/pnas.1715182115

Masters BSS et al (1996) Neuronal nitric oxide synthase, a modular enzyme formed by convergent evolution: structure studies of a cysteine thiolate-liganded heme protein that hydroxylates L-arginine to produce NO as a cellular signal. FASEB J:552–558. https://doi.org/10.1096/fasebj.10.5.8621055

Matsumoto M (2013) Faculty opinions recommendation of safety and efficacy of NA-1 in patients with iatrogenic stroke after endovascular aneurysm repair (ENACT): a phase 2, randomised, double-blind, placebo-controlled trial. Faculty Opin Post Publ Peer Rev Biomed Lit. https://doi.org/10.3410/f.717961804.793478082

Matter H et al (2002) Structural requirements for inhibition of the neuronal nitric oxide synthase (NOS-I): 3D-QSAR analysis of 4-oxo- and 4-amino-pteridine-based inhibitors. J Med Chem 45 (14):2923–2941

Matter H et al (2005) Structural analysis of isoform-specific inhibitors targeting the tetrahydrobiopterin binding site of human nitric oxide synthases. J Med Chem 48 (15):4783–4792

Mattila JT, Thomas AC (2014) Nitric oxide synthase: non-canonical expression patterns. Front Immunol 5:478

Melikian N et al (2009) Neuronal nitric oxide synthase and human vascular regulation. Trends Cardiovasc Med 19(8):256–262

Mendes AF et al (2002) Role of nitric oxide in the activation of NF-kappaB, AP-1 and NOS II expression in articular chondrocytes. Inflam Res 51(7):369–375

Mercer PF et al (2015) Translational models of lung disease. Clin Sci:235–256. https://doi.org/10.1042/cs20140373

Nanri K et al (1998) The selective inhibitor of neuronal nitric oxide synthase, 7-nitroindazole, reduces the delayed neuronal damage due to forebrain ischemia in rats. Stroke 29 (6):1248–1253; discussion 1253–4

Nathan C (2006) Role of iNOS in human host defense. Science:1874b–1875b. https://doi.org/10.1126/science.312.5782.1874b

Nedvetsky PI, Sessa WC, Schmidt HHHW (2002) There's NO binding like NOS binding: protein-protein interactions in NO/cGMP signaling. Proc Natl Acad Sci U S A:16510–16512

Nishida K et al (1992) Molecular cloning and characterization of the constitutive bovine aortic endothelial cell nitric oxide synthase. J Clin Invest 90(5):2092–2096

O'Dell T et al (1994) Endothelial NOS and the blockade of LTP by NOS inhibitors in mice lacking neuronal NOS. Science:542–546. https://doi.org/10.1126/science.7518615

O'Neill MJ et al (2000) ARL 17477, a selective nitric oxide synthase inhibitor, with neuroprotective effects in animal models of global and focal cerebral ischaemia. Brain Res 871(2):234–244

Ohtsuka M et al (2002) PPA250 [3-(2,4-Difluorophenyl)-6–2-[4-(1H-imidazol-1-ylmethyl) Phenoxy]ethoxy-2-phenylpyridine], a novel orally effective inhibitor of the dimerization of inducible nitric-oxide synthase, exhibits an anti-inflammatory effect in animal models of chronic arthritis. J Pharmacol Exp Ther:52–57. https://doi.org/10.1124/jpet.102.035857

Ota KT et al (2010) Synaptic plasticity and NO-cGMP-PKG signaling coordinately regulate ERK-driven gene expression in the lateral amygdala and in the auditory thalamus following Pavlovian fear conditioning. Learn Mem:221–235. https://doi.org/10.1101/lm.1592510

Ott C et al (2019) Effects of the nitric oxide synthase inhibitor ronopterin (VAS203) on renal function in healthy volunteers. Br J Clin Pharmacol 85(5):900–907

Ouyang J et al (2005) Effect of a novel inducible nitric oxide synthase inhibitor in prevention of rat chronic aortic rejections. Transplantation:1386–1392. https://doi.org/10.1097/01.tp.0000159144.08519.e2

Pacher P, Beckman JS, Liaudet L (2007) Nitric oxide and peroxynitrite in health and disease. Physiol Rev 87(1):315–424

Pan J et al (1996) Tyrosine phosphorylation of inducible nitric oxide synthase: implications for potential post-translational regulation. Biochem J:889–894. https://doi.org/10.1042/bj3140889

Pantke MM et al (2001) Pterin interactions with distinct reductase activities of NO synthase. Biochem J 356(Pt 1):43–51

Peng H-M et al (2012) Modulation of Heme/substrate binding cleft of neuronal nitric-oxide synthase (nNOS) regulates binding of Hsp90 and Hsp70 proteins and nNOS Ubiquitination. J Biol Chem:1556–1565. https://doi.org/10.1074/jbc.m111.323295

Percival JM (2011) nNOS regulation of skeletal muscle fatigue and exercise performance. Biophys Rev:209–217. https://doi.org/10.1007/s12551-011-0060-9

Pérez-Asensio FJ et al (2005) Inhibition of iNOS activity by 1400W decreases glutamate release and ameliorates stroke outcome after experimental ischemia. Neurobiol Dis 18(2):375–384

Petros A et al (1994) Effects of a nitric oxide synthase inhibitor in humans with septic shock. Cardiovasc Res 28(1):34–39

Pfeiffer S et al (1996) Inhibition of nitric oxide synthesis by NG-nitro-L-arginine methyl ester (L-NAME): requirement for bioactivation to the free acid, NG-nitro-L-arginine. Br J Pharmacol 118(6):1433–1440

Raman CS et al (1998) Crystal structure of constitutive endothelial nitric oxide synthase. Cell:939–950. https://doi.org/10.1016/s0092-8674(00)81718-3

Ramasamy S et al (2014) Probing the mechanism of tetrahydrobiopterin radical reduction within NO synthases. Nitric Oxide:138–139. https://doi.org/10.1016/j.niox.2014.09.118

Reif A et al (1999) Tetrahydrobiopterin inhibits monomerization and is consumed during catalysis in neuronal NO synthase. J Biol Chem 274(35):24921–24929

Reif DW et al (2000) Discovery and development of neuronal nitric oxide synthase inhibitors. Free Rad Biol Med:1470–1477. https://doi.org/10.1016/s0891-5849(00)00250-1

Rico D et al (2007) Identification of conserved domains in the promoter regions of nitric oxide synthase 2: implications for the species-specific transcription and evolutionary differences. BMC Genom 8:271

Rosenfeld RJ et al (2002a) Conformational changes in nitric oxide synthases induced by chlorzoxazone and nitroindazoles: crystallographic and computational analyses of inhibitor potency. Biochemistry 41(47):13915–13925

Rosenfeld RJ et al (2002b) Human endothelial nitric oxide synthase with 5-nitroindazole bound. https://doi.org/10.2210/pdb1m9q/pdb

Rosenfeld RJ et al (2002c) Human endothelial nitric oxide synthase with 6-nitroindazole bound. https://doi.org/10.2210/pdb1m9m/pdb

Rosenfeld RJ et al (2002d) Human endothelial nitric oxide synthase with 7-Nitroindazole bound. https://doi.org/10.2210/pdb1m9k/pdb

Rosenfeld RJ et al (2002e) Inducible nitric oxide synthase with 5-nitroindazole bound. https://doi.org/10.2210/pdb1m8i/pdb

Salem R, Mebazaa A (2007) Nitric oxide inhibition rapidly increases blood pressure with no change in outcome in cardiogenic shock: the TRIUMPH trial. Crit Care Soc Crit Care Med 11(3):136

Salerno JC et al (1997) An autoinhibitory control element defines calcium-regulated isoforms of nitric oxide synthase. J Biol Chem:29769–29777. https://doi.org/10.1074/jbc.272.47.29769

Sandstrom P et al (2004) Highly selective iNOS inhibition and sphincter of Oddi motility in the Australian possum. Acta Physiol Scand 181(3):321–331

Sandstrom P et al (2005) Highly selective inhibition of inducible nitric oxide synthase ameliorates experimental acute pancreatitis. Pancreas 30(1):e10–e15

Sattler R, Tymianski M (2000) Molecular mechanisms of calcium-dependent excitotoxicity. J Mol Med:3–13. https://doi.org/10.1007/s001090000077

Scheschowitsch K et al (2015) Rapid NOS-1-derived nitric oxide and peroxynitrite formation act as signaling agents for inducible NOS-2 expression in vascular smooth muscle cells. Pharmacol Res 100:73–84

Schinzel R, Tegtmeier F (2017) Nitric oxide synthase inhibitors in traumatic brain injury. New Ther Traumatic Brain Injury:133–144. https://doi.org/10.1016/b978-0-12-802686-1.00008-0

Schleicher M et al (2009) The Akt1-eNOS Axis illustrates the specificity of kinase-substrate relationships in vivo. Sci Signal:ra41. https://doi.org/10.1126/scisignal.2000343

Schmidt HH et al (1988) Arginine is a physiological precursor of endothelium-derived nitric oxide. Eur J Pharmacol 154(2):213–216

Schmidt HH et al (1991) Purification of a soluble isoform of guanylyl cyclase-activating-factor synthase. Proc Natl Acad Sci U S A 88(2):365–369

Schmidt HHHW et al (1992) Calcium/calmodulin-dependent nitric oxide synthase type I: a biopteroflavoprotein with calcium/calmodulin-independent diaphorase and reductase activities. Biochemistry:3243–3249. https://doi.org/10.1021/bi00127a028

Schmidt HHHW, Harald HH, Walter U (1994) NO at work. Cell:919–925. https://doi.org/10.1016/0092-8674(94)90267-4

Searles CD (2006) Transcriptional and posttranscriptional regulation of endothelial nitric oxide synthase expression. Am J Physiol Cell Physiol 291(5):C803–C816

Seddon MD et al (2008) Neuronal nitric oxide synthase regulates basal microvascular tone in humans in vivo. Circulation 117(15):1991–1996

Seddon M et al (2009) Effects of neuronal nitric oxide synthase on human coronary artery diameter and blood flow in vivo. Circulation 119(20):2656–2662

Sennequier N, Wolan D, Stuehr DJ (1999) Antifungal imidazoles block assembly of inducible NO synthase into an active dimer. J Biol Chem 274(2):930–938

Sessa WC et al (1994) Chronic exercise in dogs increases coronary vascular nitric oxide production and endothelial cell nitric oxide synthase gene expression. Circ Res 74(2):349–353

Seymour M et al (2012) Ultrasonographic measures of synovitis in an early phase clinical trial: a double-blind, randomised, placebo and comparator controlled phase IIa trial of GW274150 (a selective inducible nitric oxide synthase inhibitor) in rheumatoid arthritis. Clin Exp Rheumatol 30(2):254–261

Shabeeh H et al (2013) Differential role of endothelial versus neuronal nitric oxide synthase in the regulation of coronary blood flow during pacing-induced increases in cardiac workload. Am J Physiol Heart Circulat Physiol 304(9):H1277–H1282

Shabeeh H et al (2017) Blood pressure in healthy humans is regulated by neuronal NO synthase. Hypertension 69(5):970–976

Shahani N, Sawa A (2012) Protein S-nitrosylation: role for nitric oxide signaling in neuronal death. Biochim Biophys Acta 1820(6):736–742

Sherman PA et al (1993) Purification and cDNA sequence of an inducible nitric oxide synthase from a human tumor cell line. Biochemistry:11600–11605. https://doi.org/10.1021/bi00094a017

Singh D et al (2007) Selective inducible nitric oxide synthase inhibition has no effect on allergen challenge in asthma. Am J Respir Crit Care Med 176(10):988–993

Smith AD et al (2005) Use of exhaled nitric oxide measurements to guide treatment in chronic asthma. New Engl J Med:2163–2173. https://doi.org/10.1056/nejmoa043596

Smith BC et al (2013) Nitric oxide synthase domain interfaces regulate electron transfer and calmodulin activation. Proc Natl Acad Sci:E3577–E3586. https://doi.org/10.1073/pnas.1313331110

Song Y, Zweier JL, Yong XIA (2001) Heat-shock protein 90 augments neuronal nitric oxide synthase activity by enhancing Ca2 /calmodulin binding. Biochem J 357. https://doi.org/10.1042/0264-6021:3550357

Stover JF et al (2014) Nitric oxide synthase inhibition with the antipterin VAS203 improves outcome in moderate and severe traumatic brain injury: a placebo-controlled randomized phase IIa trial (NOSTRA). J Neurotrauma 31(19):1599–1606

Stuehr DJ, Haque MM (2019) Nitric oxide synthase enzymology in the 20 years after the Nobel prize. Br J Pharmacol 176(2):177–188

Stuehr D, Pou S, Rosen GM (2001) Oxygen reduction by nitric-oxide synthases. J Biol Chem:14533–14536. https://doi.org/10.1074/jbc.r100011200

Szabolcs MJ et al (2002) Effects of selective inhibitors of nitric oxide synthase-2 dimerization on acute cardiac allograft rejection. Circulation 106(18):2392–2396

Teerlink T (2007) Tilarginine in patients with acute myocardial infarction and cardiogenic shock. JAMA:969. https://doi.org/10.1001/jama.298.9.971-b

Tegtmeier F et al (2020a) Correction to: efficacy of Ronopterin (VAS203) in patients with moderate and severe traumatic brain injury (NOSTRA phase III trial): study protocol of a confirmatory, placebocontrolled, randomised, double blind, multi-Centre study. Trials 21(1):172

Tegtmeier F et al (2020b) Efficacy of Ronopterin (VAS203) in patients with moderate and severe traumatic brain injury (NOSTRA phase III trial): study protocol of a confirmatory, placebo-controlled, randomised, double blind, multi-centre study. Trials 21(1):80

Thomas SR, Chen K, Keaney JF (2002) Hydrogen peroxide activates endothelial nitric-oxide synthase through coordinated phosphorylation and dephosphorylation via a phosphoinositide 3-kinase-dependent signaling pathway. J Biol Chem:6017–6024. https://doi.org/10.1074/jbc.m109107200

Thomsen LL et al (1997) Selective inhibition of inducible nitric oxide synthase inhibits tumor growth in vivo: studies with 1400W, a novel inhibitor. Cancer Res 57(15):3300–3304

Tinker A, Wallace A (2006) Selective inhibitors of inducible nitric oxide synthase: potential agents for the treatment of inflammatory diseases? Curr Top Med Chem:77–92. https://doi.org/10.2174/156802606775270297

Tinker AC et al (2003) 1,2-dihydro-4-quinazolinamines: potent, highly selective inhibitors of inducible nitric oxide synthase which show antiinflammatory activity in vivo. J Med Chem:913–916. https://doi.org/10.1021/jm0255926

Tiso M et al (2008) BYK191023 (2-[2-(4-methoxy-pyridin-2-yl)-ethyl]-3h-imidazo[4,5-b]pyridine) is an NADPH- and time-dependent irreversible inhibitor of inducible nitric-oxide synthase. Mol Pharmacol 73(4):1244–1253

TRIUMPH Investigators et al (2007) Effect of tilarginine acetate in patients with acute myocardial infarction and cardiogenic shock: the TRIUMPH randomized controlled trial. JAMA 297 (15):1657–1666

Tseng EE et al (1999) Nitric oxide mediates neurologic injury after hypothermic circulatory arrest. Ann Thorac Surg 67(1):65–71

Vamvakas S, Schmidt HH (1997) Just say NO to cancer? J Natl Cancer Inst:406–407

Van der Schueren BJ et al (2009) Does the unfavorable pharmacokinetic and pharmacodynamic profile of the iNOS inhibitor GW273629 lead to inefficacy in acute migraine? J Clin Pharmacol 49(3):281–290

van Hoogdalem E-J et al (2019) First-in-human study of the safety, tolerability, pharmacokinetics and - preliminary dynamics of neuroprotectant 2-iminobiotin in healthy subjects. Curr Clin Pharmacol. https://doi.org/10.2174/1574884714666191017111109

Vaughan D et al (2010) Safety and pharmacokinetics of NXN-188 after single and multiple doses in five phase I, randomized, double-blind, parallel studies in healthy adult volunteers. Clin Ther 32 (1):146–160

Vuolteenaho K et al (2001) Regulation of the nitric oxide production resulting from the glucocorticoid-insensitive expression of iNOS in human osteoarthritic cartilage. Osteoarthritis Cartilage OARS Osteoarthritis Res Soc 9(7):597–605

Wei X et al (2020) Direct current electric field stimulates nitric oxide production and promotes NO-dependent angiogenesis: involvement of the PI3K/Akt signaling pathway. Journal of vascular research, pp.:1–11

Weinberg JB et al (1995) Human mononuclear phagocyte inducible nitric oxide synthase (iNOS): analysis of iNOS mRNA, iNOS protein, biopterin, and nitric oxide production by blood monocytes and peritoneal macrophages. Blood 86(3):1184–1195

Werner ER, Schmidt HHW (2000) Nitric-oxide-synthase inhibitors II — Pterin antagonists/anti-pterins. Nitric Oxide:137–158. https://doi.org/10.1007/978-3-642-57077-3_7

Willmot M et al (2005) Nitric oxide synthase inhibitors in experimental ischemic stroke and their effects on infarct size and cerebral blood flow: a systematic review. Free Radic Biol Med 39 (3):412–425

Woods CM et al (2007) Selective iNOS inhibition enhances spontaneous gallbladder motility in the Australian possum. Neurogastroenterol Mot 19(6):497–503

Wynia-Smith SL, Smith BC (2017) Nitrosothiol formation and S-nitrosation signaling through nitric oxide synthases. Nitric Oxide:52–60. https://doi.org/10.1016/j.niox.2016.10.001

Yuan Z et al (2004) Evidence of nuclear localization of neuronal nitric oxide synthase in cultured astrocytes of rats. Life Sci:3199–3209. https://doi.org/10.1016/j.lfs.2003.10.037

Zamora R, Vodovotz Y, Billiar TR (2000) Inducible nitric oxide synthase and inflammatory diseases. Mol Med:347–373. https://doi.org/10.1007/bf03401781

Zhang ZG et al (1996) ARL 17477, a potent and selective neuronal NOS inhibitor decreases infarct volume after transient middle cerebral artery occlusion in rats. J Cereb Blood Flow Metab 16 (4):599–604

Zhou L, Zhu D-Y (2009) Neuronal nitric oxide synthase: structure, subcellular localization, regulation, and clinical implications. Nitric Oxide Biol Chem 20(4):223–230

Zhou L et al (2010) Treatment of cerebral ischemia by disrupting ischemia-induced interaction of nNOS with PSD-95. Nat Med 16(12):1439–1443

Zitta K et al (2017) 2-Iminobiotin superimposed on hypothermia protects human neuronal cells from hypoxia-induced cell damage: an study. Front Pharmacol 8:971

Xanthine Oxidoreductase Inhibitors

Keeran Vickneson and Jacob George

Contents

Abstract

Xanthine oxidase inhibitors are primarily used in the clinical prevention and treatment of gout associated with hyperuricemia. The archetypal xanthine oxidase inhibitor, Allopurinol has been shown to have other beneficial effects such as a reduction in vascular reactive oxygen species and mechano-energetic uncoupling. This chapter discusses these properties and their relevance to human pathophysiology with a focus on Allopurinol as well as newer xanthine oxidase inhibitors such as Febuxostat and Topiroxostat.

K. Vickneson · J. George (✉)
Ninewells Hospital, Dundee, Scotland, UK
e-mail: j.george@dundee.ac.uk

H. H. H. W. Schmidt et al. (eds.), *Reactive Oxygen Species*,
Handbook of Experimental Pharmacology 264, https://doi.org/10.1007/164_2020_383

Graphical Abstract

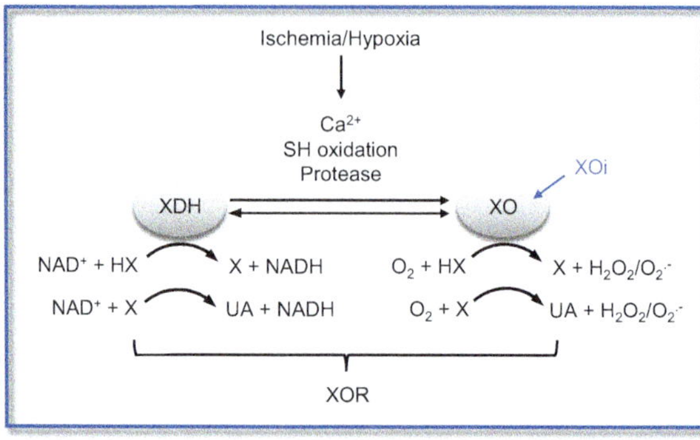

Xanthine oxidase (XO) and xanthine dehydrogenase (XDH) are collectively referred to as xanthine oxidoreductase (XOR). XDH is initially synthesised as a 150-kDa protein from which XO is derived, e.g. under conditions of ischemia/hypoxia either reversibly by conformational changes (calcium or SH oxidation) or irreversibly by proteolysis, the latter leading to formation of a 130-kDa form of XO. Both, XO and XDH, catalyse the conversion of hypoxanthine via xanthine to uric acid, the former by using oxygen forming superoxide and hydrogen peroxide and the latter NAD^+. However, XDH is in principle also able to generate ROS.

Keywords

Antioxidants · Endothelial dysfunction · Oxidative stress · Uric acid · Xanthine oxidoreductase

1 Urate and Xanthine Oxidoreductase

Urate is a heterocyclic purine derivative. In humans and some primates, it is the final product of purine breakdown. The majority of urate is filtered through the kidney (60–70%) and at least 90% of this is re-absorbed through the GLUT9 and URAT1 anion transporters. Humans lack a functional uricase gene that is present in most other lower mammals. Uricase further oxidises urate into water-soluble allantoin (Chen et al. 2016).

Urate is thought to exert opposing actions on ROS extracellularly versus intracellularly. Circulating urate is thought to contribute to 70% of all free radical scavenging capability of plasma. It is an effective scavenger of carbon-centred radicals and peroxylradicals (Waring 2002). For example, it reacts with peroxynitrite ($ONOO^-$) to release nitric oxide (NO) and therefore induce vasodilatation (Skinner et al. 1998). However, under hydrophobic conditions, it can accelerate the oxidation of LDL, increased monocyte-chemoattractant protein (MCP)-1, high sensitivity C-reactive protein and inflammatory interleukins (Bagnati et al. 1999). Therefore, it is thought that overall, hyperuricaemia contributes to the progression of CV disease because of the overwhelming oxidant property of urate (Chen et al. 2016) as well as the free radical by-products of its formation (George and Struthers 2008) (see below).

Xanthine oxidoreductase (XOR) is part of a group of enzymes known as the molybdenum iron-sulphur flavin hydroxylases. It was first discovered in milk by Schardinger in 1902 (Berry and Hare 2004) and is thought to be involved in reactions that produce reactive oxygen species (ROS) such as nitrite which enable newborn infants to overcome gut-associated bacterial gastroenteritis (Hancock et al. 2002; Stevens et al. 2000). XOR is widely distributed throughout various organs including the liver, gut, lung, kidney, heart, brain and plasma (Pacher et al. 2006) with the highest levels being found in the gut and the liver (Parks and Granger 1986). In the myocardium, it is localised to the capillary endothelial cells (Cicoira et al. 2002). The gene encoding for XOR is located at the short arm of chromosome 2 (Ichida et al. 1993). It exists in two interconvertible forms known as xanthine oxidase (EC 1.1.3.22) and xanthine dehydrogenase (EC 1.17.1.4) (Della Corte et al. 1969). Both enzymes consist of two identical subunits of 145 kDa.

The mechanism by which XOR catalyses hypoxanthine and xanthine conversion is complex and has been previously described in detail (Berry and Hare 2004; Hille and Massey 1981). A fully reduced XO contains six electrons, and its re-oxidation involves electron transfer to oxygen molecules which generates two H_2O_2 and two O_2^- species (Hille and Massey 1981) for every fully reduced XO molecule. It is interesting to note that XDH can theoretically produce more O_2^- per mole of oxygen during NADH oxidisation than XO. Along with NADPH oxidase, it is a major generator of ROS in the human.

2 Allopurinol

Allopurinol ($C_5H_4N_4O$) is the archetypal and longest established XO inhibitor in clinical use. It is a weak acid with a dissociation constant (pKa) of 9.4. It has a molecular mass of 136.11 g/mol. It is rapidly converted to oxypurinol by aldehyde oxidoreductase. While allopurinol is an analogue of hypoxanthine, oxypurinol (or alloxanthine) is an analogue of xanthine (Day et al. 2007). Oxypurinol is much more lipid soluble than allopurinol (octanol/water partition coefficient 14 vs. 0.28 for allopurinol) (Day et al. 2007).

2.1 Pharmacokinetics

Approximately 90% of allopurinol is absorbed from the gastrointestinal tract. It is rapidly absorbed and reaches peak plasma concentrations within 30–60 min (Pea 2005) following oral administration. It is rapidly metabolised to its active metabolite oxypurinol. For every 100 mg oral dose of allopurinol, 90 mg of oxypurinol is formed (Day et al. 2007). When given orally, oxypurinol has a lower bioavailability than allopurinol. An early pharmacokinetics study showed that 300 mg of allopurinol produced a slightly higher plasma levels than 600 mg of oral oxypurinol (Elion et al. 1968). Allopurinol has a short half-life in plasma between 2 and 3 h and has negligible protein binding. It has an apparent volume of distribution of 1.2 to 2.2 L/kg in healthy volunteers (Day et al. 2007). The renal clearance of allopurinol is 1.54 mL/min/kg (Day et al. 2007). Oxypurinol is detected in plasma 15–20 min after an oral dose of allopurinol, reaches peak plasma levels in 3–4 h (Guerra et al. 2001) and has a much longer plasma half-life between 14 and 30 h because it is reabsorbed in the proximal tubule of the kidney (Pea 2005). It is responsible for much of the hyperuricaemic action of allopurinol. It is excreted almost entirely unchanged in urine. Therefore the renal clearance of oxypurinol is therefore the most important aspect of the clinical pharmacokinetics of allopurinol (Day et al. 2007). Patients treated with allopurinol excrete 70% of the dose as oxypurinol and only 10% as allopurinol which indicates that the vast majority of allopurinol is converted to oxypurinol (Elion et al. 1966). Thus, the major route of allopurinol elimination is via oxidation to oxypurinol (Turnheim 1999). Turnheim et al. showed that although allopurinol elimination is not reduced in the aged, the elimination of its metabolite oxypurinol is reduced due to age-related reduction in renal function (Turnheim 1999).

The mechanism of XO inactivation by oxypurinol was determined by Massey et al. (Massey et al. 1970). Oxypurinol strongly binds at the active site of XO, resulting in the reduction of Mo^{VI} to Mo^{IV}. Its inhibition is time-dependent, and it is important to maintain an effective concentration of the inhibitor as spontaneous oxidation back from Mo^{IV} to Mo^{VI} will result in concomitant recovery of enzyme activity. It is an excellent substrate with a V_{max} which is sixfold faster than xanthine. The mechanism of oxypurinol inhibition of its own production has been termed "suicidal" (Spector 1977). In a healthy individual with a creatinine clearance of 120 mL/min, the clearance of oxypurinol is 23 mL/min. As oxypurinol is a small molecule that is not bound to plasma proteins, it is freely filtered at the glomerulus (Elion et al. 1968)

2.2 Dose-Response Studies

Dose-response studies of the hyperuricaemic effect of allopurinol suggest that this increases relatively little with increasing doses of the drug (Day et al. 2007). The

Table 1 Steady-state oxypurinol concentration in healthy volunteers (adapted from Ref. (Graham et al. 1996))

Allopurinol dose (mg)	Oxypurinol concentration (mg/L) [approx concentration in mmol/L]
50	1.77 ± 1.59 [0.01]
100	2.67 ± 1.59 [0.02]
300	5.59 ± 1.5 [0.04]
600	9.56 ± 1.92 [0.07]
900	12.21 ± 2.13 [0.09]

EC_{50} for allopurinol has been calculated as 5.6 ± 1.3 mg/L, which is identical to the trough level for the 300 mg/day dose (5.6 ± 0.6 mg/L). The steady-state oxypurinol concentration over a dose range of 50-900 mg allopurinol/day is shown in Table 1.

The oxypurinol concentration in steady-state was found to increase in a linear fashion up to the 600 mg/day dose of allopurinol. The concentration of oxypurinol did not increase proportionally between 600 mg/day and 900 mg/day (Graham et al. 1996) suggesting that tubular reabsorption of oxypurinol may be saturated at higher doses. The other possible explanation could be the saturation of xanthine oxidase, but this is unlikely as Spiekermann et al. (Spiekermann et al. 2003) have recently shown that complete inhibition of plasma XO activity in vivo requires an oxypurinol concentration of 1 mmol/L. Beneficial effects seen with higher dose could then be related to other effects of allopurinol. Although it is tempting to relate the efficacy of allopurinol to the degree of urate lowering it produces, there are many more factors that contribute to urate levels such as exogenous contributions from diet, endogenous production as well as renal function.

2.3 Pharmacodynamics

Allopurinol is generally safe and well tolerated since it was introduced into clinical practice 40 years ago. By the end of the 1980s, more than five million patient years of treatment and over 240 million doses had been prescribed (Vazquez-Mellado et al. 2001). Common adverse effects of allopurinol are gastrointestinal disturbance, hypersensitivity reactions (up to 8% of patients, sometimes occurring months to years after commencing medication) and skin rash (Committee 2007). In a study by McInnes et al., 6.2% of hospitalised patients monitored in a drug surveillance programme received allopurinol. After the exclusion of skin reactions, the most frequent reactions found were haematological abnormalities (0.6%), diarrhoea (0.3%) and pyrexia (0.3%). These adverse effects were found to be dose-related (McInnes et al. 1981). The rare allopurinol hypersensitivity syndrome (fever, rash, vasculitis, eosinophilia and renal failure) occurs in 0.4% of patients but can have a mortality of up to 25% (Gutierrez-Macias et al. 2005). It has been recently discovered that the HLA-B*5801 allele is a very significant risk factor for the allopurinol hypersensitivity syndrome (Hung et al. 2005) in Chinese patients. Whether or not this allele confers the same risk to other populations is yet to be known.

Clinically significant interactions between allopurinol and the endogenous purines mercaptopurine and azathioprine have been reported. As mentioned earlier, the initial discovery of allopurinol was as an agent to potentiate the anti-tumour effects of mercaptopurine so it is unsurprising that mercaptopurine levels are augmented by allopurinol because it is metabolised by XO into inactive metabolites (Pea 2005). Allopurinol enhances the anticoagulant effect of warfarin and increases the plasma concentration of didanosine, ciclosporin and theophylline (Committee 2007). The risk of allopurinol hypersensitivity syndrome is increased in elderly patients on thiazide diuretics (Schlesinger 2004) and ampicillin (Vazquez-Mellado et al. 2001) Both allopurinol and urate are removed by dialysis (Day et al. 2007).

2.4 Indirect Antioxidant Action (XO-Inhibition Mediated)

Allopurinol has also been shown to normalise endothelial dysfunction in individuals with Type 2 diabetes with mild hypertension and reduced plasma malondialdehyde (MDA) levels (Butler et al. 2000). MDA results from acid hydrolysis of lipid peroxides which are formed by free radical attack on plasma lipoproteins. It is therefore used as an indirect measure of oxidised LDL.

In the experimental murine myocardial infarction model, allopurinol significantly attenuated LV dilatation, hypertrophy, fibrosis and dysfunction. Once again, XO expression (as determined by electron spin resonance spectroscopy) and myocardial ROS generation were markedly increased in the post-MI ischemic model (Engberding et al. 2004). This suggests a role for allopurinol in LV remodelling, a possibility that we are investigating at present in our unit. Allopurinol has also been shown to be beneficial in conditions such as post coronary artery bypass surgery where it reduced ischemic events and produced less ST segment depression (Sisto et al. 1995) as well as in hypercholesterolaemic patients (Cardillo et al. 1997). Animal studies in other conditions such as diabetic retinopathy have yielded similar results both in terms of indirect and direct (see below) action of allopurinol. Allopurinol significantly improved the b-wave amplitude on electroretinography as well as 8-isoprostanes, a biomarker of ROS formation. Despite lowering urate to a similar degree, Benzobromarone did not result in any beneficial effect (Goharinia et al. 2017).

A recent placebo-controlled clinical trial ($n = 100$) in patients with acute coronary syndrome (ACS) showed early (1 month) reduction of markers of oxidative stress malondialdehyde (MDA), oxidised LDL. This reduction was sustained for up to 2 years (Huang et al. 2017).

Allopurinol in chronic heart failure (CHF) was assessed by Doehner et al. (Doehner et al. 2002) and by Farquharson et al. (Farquharson et al. 2002). Doehner et al. showed that the degree of improvement in forearm blood flow correlated with the degree of urate lowering. Interestingly, they also measured allantoin, a marker of oxygen free radical generation, which was reduced by 20% following 300 mg/day allopurinol. Farquharson et al. (Farquharson et al. 2002) from our unit found a 181% change in forearm blood flow with 300 mg allopurinol. They also found a 33% reduction in plasma MDA levels in patients treated with 300 mg allopurinol suggesting that the improvement in endothelial function and NO bioavailability seen was due (at least in part) to a reduction of ROS. Allopurinol also reduced B-type Natriuretic peptide (BNP) in stable CHF patients, although the reduction did not correlate with the fall in urate (Gavin and Struthers 2005). Uric acid also directly inactivates NO (Gersch et al. 2008), and therefore allopurinol may increase NO bioavailability through this indirect pathway also.

Our group demonstrated, for the first time, the antioxidant effect of high-dose allopurinol in reducing vascular oxidative stress. We studied patients with chronic heart failure and found that the effect of allopurinol on endothelial vascular function was due to xanthine oxidase inhibition and not urate lowering (George et al. 2006). We also demonstrated that there was a steep dose-response curve with high dose

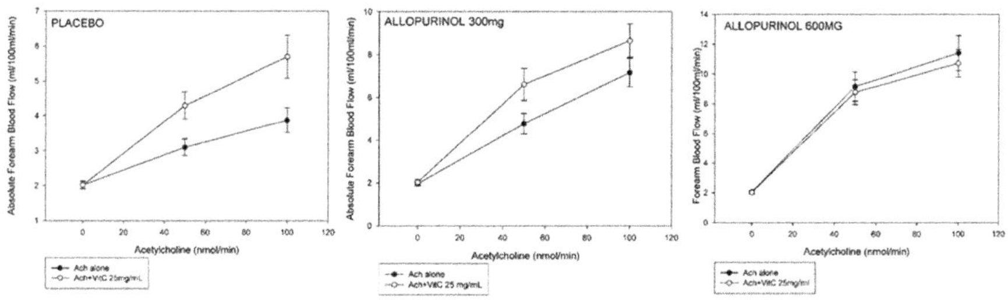

Fig. 1 Absolute forearm blood flow data for acetylcholine (50, 100 nmol/min) + vitamin C 25 mg/mL vs acetylcholine alone for – placebo, 300 mg allopurinol and 600 mg allopurinol (mean ± SEM)

allopurinol (600 mg/day) significantly better than standard dose (300 mg/day) in this respect. At high dose, allopurinol completely negated the benefits seen with high dose intra-arterial vitamin C infusions (Fig. 1). This is further strengthened by evidence that the beneficial effect of vitamin C co-infusion in patients with CHF was greatest in patients with the highest levels of oxidative stress as measured by extracellular SOD (ecSOD) (Landmesser et al. 2002) and XO activity.

This finding has been further confirmed by two other studies in coronary artery disease (Rajendra et al. 2011) and heart failure (Ogino et al. 2009) and is now widely accepted as a possible mechanism for the benefits seen with allopurinol. Our group has also previously demonstrated that high-dose (but not low-dose) allopurinol reduced oxidised LDL, further confirming allopurinol's antioxidant impact. However, as xanthine oxidase is significantly upregulated in acute ischemia or inflammation (Spiekermann et al. 2003) and otherwise constitutionally is expressed at low levels (Panus et al. 1992), there remains doubt that treating high-risk stable patients with long-term high-dose allopurinol will provide any benefit at all. Large ongoing trials such as ALL-HEART are seeking to address this question (Mackenzie et al. 2016). The biology of XO suggests that allopurinol is most beneficial in the acute ischaemia/reperfusion/inflammation setting rather than the chronic stable setting.

The other possible explanation which also relates to urate and superoxide formation is that in patients with low baseline oxidative stress, there are proportionately more urate (a known antioxidant) molecules to combat oxidative stress. In patients with high background or ischemia-induced oxidative stress however, inhibition of XO will reduce proportionately more superoxide (due to the cascade formation of superoxide). The reduction in urate with allopurinol may be an unfortunate price to pay, and the system may already be overwhelmed at this stage. This is supported by evidence in rat myocardium where the magnitude of functional improvement seen with XO inhibitors was dependent on the initial level of XO activity (Kogler et al. 2003). In chronic diseases such as CHF, sustained high levels of ROS may exceed the capacity of cellular enzymatic and non-enzymatic antioxidants (Deanfield et al. 2007) to counter its effects. Using electron spin resonance, Spiekermann et al. demonstrated that both NADPH oxidase and xanthine oxidase are upregulated in patients with coronary artery disease (Spiekermann et al. 2003). Others have demonstrated increased levels in CHF (Landmesser et al. 2002; Amado et al. 2005).

2.5 Direct Antioxidant Action

Allopurinol directly scavenges free radicals as demonstrated by Das et al. and others (Das et al. 1987; Hoey et al. 1988; Ricardo et al. 1995) in in vitro hearts where evidence of free radical scavenging occurred in the absence of XO activity. Further evidence for a possible direct antioxidant effect of allopurinol comes from models of experimental colitis where tungsten (a potent XO inhibitor) failed to improve symptoms whereas allopurinol did (Keshavarzian et al. 1990). Augustin et al. suggested that this direct effect was only seen at higher doses (Augustin et al. 1994). This was also seen in mice paracetamol toxicity models where lower doses (sufficient to block XO activity) of allopurinol failed to show antioxidant protection but higher doses did (Knight et al. 2001). There have been other non-XO effects of allopurinol suggested such as copper chelation, preventing LDL oxidation as described above (Malkiel et al. 1993), inhibition of heat shock protein (hsp) expression (Nishizawa et al. 1999) and calcium sensitisation (below). Allopurinol treatment reduces early changes in inflammation such as leukocyte activation by reducing adherence, rolling and extravasation (Granger et al. 1989). Similarly, animal studies in global cerebral ischemia-reperfusion have demonstrated that the ROS lowering effect of allopurinol was not related to its XO inhibition activity but rather due to its direct free radical scavenging activity. This was not evident with febuxostat (Yamaguchi et al. 2015).

Animal studies in experimentally induced uveitis show that at very high doses (up to 50 mg/kg), allopurinol behaves as a free radical scavenger with intrinsic antioxidant properties. Crucially, this was only achieved far beyond the XO inhibition dose of 10 mg/kg and not at that dose itself.

2.6 Mechano-energetic Uncoupling

This phenomenon refers to an imbalance between left ventricular performance and myocardial energy consumption (Kittleson and Hare 2005). The role of XO inhibition may either be to maintain cardiac output while reducing myocardial oxygen consumption or even to increase cardiac output without increasing myocardial oxygen consumption. In dogs with pacing-induced heart failure, allopurinol improved myocardial contractility, efficiency in oxygen utilisation, prevented increases in systemic vasoconstriction and ameliorates reductions in myocardial contractility (Amado et al. 2005; Ekelund et al. 1999; Saavedra et al. 2002). In murine post-ischaemic cardiomyopathy models, allopurinol attenuated the increase in end-systolic and end-diastolic volumes (Naumova et al. 2006), increased survival, augmented ventricular function as well as reduced products of lipid peroxidation (Stull et al. 2004).

Khan et al. found a direct protein-protein interaction between XO and neuronal NOS in the sarcoplasmic reticulum of cardiac myocytes (Khan et al. 2004). Allopurinol improved myofilament calcium sensitivity as contraction force increases without a concomitant rise in systolic Ca^{2+} influx. The effects were not seen in endothelial NOS-deficient mice suggesting a role for neuronal NOS preventing XO inhibition of cardiac excitation-contraction coupling (Khan et al. 2004). The finding that allopurinol is a potent myofilament Ca^{2+} sensitizer, particularly in the setting of ischaemia, is thought to be due to the inhibition of basal XO production. As with the previous study by Khan et al., Perez et al. found an almost exclusive increase in force generation without a lowering of inward transient Ca^{2+} (Perez et al. 1998).

Despite the small sample size ($n = 9$), Cappola et al. showed using cardiac catheterisation that direct intra-coronary infusions of allopurinol in these patients resulted in a marked decrease in myocardial oxygen consumption (MVO2) with no decrease in the rate of left ventricular pressure rise (dP/dT), stroke work or ventricular load (Cappola et al. 2001). Patients post-CABG given allopurinol have also been shown to require less inotropic support (Sisto et al. 1995).

As alluded to earlier, the most potent ROS-generating systems are the NADPH oxidase and xanthine oxidase enzymes, and angiotensin II is the most potent inducer of NADPH oxidase (Griendling et al. 1994; Harrison et al. 2003). However, as we have previously demonstrated, patients already on an ACE inhibitor or an AT1 receptor blocker still derive improvement in vascular function from XO inhibition suggesting that there is still a significant level of oxidative stress present even in patients who are optimally treated with current evidence-based treatments (George et al. 2006). These actions are summarised in Fig. 2.

3 Febuxostat

3.1 Pharmacokinetics

Febuxostat (2-(3-cyano-4-[2-methylpropoxyl]phenyl)-4-methylthiazole-5-carboxylic acid) is a thiazolecarboxylic acid derivative, selective for inhibition of both the oxidised and reduced forms of xanthine oxidase, and does not resemble a purine or pyrimidine (Ernst and Fravel 2009). Febuxostat has selective affinity for both the oxidised and reduced forms of xanthine oxidase, with an in vitro inhibition (*Ki*) value of <1 nM (mean [SD], 1.2 [0.05] $\times 10^{-10}$) (Takano et al. 2005). The drug has an oral bioavailability of 85% (Kamel et al. 2017), achieves maximum plasma concentration in approximately 1.5 h and has a mean elimination half-life varying between 1.3 and 15.8 h (Khosravan et al. 2006). Febuxostat is mainly metabolised via glucuronidation (22–44% of the dose) and oxidation (2–8%) with only 1–6% of the dose being excreted unchanged via the kidneys (Khosravan et al. 2006). Therefore renal function is not a key determinant in its use. It is now recognised that Febuxostat is at least as effective as allopurinol in urate reduction (Faruque et al. 2013).

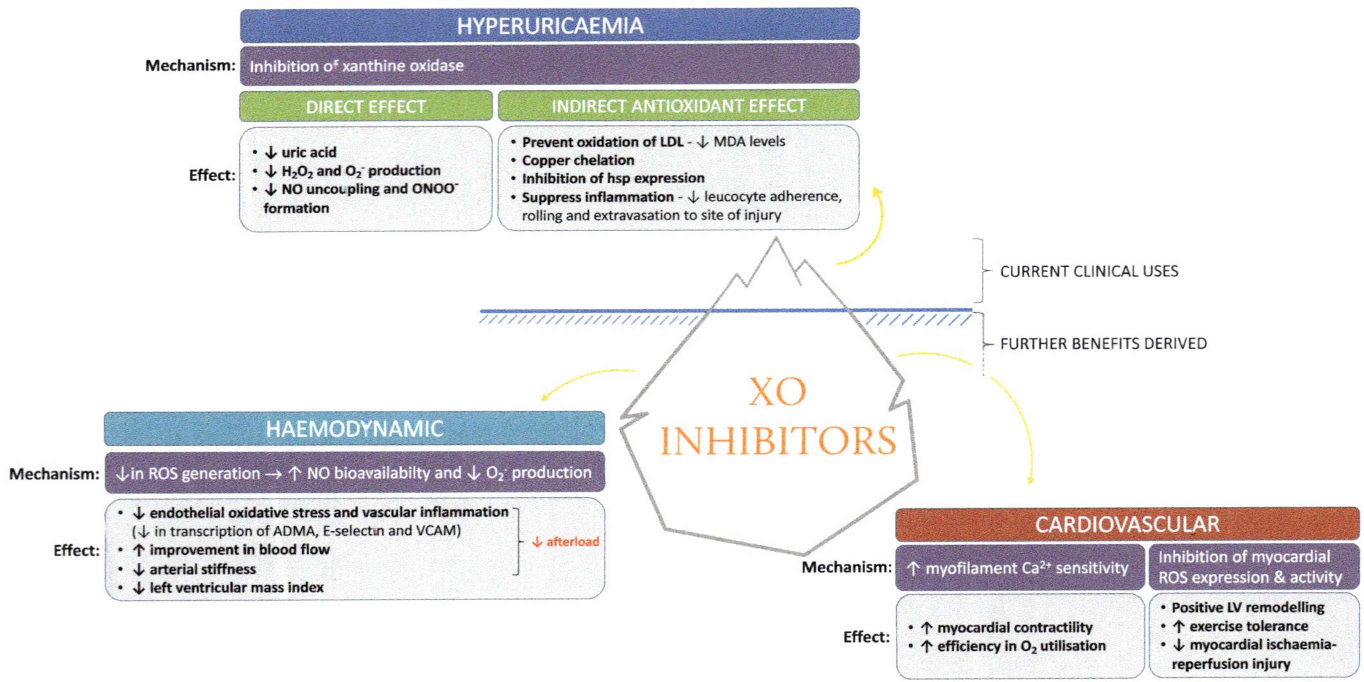

Fig. 2 Summary of effects of xanthine oxidase inhibitors on hyperuricaemia, cardiac function and cardiovascyular haemodynamics

Data on the antioxidant effects of febuxostat are conflicting. Theoretically, it is a better antioxidant agent as inflammatory/hypoxic conditions upregulate tissue XO expression which results in sequestration and immobilisation of XO by endothelial glycosaminoglycans (GAG) . Immobilised GAG-bound XO is resistant to allopurinol but not febuxostat (Malik et al. 2011). Detailed crystallography studies revealed that febuxostat reaction with XO is confined to critical amino acid residues in the tunnel leading to the Mo cofactor, where it effectively blocks substrate access to the active site (Okamoto and Nishino 2008). Thus, febuxostat should not be affected by enzyme redox state and interaction with XO should not induce ROS formation (Malik et al. 2011).

Animal models of renal ischemia-reperfusion have demonstrated amelioration of ROS and therefore tubular injury and interstitial fibrosis by febuxostat (Tsuda et al. 2012). In studies using streptozocin-induced diabetic rat model, febuxostat reduced both urinary 8-OHdG, significantly decreased renal infiltration of macrophages resulting in reduced oxidative stress, transcription levels of inflammatory genes (E-selectin and VCAM), inflammation-induced enzymes (COX-2), inflammatory mediators (NF-kB) and renal cortical nitrotyrosine. This suggests a possible therapeutic effect for febuxostat in slowing deterioration of kidney function in the setting of diabetic nephropathy (Lee et al. 2014).

In small studies of patients with gout, febuxostat has been shown to have superior effects on oxidative stress and pulse wave velocity compared to low-moderate dose allopurinol (Tausche et al. 2014). In a study of haemodialysis patients, febuxostat was shown to significantly reduce high-sensitivity CRP and asymmetric dimethylarginine (ADMA) levels and improve endothelial dysfunction (reduced ADMA-mediated eNOS inhibition) compared to placebo (Alshahawey et al. 2017). Furthermore, the increase in ADMA levels and inhibition of nitric oxide production seen with proton pump inhibitors (PPI's) has been shown to be blunted by febuxostat (Pinheiro et al. 2016). Febuxostat has also been shown to reduce other markers of oxidative stress such as oxidised LDL and EPA/AA (eicosapentaenoic acid/arachidonic acid) ratio (Sezai et al. 2013).

However when the impact of antioxidant defence is studied, febuxostat seems to reduce both biological antioxidant potential (BAP) and ROS metabolites (derivatives of reactive oxygen metabolites (d-ROMs) in equal measure (Fukui et al. 2015). In a study of obese adults with Type 2 diabetic nephropathy, febuxostat showed no effect on adipose tissue thiobarbituric acid reducing substances (TBARS) and adiponectin concentrations (Beddhu et al. 2016).

Recent large multi-centre trials of febuxostat have been reported showing mixed effects. The febuxostat for Cerebral and Cardiorenovascular Events Prevention Study (FREED) trial (Kojima et al. 2019) met its primary composite cardiovascular endpoint, but this was driven by a reduction in progression to renal dysfunction. There was no evidence of cardiovascular or cerebrovascular event rate reduction with febuxostat. This latter finding is consistent with the findings of the CARES (Cardiovascular Safety with Febuxostat vs Allopurinol) trial (White et al. 2018) which demonstrated no beneficial effect of febuxostat on cardiovascular events and in fact demonstrated that all-cause mortality and cardiovascular mortality were higher with febuxostat than with allopurinol (hazard ratio for death from any cause, 1.22 [95% CI, 1.01 to 1.47]; hazard ratio for cardiovascular death, 1.34 [95% CI, 1.03 to 1.73]). However, as Choi et al. point out, the use of non-XOI or placebo group is needed to determine whether the results of the CARES trial were due to the beneficial effects of allopurinol or the deleterious effects of febuxostat (Choi et al. 2018). Other ongoing trials such as Febuxostat versus Allopurinol Streamlined Trial (FAST) may provide some clarity to this issue.

3.2 Mechano-energetic Uncoupling

In a trial of hyperuricaemic patients undergoing cardiac surgery (NU-FLASH trial) comparing febuxostat with low-dose (300 mg) allopurinol, patients in the febuxostat arm showed significant reductions in systolic blood pressure, pulse-wave velocity and LV mass index compared to allopurinol (Sezai et al. 2013).

4 Topiroxostat

Topiroxostat [4-[5-(4-Pyridinyl)-1H-1,2,4-triazol-3-yl]-2- pyridinecarbonitrile] is a non-purine XOR inhibitor, approved in Japan in 2013 for the treatment of patients with hyperuricaemia. There is limited experience internationally with this agent. Topiroxostat behaves initially as a competitive type inhibitor to xanthine oxidoreductase before forming a strong covalent linkage to molybdenum via oxygen in the hydroxylation reaction intermediate (Chen et al. 2016). It also displays a potent non-covalent competitive type inhibition of XOR with a K_i value of 5.7×10^{-9} M (Matsumoto et al. 2011). Topiroxostat has good oral bioavailability with a half-life of up to 7.5 h after oral administration. It is predominantly eliminated in the urine. It is a strong inhibitor of Cyp 2C9 and has no inducing effect on CYP enzymes. Topiroxostat has a greater inhibitory effect on plasma XOR compared to tissue XOR (the opposite is observed with febuxostat) (Nakamura et al. 2016).

Mouse models of minimal change nephrotic syndrome demonstrated that nitrotyrosine and 8-hydroxy-2-deoxyguanosine (8-OHdG) were significantly ameliorated by topiroxostat (Kawamorita et al. 2017). The recently reported TROFEO trial (Sezai et al. 2017) in hyperuricaemic patients with cardiovascular disease comparing the effects of febuxostat and topiroxostat showed similar urate, antioxidant, anti-inflammatory and reno-protective effects for both drugs. The renoprotective effects of topiroxostat for hyperuricemic patients with overt diabetic nephropathy (ETUDE) study concluded that high-dose topiroxostat (160 mg/day) significantly reduced L-Fatty Acid Binding Protein (FABP), a validated biomarker of tubulointestitial damage and oxidative stress (Mizukoshi et al. 2018). There has not been any direct head-to-head antioxidant effect comparison between allopurinol and topiroxostat.

Table 2 summarises the current clinical trial evidence using XO inhibitors on ROS and other CV outcomes.

Table 2 Summary of clinical trials of xanthine oxidase inhibitors and impact on oxidative stress

Study, year [Ref]	Species [human (+ disease area)/ mouse/rat]	Intervention (drug, dose)	Biomarkers measured	Brief results
Butler et al. (2000)	Human (type II diabetes)	Allopurinol 300 mg/day	MDA	– Improved endothelial function through significant reduction in MDA levels (marker of lipid peroxidation, indirect measure of oxidised LDL), in diabetic patients receiving allopurinol
			Forearm blood flow ratio (infused/control arm)	– Increased forearm blood flow response in response to acetylcholine by approximately 30% ($p = 0.0012$)
Engberding et al. (2004)	Mouse [myocardial infarction]	Allopurinol (20 mg/kg/ day)	Myocardial superoxide	– Significant reduction in ROS generation by the myocardium after MI following treatment with allopurinol – Positive remodelling of the LV myocardium as evidenced by improvement in LV ejection fraction and attenuation of myocardial hypertrophy and fibrosis
Goharinia et al. (2017)	Rat [diabetic rat model]	Allopurinol (50 mg/kg/ day) vs benzbromarone (10 mg/kg/day)	8-iso-prostaglandin F2α, MDA	– Allopurinol attenuated increased oxidative stress in diabetic rates through significant reductions in plasma 8-Iso-F2α levels in vivo – Retinal MDA levels were also reduced by allopurinol but this effect was non-significant

(continued)

Table 2 (continued)

Study, year [Ref]	Species [human (+ disease area)/ mouse/rat]	Intervention (drug, dose)	Biomarkers measured	Brief results
				– Benzbromarone showed similar efficacy to allopurinol in lowering plasma urate levels but the same effect on oxidative stress was not seen
Huang et al. (2017)	Human [acute coronary syndrome]	Allopurinol (600 mg for 14 days followed by 200 mg daily)	MDA, ox-LDL	– Early reduction in oxidative stress markers and lipid peroxidation were observed with allopurinol, which was sustained during the 24-month follow-up ($p < 0.05$) – Increased endothelium dependent vasodilation with increased bioavailability of NO in the allopurinol group
			Hs-CRP, TNF-α	– Significantly steeper decline in systemic inflammatory markers with allopurinol intervention compared to control
Doehner et al. (2002)	Human [chronic heart failure]	Allopurinol	Allantoin	– Reduction in allantoin levels from 26.1 ± 1.2 μmol/L to 20.8 ± 0.6 μmol/L ($p < 0.001$) – Significant carryover effect (patients receiving placebo after crossover from allopurinol treatment arm) with reduction in allantoin levels (<0.05) but no similar effect was seen with uric acid levels

Farquharson et al. (2002)	Human [NYHA class II-III chronic heart failure]	Allopurinol (300 mg)	MDA	– Plasma MDA levels were significantly reduced with allopurinol (346 ± 128 nmol/L) compared to placebo (461 ± 101 nmol/L) ($p = 0.03$)
			Forearm blood flow	– Endothelium-dependent vasodilation in response to acetylcholine was significantly increased with allopurinol, which translated to a 181% change in blood flow compared to 120% in placebo treatment arm ($p = 0.003$)
George et al. (2006)	Human [NYHA class II-III chronic heart failure]	Allopurinol (300 mg, 600 mg)	Forearm blood flow	– A steep dose-response relationship between allopurinol and endothelial function: 43% improvement in blood flow per 0.1 mmol/L fall in urate between placebo and 300 mg allopurinol but a 129% improvement in blood flow per 0.1 mmol/L fall in urate between 300 mg and 600 mg allopurinol – High-dose allopurinol suppressed vit-C sensitive component of oxidative stress on endothelial function
Rajendra et al. (2011)	Human [coronary artery disease]	Allopurinol (300 mg, 600 mg)	Forearm blood flow ratio (FBFR)	– Co-infusion of vit C with acetylcholine resulted in highly significant increase in FBFR, marker of vascular oxidative stress, but this effect was ameliorated by high dose allopurinol

(continued)

Table 2 (continued)

Study, year [Ref]	Species [human (+ disease area)/ mouse/rat]	Intervention (drug, dose)	Biomarkers measured	Brief results
			Ox-LDL, F2-isoprostanes	– Allopurinol reduced ox-LDL levels (48.9 ± 1.8 U/L vs 57.3 ± 4 U/L; $p = 0.01$) and non-significantly reduced F2 isoprostanes (240 ± 93 pg/mL vs 259 ± 113 pg/mL; $p = 0.09$)
Kogler et al. (2003)	Rat [spontaneous hypertensive/ heart failure (SHHF) vs control]	Oxypurinol	Ca^{2+}-activated tension	– Total XOR and XO activity was significantly enhanced in failing vs non-failing myocardium ($p = 0.044$ and $p = 0.07$ respectively) – In the presence of oxypurinol, increase in inotropic Ca^{2+}-activated tension was only 25% in non-failing myocardium but nearly 3× higher (75%) in failing myocardium
Guan et al. (2003)	Human [percutaneous transluminal coronary angioplasty in patients with acute myocardial infarction]	Allopurinol	8-epi-prostaglandin $F_{2\alpha}$ (8-epi-$PGF_{2\alpha}$)	– Allopurinol pre-treatment was effective in inhibiting free radical generation (8-epi-$PGF_{2\alpha}$) during reperfusion and in recovery of LV function
Yamaguchi et al. (2015)	Mouse [global brain ischaemia reperfusion model]	Allopurinol, febuxostat	IL-1β, TNF-α, ICAM-1, MMP-9 mRNA expression	– Downstream target genes regulated by NF-kB, a pro-inflammatory transcription factor, are upregulated following cerebral ischaemia reperfusion injury

				– Neuroprotective effect of allopurinol derived from direct free-radical scavenging effect and not through inhibition of ROS production by XOR. This is evidenced by sustained high levels of inflammatory gene mRNA expression in the febuxostat group but not in the allopurinol group.
Malik et al. (2011)	Bovine aortic endothelial cells	Allopiurinol vs febuxostat	Urate and O_2^-	– Soluble XO: Febuxostat was >1,000 times more effective than allopurinol in inhibition of urate formation and more potently inhibited ROS formation – Endothelial bound XO: Complete inhibition of uric acid formation was observed with febuxostat but not with allopurinol [even at much higher concentrations seen clinically]
Tsuda et al. (2012)	Rat [ischaemia-reperfusion (I/R) injured kidneys]	Febuxostat vs control	Nitrotyrosine, thiobarbituric acid-reactive substances (TBARS), Urine-8-isoprostane	– Nitro-oxidative stress, lipid peroxidation and urinary excretion of 8-isoprostane were all significantly lower with febuxostat treatment compared to control (all $p < 0.05$)
			mRNA levels of GRP-78, ATF4 and CHOP	– Expression of endoplasmic reticulum (ER) stress-related genes, mediated by ROS, was significantly suppressed in febuxostat-treated rats (all $p < 0.05$)

(continued)

Table 2 (continued)

Study, year [Ref]	Species [human (+ disease area)/mouse/rat]	Intervention (drug, dose)	Biomarkers measured	Brief results
Lee et al. (2014)	Rat (diabetic rat model)	Febuxostat 5 mg/kg/day	8-hydroxy-2-deoxyguanosine (8-OHdG), nitrotyrosine	– Oxidative DNA damage in the kidneys (8-OHDG) and nitro-oxidative stress in the renal cortex due to diabetic nephropathy were significantly reduced with febuxostat treatment (all $p < 0.05$)
			E-selectin and VCAM-1	– Significantly reduced renal cortical mRNA expression of inflammatory genes (E-selectin and VCAM-1) and inflammatory mediators (ED-1 and NF-kB) with Febuxostat treatment ($p < 0.05$)
Tausche et al. (2014)	Human (chronic tophaceous gout)	Febuxostat vs allopurinol	Inflammation (CRP and TNFα) and oxidative stress (NADPH oxidase activity)	– Febuxostat significantly reduced serum TNFα levels ($p = 0.017$) and NADPH oxidase activity ($p = 0.009$)
			Carotid-femoral pulse wave velocity (cfPWV)	– Significant increase in cfPWV from baseline in allopurinol group after 1 year of therapy (16.8 ± 4.3 m/s, $p = 0.001$) but this trend was not observed in febuxostat patients
Alshahawey et al. (2017)	Human (haemodialysis)	Febuxostat 40 mg thrice weekly	High sensitivity CRP (hsCRP) and asymmetric dimethylarginine (ADMA)	– Improvement in endothelial dysfunction (ADMA) and reduction in inflammatory markers (hsCRP) in haemodialysis patients treated with febuxostat compared to placebo ($p < 0.001$)

Sezai et al. (2013)	Human (cardiac surgery patients with hyperuricaemia)	Febuxostat vs allopurinol	Ox-LDL	– Febuxostat significantly inhibited oxidative stress, through sustained reduction of ox-LDL levels over 6 months, as compared to low-moderate dose allopurinol ($p = 0.0007$)
			Eicosapentaenoic acid/ arachidonic acid (EPA-AA) ratio	– EPA/AA ratio (index marker of cellular inflammation and arteriosclerosis) was significantly higher with febuxostat treatment, which manifested its effect through reduced pulse wave velocity and left ventricular mass index
Kawamorita et al. (2017)	Rat (minimal change nephrotic syndrome)	Topiroxostat 1 mg/kg/ day	Nitrotyrosine	– Significant reduction in nitrotyrosine and 8-OHDG levels in rat model – Significantly reduced expression of XO and NADPH oxidase 4, known inducers of oxidative stress, in rat model
			8-hydroxy-2-deoxyguanosine (8-OHdG), NADPH	
Mizukoshi et al. (2018)	Human (hyperuricaemia and diabetic nephropathy)	Topiroxostat 40 mg/day Topiroxostat 160 mg/ day	Urine albumin-to-creatinine ratio (UACR)	– UACR reduction of -122 mg/gCR ($p = 0.041$) in the high-dose topiroxostat
			Urinary L-fatty acid binding protein (FABP)	– Urinary L-FABP reduction of 10.13 ± 3.2 ($p = 0.0021$) and 7.8 ± 1.7 μg/gCr ($p < 0.0001$) after 12 and 24 weeks respectively in the high-dose topiroxostat group

References

Alshahawey M, Shahin SM, Elsaid TW, Sabri NA (2017) Effect of febuxostat on the endothelial dysfunction in hemodialysis patients: a randomized, placebo-controlled, double-blinded study. Am J Nephrol 45:452–459

Amado LC, Saliaris AP, Raju SV, Lehrke S, St John M, Xie J, Stewart G, Fitton T, Minhas KM, Brawn J, Hare JM (2005) Xanthine oxidase inhibition ameliorates cardiovascular dysfunction in dogs with pacing-induced heart failure. J Mol Cell Cardiol 39:531–536

Augustin AJ, Boker T, Blumenroder SH, Lutz J, Spitznas M (1994) Free radical scavenging and antioxidant activity of allopurinol and oxypurinol in experimental lens-induced uveitis. Invest Ophthalmol Vis Sci 35:3897–3904

Bagnati M, Perugini C, Cau C, Bordone R, Albano E, Bellomo G (1999) When and why a water-soluble antioxidant becomes pro-oxidant during copper-induced low-density lipoprotein oxidation: a study using uric acid. Biochem J 340(Pt 1):143–152

Beddhu S, Filipowicz R, Wang B, Wei G, Chen X, Roy AC, DuVall SL, Farrukh H, Habib AN, Bjordahl T, Simmons DL, Munger M, Stoddard G, Kohan DE, Greene T, Huang Y (2016) A randomized controlled trial of the effects of febuxostat therapy on adipokines and markers of kidney fibrosis in asymptomatic hyperuricemic patients with diabetic nephropathy. Can J Kidney Health Dis 3:2054358116675343

Berry CE, Hare JM (2004) Xanthine oxidoreductase and cardiovascular disease: molecular mechanisms and pathophysiological implications. J Physiol 555:589–606

Butler R, Morris AD, Belch JJ, Hill A, Struthers AD (2000) Allopurinol normalizes endothelial dysfunction in type 2 diabetics with mild hypertension. Hypertension 35:746–751

Cappola TP, Kass DA, Nelson GS, Berger RD, Rosas GO, Kobeissi ZA, Marban E, Hare JM (2001) Allopurinol improves myocardial efficiency in patients with idiopathic dilated cardiomyopathy. Circulation 104:2407–2411

Cardillo C, Kilcoyne CM, Cannon RO 3rd, Quyyumi AA, Panza JA (1997) Xanthine oxidase inhibition with oxypurinol improves endothelial vasodilator function in hypercholesterolemic but not in hypertensive patients. Hypertension 30:57–63

Chen C, Lu JM, Yao Q (2016) Hyperuricemia-related diseases and xanthine oxidoreductase (xor) inhibitors: an overview. Med Sci Monit 22:2501–2512

Choi H, Neogi T, Stamp L, Dalbeth N, Terkeltaub R (2018) New perspectives in rheumatology: implications of the cardiovascular safety of febuxostat and allopurinol in patients with gout and cardiovascular morbidities trial and the associated food and drug administration public safety alert. Arthritis Rheumatol 70:1702–1709

Cicoira M, Zanolla L, Rossi A, Golia G, Franceschini L, Brighetti G, Zeni P, Zardini P (2002) Elevated serum uric acid levels are associated with diastolic dysfunction in patients with dilated cardiomyopathy. Am Heart J 143:1107–1111

Committee JF (2007) British national formulary. British Medical Association and Royal Pharmaceutical Society of Great Britain, London

Das DK, Engelman RM, Clement R, Otani H, Prasad MR, Rao PS (1987) Role of xanthine oxidase inhibitor as free radical scavenger: a novel mechanism of action of allopurinol and oxypurinol in myocardial salvage. Biochem Biophys Res Commun 148:314–319

Day RO, Graham GG, Hicks M, McLachlan AJ, Stocker SL, Williams KM (2007) Clinical pharmacokinetics and pharmacodynamics of allopurinol and oxypurinol. Clin Pharmacokinet 46:623–644

Deanfield JE, Halcox JP, Rabelink TJ (2007) Endothelial function and dysfunction: testing and clinical relevance. Circulation 115:1285–1295

Della Corte E, Gozzetti G, Novello F, Stirpe F (1969) Properties of the xanthine oxidase from human liver. Biochim Biophys Acta 191:164–166

Doehner W, Schoene N, Rauchhaus M, Leyva-Leon F, Pavitt DV, Reaveley DA, Schuler G, Coats AJS, Anker SD, Hambrecht R (2002) Effects of xanthine oxidase inhibition with allopurinol on endothelial function and peripheral blood flow in hyperuricemic patients with chronic heart failure: results from 2 placebo-controlled studies. Circulation 105:2619–2624

Ekelund UE, Harrison RW, Shokek O, Thakkar RN, Tunin RS, Senzaki H, Kass DA, Marban E, Hare JM (1999) Intravenous allopurinol decreases myocardial oxygen consumption and increases mechanical efficiency in dogs with pacing-induced heart failure. Circ Res 85:437–445

Elion GB, Kovensky A, Hitchings GH (1966) Metabolic studies of allopurinol, an inhibitor of xanthine oxidase. Biochem Pharmacol 15:863–880

Elion GB, Yu TF, Gutman AB, Hitchings GH (1968) Renal clearance of oxipurinol, the chief metabolite of allopurinol. Am J Med 45:69–77

Engberding N, Spiekermann S, Schaefer A, Heineke A, Wiencke A, Muller M, Fuchs M, Hilfiker-Kleiner D, Hornig B, Drexler H, Landmesser U (2004) Allopurinol attenuates left ventricular remodeling and dysfunction after experimental myocardial infarction: a new action for an old drug? Circulation 110:2175–2179

Ernst ME, Fravel MA (2009) Febuxostat: a selective xanthine-oxidase/xanthine-dehydrogenase inhibitor for the management of hyperuricemia in adults with gout. Clin Ther 31:2503–2518

Farquharson CAJ, Butler R, Hill A, Belch JJF, Struthers AD (2002) Allopurinol improves endothelial dysfunction in chronic heart failure. Circulation 106:221–226

Faruque LI, Ehteshami-Afshar A, Wiebe N, Tjosvold L, Homik J, Tonelli M (2013) A systematic review and meta-analysis on the safety and efficacy of febuxostat versus allopurinol in chronic gout. Semin Arthritis Rheum 43:367–375

Fukui T, Maruyama M, Yamauchi K, Yoshitaka S, Yasuda T, Abe Y (2015) Effects of febuxostat on oxidative stress. Clin Ther 37:1396–1401

Gavin AD, Struthers AD (2005) Allopurinol reduces b-type natriuretic peptide concentrations and haemoglobin but does not alter exercise capacity in chronic heart failure. Heart 91:749–753

George J, Struthers AD (2008) The role of urate and xanthine oxidase inhibitors in cardiovascular disease. Cardiovasc Ther 26:59–64

George J, Carr E, Davies J, Belch JJ, Struthers A (2006) High-dose allopurinol improves endothelial function by profoundly reducing vascular oxidative stress and not by lowering uric acid. Circulation 114:2508–2516

Gersch C, Palii SP, Kim KM, Angerhofer A, Johnson RJ, Henderson GN (2008) Inactivation of nitric oxide by uric acid. Nucleosides Nucleotides Nucleic Acids 27:967–978

Goharinia M, Zareei A, Rahimi M, Mirkhani H (2017) Can allopurinol improve retinopathy in diabetic rats? Oxidative stress or uric acid; which one is the culprit? Res Pharm Sci 12:401–408

Graham S, Day RO, Wong H, McLachlan AJ, Bergendal L, Miners JO, Birkett DJ (1996) Pharmacodynamics of oxypurinol after administration of allopurinol to healthy subjects. Br J Clin Pharmacol 41:299–304

Granger DN, Benoit JN, Suzuki M, Grisham MB (1989) Leukocyte adherence to venular endothelium during ischemia-reperfusion. Am J Phys 257:G683–G688

Griendling KK, Minieri CA, Ollerenshaw JD, Alexander RW (1994) Angiotensin ii stimulates nadh and nadph oxidase activity in cultured vascular smooth muscle cells. Circ Res 74:1141–1148

Guan W, Osanai T, Kamada T, Hanada H, Ishizaka H, Onodera H, Iwasa A, Fujita N, Kudo S, Ohkubo T, Okumura K (2003) Effect of allopurinol pretreatment on free radical generation after primary coronary angioplasty for acute myocardial infarction. J Cardiovasc Pharmacol 41 (5):699–705

Guerra P, Frias J, Ruiz B, Soto A, Carcas A, Govantes C, Montuenga C, Fernandez A (2001) Bioequivalence of allopurinol and its metabolite oxipurinol in two tablet formulations. J Clin Pharm Ther 26:113–119

Gutierrez-Macias A, Lizarralde-Palacios E, Martinez-Odriozola P, Miguel-De la Villa F (2005) Fatal allopurinol hypersensitivity syndrome after treatment of asymptomatic hyperuricaemia. BMJ 331:623–624

Hancock JT, Salisbury V, Ovejero-Boglione MC, Cherry R, Hoare C, Eisenthal R, Harrison R (2002) Antimicrobial properties of milk: dependence on presence of xanthine oxidase and nitrite. Antimicrob Agents Chemother 46:3308–3310

Harrison D, Griendling KK, Landmesser U, Hornig B, Drexler H (2003) Role of oxidative stress in atherosclerosis. Am J Cardiol 91:7A–11A

Hille R, Massey V (1981) Studies on the oxidative half-reaction of xanthine oxidase. J Biol Chem 256:9090–9095

Hoey BM, Butler J, Halliwell B (1988) On the specificity of allopurinol and oxypurinol as inhibitors of xanthine oxidase. A pulse radiolysis determination of rate constants for reaction of allopurinol and oxypurinol with hydroxyl radicals. Free Radic Res Commun 4:259–263

Huang Y, Zhang C, Xu Z, Shen J, Zhang X, Du H, Zhang K, Zhang D (2017) Clinical study on efficacy of allopurinol in patients with acute coronary syndrome and its functional mechanism. Hell J Cardiol 58:360–365

Hung SI, Chung WH, Liou LB, Chu CC, Lin M, Huang HP, Lin YL, Lan JL, Yang LC, Hong HS, Chen MJ, Lai PC, Wu MS, Chu CY, Wang KH, Chen CH, Fann CS, Wu JY, Chen YT (2005) Hla-b*5801 allele as a genetic marker for severe cutaneous adverse reactions caused by allopurinol. Proc Natl Acad Sci U S A 102:4134–4139

Ichida K, Amaya Y, Noda K, Minoshima S, Hosoya T, Sakai O, Shimizu N, Nishino T (1993) Cloning of the cdna encoding human xanthine dehydrogenase (oxidase): structural analysis of the protein and chromosomal location of the gene. Gene 133:279–284

Kamel B, Graham GG, Williams KM, Pile KD, Day RO (2017) Clinical pharmacokinetics and pharmacodynamics of febuxostat. Clin Pharmacokinet 56:459–475

Kawamorita Y, Shiraishi T, Tamura Y, Kumagai T, Shibata S, Fujigaki Y, Hosoyamada M, Nakagawa T, Uchida S (2017) Renoprotective effect of topiroxostat via antioxidant activity in puromycin aminonucleoside nephrosis rats. Physiol Rep 5(15):e13358

Keshavarzian A, Morgan G, Sedghi S, Gordon JH, Doria M (1990) Role of reactive oxygen metabolites in experimental colitis. Gut 31:786–790

Khan SA, Lee K, Minhas KM, Gonzalez DR, Raju SV, Tejani AD, Li D, Berkowitz DE, Hare JM (2004) Neuronal nitric oxide synthase negatively regulates xanthine oxidoreductase inhibition of cardiac excitation-contraction coupling. Proc Natl Acad Sci U S A 101:15944–15948

Khosravan R, Grabowski BA, Wu JT, Joseph-Ridge N, Vernillet L (2006) Pharmacokinetics, pharmacodynamics and safety of febuxostat, a non-purine selective inhibitor of xanthine oxidase, in a dose escalation study in healthy subjects. Clin Pharmacokinet 45:821–841

Kittleson MM, Hare JM (2005) Xanthine oxidase inhibitors: an emerging class of drugs for heart failure. Eur Heart J 26:1458–1460

Knight TR, Kurtz A, Bajt ML, Hinson JA, Jaeschke H (2001) Vascular and hepatocellular peroxynitrite formation during acetaminophen toxicity: role of mitochondrial oxidant stress. Toxicol Sci 62(2):212–220

Kogler H, Fraser H, McCune S, Altschuld R, Marban E (2003) Disproportionate enhancement of myocardial contractility by the xanthine oxidase inhibitor oxypurinol in failing rat myocardium. Cardiovasc Res 59:582–592

Kojima S, Matsui K, Hiramitsu S, Hisatome I, Waki M, Uchiyama K et al (2019) Febuxostat for cerebral and cardiorenovascular events prevention study. Eur Heart J 40(22):1778–1786

Landmesser U, Spiekermann S, Dikalov S, Tatge H, Wilke R, Kohler C, Harrison DG, Hornig B, Drexler H (2002) Vascular oxidative stress and endothelial dysfunction in patients with chronic heart failure: role of xanthine-oxidase and extracellular superoxide dismutase. Circulation 106:3073–3078

Lee HJ, Jeong KH, Kim YG, Moon JY, Lee SH, Ihm CG, Sung JY, Lee TW (2014) Febuxostat ameliorates diabetic renal injury in a streptozotocin-induced diabetic rat model. Am J Nephrol 40:56–63

Mackenzie IS, Ford I, Walker A, Hawkey C, Begg A, Avery A, Taggar J, Wei L, Struthers AD, MacDonald TM (2016) Multicentre, prospective, randomised, open-label, blinded end point trial of the efficacy of allopurinol therapy in improving cardiovascular outcomes in patients with ischaemic heart disease: protocol of the all-heart study. BMJ Open 6:e013774

Malik UZ, Hundley NJ, Romero G, Radi R, Freeman BA, Tarpey MM, Kelley EE (2011) Febuxostat inhibition of endothelial-bound xo: implications for targeting vascular ros production. Free Radic Biol Med 51:179–184

Malkiel S, Har-el R, Schwalb H, Uretzky G, Borman JB, Chevion M (1993) Interaction between allopurinol and copper: possible role in myocardial protection. Free Radic Res Commun 18:7–15

Massey V, Komai H, Palmer G, Elion GB (1970) On the mechanism of inactivation of xanthine oxidase by allopurinol and other pyrazolo[3,4-d]pyrimidines. J Biol Chem 245:2837–2844

Matsumoto K, Okamoto K, Ashizawa N, Nishino T (2011) Fyx-051: a novel and potent hybrid-type inhibitor of xanthine oxidoreductase. J Pharmacol Exp Ther 336:95–103

McInnes GT, Lawson DH, Jick H (1981) Acute adverse reactions attributed to allopurinol in hospitalised patients. Ann Rheum Dis 40:245–249

Mizukoshi T, Kato S, Ando M, Sobajima H, Ohashi N, Naruse T, Saka Y, Shimizu H, Nagata T, Maruyama S (2018) Renoprotective effects of topiroxostat for hyperuricaemic patients with overt diabetic nephropathy study (ETUDE study): a prospective, randomized, multicentre clinical trial. Nephrology (Carlton, Vic) 23(11):1023–1030

Nakamura T, Murase T, Nampei M, Morimoto N, Ashizawa N, Iwanaga T, Sakamoto R (2016) Effects of topiroxostat and febuxostat on urinary albumin excretion and plasma xanthine oxidoreductase activity in db/db mice. Eur J Pharmacol 780:224–231

Naumova AV, Chacko VP, Ouwerkerk R, Stull L, Marban E, Weiss RG (2006) Xanthine oxidase inhibitors improve energetics and function after infarction in failing mouse hearts. Am J Physiol Heart Circ Physiol 290:H837–H843

Nishizawa J, Nakai A, Matsuda K, Komeda M, Ban T, Nagata K (1999) Reactive oxygen species play an important role in the activation of heat shock factor 1 in ischemic-reperfused heart. Circulation 99:934–941

Ogino K, Kato M, Furuse Y, Kinugasa Y, Ishida K, Osaki S, Kinugawa T, Igawa O, Hisatome I, Shigemasa C, Anker SD, Doehner W (2009) Uric acid lowering treatment with benzbromarone in patients with heart failure: a double-blind placebo-controlled cross-over preliminary study. Circ Heart Fail 3:73–81

Okamoto K, Nishino T (2008) Crystal structures of mammalian xanthine oxidoreductase bound with various inhibitors: allopurinol, febuxostat, and fyx-051. J Nippon Med Sch 75:2–3

Pacher P, Nivorozhkin A, Szabo C (2006) Therapeutic effects of xanthine oxidase inhibitors: renaissance half a century after the discovery of allopurinol. Pharmacol Rev 58:87–114

Panus PC, Wright SA, Chumley PH, Radi R, Freeman BA (1992) The contribution of vascular endothelial xanthine dehydrogenase/oxidase to oxygen-mediated cell injury. Arch Biochem Biophys 294:695–702

Parks DA, Granger DN (1986) Xanthine oxidase: biochemistry, distribution and physiology. Acta Physiol Scand Suppl 548:87–99

Pea F (2005) Pharmacology of drugs for hyperuricemia. Mechanisms, kinetics and interactions. Contrib Nephrol 147:35–46

Perez NG, Gao WD, Marban E (1998) Novel myofilament ca2+-sensitizing property of xanthine oxidase inhibitors. Circ Res 83:423–430

Pinheiro LC, Oliveira-Paula GH, Portella RL, Guimaraes DA, de Angelis CD, Tanus-Santos JE (2016) Omeprazole impairs vascular redox biology and causes xanthine oxidoreductase-mediated endothelial dysfunction. Redox Biol 9:134–143

Rajendra NS, Ireland S, George J, Belch JJ, Lang CC, Struthers AD (2011) Mechanistic insights into the therapeutic use of high-dose allopurinol in angina pectoris. J Am Coll Cardiol 58:820–828

Ricardo SD, Bertram JF, Ryan GB (1995) Podocyte architecture in puromycin aminonucleoside-treated rats administered tungsten or allopurinol. Exp Nephrol 3:270–279

Saavedra WF, Paolocci N, St John ME, Skaf MW, Stewart GC, Xie JS, Harrison RW, Zeichner J, Mudrick D, Marban E, Kass DA, Hare JM (2002) Imbalance between xanthine oxidase and nitric oxide synthase signaling pathways underlies mechanoenergetic uncoupling in the failing heart. Circ Res 90:297–304

Schlesinger N (2004) Management of acute and chronic gouty arthritis: present state-of-the-art. Drugs 64:2399–2416

Sezai A, Soma M, Nakata K, Hata M, Yoshitake I, Wakui S, Hata H, Shiono M (2013) Comparison of febuxostat and allopurinol for hyperuricemia in cardiac surgery patients (nu-flash trial). Circ J 77:2043–2049

Sezai A, Obata K, Abe K, Kanno S, Sekino H (2017) Cross-over trial of febuxostat and topiroxostat for hyperuricemia with cardiovascular disease (TROFEO trial). Circ J 81(11):1707–1712

Sisto T, Paajanen H, Metsa-Ketela T, Harmoinen A, Nordback I, Tarkka M (1995) Pretreatment with antioxidants and allopurinol diminishes cardiac onset events in coronary artery bypass grafting. Ann Thorac Surg 59:1519–1523

Skinner KA, White CR, Patel R, Tan S, Barnes S, Kirk M, Darley-Usmar V, Parks DA (1998) Nitrosation of uric acid by peroxynitrite: formation of a vasoactive nitric oxide donor. J Biol Chem 273:24491–24497

Spector T (1977) Inhibition of urate production by allopurinol. Biochem Pharmacol 26:355–358

Spiekermann S, Landmesser U, Dikalov S, Bredt M, Gamez G, Tatge H, Reepschlager N, Hornig B, Drexler H, Harrison DG (2003) Electron spin resonance characterization of vascular xanthine and nad(p)h oxidase activity in patients with coronary artery disease: relation to endothelium-dependent vasodilation. Circulation 107:1383–1389

Stevens CR, Millar TM, Clinch JG, Kanczler JM, Bodamyali T, Blake DR (2000) Antibacterial properties of xanthine oxidase in human milk. Lancet 356:829–830

Stull LB, Leppo MK, Szweda L, Gao WD, Marban E (2004) Chronic treatment with allopurinol boosts survival and cardiac contractility in murine postischemic cardiomyopathy. Circ Res 95:1005–1011

Takano Y, Hase-Aoki K, Horiuchi H, Zhao L, Kasahara Y, Kondo S, Becker MA (2005) Selectivity of febuxostat, a novel non-purine inhibitor of xanthine oxidase/xanthine dehydrogenase. Life Sci 76:1835–1847

Tausche AK, Christoph M, Forkmann M, Richter U, Kopprasch S, Bielitz C, Aringer M, Wunderlich C (2014) As compared to allopurinol, urate-lowering therapy with febuxostat has superior effects on oxidative stress and pulse wave velocity in patients with severe chronic tophaceous gout. Rheumatol Int 34:101–109

Tsuda H, Kawada N, Kaimori JY, Kitamura H, Moriyama T, Rakugi H, Takahara S, Isaka Y (2012) Febuxostat suppressed renal ischemia-reperfusion injury via reduced oxidative stress. Biochem Biophys Res Commun 427:266–272

Turnheim K (1999) Oberbauer. Pharmacokinetics and pharmacodynamics of allopurinol in elderly and young subjects. Br J Clin Pharmacol 48:501–509

Vazquez-Mellado J, Morales EM, Pacheco-Tena C, Burgos-Vargas R (2001) Relation between adverse events associated with allopurinol and renal function in patients with gout. Ann Rheum Dis 60:981–983

Waring WS (2002) Uric acid: an important antioxidant in acute ischaemic stroke. QJM 95:691–693

White WB, Saag KG, Becker MA, Borer JS, Gorelick PB, Whelton A, Hunt B, Castillo M, Gunawardhana L, CARES Investigators (2018) Cardiovascular safety of febuxostat or allopurinol in patients with gout. N Engl J Med 378:1200–1210

Yamaguchi M, Okamoto K, Kusano T, Matsuda Y, Suzuki G, Fuse A, Yokota H (2015) The effects of xanthine oxidoreductase inhibitors on oxidative stress markers following global brain ischemia reperfusion injury in c57bl/6 mice. PLoS One 10:e0133980

Monoamine Oxidase Inhibitors: From Classic to New Clinical Approaches

Pablo Duarte, Antonio Cuadrado, and Rafael León

Contents

P. Duarte
Instituto Teófilo Hernando y Departamento de Farmacología y Terapéutica, Facultad de Medicina, Universidad Autónoma de Madrid, Madrid, Spain

Instituto de Investigación Sanitaria, Servicio de Farmacología Clínica, Hospital Universitario de la Princesa, Madrid, Spain

A. Cuadrado
Departmento de Bioquímica, Facultad de Medicina, Centro de Investigación Biomédica en Red Sobre Enfermedades Neurodegenerativas (CIBERNED), Instituto de Investigación Sanitaria La Paz (IdiPaz), Instituto de Investigaciones Biomédicas 'Alberto Sols' UAM-CSIC, Universidad Autónoma de Madrid, Madrid, Spain

R. León (✉)
Instituto Teófilo Hernando y Departamento de Farmacología y Terapéutica, Facultad de Medicina, Universidad Autónoma de Madrid, Madrid, Spain

Instituto de Investigación Sanitaria, Servicio de Farmacología Clínica, Hospital Universitario de la Princesa, Madrid, Spain

Instituto de Química Médica, Consejo Superior de Investigaciones Científicas (IQM-CSIC), Madrid, Spain
e-mail: rafael.leon@inv.uam.es

H. H. H. W. Schmidt et al. (eds.), *Reactive Oxygen Species*,
Handbook of Experimental Pharmacology 264, https://doi.org/10.1007/164_2020_384

Abstract

Monoamine oxidases (MAOs) are involved in the oxidative deamination of different amines and neurotransmitters. This pointed them as potential targets for several disorders and along the last 70 years a wide variety of MAO inhibitors have been developed as successful drugs for the treatment of complex diseases, being the first drugs approved for depression in the late 1950s. The discovery of two MAO isozymes (MAO-A and B) with different substrate selectivity and tissue expression patterns led to novel therapeutic approaches and to the development of new classes of inhibitors, such as selective irreversible and reversible MAO-B inhibitors and reversible MAO-A inhibitors. Significantly, MAO-B inhibitors constitute a widely studied group of compounds, some of them approved for the treatment of Parkinson's disease. Further applications are under development for the treatment of Alzheimer's disease, amyotrophic lateral sclerosis, and cardiovascular diseases, among others. This review summarizes the most important aspects regarding the development and clinical use of MAO inhibitors, going through mechanistic and structural details, new indications, and future perspectives.

Graphical Abstract

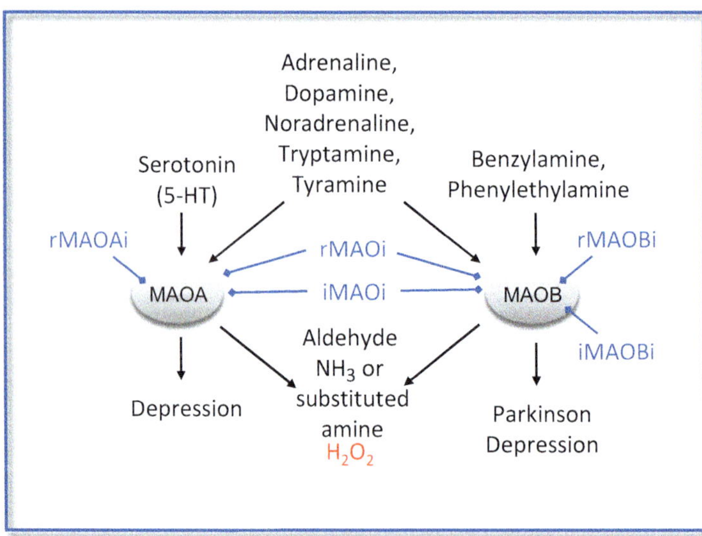

Monoamine oxidases (MAOs) catalyze the oxidative deamination of different amines and neurotransmitters. The two different isozymes, MAO-A and MAO-B, are located at the outer mitochondrial membrane in different tissues. The enzymatic reaction involves formation of the corresponding aldehyde and releasing hydrogen peroxide (H_2O_2) and ammonia or a substituted amine depending on the substrate. MAO's role in neurotransmitter metabolism made them targets for major depression and Parkinson's disease, among other neurodegenerative diseases. Currently, these compounds are being studied for other diseases such as cardiovascular ones.

Keywords

Depression · MAO-A · MAO-B · Monoamine oxidases · Neurodegeneration · Oxidative stress · Parkinson

1 Overview

Monoamine oxidases (MAOs) are part of the family of flavin adenine dinucleotide (FAD)-dependent enzymes that catalyze the oxidative deamination of different amines and neurotransmitters. There are two different isozymes that are well characterized (MAO-A and MAO-B), and they are located at the outer mitochondrial membrane in most mammalian tissues with different expression patterns. MAOs are responsible of neurotransmitters regulation by their degradation in which the enzymatic reaction involves formation of the corresponding aldehyde from the neurotransmitter to form the oxidized FAD and releasing hydrogen peroxide resulting from reduction of molecular oxygen and ammonia or a substituted amine depending on the substrate. The production of H_2O_2 by these enzymes might lead to the generation of free radicals that can exert cytotoxicity under pathological conditions. The relationship of MAOs with the neurotransmitter metabolism pointed them as potential target for different disorders including major depression, Parkinson's disease (PD), Alzheimer's disease (AD), Lewy body diseases with dementia, and amyotrophic lateral sclerosis (ALS), and currently they are used in clinics for some of these disorders.

In general, MAO inhibition has a long history since first drug was approved in the late 1950s. Initially, iproniazid was tested as a treatment for tuberculosis; although it was not effective, it showed an improvement in patient emotional state. It would be later described as a non-selective irreversible MAO inhibitor of the hydrazine group and approved in 1958 for depression. Nevertheless, those drugs showed high liver toxicity and were associated to hypertensive crisis, secondary effects that resulted in market removal of some of them. Thereafter, non-hydrazine MAO inhibitors were developed as an alternative to avoid liver toxicity (i.e., tranylcypromine, early 1960s); however, hypertensive crisis as secondary effect was still present. Therefore, some initially approved MAO inhibitors were retired as antidepressants and substituted by other drugs directed to novel targets (Shulman et al. 2013; Youdim et al. 2006; Edmondson and Binda 2018).

The discovery of two different MAO isozymes (MAO-A and B) with different substrate selectivity and expression pattern in different tissues led to novel therapeutic approaches and the development of new classes of inhibitors, such as selective irreversible and reversible MAO-B inhibitors and reversible MAO-A inhibitors. Typically, MAO-A is related to the oxidative deamination of serotonin, while MAO-B catalyzes preferentially benzylamine and 2-Phenethylamine. Dopamine, noradrenaline, adrenaline, tryptamine and tyramine exhibit similar substrate specificities for both isozymes. Hypertensive crises suffered by patients treated with MAO inhibitors were then associated with a strong MAO-A inhibition, since this enzyme metabolizes tyramine in the small intestine due to its higher expression compared to MAO-B in that organ. The toxic effect is related with the ingestion of food containing tyramine coupled to MAO-A inhibitors. This combination induces tyramine accumulation leading to high blood pressure due to the displacement of norepinephrine from neuronal storage, extremely increasing its concentration in the bloodstream (Anderson et al. 1993). The related toxic effect is known as the "cheese effect" (tyramine-induced hypertensive crisis of MAO-A inhibitors). In this sense, selective MAO-B inhibitors provide an improved pharmacological profile avoiding this effects; also reversible inhibitors of MAO-A (RIMAs) obtaining in some cases a sufficient blockade without toxicity (Da Prada et al. 1988).

MAO-B inhibitors were extensively studied after the case of intoxication with 1-methyl-4-phenyl-1,2,3,6-tetrahydropyridine (MPTP) in 1980. It was discovered that the toxicity induced by MPTP was related to its conversion to 1-methyl-4-phenylpyridinium (MPP+) by MAO-B, a metabolite that induced Parkinsonian symptoms. These symptoms were reversed by pargyline, a selective and irreversible MAO-B inhibitor (Langston et al. 1984). This correlation induced an intensive investigation of MAO-B inhibitors as potential therapeutics for the treatment PD. Nowadays, three selective MAO-B inhibitors are approved for the treatment of this disease: selegiline, rasagiline, and safinamide.

2 Structure and Binding Site of MAO

Human MAO proteins are flavoenzymes constituted by 527 and 520 amino acids for MAO-A and MAO-B isozymes, respectively. They are attached to the mitochondrial outer membrane as dimer structures (Fig. 1a) (Upadhyay et al. 2008). Both isozymes share about 70% aminoacidic sequence with a flavin adenine dinucleotide (FAD) coenzyme covalently bonded by 8α-methylene to Cys397 in MAO-B and Cys406 in MAO-A (Hubalek et al. 2003). Studies on their crystal structures suggest that the C-terminal α-helix would be related to the anchoring at the outer mitochondrial membrane of each monomer of the dimer. Substrate entry to the catalytic site of each monomer is independent from each other (Binda et al. 2002; Edmondson et al. 2009). Human MAO-A is demonstrated to be also a dimer, but it crystallizes as a monomer due to its higher instability (Edmondson and Binda 2018).

The catalytic mechanism is similar in both enzymes; however, different binding site structures revealed important aspects for substrate selectivity and inhibitors binding. One of the most important differences is a unique 550 $Å^3$ substrate cavity present in MAO-A opposite to FAD coenzyme. MAO-B isozyme presents two cavities, an entry cavity and a reactive site cavity, with total combined volume of about 700 $Å^3$ (Fig. 1b). MAO-B cavity exhibits an entrance cavity (290 $Å^3$) that precedes the flat hydrophobic substrate cavity (490 $Å^3$) (Binda et al. 2002; De Colibus et al. 2005). Open or closed conformation of Ile199 side chain in MAO-B determines the connectivity between cavities, flanked by other important residues as

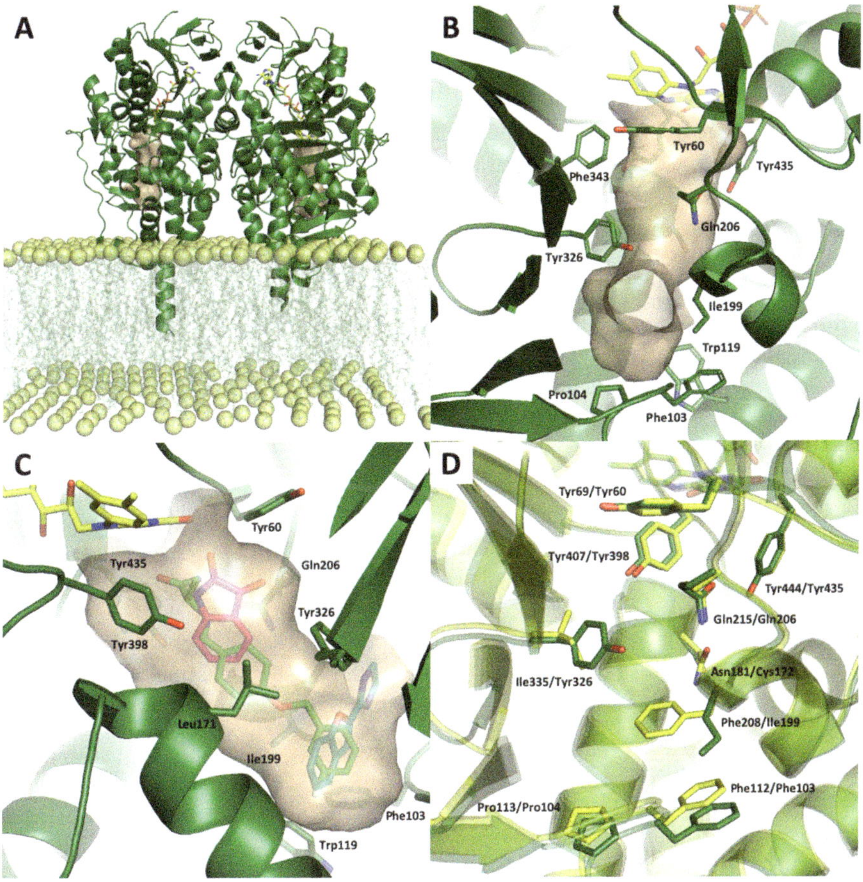

Fig. 1 Structural details about MAO enzymes. (**a**) Schematic model of the human MAO-B crystal structure PDB-ID *2V5Z* (Binda et al. 2007) as a dimer inserted in a phospholipid bilayer shown as sphere-stick representation. MAO-B protein is represented as dark green cartoon with FAD coenzyme of each monomer as yellow sticks and binding site cavity as light orange surface. (**b**) Zoom of the human MAO-B binding site with key residues as dark green sticks. (**c**) Detail of different inhibitor binding modes. Isatin represented as pink sticks is located in the substrate cavity near FAD coenzyme, 2-(2-benzofuranyl)-2-imidazoline represented as blue sticks is located in the entrance cavity at the other end of the binding site, and safinamide represented as light green sticks fits along the bipartite cavity. Isatin and 2-(2-benzofuranyl)-2-imidazoline come from crystal structures (PDB-ID *1OJA* and *2XFN*) (Bonivento et al. 2010; Binda et al. 2003) aligned to *2V5Z* structure in complex with safinamide. (**d**) Comparison between binding site key residues of human MAO-A (PDB-ID *2Z5X*) (Son et al. 2008) and MAO-B (PDB-ID *2V5Z*) isozymes represented as yellow and dark green cartoon, respectively. Most of the residues are conserved and show equivalent positions of the side chains, except for Phe208/Ile199 and Ile335/Tyr326 (MAO-A/MAO-B), key residues for substrate and inhibitor specificity. All images were constructed using PyMOL software [*The PyMOL Molecular Graphics System, Version 2.2 Schrödinger, LLC*]

Tyr326 or Phe168 (Hubalek et al. 2005). In spite of the higher total pocket volume for MAO-B, narrowness and stiffness of the bipartite cavity limit size of ligands in comparison with the A isozyme, and depending on its nature, they will bind in different positions along the cavity (Fig. 1c, discussed in the next section). The main differences of both isozymes can be found in these residues: Ile199 and Tyr326 of MAO-B instead of Phe208 and Ile335 of MAO-A. In this sense, MAO-A Phe208 side chain prevents the possibility of stablishing a double cavity. The human mutant I199F MAO-B protein showed no binding for some selective MAO-B inhibitors stating the importance of Ile199 gate as a determinant for MAO-B specificity (Hubalek et al. 2005). Likewise, double I199A/Y326A human MAO-B mutation led to a protein that exhibits binding properties closer to MAO-A, highlighting the importance of Ile199 and Tyr326 side chains for selectivity (Milczek et al. 2011). In addition, the hydrophobic environment near FAD and some polar residues as Gln206 in MAO-B are important for substrate recognition and, therefore, inhibitor orientation at the coenzyme surroundings (Bonivento et al. 2010; Dasgupta et al. 2018). Furthermore, the presence of aromatic amino acid residues such as Tyr398 and Tyr435 in MAO-B oriented perpendicular to the flavin ring has been suggested as important for catalysis (Tyr407 and Tyr444 form a similar aromatic cavity in MAO-A, see Fig. 1d). Mutations in Tyr435 residue revealed differences in catalysis and function of this aromatic cage for substrate specificity (Li et al. 2006). Besides that, Y407F/Y444F mutant MAO-A enzymes also led to modified catalytic properties (Nandigama et al. 2001). Altogether, there are many structural evidences and information about what are the most influential residues in relationship with MAO activity and, therefore, crucial for inhibitor performance and design (for review, see Edmondson and Binda (2018)). Structural knowledge and binding site details are summarized in Fig. 1.

2.1 Mechanisms of MAO Catalysis and Inhibition

Compounds targeting MAO inhibition can be generally classified as irreversible or reversible inhibitors. This refers to the ability to react covalently or not with the enzyme, in particular with FAD coenzyme involved in catalysis. Irreversible inhibitors block the enzyme, and they are not released from binding site over time as a covalent bond is stablished. Conversely, reversible inhibitors are stabilized by weaker interactions in the pocket for inhibition. Depending on the therapeutic approach, reversible or irreversible inhibition will be desired. Classically, irreversible MAO inhibitors were associated with higher toxicity, although the development of specific isozyme selective compounds would reduce these complications (Youdim et al. 2006; Edmondson and Binda 2018; Anderson et al. 1993). In general, irreversible inhibitors can be classified into different classes attending to its chemical structure: hydrazines, cyclopropylamines, and propargylamines, among others. Irreversible inhibitors permanently deactivate the enzyme, and its action can only be recovered by the expression of new enzyme. The formation of the enzyme-inhibitor adduct has been related to increased secondary effects and potential immunogenicity of these adducts.

From a structural point of view, there are some important details to mention, particularly for MAO-B inhibitors. Related to the well-known architecture of MAO-B cavities, nature of the inhibitor will determine different binding site conformations. Larger compounds will occupy both, entrance and substrate cavity, inducing an open conformation of Ile199 side chain (i.e., safinamide, approved drug for the treatment of PD). In this case, this compound guides its amine moieties to flavin ring mimicking natural substrate orientation and stablishing hydrogen bonds with Gln206 (Binda et al. 2007). This situation, with both cavities engaged, is found also with other compounds (Binda et al. 2003). In turn, smaller compounds can bind in the substrate cavity with a close conformation of Ile199 gate. Thus, depending on the inhibitor size, connection between cavities will vary, regardless of nature of inhibition (Fig. 1c) (Binda et al. 2003). Considering mechanistic details, known irreversible inhibitors such as rasagiline stablish a covalent bond with N5 atom of the flavin ring (Binda et al. 2004, 2005). Other mechanisms have been described for other compounds as tranylcypromine, one of the first antidepressants, consisting in covalent linkage in this case with C4A atom of the flavin ring (Bonivento et al. 2010).

Considering selectivity, compounds can exhibit different levels of inhibition towards A or B isozyme. Apart from this, it is important to mention brain selectivity. Some compounds, regardless its affinity for A or B enzymes, can exhibit preferential brain activity and no peripheral MAO activities. Ladostigil, a compound that has completed phase II clinical trials for mild cognitive impairment as dual neuroprotective agent with cholinesterase and brain-selective MAO activity (Schneider et al. 2019) is a good example of this situation. This compound did not show intestinal MAO-A inhibition upon oral administration in rabbits, exhibiting brain selectivity for MAO inhibition (Weinreb et al. 2012).

3 Therapeutic Value of MAO Inhibitors

Several chemical structures are described as MAO inhibitors, and many new compounds are being developed. An extensive review about privileged scaffolds as MAO inhibitors is described in Tripathi et al. (Tripathi et al. 2018). Here we focused on preclinical and clinical studies with MAO inhibitors targeting several diseases from classical to new therapeutic approaches.

3.1 Affective Diseases

Classically, the effectiveness of MAO inhibitors as antidepressants raised the hypothesis of a potential overexpression of MAO enzymes as the cause of some forms of depression, and on the opposite, reduced MAO activity has been related to violent behavior (Alia-Klein et al. 2008). MAO inhibitors have been typically used for the treatment of depression and related disorders, and they were the first antidepressant drugs developed (Ramachandraih et al. 2011). The therapeutic value for this and other related affective diseases is, in general, related to MAO-A inhibition in the central nervous system (CNS), leading to increased levels of serotonin among other neurotransmitters such as noradrenaline (Youdim et al. 2006; Finberg 2014; Finberg and Rabey 2016). In addition, the selective MAO-A inhibitor ^{11}C-harmine was used for brain imaging and gave evidence of increased MAO-A levels in striatal, midbrain, and cortical locations of major depressive disorder patients (Meyer et al. 2006). In this line, MAO-A activity is potentiated upon chronic glucocorticoid treatment in several experimental models, thus indicating relationship between stress and increased MAO-A activity (Soliman et al. 2012). MAO-A overactivity results in a decrease in monoamine neurotransmitters supporting the idea of using MAO inhibitors for the treatment of this kind of diseases, in line with classic biogenic amine hypothesis of depression (Ramachandraih et al. 2011; Finberg 2014). In addition, different MAO-A gene polymorphisms have been connected with behavioral traits, associating reduced MAO-A expression to aggressive mood and overexpression to depression (Alia-Klein et al. 2008). Most relevant MAO inhibitors for the treatment of affective disorders are summarized in Table 1.

Some of the first non-selective irreversible MAO inhibitors are still being used in clinic (i.e., phenelzine and tranylcypromine). Nevertheless, the abovemention medical complications lead to new compound profiles. In this sense, reversible and selective MAO-A inhibitors (RIMAs) appeared from 1980s to deal with "cheese effect" among other complications. In relation with it, the safety of these compounds was assessed by several clinical studies with tyramine combination (Finberg 2014; Finberg and Rabey 2016). Also, a potential secondary effect is related to the combination of irreversible MAO inhibitors and drugs with the ability to elevate 5-HT, such us serotonin reuptake inhibitors (SSRIs), leading to the serotonin toxicity syndrome (Gillman 2006). Thus, in the case of needing a serotonin reuptake inhibitor, it is necessary a washout period between 7 and 10 days for complete recovery of MAO activity. The reversible mechanism of action facilitates the competition of concentrated substrate with the inhibitor; thus, in case of high inhibition, the substrate is able to displace the inhibitor from the catalytic site limiting the secondary effects.

Moclobemide is the most important RIMA available for clinical use (approved in several western countries such as United Kingdom) also pirlindole (available in Russia) (Fasipe 2019; Lotufo-Neto et al. 1999). Other RIMAs were marketed; however, most of them were rejected. They are valuable in the treatment of depression associated to aged people and also other affective disorders such as bulimia or hypersomnia (Zisook 1985). In this line, compound CX157 was developed to achieve higher levels of brain MAO-A inhibition, and it is the first RIMA with documented reversible brain MAO-A inhibition which correlates to its plasma concentration (Fowler et al. 2010). It ended phase II clinical trials for major depressive disorder (NCT00739908); however, no more clinical studies have been documented to date. Novel compounds targeting MAO-A are still being developed, i.e., pyrazoline and hydrazone derivatives, that led to more potent MAO-A inhibitors

Table 1 MAO inhibitors for the treatment of affective disorders

Compound	MAO inhibition and state of development	Comments	Structure
Befloxatone	Reversible A; Not approved	Research and clinical use for brain positron emission tomography (PET) imaging of MAO-A with [^{11}C]befloxatone radioligand (Curet et al. 1996; Zanotti-Fregonara and Bottlaender 2014; Zanotti-Fregonara et al. 2014)	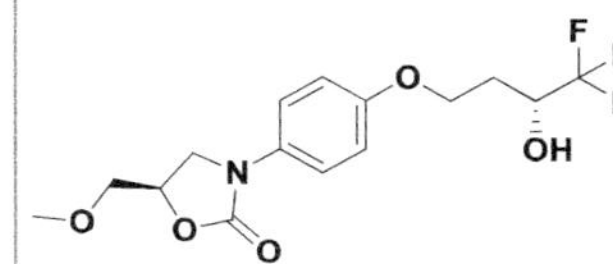
Brofaromine	Reversible A; Not approved	It is also a modest inhibitor of serotonin reuptake. Broadly studied and clinical trials although not approved, probably due to the limited market (Lotufo-Neto et al. 1999; Chouinard et al. 1993)	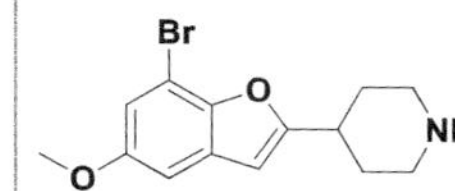
Clorgyline	Irreversible A; Not approved	Crystal structure of MAO-A with clorgyline is one of the only two available for this isozyme (De Colibus et al. 2005). Improved affective phenotypes in a mouse model of Huntington's disease (HD) and antidepressant effects demonstrated in humans (Finberg and Rabey 2016; Garcia-Miralles et al. 2016)	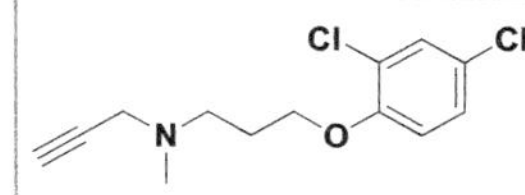
CX157	Reversible A; Not approved	Potent brain MAO-A inhibition observed in humans and first RIMA with documented reversible brain MAO-A inhibition which correlates to its plasma concentration (Fowler et al. 2010). Completed phase II clinical trials for major depressive disorder (NCT00739908)	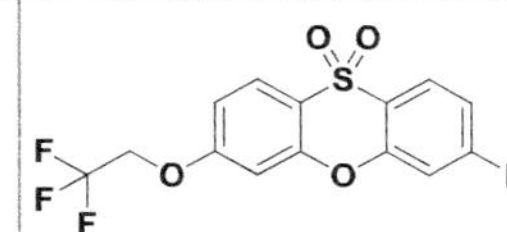
Harmine	Reversible A; Not approved	Research and clinical use for brain PET imaging of MAO-A with [^{11}C]harmine radioligand (Zanderigo et al. 2018). Crystal structure of MAO-A in complex with harmine (Son et al. 2008). Harmine is related with stimulation of human neural progenitors and restoration of astrocytic functions, which could be linked with its antidepressant potential (Dakic et al. 2016; Liu et al. 2017a). Clinical trials for resistant depression with Ayahuasca, a botanical hallucinogenic brew (NCT02914769) (Sanches et al. 2016)	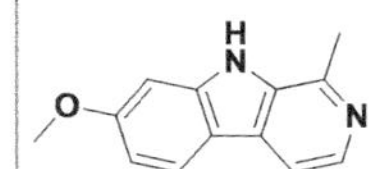

(continued)

Table 1 (continued)

Compound	MAO inhibition and state of development	Comments	Structure
Iproniazid	Irreversible A and B; Not approved (removed from market)	Strong liver toxicity associated to hydrazine compounds (Youdim et al. 2006; Edmondson and Binda 2018)	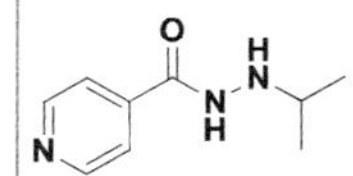
Isocarboxacid	Irreversible A and B; Approved	Rarely use due to dietary restrictions and toxicity associated to hydrazine compounds (Youdim et al. 2006; Edmondson and Binda 2018)	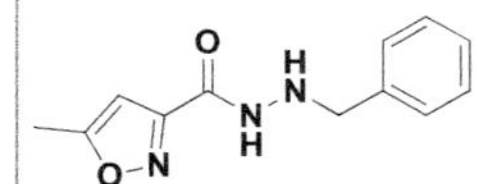
Ladostigil	Irreversible A and B; Not approved	Potential antidepressant activity showed in rats. Brain selectivity towards MAO inhibition, avoiding dietary restrictions (Weinreb et al. 2012)	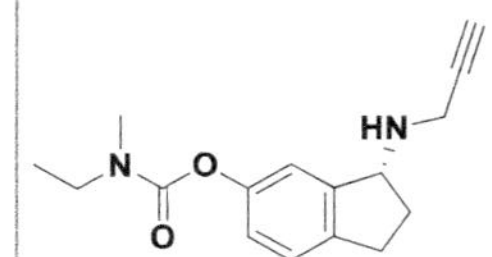
Methylene blue	Reversible A; Approved for the treatment of methemoglobinemia	It is also a non-selective inhibitor of nitric oxide synthase, guanylate cyclase, and selective reversible MAO-A inhibitor. It has completed phase III clinical trials for bipolar disorder showing improved symptoms of depression and anxiety (NCT00214877) (Alda et al. 2017)	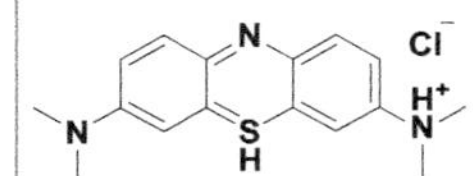
Moclobemide	Reversible A; Approved	Evaluated for nicotine dependence (Berlin et al. 1995). Compared efficacy to SSRIs for the treatment of depression. The first RIMA approved in Europe (Lotufo-Neto et al. 1999)	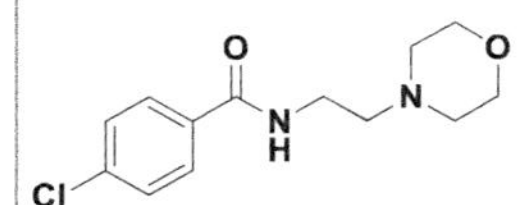
Pargyline	Irreversible B; Not approved	Antidepressant and antihypertensive activities (Finberg and Rabey 2016)	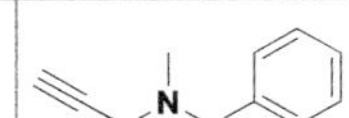

Pirlindole	Reversible A; Approved	Promotes hippocampal adult neurogenesis and prevent stress-associated dendritic atrophy of granule neurons in a chronic stress rat model (Morais et al. 2014)	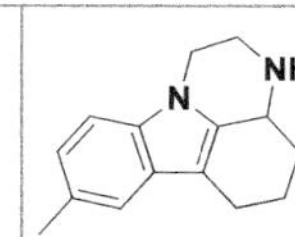
Phenelzine	Irreversible A and B; Approved	One of the first developed antidepressants. Needs dietary control for tyramine ingestion, hepatotoxicity (Finberg and Rabey 2016)	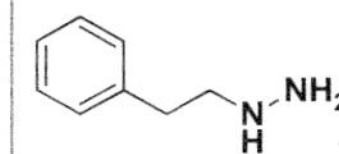
Selegiline	Irreversible B; Approved	Approved for the treatment of major depressive disorder in a transdermal patch to avoid "cheese effect" (Lee and Chen 2007)	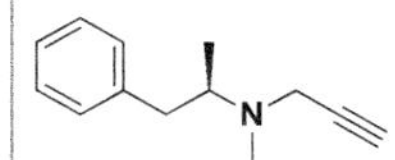
Tranylcypromine	Irreversible A and B; Approved	Needs dietary control for tyramine ingestion (Finberg and Rabey 2016)	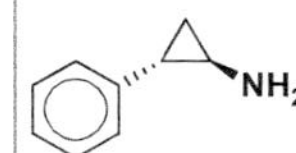 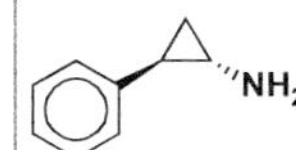
Zonisamide	Reversible B; Approved for the treatment of epilepsy	Clinical trials conducted for sleep apnea (NCT01765608), alcohol use (several clinical trials available; one phase IV study for the treatment of alcohol dependence NCT00595556), bipolar disorder (NCT00047567), post-traumatic stress disorder (NCT01847469, NCT03376139), and cocaine dependence (NCT01137890)	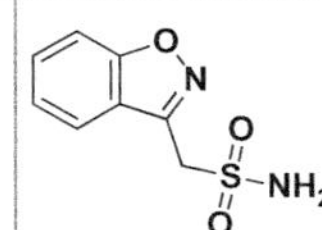

than moclobemide that exerted antidepressant properties in behavioral in vivo tests (Evranos-Aksoz et al. 2017).

Development of selective MAO-B inhibitors, accelerated for their use in PD, leads to their evaluation in depression, considering that they do not induce hypertensive crisis ("cheese effect"). A prominent example was selegiline, a selective irreversible MAO-B inhibitor, approved for major depressive disorder (Finberg and Rabey 2016; Lee and Chen 2007). In order to increase CNS levels, it was administrated as transdermal patch, a formulation that increased its brain distribution. Its high brain concentration inhibited both isozymes (A and B), since it is a selective MAO-B inhibitor only at low concentrations. Therefore, under this administration regime, MAO-A inhibition was observed at CNS level without hepatic or intestinal effects (Mawhinney et al. 2003). Seleginine was further evaluated clinically for the treatment of attention deficit hyperactivity disorder (ADHD) in children where it showed increased levels of attention similar to those achieved with methylphenidate (first-line treatment) (Akhondzadeh et al. 2003; Rubinstein et al. 2006). Rasagiline and safinamide, other selective irreversible and reversible MAO-B inhibitors, respectively, were also found to be effective in the treatment of depression associated to PD patients, as shown in different clinical trials (Barone et al. 2015; Korchounov et al. 2012; Ryan et al. 2019; Smith et al. 2015) (Table 1).

3.2 Parkinson's Disease

PD is a chronic and fatal neurodegenerative disease that affects mainly the nigrostriatal dopaminergic neurons, resulting in tremor, muscle rigidity, bradykinesia, and instability. Following the development of MAO inhibitors as treatments for depression and other affective disorders, and considering the implication of MAO enzymes in the metabolism of dopamine, their application for the treatment of PD started to be clinically tested. Furthermore, the target in this case was MAO-B inhibition that also led to reduced toxicity risk. In general, MAO-B inhibitors are indicated for patients showing mild motor deficits as results in clinical trials in early PD demonstrated mobility scores improvement compared to levodopa. These compounds showed benefits in the rate of motor fluctuations with reduced adverse effects compared to other dopaminergic agents (Caslake et al. 2009). In that sense, the application of MAO-B inhibitors for PD relies on several observations: (1) both MAO-A and MAO-B isozymes show similar rates for dopamine metabolism (Youdim et al. 2006); (2) levels of MAO-B are increased in PD conditions as consequence of gliosis, considering that this enzyme is mainly present in glial cells (Nagatsu and Sawada 2006); (3) human basal ganglia dysfunction is closely related to PD, and these structures contain higher levels of MAO-B isozyme. In this sense, MAO-B activity has been found to be increased in PD affected areas as *substantia nigra* (Youdim et al. 2006; Mallajosyula et al. 2008); (4) aging is the most important PD risk factor, and MAO-B levels increase with age, as observed in human brain postmortem studies. Thus, MAO-B overactivity might be related to neurodegeneration as a consequence increased ROS levels (Mallajosyula et al. 2008; Fowler et al. 1997).

ROS are considered to play a central role on the onset and progression of the disease. In PD there is a selective loss of dopaminergic neurons in the *substantia nigra* pars compacta, and this neuronal death is considered to be linked to oxidative stress due to dopamine metabolism (Segura-Aguilar et al. 2014). In brief, tyrosine is converted into L-DOPA by tyrosine hydroxylase (TH) and then decarboxylated by the DOPA-decarboxylase to generate dopamine in the brain (Meiser et al. 2013). The metabolism of dopamine by MAO enzymes, or its auto-oxidation, leads to the production of reactive oxygen species (ROS) (Blesa et al. 2015). On the one hand, the auto-oxidized form of dopamine, a quinone-related structure, is a powerful electrophile that exerts high toxicity to the cell. On the other hand, MAO metabolism of dopamine releases hydrogen peroxide that decomposes to generate ROS, thus, exacerbating cellular damage. In this line, dopaminergic neurons present in *substantia nigra* are surrounded by astrocytes containing high levels of MAO-B enzyme; thus, they are exposed to increased oxidative stress, as previously described. In general, neurons contain lower levels of glutathione (GSH) than astrocytes to control oxidative stress; therefore, they are highly susceptible to this toxic stimulus (Mallajosyula et al. 2008; Sian et al. 1994). In fact, decreased levels of GSH increase neuronal vulnerability to free radicals that induce mitochondrial dysfunction and maintain α-synuclein in a protofibril state. α-Synuclein protofibril finally aggregates to form Lewy bodies observed in postmortem brain of PD patients (Croisier et al. 2005; Poewe et al. 2017). Most relevant MAO inhibitors for PD treatment are summarized in Table 2.

The first MAO inhibitor approved for the treatment of PD was selegiline (L-deprenyl), a selective irreversible MAO-B inhibitor. This compound was marketed on the 1980s and used in combination with L-DOPA or levodopa, a dopamine precursor that was the main treatment for PD symptoms. This inhibitor belongs to propargylamine class of compounds, in general, acting through covalent bonding to N5 atom of flavin ring of FAD coenzyme, as previously mentioned. It suffers first-pass metabolism to generate amphetamine and methamphetamine (Yasar et al. 2006), related with potential CNS and cardiovascular secondary effects (Gal et al. 2005). Nevertheless, there is no reported toxicity derived from selegiline metabolites probably due to the safer clinical doses used, although it could contribute to sleep disturbances (Muller et al. 2013; Yasar et al. 1996). Conversely, it has been reported a psychostimulant-like behavioral effects related with production of selegiline-derived amphetamine metabolites at higher doses (Yasar et al. 2006). Although it can be used in monotherapy (Youdim et al. 2006), clinical trials demonstrated that levodopa dosing can be reduced 30–40% when combined with selegiline (Myllyla et al. 1997); additionally, the DATATOP study demonstrated that only 26% of selegiline treated patients needed levodopa combination after 12 months compared to 47% of placebo group (Parkinson Study 1989). Thereafter, the SELEDO study demonstrated that the mean time delay for an increase of 50% of levodopa dosage in selegiline treated patients was 4.9 years, compared to 2.6 years

Table 2 MAO inhibitors for the treatment of PD

Compound	MAO inhibition and state of development	Comments	Structure
2,3,6-Trimethyl-1,4-naphthoquinone (TMN)	Reversible A and B; Not approved	Present in tobacco plant. Increased dopamine levels in the striatum observed in TMN pre-treated MPTP mice (Castagnoli et al. 2003). Similar structures such as 1,4-naphthoquinones also showed MAO inhibition (Coelho Cerqueira et al. 2011)	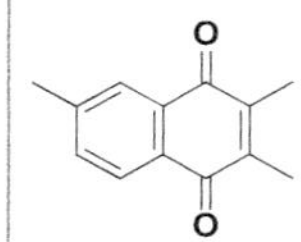
2-HMP (N-(2-heptyl)-N-methylpropargylamine)	Irreversible B; Not approved	Showed in vivo selective and brain specific MAO-B inhibition. Neuroprotection and antiapoptotic activity, MAO-B independent (Berry 1999)	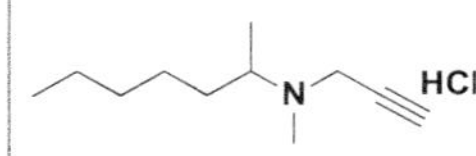
(E)-8-(3-Chlorostyryl)caffeine (CSC)	Reversible B; Not approved	Dual adenosine A_{2A} receptor antagonist and MAO-B inhibitor developed as xanthine derivative. Protective profile against 6-hydroxydopamine (6-OHDA) in rats (Aguiar et al. 2008)	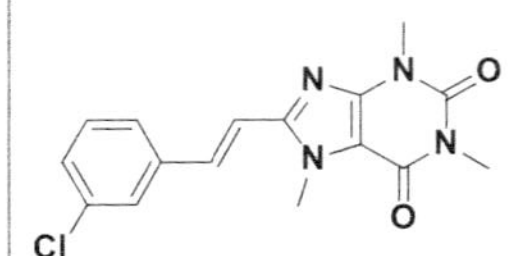
Farnesol	Reversible B; Not approved	Showed in vivo rat brain MAO-B inhibition (Khalil et al. 2006). Reported neuroprotective activities in LPS mice model by regulating apoptotic cascade through antioxidant effects (Santhanasabapathy and Sudhandiran 2015)	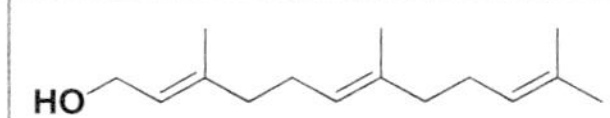
Isatin	Reversible B; Not approved	An endogenous indole with MAO inhibitory properties, more selective for B isozyme. Isatin is able to increase brain dopamine levels upon striatal administration in rat (Justo et al. 2016). It improves bradykinesia and striatum dopamine levels in rat model of PD induced by the Japanese encephalitis virus (Ogata et al. 2003). It improves apomorphine (APO)-induced rotations	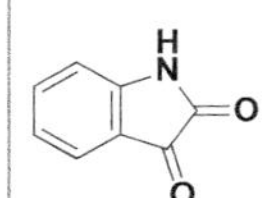

		in a 6-OHDA rat model, although this effect seems to be MAO independent as caudate putamen dopamine levels do not change (Zhou et al. 2001). Isatin pre-treatment in MPTP mice improved locomotor activity and profile of ubiquitinated brain mitochondrial proteins, related with mitochondrial dysfunction elicited by the toxin (Buneeva et al. 2018)	
Ladostigil	Irreversible A and B; Not approved	Ladostigil showed brain-selective MAO inhibition. It inhibits hippocampal and striatal MAO isozymes in rats upon chronic treatment, increases striatal dopamine levels, and inhibits behavioral hyperactivity following L-DOPA treatment (Sagi et al. 2005). Ladostigil also avoids nigrostriatal neurodegeneration in MPTP mice model (Youdim 2013)	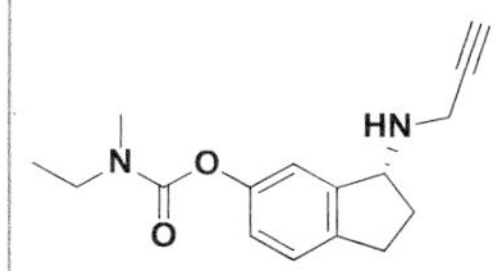
Lazabemide	Reversible B; Not approved	Shows higher potency than selegiline (Henriot et al. 1994). Clinical trials in levodopa-treated PD patients were carried out without improvement in clinical features of the disease (Parkinson Study 1994). Some lazabemide derivatives are currently being developed for further research (i.e., an L-DOPA-lazabemide prodrug composed by L-DOPA linked with lazabemide via an amide) (Hoon et al. 2017; Zhou et al. 2018)	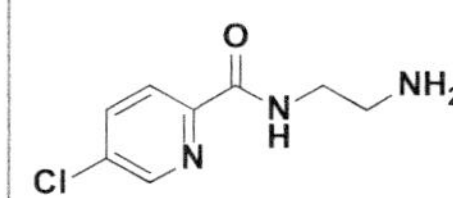

(continued)

Table 2 (continued)

Compound	MAO inhibition and state of development	Comments	Structure
M30	Irreversible A and B; Not approved	The propargyl moiety of the compound is thought to be responsible of neuroprotective properties observed, related to interaction with Bcl-2 and Bcl-xl apoptotic proteins (Youdim 2013; Zheng et al. 2005). Limited actuation of M30 on cardiovascular effect of oral tyramine in rats in comparison with tranylcypromine as control, related to its brain selectivity (Gal et al. 2010)	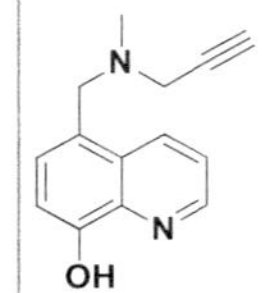
Pioglitazone	Reversible B; Not approved	Antidiabetic drug under repurposing studies for PD treatment. Phase II clinical trials in early PD patients was carried out although pioglitazone was not found to modify progression of the disease (NCT01280123) (Neurol 2015)	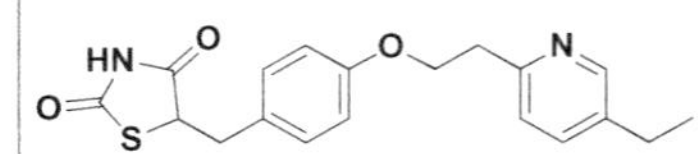
PF9601N (N-(2-propynyl)-2-(5-benzyloxy-indolyl) methylamine)	Irreversible B; Not approved	Tryptamine derivative (Pérez et al. 1999). It showed nigrostriatal dopamine neurons protection in MPTP mice and rats under 6-OHDA striatal lesion (Cutillas et al. 2002; Perez and Unzeta 2003). As other propargylamine inhibitors, neuroprotective effects are in part related with antiapoptotic properties MAO-independent (Sanz et al. 2008)	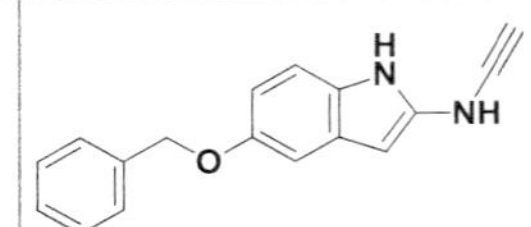
Rasagiline	Irreversible B; Approved	Approved for the treatment of PD	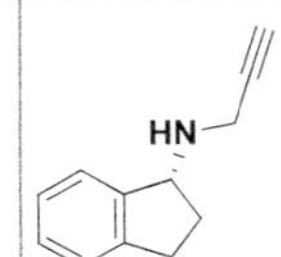

Rosiglitazone	Reversible A and B; Not approved	Antidiabetic drug under repurposing studies for PD treatment. Low selectivity for MAO-B inhibition (Binda et al. 2011a)	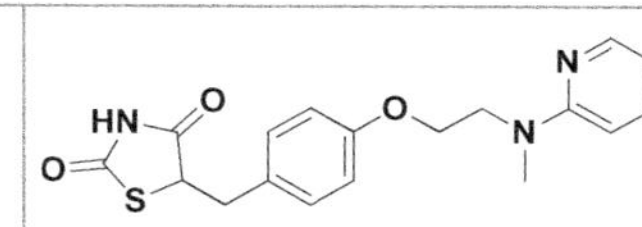
Safinamide	Reversible B; Approved	Approved for the treatment of PD	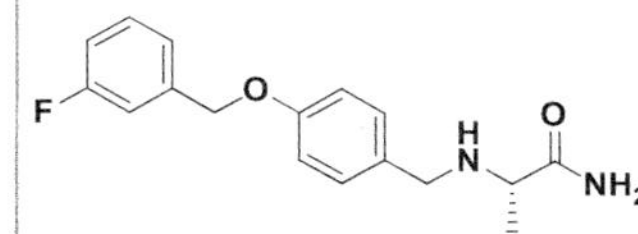
Selegiline	Irreversible B; Approved	Approved for the treatment of PD	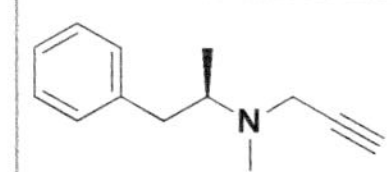
Zonisamide	Reversible B; Approved (not in United States, but available in Japan)	Antiepileptic drug repurposed for PD treatment and approved in 2009 in Japan	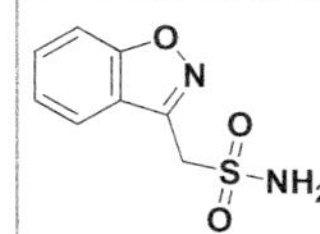
VAR103039	Irreversible A and B; Not approved	Multitarget compound that achieves iron chelation, iron-induced lipid-peroxidation inhibition, and brain-selective MAO inhibition, structurally related with M30. It showed neuroprotective profile in MPTP mice and 6-OHDA rats, attenuating striatal dopamine loss and increasing tyrosine hydroxylase levels. Chronic treatment in aged rats led to an increase in neurotrophic factors (BDNF, GDNF) and Bcl-2 family proteins (Bar-Am et al. 2015)	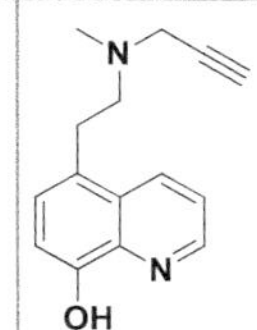

of placebo patients (Przuntek et al. 1999). This is important for reducing levodopa-induced dyskinesia observed in levodopa-treated PD patients.

Another propargylamine potent selective irreversible MAO-B inhibitor, rasagiline, was then developed on the 2000s and approved for the treatment of PD. This compound is not metabolized into amphetamine as selegiline, offering an improved profile in terms of potential neurotoxicity (Edmondson and Binda 2018). Rasagiline can improve motor behavior, motor complications, mood, and sleep disorders due to its additional glutamate antagonizing properties (Muller et al. 2013; Dong et al. 2016). After satisfactory safety clinical trials, rasagiline showed reduced PD advance (Rabey et al. 2000), and this observation was further evaluated in the TEMPO study. Results from this study suggested a potential disease modifying activity of rasagiline that reduced the disease progression and increased patient quality of life (Parkinson Study 2002, 2004). Considering these encouraging results, a new clinical trial was designed to evaluate the effectiveness of rasagiline in combination with levodopa in advanced PD patients, the LARGO study (Rascol et al. 2005). In this case, rasagiline showed reduced off-time duration intraday and improved clinical global improvement and motor function on patients. More recently, a new double-blind clinical trial was conducted to further evaluate the potential disease modifying ability of rasagiline in PD, the ADAGIO study (Olanow et al. 2009). The results showed a possible disease modifying effect of rasagiline given at 1 mg daily dosage; however, it did not show this effect at 2 mg daily dosage; thus, these divergent results are not conclusive (Rascol et al. 2011; Jankovic et al. 2014).

As previously depicted, irreversible inhibitors led to strong and long-lasting inhibition, and new enzyme has to be synthesized. In that sense, reversible inhibition appears to be more manageable, maintaining selectivity to the B isozyme. Safinamide, a reversible and highly selective MAO-B inhibitor, was then developed following this idea and has been recently approved as an adjunctive treatment to L-DOPA for mild- to late-stage PD (Borgohain et al. 2014). Separately from selective reversible MAO-B inhibition, this drug acts through multiple mechanisms of action. It is able to block sodium and calcium channels and reduce excessive glutamate release (Stocchi et al. 2006). It has shown a safer profile, and there is evidence of improved motor symptoms in early PD patients (Edmondson and Binda 2018; Bette et al. 2018; Fabbri et al. 2015). Initial clinical studies (study 015 and continued in study 017) showed improved motor scores in safinamide treated patients as well as improved quality of life (Schapira et al. 2013). Thereafter, the MOTION study also demonstrated benefits in the ADL score, heath-related quality of life, better cognition, and increased rate of responders (Stocchi and Torti 2016). Thereafter, the SETTLE trial (in combination with levodopa) demonstrated a significant increase in the on-time and improved motor symptoms (Schapira et al. 2013) (For a critical review of safinamide clinical evidence in PD, see Bette et al. (2018)). In addition, a recent post hoc analysis of the 2-year study 018 revealed a reduction on chronic pain suffered by PD patients with a reduction on pain drugs usage (Cattaneo et al. 2018). Finally, the recent SIN-DEP-PAR and SELEDO clinical studies further confirm the beneficial effects of seleginine for the management of PD (Muller and Mohr 2019).

Apart from MAO-B inhibition-related aspects, this kind of inhibitors exhibit MAO-independent activities, also responsible of its beneficial effects as mentioned for safinamide. It has been demonstrated that selegiline exerts neuroprotective activities against MPTP toxicity also in cell lines lacking MAO-B (Le et al. 1997). In this sense, rasagiline and selegiline have been effective in several neuroprotection models of disease, and they have demonstrated antiapoptotic properties, an effect attributed to the propargylamine moiety (Szoko et al. 2018).

During the last years, new multitarget drugs derived from rasagiline have been developed for the treatment of PD among other neurodegenerative diseases. Ladostigil, an example of rasagiline multitarget derivative, inhibits brain acetylcholinesterase, butyrylcholinesterase, and brain MAO-A and MAO-B isozymes after chronic treatment in rats. Furthermore, it exerts neuroprotection against MPTP toxicity in mice, among other activities (Youdim 2013). Related to its neuroprotective profile in preclinical models, ladostigil has completed phase IIb clinical trials for mild cognitive impairment and Alzheimer's disease (AD) (NCT01354691, NCT01429623) (discussed below). M30, another rasagiline derivative, is a multitarget iron chelator that exhibits brain-selective MAO inhibition and neuroprotection against MPTP, lacatcystin, and 6-hydroxydopamine in animal models of PD. The protective properties of M30 were also dependent to the ability of the drug to activate the hypoxia-induced factor (HIF) and derived induction of neurotrophins as brain-derived neurotrophic factor (BDNF), vascular endothelial growth factor (VEGF), erythropoietin, and glia-derived neurotrophic factor (GDNF) (Youdim 2013). Finally, it was demonstrated that M30 elevated dopaminergic and transferrin receptor cell count in the *substantia nigra* of MPTP-treated mice (Youdim 2012). Apart from the previously mentioned increase in MAO activity with PD and ageing, iron was also observed to increase in such conditions stating the promising future of multitarget iron chelators (Youdim et al. 2014).

Last years, repurposing programs are being used for the development of new treatments for complex diseases, due to their interesting cost reduction. In that sense, glitazones, used as antidiabetic drugs, were evaluated for PD as they showed MAO-B inhibition properties. This is the case of pioglitazone and rosiglitazone. Crystal structure of pioglitazone in complex with MAO-B explained binding mode through both the entrance and substrate cavities (Binda et al. 2011a). Several studies position this compound for potential PD treatment: (1) acute treatment of MPTP mice leaded to the absence of striatal dopamine reduction and improved motor behaviors (Barbiero et al. 2011); (2) beneficial effects on rotenone model of PD in rats related to recovery of striatal dopamine levels and locomotion (Ulusoy et al. 2011); (3) improvements in the 6-OHDA model in rats: reduced mortality rate, attenuated microglial activation, exerted antidepressant-like effects, induction of hippocampal neurogenesis (often associated with depression in PD patients), and protection against hypolocomotion (Bonato et al. 2018; Machado et al. 2019); and (4) decreased neuroinflammation in the midbrain and striatum and improved motor phenotype in Cox10/DAT-cre mice (dopaminergic neuron knockout of Cox10 essential for maturation of COXI, catalytic subunit of Complex IV of mitochondrial respiratory chain) (Pinto et al. 2016). Phase II clinical trials in early PD patients was carried out although pioglitazone was not found to modify progression of the disease (NCT01280123) (Neurol 2015). On the other hand, rosiglitazone showed low selectivity for MAO-B inhibition (Binda et al. 2011a) and also reported properties of this drug in PD models: (1) pre-treatment with rosiglitazone prior to 6-OHDA insult in rats leads to protection against striatal dopaminergic death and decrease in some pro-inflammatory markers as TNF-α (Lee et al. 2012) and (2) partial recovery of striatal dopamine content and dopamine neuron degeneration in *substantia nigra* after rosiglitazone treatment in MPTP mice (Carta et al. 2011). The protective effects of this kind of compounds may be MAO-independent, at least in part, as demonstrated by importance of peroxisome proliferator-activated receptor γ (PPARγ) activation by glitazones for neuroprotection in *substantia nigra* and its related anti-inflammatory actions (Martin et al. 2012).

Another important example of repurposing is the case of the antiepileptic drug zonisamide, compound approved for the treatment of PD in Japan. Its binding mode was elucidated in the substrate cavity in complex with MAO-B (Binda et al. 2011b). Among several results in PD-related models, it has been demonstrated implication of zonisamide as MAO inhibitor, showing an attenuation of striatal dopamine and tyrosine hydroxylase reduction in MPTP mice (Sonsalla et al. 2010). Zonisamide has also showed a neuroprotective profile in a genetic mouse model of PD, reducing motor symptoms (Sano et al. 2015), enhancing L-DOPA treatment (Nishijima et al. 2018), and exhibiting a reduction in associated dyskinesia in 6-OHDA rats (Oki et al. 2017). In this line, zonisamide has been evaluated in several clinical trials, most of them in Japan (Murata et al. 2015, 2016, 2018), and there is one recent open study in advanced PD in Egyptian population (NCT04182399). As previously mentioned, it was approved in Japan for use as anti-PD agent in 2009, in patients treated with other drugs in combination with L-DOPA that show inadequate responses (Grover et al. 2013) (Table 2).

3.3 Other Neurodegenerative Disorders

Given the neuroprotective profile of several MAO inhibitors, they could be useful for treatment of other neurodegenerative diseases. These complex diseases share many physiopathological mechanisms, in which MAO enzymes are involved, such as oxidative stress and neuroinflammation.

In this sense, several MAO-B inhibitors have been evaluated as potential treatments of AD. This is the most prevalent neurodegenerative disease, and it is characterized by progressive memory loss and incapacitation for daily task. AD major symptoms are accompanied by depression, agitation, delusions, and hallucinations. AD patients showed higher levels of MAO-B activity, and this may lead to an exacerbated oxidative stress condition in this disease (Kennedy et al. 2003). MAO activation has also been related to cognitive impairment, altering balance of neurotransmitters, and the formation of amyloid plaques through the modulation of amyloid precursor protein (APP) processing by MAO (for review, see Cai (2014)). These evidences prompted the application of MAO inhibitors as potential treatments of AD. In this line, selegiline and rasagiline have been evaluated in clinical trials for AD (NCT01701089, NCT00104273, NCT02359552). Initial clinical trials with seleginine pointed to a potential cognitive and behavioral improvement after 3–6 months treatment (Campi et al. 1990; Filip and Kolibas 1999); however, these results were not corroborated in other cases (Tariot et al. 1987; Burke et al. 1993). Further post hoc analysis determined that selegiline might have a short-term beneficial effect with no clinical relevance.

Given the potential protective effects of these compounds, also novel rasagiline derivatives have been tested. This is the case of M30 and ladostigil, previously introduced in this review for the treatment of PD. These compounds were developed as multitarget drugs combining interesting activities for the treatment of neurodegenerative diseases (see Sect. 3.2). M30 has been observed to be involved in regulation of amyloid β (Aβ), inhibiting its accumulation in APP/presenilin 1 mice, neurogenesis, suppression of oxidative stress, pro-cognitive, and anti-inflammatory effects (improvement in a rat model of AD with amyloid pathology McGill-R-Thy1-APP transgenic rats) (Cai 2014; Kupershmidt et al. 2012; Pimentel et al. 2015; Zheng et al. 2010). Ladostigil was shown to be related with APP translation and processing (Yogev-Falach et al. 2006) and neuroinflammation in a model of aged rats (Panarsky et al. 2012). These among other properties such as its acetylcholinesterase inhibition point out ladostigil as a new candidate for AD treatment, and phase II clinical trials has been conducted (NCT01354691, NCT01429623). Although it was safe and well tolerated, ladostigil did not delay progression to dementia. In spite of its failure, ladostigil treatment was associated with reduced brain and hippocampus volume loss (Schneider et al. 2019).

Apart from the abovementioned compounds, many others have shown potential therapeutic value for AD: (1) *harmine* improved memory and learning in animal models and exerted anti-inflammatory properties in LPS-mouse model (Dos Santos and Hallak 2017; Liu et al. 2017b); (2) *methylene blue* relationship between serotonin levels and psychological symptoms of AD could be an approach for this compound. Apart from this, methylene blue has shown attenuation of the formation of amyloid plaques, neurofibrillary tangles of tau protein, and involvement in mitochondrial function restoration, among others (Oz et al. 2009). It is currently under clinical trials (NCT02380573); (3) *KDS2010* is a novel compound highly potent and selective reversible MAO-B inhibitor (Fig. 2). It has demonstrated learning and memory improvements, promotion of synaptic transmission, and reduction of astrogliosis and astrocytic GABA levels in APP/presenilin 1 mice (Park et al. 2019). Importantly, abundant GABA production by MAO-B in reactive astrocytes has been linked to AD-like pathology in animal models of AD (Jo et al. 2014); (4) *ASS234* is a multitarget compound that inhibits acetylcholinesterase, butyrylcholinesterase, and MAO enzymes (Fig. 2). From crystal structure in complex with MAO-B, it is known that it acts as irreversible inhibitor binding to the FAD coenzyme (Esteban et al. 2014). It has been shown that ASS234 is able to restore scopolamine-induced cognitive impairment and prevent Aβ aggregation in animal

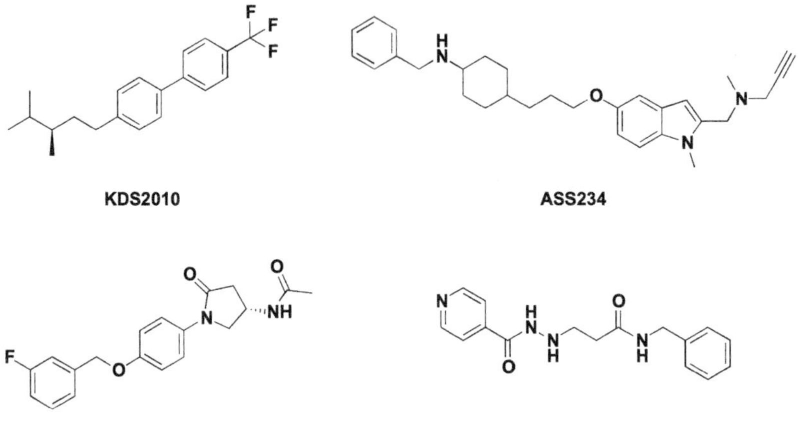

KDS2010 ASS234

Sembragiline Nialamide

Fig. 2 Chemical structure of novel MAO inhibitors for the treatment of AD (KDS2010, ASS234, sembragiline) and ischemic stroke (nialamide)

models of AD (Marco-Contelles et al. 2016); (5) *sembragiline* has shown potent and reversible MAO-B inhibition with an improved safety profile, regarding serotonin syndrome and the "cheese effect" observed with other MAO inhibitors (Fig. 2). Interestingly, this compound was protective against oxidative stress and astrogliosis in transgenic mice overexpressing MAO-B in astroglia (Borroni et al. 2017). In accordance with these promising results and after phase I trials, sembragiline was evaluated in phase II clinical trials for AD (MAyflOwer RoAD study) suggesting potential benefits on behavioral symptoms in some subpopulations (NCT01677754) (Nave et al. 2017). No differences were found between treated groups and placebo in ADAS-Cog11 scale performance, missing the primary endpoint. Positron emission tomography measurements have demonstrated that same dose used in phase II clinical trial (NCT01677754) achieved near-complete brain MAO-B inhibition in AD patients daily treated with sembragiline (NCT01701089) (Sturm et al. 2017).

Amyotrophic lateral sclerosis (ALS) and Huntington's disease (HD) share many of the pathophysiological characteristics of PD and AD. Although selegiline was shown to be ineffective in clinical trials for ALS (Lange et al. 1998), other MAO inhibitors have demonstrated potential properties. Previously mentioned iron chelator M30 was evaluated in SOD1-G93A transgenic mouse model of ALS in combination with high-calorie energy supplemented diet, and it has been related with protective effects on motor performance and increased survival, with brain MAO inhibition and reduction in dopamine metabolism (Golko-Perez et al. 2016). Other MAO inhibitors such as rasagiline have been evaluated in clinical trials for ALS. This compound was safe in ALS patients, also suggesting potential disease modifying properties in some subpopulations of the study when co-administered with riluzole (NCT01879241) (Ludolph et al. 2018). However, when administered alone it did not show any alteration of the disease progression when compared to control after 12 months (Statland et al. 2019).

Regarding dopamine imbalance present in HD (Chen et al. 2013), MAO inhibitors have been evaluated for this condition. MAO activity is increased in brain tissue from HD patients and mouse striatal neural cells expressing huntingtin. Treatment with MAO inhibitors (clorgyline, phenelzine, selegiline) improved oxidative stress condition and survival of these cells (Ooi et al. 2015). In this line, clorgyline treatment of YAC128 HD mice restored dopamine, serotonin, and norepinephrine levels in the striatum (Garcia-Miralles et al. 2016).

Other related diseases such as cerebral ischemia have been postulated for MAO inhibition treatment. Probably related to its neuroprotective activities, selegeline treatment within 48 h after stroke improved patient recovery in a phase II study (Sivenius et al. 2001; Bartolo et al. 2015). Recently, the MAO inhibitor nialamide has shown to be protective in terms of improved functional recovery and survival after post-ischemic administration in a stroke mouse model (Liu et al. 2019) (Fig. 2).

3.4 Cardiac Diseases

Apart from the deeply discussed role of oxidative stress in neurodegenerative diseases, it also appears as an important factor for cardiovascular diseases. Serotonin and norepinephrine, which are MAO-A substrates, are related with cardiac function. On the one hand, variations in serotonin levels can lead to cardiovascular dysfunction, being important regulator of heart function (Cote et al. 2004). On the other hand, norepinephrine metabolism by MAO-A is increased in mice subjected to hemodynamic stress leading to enhanced oxidative stress, hypertrophy, chamber dilation, and reduced systolic function; clorgyline treatment of these animals showed beneficial effects on cardiac function (Kaludercic et al. 2010). Levels of MAO-A are increased in several models of cardiomyopathies, and cardiac-specific MAO-A overexpression in transgenic mice was related with oxidative mitochondrial damage, cardiomyocyte necrosis, and chronic ventricular dysfunction (Villeneuve et al. 2013). Recent studies with reversible MAO-A inhibitor moclobemide exhibited myocardium protection in rats submitted to acute volume overload (Huuskonen et al. 2019). There are some studies in the same way with MAO-B demonstrating a role of enhance activity of this isozyme in cardiac injuries after chronic hemodynamic stress (Kaludercic et al. 2014). All this shows the potential for the future clinical use of MAO inhibitors for the treatment of heart disease.

4 Summary and Future Prospects

MAO inhibitors have provided a number of successful drugs for the treatment of complex diseases, being the first drugs approved for depression. During the last 70 years, the advances on the knowledge of the crystal structures of both isozymes, MAO-A and MAO-B, tissue distribution, and substrate selectivity have led to the development of a plethora of highly selective reversible and irreversible inhibitors with clinical application. These advances overcome the deleterious side effects improving the efficacy and security of these classes of advanced drugs. Novel therapeutics as MAO inhibitors included classical applications such as major depressive disorder, transient resistant depression, bipolar depression, etc., and it is expected as exponential growth of their application in this therapeutic area. Nonetheless, novel therapeutic areas are under use due to the development of selective MAO-B inhibitors such us the treatment of neurodegenerative diseases. In particular, the treatment of Parkinson's disease is a prominent example with four compounds in clinical use. Further applications are under development for the treatment of Alzheimer's disease, amyotrophic lateral sclerosis, and cardiovascular diseases. The ubiquitous distribution of MAO enzymes ensures future exciting development of novel therapeutics including the development of novel and safer formulations.

References

Aguiar LM et al (2008) CSC, an adenosine A(2A) receptor antagonist and MAO B inhibitor, reverses behavior, monoamine neurotransmission, and amino acid alterations in the 6-OHDA-lesioned rats. Brain Res 1191:192–199

Akhondzadeh S et al (2003) Selegiline in the treatment of attention deficit hyperactivity disorder in children: a double blind and randomized trial. Prog Neuropsychopharmacol Biol Psychiatry 27 (5):841–845

Alda M et al (2017) Methylene blue treatment for residual symptoms of bipolar disorder: randomised crossover study. Br J Psychiatry 210(1):54–60

Alia-Klein N et al (2008) Brain monoamine oxidase A activity predicts trait aggression. J Neurosci 28(19):5099–5104

Anderson MC et al (1993) Monoamine oxidase inhibitors and the cheese effect. Neurochem Res 18 (11):1145–1149

Bar-Am O et al (2015) Neuroprotective and neurorestorative activities of a novel iron chelator-brain selective monoamine oxidase-A/monoamine oxidase-B inhibitor in animal models of Parkinson's disease and aging. Neurobiol Aging 36(3):1529–1542

Barbiero JK et al (2011) Acute but not chronic administration of pioglitazone promoted behavioral and neurochemical protective effects in the MPTP model of Parkinson's disease. Behav Brain Res 216(1):186–192

Barone P et al (2015) A randomized clinical trial to evaluate the effects of rasagiline on depressive symptoms in non-demented Parkinson's disease patients. Eur J Neurol 22(8):1184–1191

Bartolo M et al (2015) An explorative study regarding the effect of l-deprenyl on cognitive and functional recovery in patients after stroke. J Neurol Sci 349(1-2):117–123

Berlin I et al (1995) A reversible monoamine oxidase A inhibitor (moclobemide) facilitates smoking cessation and abstinence in heavy, dependent smokers. Clin Pharmacol Ther 58 (4):444–452

Berry MD (1999) R-2HMP: an orally active agent combining independent antiapoptotic and MAO-B-inhibitory activities. CNS Drug Rev 5(2):105–124

Bette S et al (2018) Safinamide in the management of patients with Parkinson's disease not stabilized on levodopa: a review of the current clinical evidence. Ther Clin Risk Manag 14:1737–1745

Binda C et al (2002) Structure of human monoamine oxidase B, a drug target for the treatment of neurological disorders. Nat Struct Biol 9(1):22–26

Binda C et al (2003) Insights into the mode of inhibition of human mitochondrial monoamine oxidase B from high-resolution crystal structures. Proc Natl Acad Sci U S A 100(17):9750–9755

Binda C et al (2004) Crystal structures of monoamine oxidase B in complex with four inhibitors of the N-propargylaminoindan class. J Med Chem 47(7):1767–1774

Binda C et al (2005) Binding of rasagiline-related inhibitors to human monoamine oxidases: a kinetic and crystallographic analysis. J Med Chem 48(26):8148–8154

Binda C et al (2007) Structures of human monoamine oxidase B complexes with selective noncovalent inhibitors: safinamide and coumarin analogs. J Med Chem 50(23):5848–5852

Binda C et al (2011a) Molecular insights into human monoamine oxidase B inhibition by the glitazone anti-diabetes drugs. ACS Med Chem Lett 3(1):39–42

Binda C et al (2011b) Interactions of monoamine oxidases with the antiepileptic drug zonisamide: specificity of inhibition and structure of the human monoamine oxidase B complex. J Med Chem 54(3):909–912

Blesa J et al (2015) Oxidative stress and Parkinson's disease. Front Neuroanat 9:91

Bonato JM et al (2018) Pioglitazone reduces mortality, prevents depressive-like behavior, and impacts hippocampal neurogenesis in the 6-OHDA model of Parkinson's disease in rats. Exp Neurol 300:188–200

Bonivento D et al (2010) Potentiation of ligand binding through cooperative effects in monoamine oxidase B. J Biol Chem 285(47):36849–36856

Borgohain R et al (2014) Two-year, randomized, controlled study of safinamide as add-on to levodopa in mid to late Parkinson's disease. Mov Disord 29(10):1273–1280

Borroni E et al (2017) Sembragiline: a novel, selective monoamine oxidase type B inhibitor for the treatment of Alzheimer's disease. J Pharmacol Exp Ther 362(3):413–423

Buneeva O et al (2018) The effect of neurotoxin MPTP and neuroprotector isatin on the profile of ubiquitinated brain mitochondrial proteins. Cell 7(8):91

Burke WJ et al (1993) L-deprenyl in the treatment of mild dementia of the Alzheimer type: results of a 15-month trial. J Am Geriatr Soc 41(11):1219–1225

Cai Z (2014) Monoamine oxidase inhibitors: promising therapeutic agents for Alzheimer's disease (Review). Mol Med Rep 9(5):1533–1541

Campi N, Todeschini GP, Scarzella L (1990) Selegiline versus L-acetylcarnitine in the treatment of Alzheimer-type dementia. Clin Ther 12(4):306–314

Carta AR et al (2011) Rosiglitazone decreases peroxisome proliferator receptor-gamma levels in microglia and inhibits TNF-alpha production: new evidences on neuroprotection in a progressive Parkinson's disease model. Neuroscience 194:250–261

Caslake R et al (2009) Monoamine oxidase B inhibitors versus other dopaminergic agents in early Parkinson's disease. Cochrane Database Syst Rev 4:CD006661

Castagnoli K et al (2003) Inhibition of human MAO-A and MAO-B by a compound isolated from flue-cured tobacco leaves and its neuroprotective properties in the MPTP mouse model of neurodegeneration. Inflammopharmacology 11(2):183–188

Cattaneo C et al (2018) Long-term efficacy of safinamide on Parkinson's disease chronic pain. Adv Ther 35(4):515–522

Chen JY et al (2013) Dopamine imbalance in Huntington's disease: a mechanism for the lack of behavioral flexibility. Front Neurosci 7:114

Chouinard G et al (1993) Brofaromine in depression: a Canadian multicenter placebo trial and a review of standard drug comparative studies. Clin Neuropharmacol 16(Suppl 2):S51–S54

Coelho Cerqueira E et al (2011) Molecular insights into human monoamine oxidase (MAO) inhibition by 1,4-naphthoquinone: evidences for menadione (vitamin K3) acting as a competitive and reversible inhibitor of MAO. Bioorg Med Chem 19(24):7416–7424

Cote F et al (2004) Recent advances in understanding serotonin regulation of cardiovascular function. Trends Mol Med 10(5):232–238

Croisier E et al (2005) Microglial inflammation in the Parkinsonian substantia nigra: relationship to alpha-synuclein deposition. J Neuroinflammation 2:14

Curet O et al (1996) Befloxatone, a new reversible and selective monoamine oxidase-A inhibitor. I. Biochemical profile. J Pharmacol Exp Ther 277(1):253–264

Cutillas B, Ambrosio S, Unzeta M (2002) Neuroprotective effect of the monoamine oxidase inhibitor PF 9601N [N-(2-propynyl)-2-(5-benzyloxy-indolyl) methylamine] on rat nigral neurons after 6-hydroxydopamine-striatal lesion. Neurosci Lett 329(2):165–168

Da Prada M et al (1988) On tyramine, food, beverages and the reversible MAO inhibitor moclobemide. J Neural Transm Suppl 26:31–56
Dakic V et al (2016) Harmine stimulates proliferation of human neural progenitors. PeerJ 4:e2727
Dasgupta S et al (2018) Recognition dynamics of dopamine to human monoamine oxidase B: role of Leu171/Gln206 and conserved water molecules in the active site cavity. J Biomol Struct Dyn 36(6):1439–1462
De Colibus L et al (2005) Three-dimensional structure of human monoamine oxidase A (MAO A): relation to the structures of rat MAO A and human MAO B. Proc Natl Acad Sci U S A 102 (36):12684–12689
Dong J et al (2016) Current pharmaceutical treatments and alternative therapies of Parkinson's disease. Curr Neuropharmacol 14(4):339–355
Dos Santos RG, Hallak JE (2017) Effects of the natural beta-carboline alkaloid harmine, a main constituent of Ayahuasca, in memory and in the hippocampus: a systematic literature review of preclinical studies. J Psychoactive Drugs 49(1):1–10
Edmondson DE, Binda C (2018) Monoamine oxidases. Subcell Biochem 87:117–139
Edmondson DE et al (2009) Molecular and mechanistic properties of the membrane-bound mitochondrial monoamine oxidases. Biochemistry 48(20):4220–4230
Esteban G et al (2014) Kinetic and structural analysis of the irreversible inhibition of human monoamine oxidases by ASS234, a multi-target compound designed for use in Alzheimer's disease. Biochim Biophys Acta 1844(6):1104–1110
Evranos-Aksoz B et al (2017) New human monoamine oxidase A inhibitors with potential anti-depressant activity: design, synthesis, biological screening and evaluation of pharmacological activity. Comb Chem High Throughput Screen 20(6):461–473
Fabbri M et al (2015) Clinical pharmacology review of safinamide for the treatment of Parkinson's disease. Neurodegener Dis Manag 5(6):481–496
Fasipe OJ (2019) The emergence of new antidepressants for clinical use: agomelatine paradox versus other novel agents. IBRO Rep 6:95–110
Filip V, Kolibas E (1999) Selegiline in the treatment of Alzheimer's disease: a long-term randomized placebo-controlled trial. Czech and Slovak Senile Dementia of Alzheimer Type Study Group. J Psychiatry Neurosci 24(3):234–243
Finberg JP (2014) Update on the pharmacology of selective inhibitors of MAO-A and MAO-B: focus on modulation of CNS monoamine neurotransmitter release. Pharmacol Ther 143 (2):133–152
Finberg JP, Rabey JM (2016) Inhibitors of MAO-A and MAO-B in psychiatry and neurology. Front Pharmacol 7:340
Fowler JS et al (1997) Age-related increases in brain monoamine oxidase B in living healthy human subjects. Neurobiol Aging 18(4):431–435
Fowler JS et al (2010) Reversible inhibitors of monoamine oxidase-A (RIMAs): robust, reversible inhibition of human brain MAO-A by CX157. Neuropsychopharmacology 35(3):623–631
Gal S et al (2005) Novel multifunctional neuroprotective iron chelator-monoamine oxidase inhibitor drugs for neurodegenerative diseases. In vivo selective brain monoamine oxidase inhibition and prevention of MPTP-induced striatal dopamine depletion. J Neurochem 95(1):79–88
Gal S, Abassi ZA, Youdim MB (2010) Limited potentiation of blood pressure in response to oral tyramine by the anti-Parkinson brain selective multifunctional monoamine oxidase-AB inhibitor, M30. Neurotox Res 18(2):143–150
Garcia-Miralles M et al (2016) Treatment with the MAO-A inhibitor clorgyline elevates monoamine neurotransmitter levels and improves affective phenotypes in a mouse model of Huntington disease. Exp Neurol 278:4–10
Gillman PK (2006) A review of serotonin toxicity data: implications for the mechanisms of antidepressant drug action. Biol Psychiatry 59(11):1046–1051
Golko-Perez S et al (2016) Additive neuroprotective effects of the multifunctional iron chelator M30 with enriched diet in a mouse model of amyotrophic lateral sclerosis. Neurotox Res 29 (2):208–217

Grover ND et al (2013) Zonisamide: a review of the clinical and experimental evidence for its use in Parkinson's disease. Indian J Pharmacol 45(6):547–555

Henriot S et al (1994) Lazabemide (Ro 19-6327), a reversible and highly sensitive MAO-B inhibitor: preclinical and clinical findings. J Neural Transm Suppl 41:321–325

Hoon M et al (2017) The design and evaluation of an l-dopa-lazabemide prodrug for the treatment of Parkinson's disease. Molecules 22(12):2076

Hubalek F, Pohl J, Edmondson DE (2003) Structural comparison of human monoamine oxidases A and B: mass spectrometry monitoring of cysteine reactivities. J Biol Chem 278 (31):28612–28618

Hubalek F et al (2005) Demonstration of isoleucine 199 as a structural determinant for the selective inhibition of human monoamine oxidase B by specific reversible inhibitors. J Biol Chem 280 (16):15761–15766

Huuskonen C et al (2019) Monoamine oxidase A inhibition protects the myocardium after experimental acute volume overload. Anatol J Cardiol 21(1):39–45

Jankovic J et al (2014) Symptomatic efficacy of rasagiline monotherapy in early Parkinson's disease: post-hoc analyses from the ADAGIO trial. Parkinsonism Relat Disord 20(6):640–643

Jo S et al (2014) GABA from reactive astrocytes impairs memory in mouse models of Alzheimer's disease. Nat Med 20(8):886–896

Justo LA et al (2016) Effects and mechanism of action of isatin, a MAO inhibitor, on in vivo striatal dopamine release. Neurochem Int 99:147–157

Kaludercic N et al (2010) Monoamine oxidase A-mediated enhanced catabolism of norepinephrine contributes to adverse remodeling and pump failure in hearts with pressure overload. Circ Res 106(1):193–202

Kaludercic N et al (2014) Monoamine oxidase B prompts mitochondrial and cardiac dysfunction in pressure overloaded hearts. Antioxid Redox Signal 20(2):267–280

Kennedy BP et al (2003) Early and persistent alterations in prefrontal cortex MAO A and B in Alzheimer's disease. J Neural Transm 110(7):789–801

Khalil AA, Davies B, Castagnoli N Jr (2006) Isolation and characterization of a monoamine oxidase B selective inhibitor from tobacco smoke. Bioorg Med Chem 14(10):3392–3398

Korchounov A, Winter Y, Rossy W (2012) Combined beneficial effect of rasagiline on motor function and depression in de novo PD. Clin Neuropharmacol 35(3):121–124

Kupershmidt L et al (2012) Multi-target, neuroprotective and neurorestorative M30 improves cognitive impairment and reduces Alzheimer's-like neuropathology and age-related alterations in mice. Mol Neurobiol 46(1):217–220

Lange DJ et al (1998) Selegiline is ineffective in a collaborative double-blind, placebo-controlled trial for treatment of amyotrophic lateral sclerosis. Arch Neurol 55(1):93–96

Langston JW et al (1984) Pargyline prevents MPTP-induced parkinsonism in primates. Science 225 (4669):1480–1482

Le W et al (1997) (−)-Deprenyl protection of 1-methyl-4 phenylpyridinium ion (MPP+)-induced apoptosis independent of MAO-B inhibition. Neurosci Lett 224(3):197–200

Lee KC, Chen JJ (2007) Transdermal selegiline for the treatment of major depressive disorder. Neuropsychiatr Dis Treat 3(5):527–537

Lee EY et al (2012) Rosiglitazone, a PPAR-gamma agonist, protects against striatal dopaminergic neurodegeneration induced by 6-OHDA lesions in the substantia nigra of rats. Toxicol Lett 213 (3):332–344

Li M et al (2006) Functional role of the "aromatic cage" in human monoamine oxidase B: structures and catalytic properties of Tyr435 mutant proteins. Biochemistry 45(15):4775–4784

Liu F et al (2017a) Harmine produces antidepressant-like effects via restoration of astrocytic functions. Prog Neuro-Psychopharmacol Biol Psychiatry 79(Pt B):258–267

Liu X et al (2017b) Harmine is an inflammatory inhibitor through the suppression of NF-kappaB signaling. Biochem Biophys Res Commun 489(3):332–338

Liu Y et al (2019) Attenuation of ischemic stroke-caused brain injury by a monoamine oxidase inhibitor involves improved proteostasis and reduced neuroinflammation. Mol Neurobiol 57 (2):937–948

Lotufo-Neto F, Trivedi M, Thase ME (1999) Meta-analysis of the reversible inhibitors of monoamine oxidase type A moclobemide and brofaromine for the treatment of depression. Neuropsychopharmacology 20(3):226–247

Ludolph AC et al (2018) Safety and efficacy of rasagiline as an add-on therapy to riluzole in patients with amyotrophic lateral sclerosis: a randomised, double-blind, parallel-group, placebo-controlled, phase 2 trial. Lancet Neurol 17(8):681–688

Machado MMF et al (2019) PPAR-gamma agonist pioglitazone reduces microglial proliferation and NF-kappaB activation in the substantia nigra in the 6-hydroxydopamine model of Parkinson's disease. Pharmacol Rep 71(4):556–564

Mallajosyula JK et al (2008) MAO-B elevation in mouse brain astrocytes results in Parkinson's pathology. PLoS One 3(2):e1616

Marco-Contelles J et al (2016) ASS234, as a new multi-target directed propargylamine for Alzheimer's disease therapy. Front Neurosci 10:294

Martin HL et al (2012) Pharmacological manipulation of peroxisome proliferator-activated receptor gamma (PPARgamma) reveals a role for anti-oxidant protection in a model of Parkinson's disease. Exp Neurol 235(2):528–538

Mawhinney M, Cole D, Azzaro AJ (2003) Daily transdermal administration of selegiline to guinea-pigs preferentially inhibits monoamine oxidase activity in brain when compared with intestinal and hepatic tissues. J Pharm Pharmacol 55(1):27–34

Meiser J, Weindl D, Hiller K (2013) Complexity of dopamine metabolism. Cell Commun Signal 11 (1):34

Meyer JH et al (2006) Elevated monoamine oxidase a levels in the brain: an explanation for the monoamine imbalance of major depression. Arch Gen Psychiatry 63(11):1209–1216

Milczek EM et al (2011) The 'gating' residues Ile199 and Tyr326 in human monoamine oxidase B function in substrate and inhibitor recognition. FEBS J 278(24):4860–4869

Morais M et al (2014) The effects of chronic stress on hippocampal adult neurogenesis and dendritic plasticity are reversed by selective MAO-A inhibition. J Psychopharmacol 28(12):1178–1183

Muller T, Mohr JD (2019) Pharmacokinetics of monoamine oxidase B inhibitors in Parkinson's disease: current status. Expert Opin Drug Metab Toxicol 15(5):429–435

Muller T et al (2013) Switch from selegiline to rasagiline is beneficial in patients with Parkinson's disease. J Neural Transm 120(5):761–765

Murata M et al (2015) Zonisamide improves wearing-off in Parkinson's disease: a randomized, double-blind study. Mov Disord 30(10):1343–1350

Murata M et al (2016) Randomized placebo-controlled trial of zonisamide in patients with Parkinson's disease. Neurol Clin Neurosci 4(1):10–15

Murata M et al (2018) Adjunct zonisamide to levodopa for DLB parkinsonism: a randomized double-blind phase 2 study. Neurology 90(8):e664–e672

Myllyla VV et al (1997) Selegiline as the primary treatment of Parkinson's disease--a long-term double-blind study. Acta Neurol Scand 95(4):211–218

Nagatsu T, Sawada M (2006) Molecular mechanism of the relation of monoamine oxidase B and its inhibitors to Parkinson's disease: possible implications of glial cells. J Neural Transm Suppl 71:53–65

Nandigama RK, Miller JR, Edmondson DE (2001) Loss of serotonin oxidation as a component of the altered substrate specificity in the Y444F mutant of recombinant human liver MAO A. Biochemistry 40(49):14839–14846

Nave S et al (2017) Sembragiline in moderate Alzheimer's disease: results of a randomized, double-blind, placebo-controlled phase II trial (MAyflOwer RoAD). J Alzheimers Dis 58 (4):1217–1228

Neurol L (2015) Pioglitazone in early Parkinson's disease: a phase 2, multicentre, double-blind, randomised trial. Lancet Neurol 14(8):795–803

Nishijima H et al (2018) Zonisamide enhances motor effects of levodopa, not of apomorphine, in a rat model of Parkinson's disease. Parkinsons Dis 2018:8626783

Ogata A et al (2003) Isatin, an endogenous MAO inhibitor, improves bradykinesia and dopamine levels in a rat model of Parkinson's disease induced by Japanese encephalitis virus. J Neurol Sci 206(1):79–83

Oki M et al (2017) Zonisamide ameliorates levodopa-induced dyskinesia and reduces expression of striatal genes in Parkinson model rats. Neurosci Res 122:45–50

Olanow CW et al (2009) A double-blind, delayed-start trial of rasagiline in Parkinson's disease. N Engl J Med 361(13):1268–1278

Ooi J, Hayden MR, Pouladi MA (2015) Inhibition of excessive monoamine oxidase A/B activity protects against stress-induced neuronal death in Huntington disease. Mol Neurobiol 52 (3):1850–1861

Oz M, Lorke DE, Petroianu GA (2009) Methylene blue and Alzheimer's disease. Biochem Pharmacol 78(8):927–932

Panarsky R, Luques L, Weinstock M (2012) Anti-inflammatory effects of ladostigil and its metabolites in aged rat brain and in microglial cells. J Neuroimmune Pharmacol 7(2):488–498

Park JH et al (2019) Newly developed reversible MAO-B inhibitor circumvents the shortcomings of irreversible inhibitors in Alzheimer's disease. Sci Adv 5(3):eaav0316

Parkinson Study G (1989) Effect of deprenyl on the progression of disability in early Parkinson's disease. N Engl J Med 321(20):1364–1371

Parkinson Study G (1994) A controlled trial of lazabemide (Ro 19-6327) in levodopa-treated Parkinson's disease. Arch Neurol 51(4):342–347

Parkinson Study G (2002) A controlled trial of rasagiline in early Parkinson disease: the TEMPO study. Arch Neurol 59(12):1937–1943

Parkinson Study G (2004) A controlled, randomized, delayed-start study of rasagiline in early Parkinson disease. Arch Neurol 61(4):561–566

Perez V, Unzeta M (2003) PF 9601N [N-(2-propynyl)-2-(5-benzyloxy-indolyl) methylamine], a new MAO-B inhibitor, attenuates MPTP-induced depletion of striatal dopamine levels in C57/BL6 mice. Neurochem Int 42(3):221–229

Pérez V et al (1999) Relevance of benzyloxy group in 2-indolyl methylamines in the selective MAO-B inhibition. Br J Pharmacol 127(4):869–876

Pimentel LS et al (2015) The multi-target drug m30 shows pro-cognitive and anti-inflammatory effects in a rat model of Alzheimer's disease. J Alzheimers Dis 47(2):373–383

Pinto M et al (2016) Pioglitazone ameliorates the phenotype of a novel Parkinson's disease mouse model by reducing neuroinflammation. Mol Neurodegener 11:25

Poewe W et al (2017) Parkinson disease. Nat Rev Dis Primers 3:17013

Przuntek H et al (1999) SELEDO: a 5-year long-term trial on the effect of selegiline in early Parkinsonian patients treated with levodopa. Eur J Neurol 6(2):141–150

Rabey JM et al (2000) Rasagiline mesylate, a new MAO-B inhibitor for the treatment of Parkinson's disease: a double-blind study as adjunctive therapy to levodopa. Clin Neuropharmacol 23(6):324–330

Ramachandraih CT et al (2011) Antidepressants: from MAOIs to SSRIs and more. Indian J Psychiatry 53(2):180–182

Rascol O et al (2005) Rasagiline as an adjunct to levodopa in patients with Parkinson's disease and motor fluctuations (LARGO, Lasting effect in Adjunct therapy with Rasagiline Given Once daily, study): a randomised, double-blind, parallel-group trial. Lancet 365(9463):947–954

Rascol O et al (2011) A double-blind, delayed-start trial of rasagiline in Parkinson's disease (the ADAGIO study): prespecified and post-hoc analyses of the need for additional therapies, changes in UPDRS scores, and non-motor outcomes. Lancet Neurol 10(5):415–423

Rubinstein S et al (2006) Placebo-controlled study examining effects of selegiline in children with attention-deficit/hyperactivity disorder. J Child Adolesc Psychopharmacol 16(4):404–415

Ryan M, Eatmon CV, Slevin JT (2019) Drug treatment strategies for depression in Parkinson disease. Expert Opin Pharmacother 20(11):1351–1363

Sagi Y, Drigues N, Youdim MB (2005) The neurochemical and behavioral effects of the novel cholinesterase-monoamine oxidase inhibitor, ladostigil, in response to L-dopa and L-tryptophan, in rats. Br J Pharmacol 146(4):553–560

Sanches RF et al (2016) Antidepressant effects of a single dose of ayahuasca in patients with recurrent depression: a SPECT study. J Clin Psychopharmacol 36(1):77–81

Sano H, Murata M, Nambu A (2015) Zonisamide reduces nigrostriatal dopaminergic neurodegeneration in a mouse genetic model of Parkinson's disease. J Neurochem 134 (2):371–381

Santhanasabapathy R, Sudhandiran G (2015) Farnesol attenuates lipopolysaccharide-induced neurodegeneration in Swiss albino mice by regulating intrinsic apoptotic cascade. Brain Res 1620:42–56

Sanz E et al (2008) Anti-apoptotic effect of Mao-B inhibitor PF9601N [N-(2-propynyl)-2-(5-benzyloxy-indolyl) methylamine] is mediated by p53 pathway inhibition in MPP+-treated SH-SY5Y human dopaminergic cells. J Neurochem 105(6):2404–2417

Schapira AH et al (2013) Long-term efficacy and safety of safinamide as add-on therapy in early Parkinson's disease. Eur J Neurol 20(2):271–280

Schneider LS et al (2019) Low-dose ladostigil for mild cognitive impairment: a phase 2 placebo-controlled clinical trial. Neurology 93(15):e1474–e1484

Segura-Aguilar J et al (2014) Protective and toxic roles of dopamine in Parkinson's disease. J Neurochem 129(6):898–915

Shulman KI, Herrmann N, Walker SE (2013) Current place of monoamine oxidase inhibitors in the treatment of depression. CNS Drugs 27(10):789–797

Sian J et al (1994) Alterations in glutathione levels in Parkinson's disease and other neurodegenerative disorders affecting basal ganglia. Ann Neurol 36(3):348–355

Sivenius J et al (2001) Selegiline treatment facilitates recovery after stroke. Neurorehabil Neural Repair 15(3):183–190

Smith KM, Eyal E, Weintraub D (2015) Combined rasagiline and antidepressant use in Parkinson disease in the ADAGIO study: effects on nonmotor symptoms and tolerability. JAMA Neurol 72(1):88–95

Soliman A et al (2012) Convergent effects of acute stress and glucocorticoid exposure upon MAO-A in humans. J Neurosci 32(48):17120–17127

Son SY et al (2008) Structure of human monoamine oxidase A at 2.2-Å resolution: the control of opening the entry for substrates/inhibitors. Proc Natl Acad Sci U S A 105(15):5739–5744

Sonsalla PK et al (2010) The antiepileptic drug zonisamide inhibits MAO-B and attenuates MPTP toxicity in mice: clinical relevance. Exp Neurol 221(2):329–334

Statland JM et al (2019) Rasagiline for amyotrophic lateral sclerosis: a randomized, controlled trial. Muscle Nerve 59(2):201–207

Stocchi F, Torti M (2016) Adjuvant therapies for Parkinson's disease: critical evaluation of safinamide. Drug Des Devel Ther 10:609–618

Stocchi F et al (2006) Symptom relief in Parkinson disease by safinamide: biochemical and clinical evidence of efficacy beyond MAO-B inhibition. Neurology 67(7 Suppl 2):S24–S29

Sturm S et al (2017) Positron emission tomography measurement of brain MAO-B inhibition in patients with Alzheimer's disease and elderly controls after oral administration of sembragiline. Eur J Nucl Med Mol Imaging 44(3):382–391

Szoko E et al (2018) Pharmacological aspects of the neuroprotective effects of irreversible MAO-B inhibitors, selegiline and rasagiline, in Parkinson's disease. J Neural Transm (Vienna) 125 (11):1735–1749

Tariot PN et al (1987) Cognitive effects of L-deprenyl in Alzheimer's disease. Psychopharmacology 91(4):489–495

Tripathi AC et al (2018) Privileged scaffolds as MAO inhibitors: retrospect and prospects. Eur J Med Chem 145:445–497

Ulusoy GK et al (2011) Effects of pioglitazone and retinoic acid in a rotenone model of Parkinson's disease. Brain Res Bull 85(6):380–384

Upadhyay AK et al (2008) Determination of the oligomeric states of human and rat monoamine oxidases in the outer mitochondrial membrane and octyl beta-D-glucopyranoside micelles using pulsed dipolar electron spin resonance spectroscopy. Biochemistry 47(6):1554–1566

Villeneuve C et al (2013) p53-PGC-1alpha pathway mediates oxidative mitochondrial damage and cardiomyocyte necrosis induced by monoamine oxidase-A upregulation: role in chronic left ventricular dysfunction in mice. Antioxid Redox Signal 18(1):5–18

Weinreb O et al (2012) Ladostigil: a novel multimodal neuroprotective drug with cholinesterase and brain-selective monoamine oxidase inhibitory activities for Alzheimer's disease treatment. Curr Drug Targets 13(4):483–494

Yasar S, Goldberg JP, Goldberg SR (1996) Are metabolites of l-deprenyl (selegiline) useful or harmful? Indications from preclinical research. J Neural Transm Suppl 48:61–73

Yasar S et al (2006) Metabolic transformation plays a primary role in the psychostimulant-like discriminative-stimulus effects of selegiline [(R)-(−)-deprenyl]. J Pharmacol Exp Ther 317 (1):387–394

Yogev-Falach M et al (2006) A multifunctional, neuroprotective drug, ladostigil (TV3326), regulates holo-APP translation and processing. FASEB J 20(12):2177–2179

Youdim MB (2012) M30, a brain permeable multitarget neurorestorative drug in post nigrostriatal dopamine neuron lesion of parkinsonism animal models. Parkinsonism Relat Disord 18(Suppl 1):S151–S154

Youdim MB (2013) Multi target neuroprotective and neurorestorative anti-Parkinson and anti-Alzheimer drugs ladostigil and m30 derived from rasagiline. Exp Neurobiol 22(1):1–10

Youdim MB, Edmondson D, Tipton KF (2006) The therapeutic potential of monoamine oxidase inhibitors. Nat Rev Neurosci 7(4):295–309

Youdim MB et al (2014) Promises of novel multi-target neuroprotective and neurorestorative drugs for Parkinson's disease. Parkinsonism Relat Disord 20(Suppl 1):S132–S136

Zanderigo F et al (2018) [(11)C]Harmine binding to brain monoamine oxidase A: test-retest properties and noninvasive quantification. Mol Imaging Biol 20(4):667–681

Zanotti-Fregonara P, Bottlaender M (2014) [11C]befloxatone distribution is well correlated to monoamine oxidase A protein levels in the human brain. J Cereb Blood Flow Metab 34 (12):1951–1952

Zanotti-Fregonara P et al (2014) Imaging of monoamine oxidase-A in the human brain with [11C] befloxatone: quantification strategies and correlation with mRNA transcription maps. Nucl Med Commun 35(12):1254–1261

Zheng H et al (2005) Novel multifunctional neuroprotective iron chelator-monoamine oxidase inhibitor drugs for neurodegenerative diseases: in vitro studies on antioxidant activity, prevention of lipid peroxide formation and monoamine oxidase inhibition. J Neurochem 95(1):68–78

Zheng H, Fridkin M, Youdim MB (2010) Site-activated chelators derived from anti-Parkinson drug rasagiline as a potential safer and more effective approach to the treatment of Alzheimer's disease. Neurochem Res 35(12):2117–2123

Zhou Y, Zhao ZQ, Xie JX (2001) Effects of isatin on rotational behavior and DA levels in caudate putamen in Parkinsonian rats. Brain Res 917(1):127–132

Zhou S, Chen G, Huang G (2018) Design, synthesis and biological evaluation of lazabemide derivatives as inhibitors of monoamine oxidase. Bioorg Med Chem 26(17):4863–4870

Zisook S (1985) A clinical overview of monoamine oxidase inhibitors. Psychosomatics 26 (3):240–251

Inhibition of Myeloperoxidase

Jala Soubhye, Paul G. Furtmüller, Francois Dufrasne, and Christian Obinger

Contents

Abstract

Myeloperoxidase participates in innate immune defense mechanism through formation of microbicidal reactive oxidants and diffusible radical species. A unique activity is its ability to use chloride as a cosubstrate with hydrogen peroxide to generate chlorinating oxidants such as hypochlorous acid, a potent antimicrobial agent. However, chronic MPO activation can lead to indiscriminate protein modification causing tissue damage, and has been associated with chronic

J. Soubhye
Pharmacognosy, Bioanalysis and Drug Discovery, Research in Drug Development, Faculté de Pharmacie, Université Libre de Bruxelles, Brussels, Belgium

P. G. Furtmüller · C. Obinger (✉)
Department of Chemistry, Institute of Biochemistry, University of Natural Resources and Life Sciences, Vienna, Austria
e-mail: paul.furtmueller@boku.ac.at; christian.obinger@boku.ac.at

F. Dufrasne
Microbiology, Bioorganic and Macromolecular Chemistry, Research in Drug Development, Faculté de Pharmacie, Université Libre de Bruxelles, Bruxelles, Belgium
e-mail: francois.dufrasne@ulb.be

H. H. H. W. Schmidt et al. (eds.), *Reactive Oxygen Species*,
Handbook of Experimental Pharmacology 264, https://doi.org/10.1007/164_2020_388

inflammatory diseases, atherosclerosis, and acute cardiovascular events. This has attracted considerable interest in the development of therapeutically useful MPO inhibitors. Today, based on the profound knowledge of structure and function of MPO and its biochemical and biophysical differences with the other homologous human peroxidases, various rational and high-throughput screening attempts were performed in developing specific irreversible and reversible inhibitors. The most prominent candidates as well as MPO inhibitors already studied in clinical trials are introduced and discussed.

Graphical Abstract

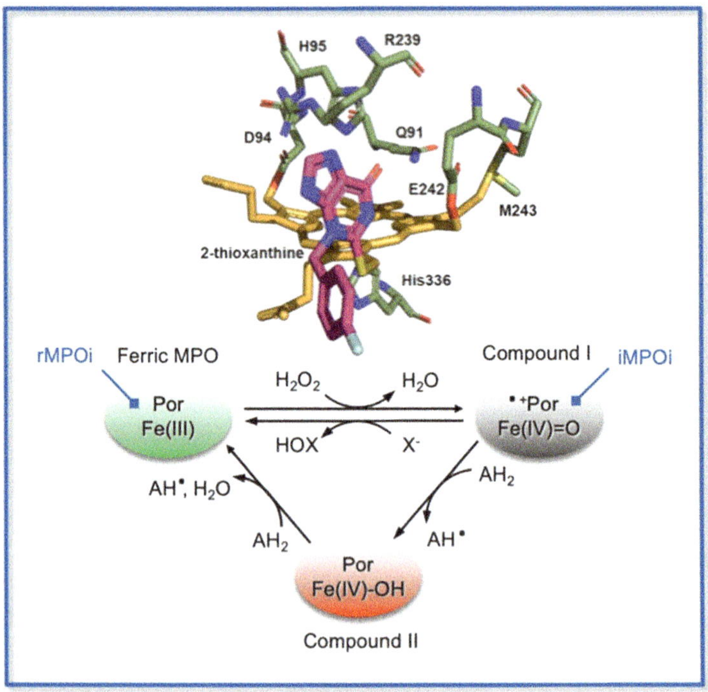

Keywords

Eosinophil peroxidase · Irreversible inhibition · Lactoperoxidase · Myeloperoxidase · Peroxidasin · Reversible inhibition · Thyroid peroxidase

1 Introduction

Myeloperoxidase (MPO) is a member of the peroxidase-cyclooxygenase superfamily (Zámocký et al. 2008, 2015) and mainly stored in azurophilic granules of neutrophils and to a lesser extent in monocytes, but its expression has also been observed in neurons (Green et al. 2004) and endothelial cells (La Rocca et al. 2009). MPO is an essential part of host defense in the innate immune system (Klebanoff 2005; Davies et al. 2008; Nauseef 2014). During neutrophil activation, MPO is released into the phagolysosome, while assembly and activation of membrane-associated NADPH oxidase result in the production of superoxide and in consequence hydrogen peroxide (Winterbourn et al. 2016). Finally, MPO catalyzes the H_2O_2-mediated oxidation of two-electron donors (chloride, bromide, thiocyanate) and one-electron donors (ascorbate, tyrosine, sertotonin, urate, etc.) to the corresponding hypohalous acids and radicals. Among others, reaction products of MPO like hypochlorous acid and hypothiocyanous acid support microbial killing within the phagolysosome (Klebanoff et al. 2013; Winterbourn and Kettle 2013). However, evidence has emerged that MPO-derived oxidants contribute to tissue damage and the initiation and propagation of acute and chronic vascular inflammatory diseases. Under pathological conditions, persistent activation of the MPO–H_2O_2 system of activated phagocytes may adversely affect tissues. Hypochlorous acid is able to initiate modification reactions targeting lipids, DNA and (lipo)proteins, including halogenation, nitration and oxidative crosslinking (Malle et al. 2007), which may cause and promote chronic inflammation and development of diseases (Davies and Hawkins 2020). As a consequence, several therapeutic intervention strategies aiming at efficient MPO inhibition were tested in the last decades. These include (1) active site blockade of MPO, (2) irreversible suicide inhibition, (3) diversion of MPO from the chlorination cycle, and (4) application of HOCl scavengers to prevent initiation and propagation of diseases with an inflammatory component. It is the aim of this review to provide an overview about potential inhibitors and their mode of action that is based on our present knowledge about structure–function relationship of MPO.

Importantly, MPO inhibitors must not affect other human peroxidases which show high structural and functional similarities with MPO (Furtmüller et al. 2006; Zámocký et al. 2015; Nicolussi et al. 2018). These include thyroid peroxidase (TPO), eosinophil peroxidase (EPO), lactoperoxidase (LPO), and peroxidasin (PXDN). Dimeric membrane-anchored TPO catalyzes iodination of tyrosine residues in thyroglobulin and, finally, the synthesis of the thyroid hormones triiodothyronine (T3) and thyroxine (T4) (Ruf and Carayon 2006). Together with MPO monomeric LPO and EPO have been shown to be an essential part of host defense in the innate immune system and comprise the frontline defense against invading pathogens (Ihalin et al. 2006; Malik and Batra 2012). Finally, PXDN is embedded in the extracellular matrix and catalyzes the formation of hypobromous acid, which mediates the formation of essential sulfilimine cross-links between methionine and hydroxylysine residues in collagen IV. This confers critical structural reinforcement to the extracellular matrix (Soudi et al. 2012, 2015; Colon et al. 2017; Paumann-Page et al. 2017, 2020). All these essential biological functions should not be affected by MPO inhibitor.

2 Structural and Functional Features of Myeloperoxidase

Myeloperoxidase is a glycosylated, soluble, cationic, and homodimeric heme enzyme (Fig. 1a) with a molar mass of around 146 kDa (Furtmüller et al. 2006; van Antwerpen et al. 2010; Nauseef 2018). The two protomers are functionally independent, crosslinked by a single disulfide bridge and composed of a heavy and light chain. Most importantly, MPO contains one posttranslational modified heme *b* per protomer. The characteristic sequence motif -X-G-**Q**-X-X-**D**-**H**-**D**-X- allows the assignment of myeloperoxidase (but also TPO, EPO, LPO, and PXDN) to the peroxidase-cyclooxygenase superfamily (Zámocký et al. 2015). The motif includes the distal catalytic histidine (H95, MPO numbering) with two adjacent aspartates, i.e. D94 being involved in ester bond formation with the prosthetic group and D96 acting as ligand in the distal Ca^{2+}-binding site that enhances the conformational stability of the heme cavity (Fiedler et al. 2000; Furtmüller et al. 2006; Carpena et al. 2009) (Fig. 1b). The fully conserved glutamine (Q91) seems to be involved in halide binding (Blair-Johnson et al. 2001). Another conserved sequence motif is -X-**R**-X-X-**E**-X-, which includes the catalytic R239 and E242 that are involved in the formation of the second ester bond with the modified prosthetic group (Fig. 1b). The proximal heme ligand H336 is hydrogen bonded to the amide side chain of fully conserved N421 (Fig. 1b). The resulting imidazolate character partially stabilizes the highly reactive redox intermediate Compound I (see below) (Carpena et al. 2009; Grishkovskaya et al. 2017). A comparison of the crystal structures of human mature MPO, partially unprocessed monomeric promyeloperoxidase (proMPO) (Grishkovskaya et al. 2017) and bovine lactoperoxidase (Singh et al. 2008)

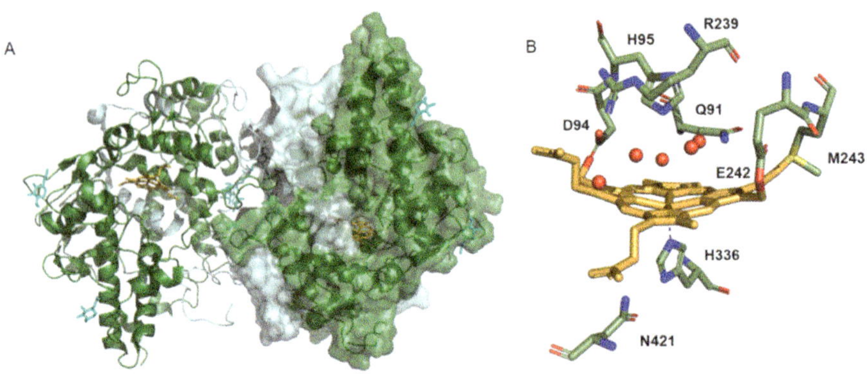

Fig. 1 Overall and active site structure of human myeloperoxidase. (**a**) Overall structure of mature homodimeric human MPO. One monomer is shown as ribbon structure, the other with transparent surface. The heavy chain is depicted in green, the light chain in pale cyan, sugars in cyan, and the prosthetic group in orange. (**b**) Conserved active site residues of MPO including the non-planar porphyrin ring and its covalent ester bond to D94 and E242, the sulfonium linkage with M243 and the catalytic residues H95, R239, and Q91, which are involved in the heterolytic cleavage of hydrogen peroxide and halide binding. The figure was constructed using the coordinates deposited in the Protein Data Bank (accession code 1CXP)

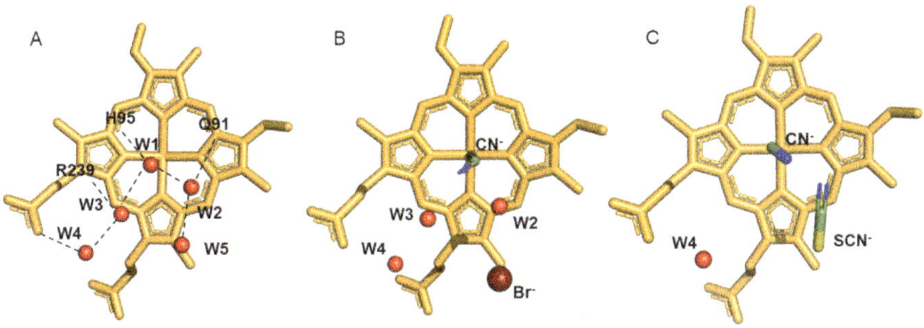

Fig. 2 Water and halide binding in the distal cavity of human myeloperoxidase. (**a**) Ferric resting state of MPO, (**b**) MPO-cyanide-bromide complex, (**c**) MPO-cyanide-thiocyanate complex. PDB accession codes 1CXP, 1D7W, and 1DNW, respectively

demonstrates an almost identical heme cavity architecture. In contrast to MPO and LPO, no crystal structures for TPO, EPO, and PXDN are available but sequence alignment and modelling suggest a similar active site.

All five human peroxidases have two heme to protein ester bonds, which are formed autocatalytically (Colas and Ortiz de Montellano 2003; Nicolussi et al. 2018). Importantly, MPO has an additional sulfonium ion linkage between the β-carbon of the vinyl group on pyrrole ring A and the sulfur atom of Met243 (Fiedler et al. 2000). As a consequence, MPO has unique biochemical and biophysical features, which can be exploited in the design of inhibitors as demonstrated below.

Figure 2a shows the distal H-bonding in the ferric resting state of MPO involving five water molecules (W1–W5). The waters are part of a pronounced hydrogen bonding network with H95, R239, Q91, and the heme pyrrole ring C propionate as well as between themselves (Fig. 2a). Distal H95 is hydrogen bonded via W1 and W2 with Q91 (Fig. 2a), whereas upon bromide binding, the halide replaces W2. No crystal structure of a MPO-Cl complex is available so far. In the MPO–CN$^-$–Br$^-$ complex (Fig. 2b), Br$^-$ substitutes W5 and does not directly interact with neither the low-spin ligand nor with H95 (Blair-Johnson et al. 2001). Thus this structure most probably does not mirror the oxoiron(IV) structure of the Compound I-chloride complex which is necessary for efficient oxygen transfer and formation of HOCl (see below). In general, at pH 7.4 chloride is weakly bound to ferric MPO (K_D = 100 mM) (Furtmüller et al. 2006). In the MPO–CN$^-$-SCN$^-$ complex (Fig. 2c), thiocyanate is arranged almost parallel with the heme plane and replaces W2, W3, and W5, respectively (Blair-Johnson et al. 2001).

Aromatic one-electron donors (as well as several inhibitors) bind at the substrate channel entry to the heme cavity as already demonstrated by Hori et al. (1994) for benzylhydroxamic acid and salicylhydroxamic acid (SHA), which both act as substrate and inhibitor of MPO. The 2.3 Å resolution X-ray crystal structure of the MPO-SHA complex (Davey and Fenna 1996) shows an interaction of the benzene ring of SHA with pyrrole ring D and the side chain of Arg239. The hydroxamic group is hydrogen bonded to distal H95 and Q91 but is not directly associated with the heme iron.

The posttranslational modification of the prosthetic group and formation of three covalent bonds (Fig. 1b) not only increases the overall conformational and thermal stability of MPO (Banerjee et al. 2011), but additionally has a strong impact on its redox properties. Typically, in (unmodified) heme *b* peroxidases, the ferric state is stabilized by a negative reduction potential of the redox couple Fe(III)/Fe(II) (Battistuzzi et al. 2001). By contrast, in MPO the corresponding $E^{\circ\prime}$ value is significantly more positive mainly due to the electron withdrawing effect of the MPO-typical sulfonium ion linkage ($E^{\circ\prime}$ is 0.005 V) (Battistuzzi et al. 2006, 2011). In addition the prosthetic group in MPO is strongly distorted thereby diminishing the interaction of the pyrrole nitrogens with the heme iron (Fig. 1b). The impact of the MPO-typical sulfonium ion bond on its redox and enzymatic feature could be demonstrated by comprehensive mutational studies (Battistuzzi et al. 2011).

The fact that the homologous human peroxidases EPO, LPO, TPO, and PXDN exhibit more negative reduction potentials of the redox couple Fe(III)/Fe(II) compared to MPO, i.e. LPO (−0.176 V), PXDN (−0.128 V), and EPO (−0.126 V) (Battistuzzi et al. 2010, 2011; Paumann-Page et al. 2017) can be explored in inhibitor design (Jantschko et al. 2005) because the molecular factors which determine $E^{\circ\prime}$ of the Fe(III)/Fe(II) also influence $E^{\circ\prime}$ of the catalytically relevant redox couples like Compound I/Fe(III) and Compound I/Compound II. Indeed, the hierarchy observed for E°'[Fe(III)/Fe(II)], namely, MPO > EPO > PXDN > LPO, is also reflected in E°'[Compound I/Fe(III)] and E°'[Compound I/Compound II] (Arnhold et al. 2003, 2006; Zederbauer et al. 2007a, b; Battistuzzi et al. 2011). Myeloperoxidase Compound I is an extremely strong oxidant with E°`values of the couples Compound I/Fe(III) of 1.160 V (Arnhold et al. 2001) and Compound I/Compound II of 1.350 V (Furtmüller et al. 2003, 2005). The respective standard reduction potentials in LPO and EPO are less positive (Arnhold et al. 2001; Furtmüller et al. 2005). Here, it is important to note that recombinant monomeric, partially unprocessed proMPO shares the same heme cavity and substrate channel architecture as well as almost identical redox and enzymatic properties with mature leukocyte MPO (Grishkovskaya et al. 2017; Furtmüller et al. 2001) and thus can be used in the design and testing of MPO inhibitors.

Figure 3 summarizes the redox interconversions of human myeloperoxidase. As already mentioned, MPO is an efficient catalyst of one- and two-electron oxidation reactions. Compound I [oxoiron(IV) porphyrin π-cation radical] is the central redox intermediate for both the halogenation cycle (Reactions 1 & 2) as well as for the peroxidase cycle (Reactions 1, 3 & 4). Compound I is rapidly formed by oxidation of the ferric resting state by hydrogen peroxide according to Reaction 1 (Fig. 3). Here, the fully conserved H95-R239 couple supports the heterolytic cleavage of H_2O_2 (Poulos and Kraut 1980; Marquez et al. 1994; Furtmüller et al. 2006.). The main physiological activity of MPO (as well as of EPO, LPO, TPO, and PXDN) includes the two-electron oxidation of (pseudo-)halides (X^-, i.e. Cl^-, Br^-, I^-, SCN^-) to the corresponding (pseudo-)hypohalous acids (HOX, i.e. HOCl, HOBr, HOI, HOSCN)

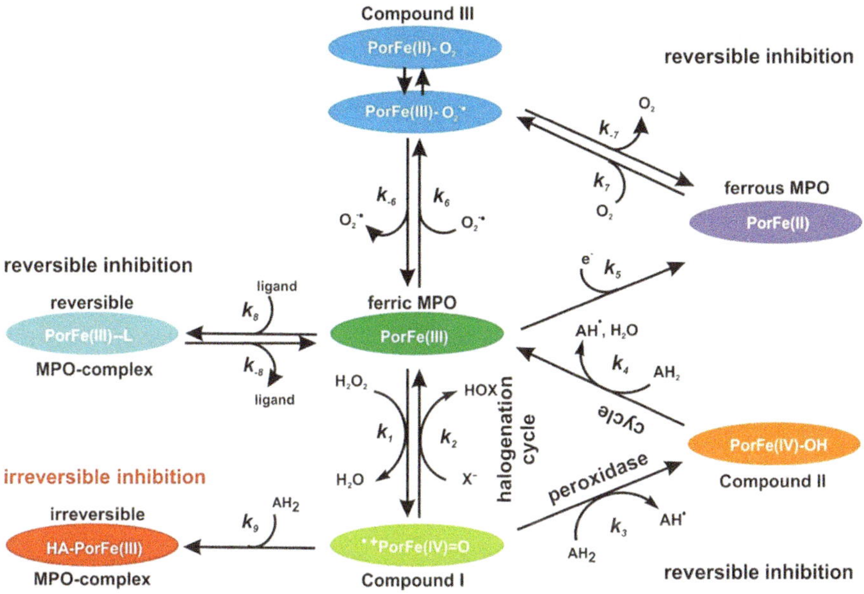

Fig. 3 Reactions of human myeloperoxidase. The halogenation cycle includes Reactions 1 and 2, whereas the peroxidase cycle involves Reactions 1, 3, & 4. In both cycles MPO Compound I acts as strong oxidant of both one- and two-electron donors. Compound III is catalytically inactive and is formed from ferric or ferrous MPO with excess of superoxide (Reaction 6) or dioxygen (Reaction 7), respectively. Reversible inhibitors follow Reactions 1 & 3 and are trapped in the Compound II state due its poor oxidation capacity. Suicide inhibitors are oxidized in the peroxidase cycle and the resulting radicals interact with either the protein or the prosthetic group thereby irreversibly inhibiting MPO. Finally, inhibitors bind (reversibly) to the heme cavity thereby blocking substrate access or product release

according to Reaction 2 (Fig. 3). However, only MPO Compound I is able to efficiently react with chloride at pH 7.0 (Furtmüller et al. 1998 ,2000), thereby reflecting the high reduction potential of the redox couple Compound I/Fe(III) (Battistuzzi et al. 2006; Arnhold et al. 2001, 2003). With decreasing pH chloride oxidation is increased (Furtmüller et al. 1998; Ramos et al. 2008). In general, oxidation of halides strictly follows the hierarchy of the $E^{\circ\prime}$ of the redox couple HOX/X^-, H_2O, i.e. 1.28 V ($X^- = Cl^-$), 1.13 V (Br^-), 0.78 V (I^-), and 0.56 V (SCN^-), respectively (Arnhold et al. 2006). As a consequence, MPO is the only HOCl source in the human body, whereas EPO and PXDN are known to efficiently oxidize Br^- to HOBr.

Besides the halogenation reaction MPO can follow the peroxidase cycle (Fig. 3). Here Compound I is reduced via Compound II [oxoiron(IV)] to the ferric resting state by two one-electron donors (AH_2, i.e. tyrosine, serotonin, ascorbate, urate, nitrite, etc.) thereby releasing the corresponding radicals ($^\bullet$AH). As mentioned above MPO Compound I is an extremely strong one-electron oxidant, which is reflected by oxidation of substrates with very positive $E^{\circ\prime}$ values (Jantschko et al. 2002, 2005).

This unique property can be exploited in the design of specific and reversible inhibitors of MPO (Jantschko et al. 2005; Soubhye et al. 2010, 2013; Aldib et al. 2012). These potential inhibiting compounds are not oxidized by the other homologous human peroxidases, but can divert MPO from the halogenation cycle to Compound II. The latter has a significantly lower oxidation capacity compared to Compound I thus trapping MPO in the Compound II state and dampening its chlorination activity. Importantly, the binding and oxidation site(s) for (aromatic) substrates are identical in Compound I and Compound II, i.e. the hydrophobic heme periphery at the entrance to the distal cavity close to the pyrrole ring D (Jantschko et al. 2005; Soubhye et al. 2010; Aldib et al. 2012).

Further reactions of MPO include complex formation between the ferric protein with low-spin or high-spin ligands (Reaction 8) or formation of the catalytically inactive Compound III. The latter can be formed either from ferric MPO in the presence of millimolar superoxide (Reaction 6) or from ferrous MPO by dioxygen binding (Reaction 7) (Jantschko et al. 2003, 2004) or during inhibition reactions via Compound III (Reaction 9) (Fig. 3).

3 Strategies of Design of Inhibitors of Human Peroxidases

As outlined above MPO may be released extracellularly during (patho)physiological conditions (Klebanoff 2005) thereby promoting the production of (highly) oxidative molecules in extracellular fluids including HOCl (Klebanoff 2005), HOSCN (Gau et al. 2016), nitrogen monoxide, peroxynitrite (Eiserich et al. 1998), and other radicals (Davies 2010). These compounds easily react with host tissue causing oxidative damage (Soubhye et al. 2016a). In addition, circulating MPO may oxidize secondary messengers like NO (Eiserich et al. 1998), hormones like thyroxin (Klebanoff and Green 1973) or estradiol (Klebanoff 1977). Moreover, MPO has been implicated in the acceleration of inflammation by stimulating the release of inflammatory mediators and activation of proteases (Clark and Klebanoff 1979). In this context, it has been suggested that MPO might be an important target to develop new anti-inflammatory agents and its inhibition might be useful to cure the inflammation unlike the known non-steroidal anti-inflammatory drugs (NSAIDs) that treat only the symptoms but not the origin of the disease (Lazarević-Pasti et al. 2015; Malle et al. 2007). According to the role of MPO in immune system, the only expected side effect of MPO could be impairing the activity of neutrophils against pathogens. However, this effect is assumed to be reduced by using relatively polar molecules which would not be able to penetrate the cell membranes and could target only the extracellular enzyme (Soubhye et al. 2016a).

Access and binding to the heme periphery of substrates and potential inhibitors are guided by the same principles. The prosthetic group of MPO is located in a crevice, about 15 Å in depth, with access to the solvent via an open channel, approximately 10 Å in diameter. Binding studies of aromatic molecules like benzylhydroxamic acid and salicylhydroxamic acid (SHA) have indicated that the respective aromatic ring binds to the hydrophobic region at the entrance to the distal

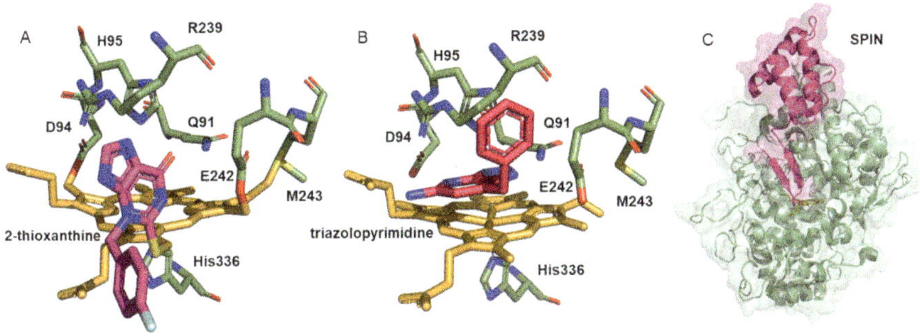

Fig. 4 Binding of MPO inhibitors at the heme periphery. (**a**) Binding site for 2-thioxantine, (**b**) 7-benzyl triazolopyridine, and (**c**) for SPIN (SPIN for *S*taphylococcal *P*eroxidase *In*hibitor). Figure 3 was constructed using the coordinates deposited in the Protein Data Bank, accession code (A) 3ZS0, (B) 6WYD, and (C) 6BMT, respectively

heme pocket between pyrrole ring D and the side chain of Arg239 (Fig. 4). The hydroxamic acid moiety is hydrogen bonded to both the distal H95 and Q91 amide group but is not coordinated to the heme iron (Ikeda-Saito et al. 1991; Hori et al. 1994; Davey and Fenna 1996). Although SHA binding displaces three water molecules from the distal cavity (W1, W2, and W3), no significant conformational differences between the active site regions of the complex and the native enzyme became apparent. Recently, the well-established interaction of SHA with MPO was used to validate the docking procedure for screening of large databases for selection of potential MPO inhibitors (Soubhye et al. 2017a, b).

Similar binding modi were observed for melatonin (N-acetyl-5-methoxytryptamine) and serotonin (5-hydroxytryptamine) binding to ferric MPO in computational docking studies (Hallingbäck et al. 2006). Both indole derivatives were shown to bind with their indole rings parallel to the heme plane and close enough to pyrrole ring D to achieve stacking. The ligand side chain was never directed toward the distal cavity, whereas the indole substituent at position 5 was close to the heme center (Hallingbäck et al. 2006). Computational docking of Compounds I and II demonstrated that both melatonin and serotonin were pushed about 1 Å away from the ferryl oxygen, which abolished the 5-substituent to point to the heme center thus favoring an alternative binding mode with a rotated indole still parallel to pyrrole ring D (Hallingbäck et al. 2006). Following studies demonstrated that most aromatic inhibitors follow similar binding modi.

Figure 4 depicts the crystal complexes of various compounds binding at the heme periphery including the irreversible MPO inhibitor 2-thioxanthine, the reversible MPO inhibitor 7-benzyl triazolopyridines (Shaw et al. 2020), and the MPO inhibiting protein SPIN from *Staphylococcus aureus* (de Jong et al. 2017).

Based on the mechanism of the inhibition of MPO, inhibitors have been divided into two main classes:

1. Irreversible inhibitors that act as substrate of MPO and are oxidized by Compound I and Compound II in the peroxidase cycle (Fig. 3). The oxidized (radicalic) reaction product attacks the heme group and establishes a covalent bond. As a consequence, the access to the heme cavity is blocked irreversibly.
2. Reversible inhibitors bind to the active site of MPO by non-covalent interactions with high affinity and low dissociation rate (Soubhye et al. 2016a). In addition, they may react easily and quickly with Compound I of MPO but slowly with Compound II. Formed radicals do interact neither with the protein nor with the prosthetic group. Such inhibitors may lose their activity in the physiological environment because of the presence of one-electron donors which can reduce Compound II to the ferric state (Soubhye et al. 2016b).

4 Irreversible MPO Inhibitors

Mechanism-based inhibitors are oxidized by MPO and the resulting products bind to the active site via covalent bond formation thereby causing irreversible inhibition. Such inhibitors prevent MPO from recovering its activity and therefore need lower doses than reversible inhibitors (Lundblad 2005). Additionally, the administration schedule is different with less doses needed. In this type of inhibition, the potency of the inhibitor is usually evaluated by the k_i of the reaction between the inhibitor and the enzyme rather than the IC_{50} value (Adam et al. 2001). In case of MPO, the inhibitor needs to be oxidized by either Compound I or Compound II (Soubhye et al. 2016a).

p-Aminobenzoic acid hydrazide (ABAH) was the first specific irreversible MPO inhibitor (Kettle et al. 1995, 1997; Burner et al. 1999). It has been shown that MPO oxidizes its hydrazide group to the corresponding hydrazyl radical which forms covalent acyl adducts with the active site (Forbes et al. 2012). Due to its commercial availability and low price, ABAH was widely used to evaluate the inhibition of MPO as a new strategy for treatment of inflammatory syndromes such as atherosclerosis in in vitro or animal models (Bekesi et al. 2005) as well as multiple sclerosis (MS), Parkinson's disease (PD), and Alzheimer's disease (AD) (Forghani et al. 2012). In vivo experiments in animals demonstrated that ABAH can prevent the development of myeloid inflammation, demyelinating diseases such as MS (Forghani et al. 2012), and blood–brain barrier (BBB) dysfunction (Üllen et al. 2013). Moreover, ABAH has been found to be useful to avert the growth of aortic atherosclerosis (Bekesi et al. 2005) and to block the formation of DNA double-strand breaks (Papież et al. 2015). It has also been reported that the inhibition of MPO by ABAH in mice with ischemic stroke can increase the neurogenesis (Kim et al. 2016). However, due to its low molecular mass that leads to a large distribution in the body and the reactive hydrazide group, ABAH may inhibit also other enzymes causing several unwanted side effects (Lipinski et al. 2001; Smith 2011). These characteristics preclude ABAH from usage as drug and in human tests.

Another interesting family of MPO inhibitors is based on the xanthine scaffold. It is well known that urate, a xanthine derivative, is a good substrate for MPO (Stamp

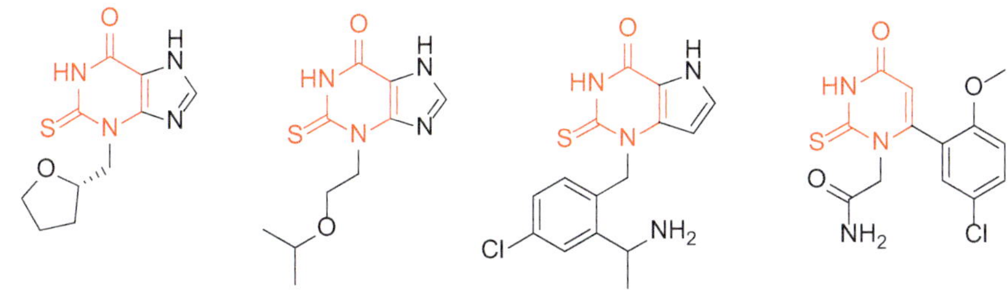

AZD-5904 AZD-3241 AECTPPO PF-06282999

Fig. 5 Structures of irreversible inhibitors of myeloperoxidase comprising a thiouracil moiety highlighted in red

et al. 2014). Therefore, xanthine has been used as a backbone for developing new inhibitors (Tiden et al. 2011). In this context, potent inhibitors derived from 2-thioxanthine have been developed by AstraZeneca. Several compounds were identified and tested in vivo and ex vivo, namely AZD-5904 and AZD-3241, and these inhibitors can be administered orally (Fig. 5) (Maiocchi et al. 2017; Jucaite et al. 2015). These compounds have been identified as irreversible MPO inhibitors. It has been proposed that thioxanthine is oxidized by Compound I to the corresponding free radical that covalently binds to the prosthetic group and blocks the substrate entry channel (Fig. 4a) (Tiden et al. 2011). Likewise, AstraZeneca has developed other MPO inhibitors with similar structures, but they have also discovered novel potent inhibitors derived from 1-[2-(aminomethyl)benzyl]-2-thioxo-1,2,3,5-tetrahydro-4H-pyrrolo[3,2-d]pyrimidin-4-one (Fig. 5). Among these inhibitors, AECTPPO has shown the best activity at low nanomolar range and also the highest selectivity. AECTPPO has been designed for the treatment or the prevention of cardiovascular diseases (Inghardt et al. 2016). However, it is not known whether these inhibitors act reversibly or irreversibly but, owing to their structural similarity with thioxanthine, it is reasonable to assume that they belong to the irreversible inhibitors (Inghardt et al. 2016). Nevertheless, not all the inhibitors carrying 2-thioxopyrimidin-4-one exhibit irreversible inhibition for MPO. Li et al. have shown that inhibitors derived from thioxodihydroquinazolinone inactivate MPO reversibly (Li et al. 2015).

New irreversible MPO inhibitors derived from thiouracil have been developed by Pfizer, based on their structural analogy to thioxanthine (Fig. 5) (Ruggeri et al. 2015). Compound PF-06282999 has shown the best potency and selectivity toward MPO. It can easily be absorbed from gastro-intestinal tract giving a plasma concentration corresponding to its IC_{50} in vitro (Ruggeri et al. 2015). As the structure of PF-06282999 is related to thioxanthine compounds, it has been proposed that the mechanism of inactivation is similar (Ruggeri et al. 2015).

From the armamentarium, two compounds have been identified as mechanism-based inhibitors of MPO, namely paroxetine (Soubhye et al. 2014) and hydralazine (Soubhye et al. 2017a). The selective serotonin transporter inhibitor paroxetine exhibits irreversible MPO inhibition at low nanomolar range (Soubhye et al. 2014). It has been suggested that one of the oxygens of benzodioxol group of paroxetine is oxidized by Compound I into a carbene which in turn reacts with the heme group causing the degradation of the enzyme. At high concentrations of paroxetine, MPO activity can be partially recovered suggesting that the antidepressant drug can inhibit MPO via both reversible and irreversible pathways (Soubhye et al. 2014). In contrast, hydralazine has been found to be oxidized by Compound I into hydralazyl radicals and subsequently to reactive oxygen-centered radicals. The latter may react irreversibly with the heme (Soubhye et al. 2017a). According to the ADME profile of paroxetine and hydralazine, it has been demonstrated that after oral administration of doses usually recommended in the clinical guidelines (between 10 and 30 mg per day and between 40 and 300 mg per day, respectively), their plasma concentrations could be sufficient to inhibit MPO irreversibly (Lima et al. 2008; Shepherd et al. 1981). In vitro experiments have demonstrated that paroxetine was able to inhibit the MPO-mediated oxidation of LDL at low nanomolar range suggesting that it can be used as a hybrid drug for preventing atherosclerosis associated with major depression (Soubhye et al. 2014), while hydralazine did not show good inhibition of LDL oxidation at the therapeutic concentrations (Soubhye et al. 2017a).

Due to the low activity of hydralazine on the oxidation of LDL, attempts were made to improve its activity. By dynamic combinatorial chemistry, a hydrazone compound was obtained from hydralazine and glycolaldehyde (HYDG). The resulting compound showed an activity 10-fold higher than hydralazine. Further study on this compound has established that HYDG acts as irreversible MPO inhibitor. Furthermore, in vivo experiments have demonstrated that HYDG can inhibit MPO for more than 24 h with one dose (Soubhye et al. 2017b).

5 Reversible MPO Inhibitors

Several compounds have been identified as reversible MPO inhibitors. All of these inhibitors are good electron donors for Compound I but poor substrates for Compound II. These inhibitors (1) bind reversibly with high affinity via non-covalent interactions to the active site (Soubhye et al. 2016b) and (2) efficiently reduce Compound I to Compound II while keeping the binding between the enzyme and the inhibitor. As a consequence, the inhibitor switches the enzyme from the chlorination cycle to the peroxidation cycle (Jantschko et al. 2005). In contrast, some reversible inhibitors have been suggested to follow other inhibition pathways. It has been proposed that derivatives of hydroxamic acid are oxidized by Compound I and promote the redox interconversion into Compound III which is catalytically inactive. Therefore hydroxamic acids are thought to inhibit MPO by pushing it out the chlorination cycle without accumulating Compound II (Ikeda-Saito et al. 1991; Forbes et al. 2013).

Naturally occurring polyphenols and flavonoids are common examples for reversible MPO inhibitors (Boufadi et al. 2014). Quercetin, kaempferol, and epicatechin showed high activity on MPO (Spalteholz et al. 2008). Such compounds can be found in high amount in plants such as capers (Inocencio et al. 2000), onions (Slimestad et al. 2007), sweet potatoes (Jiang et al. 2011), green and black teas (Langley-Evans 2000), propolis (Boufadi et al. 2014), tomatoes (Zhao et al. 2015), and cacao (Schewe and Sies 2005), for example. This widespread abundance of flavonoids in plants commonly used in alimentation gives these compounds special importance for clinical use as MPO inhibitors. Indeed, several studies have been published showing the effects of natural compounds on MPO in vivo. It has been reported that polyphenols can be absorbed from intestine easier when they are administered as crude extracts or plants than giving them as isolated compounds (Manach et al. 2004; Rasoanaivo et al. 2011; Boufadi et al. 2017). In vivo experiments have shown that the oral administration of plant polyphenols can suppress (1) lipid peroxidation of LDL caused by MPO/nitrite or peroxynitrite (Schewe and Sies 2005; Kostyuk et al. 2003), (2) MPO-mediated injury of endothelial cell (Tian et al. 2017), and (3) development of colitis (Nastase et al. 2016), and leads to muco-protective effects (Rtibi et al. 2016). It has been suggested that polyphenols including flavonoids might exhibit beneficial roles in two ways: (1) by reversible inhibition of MPO, and (2) by scavenging the oxidative reaction products of MPO (Loke et al. 2008). Kinetic studies on reactions of flavonoids with MPO Compounds I and II and with oxidative products of the enzyme suggested that these antioxidants mainly prevent oxidative damage by inhibition of MPO rather than detoxification of the oxidative reaction products (Skaff et al. 2007; Pattison and Davies 2004, 2006).

Random and rational screening of libraries comprising known drugs led to the discovery of several new reversible MPO inhibitors (Soubhye et al. 2016a). Most of these molecules belong to the NSAIDs family (e.g., nimesulide, flufenamic acid, and mefenamic acid) (Soubhye et al. 2016a). However, these anti-inflammatory agents did not show enough potency to be used as MPO inhibitor at a safe dose. Although the in vitro activity of dapsone on MPO is similar to those of the NSAIDs inhibitors, the former has shown a strong anti-inflammatory effect in humans at the therapeutic doses due to its capacity of MPO inhibition (Stendahl et al. 1978; Foye et al. 2013; Bozeman et al. 1992). Anyway, in vitro experiments have demonstrated that none of these drugs can inhibit MPO-mediated oxidation of LDL (Soubhye et al. 2016a). Recently, mesna (2-mercaptoethane sodium sulfonate), which is used as a medication for reducing hemorrhagic cystitis and hematuria associated with cancer chemotherapy, has shown inhibitory properties on MPO. It inhibits the enzyme by switching the chlorination cycle to the peroxidase cycle thus preventing the enzyme to generate HOCl. However, it is not known whether mesna is able to dampen LDL oxidation, CNS tissue damage or whether it inactivates MPO in vivo (Jeelani et al. 2017).

The endogenous indolic compounds melatonin and tryptophan have been identified as potent reversible MPO inhibitors (Galijasevic et al. 2008; Ximenes et al. 2005). They inhibit MPO by high affinity binding and efficient oxidation by Compound I but poor oxidation by Compound II (Galijasevic et al. 2008; Ximenes et al. 2005). In this context, attempts were made to develop new potent indole-derived inhibitors with high activities and low toxicity which allows them being used

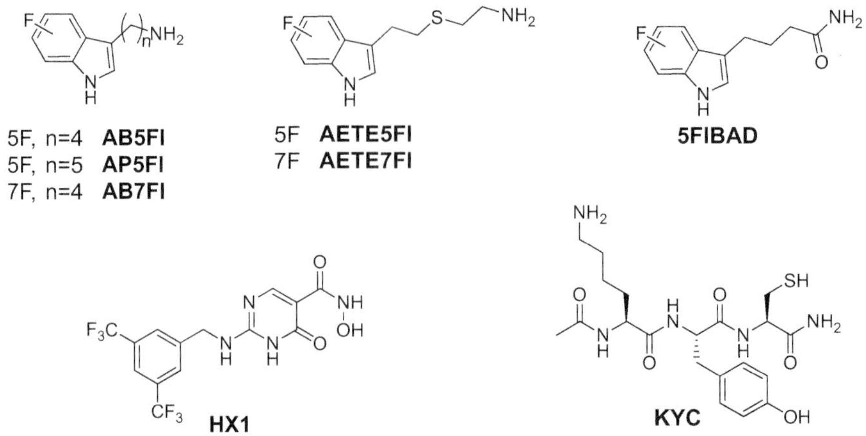

Fig. 6 Structures of reversible MPO inhibitors

in vivo (Soubhye et al. 2016a). Several potent inhibitors derived from 3-alkylfluoroindole were developed (Soubhye et al. 2010, 2013). It has been demonstrated that 3-(4-aminobutyl)-5-fluoro-*1H*-indole (AB5FI), 3-(5-aminopentyl)-5-fluoro-*1H*-indole (AP5FI), 3-(2-(2-aminoethylthio)ethyl)-5-fluoro-*1H*-indole (AETE5FI), 3-(4-aminobutyl)-7-fluoro-*1H*-indole (AB7FI), 3-(2-(2-aminoethylthio)ethyl)-7-fluoro-*1H*-indole (AETE7FI), and 4-(5-fluoro-*1H*-indol-3-yl)butanamide (5FIBAD) are potent MPO inhibitors (Fig. 6). The mechanism of inhibition by these molecules includes high affinity binding at the heme periphery followed by oxidation by Compound I thereby generating Compound II that accumulates (Soubhye et al. 2010, 2013).

All the 3-alkylfluoroindole derivatives depicted in Fig. 6 have been shown to exhibit inhibitory activities at low nanomolar level ($IC_{50} < 20$ nM) and prevent MPO-mediated LDL oxidation. The in vivo toxicological experiments have suggested that this family of inhibitors has large safety margin (Soubhye et al. 2013, 2014). However, with the exception of 5FIBAD, 3-alkylfluoroindole inhibitors have been shown to exhibit also inhibitory effects on the serotonin transporter (SERT) roughly at the same concentrations required for MPO inhibition. However, the high activity of these inhibitors on both MPO and SERT can be considered as a double-edged sword. On the one hand are safety issues due to promotion of the serotonin syndrome, but on the other hand these compounds can be used as hybrid molecules for treating and preventing atherosclerosis associated with major depressive disorder (MDD) (Soubhye et al. 2013, 2014). In any case, among the indolic inhibitors, 5FIBAD showed high potency toward MPO with high selectivity toward SERT. Indeed, the in vivo toxicological tests have demonstrated that 5FIBAD has the largest safety margin among this family of molecules (Soubhye et al. 2013).

As mentioned above, aromatic hydroxamic inhibit MPO by promoting Compound III formation via Compound I (Davies and Edwards 1989; Ikeda-Saito et al.

1991; Forbes et al. 2013). Based on this fact, very potent inhibitors were developed starting from aromatic hydroxamic acids. Among these inhibitors, HX1 showed the best activity and selectivity (Fig. 6). It has been suggested that HX1 inhibits MPO reversibly via accumulation of the nitrosyl ferrous complex of ferrous MPO [NO-Fe (II)] (Forbes et al. 2013). Despite its high activity on MPO in vitro, there are no data about the activity and toxicity of HX1 in vivo. However, ex vivo experiments demonstrated that HX1 prevents the formation of HOCl released by stimulating neutrophils at very low concentrations.

Similar to 3-alkylfluoroindole inhibitors, the tripeptide N-acetyl lysyltyrosylcysteine amide (KYC) inhibits MPO by causing accumulation of Compound II but with less potency (Zhang et al. 2013). Preclinical studies on KYC in mice showed that this safe inhibitor can reduce the migration of myeloid cells in conditions such as middle cerebral artery occlusion (MCAO) (Zhang et al. 2016). In addition, KYC has shown useful efficacy in treating brain damage after stroke (Yu et al. 2016), and vascular disease (Zhang et al. 2013).

Next, triazolopyrimidine was also shown to act as reversible MPO inhibitor. However, it suffers from poor stability in acidic media and is an irreversible inhibitor of the DNA repair protein methyl guanine methyl transferase (MGMT). Consequently, a structure-based drug design was employed to discover benzyl triazolopyridine derivatives with improved MPO potency, as well as acid stability, no reactivity with MGMT, and selectivity against thyroid peroxidase (TPO). Finally, structure–activity relationships, a crystal structure of the MPO-inhibitor complex (Fig. 4b), and acute in vivo pharmacodynamic data have been reported (Shaw et al. 2020).

Recently, with the aid of secretome phage display, a highly conserved protein that specifically binds and inhibits MPO was detected and designated as "staphylococcal peroxidase inhibitor" (SPIN). A co-crystal structure of SPIN bound to MPO suggested that SPIN blocks substrate access to the catalytic heme by inserting an N-terminal β-hairpin into the MPO active-site channel (de Jong et al. 2017) whereas the C-terminal domain specifically binds to human MPO (Fig. 4c). Further studies have to show whether SPIN can be used as starting scaffold for the design and production of a new generation of reversible and specific MPO inhibitors.

6 MPO Inhibitors in Clinical Development

Although MPO has been implicated in many chronic inflammatory syndromes, no medication targeting this enzyme is available so far. In fact, the identification of MPO as a target for developing drugs is relatively recent (Malle et al. 2007). However, several potent MPO inhibitors have been obtained in the last decade but only few of them have been subjected to preclinical trials (Ruggeri et al. 2015; Dong et al. 2016; Maiocchi et al. 2017; Jucaite et al. 2015; Zhang et al. 2013; Yu et al. 2016). It is remarkable that most compounds subjected to clinical trials are irreversible inhibitors. The Astra Zeneca compound AZD-3241 is currently in clinical trials for multiple system atrophy and phases 1 and 2 were already completed. The phases 1 and 2 of its clinical trials were already completed. The preliminary results have shown good efficacy in neuro-inflammation in patients suffering from PD and multiple system atrophy with good tolerance when administered orally at a treatment dose of 600 mg twice a day (Jucaite et al. 2015). ADME-T profile of another compound in clinical trial, PF-06282999, which has been established in mice, rats, dogs, and monkeys, indicated that the inhibitor is absorbed by intestinal tract with a bioavailability suitable for in vivo inhibition of MPO. In addition, the results of phase 1 trials suggested a good safety profile (Ruggeri et al. 2015).

Many preclinical and ex vivo experimental data also suggest that the inhibition of MPO not only prevents inflammation but also improves tissue injury. Indeed, it has been reported that the administration of MPO inhibitors in vivo can prevent endothelial cells dysfunction as well as the oxidation of LDL and therefore slow down the development of atherosclerosis (Maiocchi et al. 2017; Bekesi et al. 2005). Moreover, inhibition of MPO may be useful for treating and preventing MS (Forghani et al. 2012), BBB dysfunction (Üllen et al. 2013), chronic obstructive pulmonary disease (Churg et al. 2012), neuro-inflammation in patients suffering from PD and multiple system atrophy (Jucaite et al. 2015), and it can improve neurogenesis after ischemic stroke (Kim et al. 2016). However, MPO inhibitors cannot be given alone as a medication for treating or preventing the chronic inflammatory disorders but they must be administered with a combination of other drugs such as anti-hypertensives, diuretics, peripheral vasodilators, and lipid modifying agents (Inghardt et al. 2016). It has been recommended to use MPO inhibitors in inflammatory diseases associated with enhanced neutrophil attack such as chronic obstructive pulmonary disease, cystic fibrosis, and systemic autoimmune diseases (Forbes et al. 2013). Many authors suggested MPO inhibitors as essential medications for preventing atherosclerosis in patients with hyperlipidemia (Liu et al. 2012, 2015; Malle et al. 2007; Ali et al. 2016; Nicholls and Hazen 2008; Shao and Heinecke 2009; Nicholls 2005). In all cases, more experiments are needed to compare the benefit of the synthesized MPO inhibitors versus natural antioxidant agents since the latter can be used as safe food supplements or, more simply, through a balanced diet (Boufadi et al. 2017).

7 Conclusions

Reactive oxidants released by human myeloperoxidase are involved in the development of several diseases including chronic inflammatory diseases, atherosclerosis, and acute cardiovascular events and are considered as a major problem for human health (Malle et al. 2007; Lazarević-Pasti et al. 2015; Soubhye et al. 2016a). Many potent specific reversible and irreversible MPO inhibitors have been developed based on our profound knowledge of structure and function of this human heme peroxidase and its homologous human counterparts as well as novel (virtual) high-throughput screening procedures. But only a few drug candidates have been subjected to preclinical or clinical experiments. Despite satisfying results obtained in preclinical and ex vivo tests, only two compounds have been studied in clinical trials so far, namely AZD-3241 and PF-06282999 and the beneficial effect of MPO inhibition in the treatment and prophylaxis of inflammatory disorders is still under discussion. Nevertheless, the availability of potent MPO inhibitors found among marketed drugs, such as paroxetine, as well as the fact that these drugs are used for a long time in large populations, might allow epidemiological studies for evaluation of the impact of MPO inhibition in the improvement of inflammatory syndromes.

References

Adam GC, Cravatt BF, Sorensen EJ (2001) Profiling the specific reactivity of the proteome with non-directed activity-based probes. Chem Biol 8:81–95. https://doi.org/10.1016/s1074-5521(00)90060-7

Aldib I, Soubhye J, Zouaoui Boudjeltia K, Vanhaeverbeek M, Rousseau A, Furtmüller PG, Obinger C, Dufrasne F, Nève J, Van Antwerpen P, Prévost M (2012) Evaluation of new scaffolds of myeloperoxidase inhibitors by rational design combined with high-throughput virtual screening. J Med Chem 55:7208–7218. https://doi.org/10.1021/jm3007245

Ali M, Pulli B, Courties G, Tricot B, Sebas M, Iwamoto Y, Hilgendorf I, Schob S, Dong A, Zheng W, Skoura A, Kalgukar A, Cortes C, Ruggeri R, Swirski FK, Nahrendorf M, Buckbinder L, Chen JW (2016) Myeloperoxidase inhibition improves ventricular function and remodeling after? Experimental myocardial infarction. JACC Basic Transl Sci 1:633–643. https://doi.org/10.1016/j.jacbts.2016.09.004

Arnhold J, Furtmüller PG, Regelsberger G, Obinger C (2001) Redox properties of the couple compound I/native enzyme of myeloperoxidase and eosinophil peroxidase. Eur J Biochem 268:5142–5148. https://doi.org/10.1046/j.0014-2956.2001.02449.x

Arnhold J, Furtmüller PG, Obinger C (2003) Redox properties of myeloperoxidase. Redox Rep 8:179–186. https://doi.org/10.1179/135100003225002664

Arnhold J, Monzani E, Furtmüller PG, Zederbauer M, Casella L, Obinger C (2006) Kinetics and thermodynamics of halide and nitrite oxidation by mammalian heme peroxidases. Eur J Inorg Chem 19:3801–3811. https://doi.org/10.1002/ejic.200600436

Banerjee S, Stampler J, Furtmüller PG, Obinger C (2011) Conformational and thermal stability of mature dimeric human myeloperoxidase and a recombinant monomeric form from CHO cells. Biochim Biophys Acta 1814:375–387. https://doi.org/10.1016/j.bbapap.2010.09.015

Battistuzzi G, Borsari M, Ranieri J, Sola M (2001) Redox thermodynamic of the Fe^{3+}/Fe^{2+} couple in horseradish peroxidase and its cyanide complex. J Am Chem Soc 124:26–27. https://doi.org/10.1021/ja017188m

Battistuzzi G, Bellei M, Zederbauer M, Furtmüller PG, Sola M, Obinger C (2006) Redox thermodynamics of the Fe(III)/Fe(II) couple of human myeloperoxidase in its high-spin and low-spin forms. Biochemistry 45:12750–12755. https://doi.org/10.1074/jbc.M610685200

Battistuzzi G, Bellei M, Bortolotti CA, Sola M (2010) Redox properties of heme peroxidases. Arch Biochem Biophys 500:21–36. https://doi.org/10.1016/j.abb.2010.03.002

Battistuzzi G, Stampler J, Bellei M, Vlasits J, Soudi M, Furtmüller PG, Obinger C (2011) Influence of the covalent heme-protein bonds on the redox thermodynamics of human myeloperoxidase. Biochemistry 50:7987–7994. https://doi.org/10.1021/bi2008432

Bekesi G, Heinle H, Kakucs R, Pazmany T, Szombath D, Dinya M, Tulassay Z, Feher J, Racz K, Szekacs B (2005) Effect of inhibitors of myeloperoxidase on the development of aortic atherosclerosis in an animal model. Exp Gerontol 40:199–208. https://doi.org/10.1016/j.exger.2004.12.004

Blair-Johnson M, Fiedler T, Fenna R (2001) Human myeloperoxidase: structure of a cyanide complex and its interaction with bromide and thiocyanate substrates at 1.9 A resolution. Biochemistry 40:13990–14007. https://doi.org/10.1021/bi0111808

Boufadi YM, Soubhye J, Riazi A, Rousseau A, Vanhaeverbeek M, Nève J, Boudjeltia KZ, Van Antwerpen P (2014) Characterization and antioxidant properties of six Algerian propolis extracts: ethyl acetate extracts inhibit myeloperoxidase activity. Int J Mol Sci 15:2327–2345. https://doi.org/10.3390/ijms15022327

Boufadi Y, Van Antwerpen P, Chikh Alard I, Nève J, Djennas N, Riazi A, Soubhye J (2017) Antioxidant effects and bioavailability evaluation of propolis extract and its content of pure polyphenols. J Food Biochem 42:e12434. https://doi.org/10.1111/jfbc.12434

Bozeman PM, Learn DB, Thomas EL (1992) Inhibition of the human leukocyte enzymes myeloperoxidase and eosinophil peroxidase by dapsone. Biochem Pharmacol 44 (553–563):372. https://doi.org/10.1016/0006-2952(92)90449-s

Burner U, Obinger C, Paumann M, Furtmüller PG, Kettle AJ (1999) Transient and steady-state kinetics of the oxidation of substituted benzoic acid hydrazides by myeloperoxidase. J Biol Chem 274:9494–9502. https://doi.org/10.1074/jbc.274.14.9494

Carpena X, Vidossich P, Schroettner K, Calisto BM, Banerjee S, Stampler J, Soudi M, Furtmüller PG, Rovira C, Fita I, Obinger C (2009) Essential role of proximal histidine-asparagine interaction in mammalian peroxidases. J Biol Chem 284:25929–25937. https://doi.org/10.1074/jbc.M109.002154

Churg A, Marshall CV, Sin DD, Bolton S, Zhou S, Thain K, Cadogan EB, Maltby J, Soars MG, Mallinder PR, Wright JL (2012) Late intervention with a myeloperoxidase inhibitor stops progression of experimental chronic obstructive pulmonary disease. Am J Respir Crit Care Med 185:34–43. https://doi.org/10.1164/rccm.201103-0468OC

Clark RA, Klebanoff SJ (1979) Myeloperoxidase-mediated platelet release reaction. J Clin Invest 63:177–183. https://doi.org/10.1172/JCI109287

Colas C, Ortiz de Montellano PR (2003) Autocatalytic radical reactions in physiological prosthetic heme modification. Chem Rev 103:2305–2332. https://doi.org/10.1021/cr0204303

Colon S, Page-McCaw P, Bhave G (2017) Role of hypohalous acids in basement membrane homeostasis. Antioxid Redox Signal 27:839–854. https://doi.org/10.1089/ars.2017.7245

Davey CA, Fenna RE (1996) 2.3 Å resolution X-ray crystal structure of the bisubstrate analogue inhibitor salicylhydroxamic acid bound to human myeloperoxidase: a model for a prereaction complex with hydrogen peroxide. Biochemistry 35:10967–10973. https://doi.org/10.1021/bi960577m.

Davies MJ (2010) Myeloperoxidase-derived oxidation: mechanisms of biological damage and its prevention. J Clin Biochem Nutr 48:8–19. https://doi.org/10.3164/jcbn.11-006FR

Davies B, Edwards SW (1989) Inhibition of myeloperoxidase by salicylhydroxamic acid. Biochem J 258:801–806. https://doi.org/10.1042/bj2580801

Davies MJ, Hawkins CL (2020) The role of myeloperoxidase in biomolecule modification, chronic inflammation, and disease. Antioxid Redox Signal 32:957–981. https://doi.org/10.1089/ars.2020.8030

Davies MJ, Hawkins CL, Pattison DI, Rees MD (2008) Tryptophan residues are targets in hypothiocyanous acid-mediated protein oxidation. Antioxid Redox Signal 7:1199–1234. https://doi.org/10.1089/ars.2007.1927

de Jong NWM, Ramyar KX, Guerra FE, Nijland R, Fevre C, Voyich JM, McCarthy AJ, Garcia BL, van Kessel KPM, van Strijp JAG, Geisbrecht BV, Haas PA (2017) Immune evasion by a staphylococcal inhibitor of myeloperoxidase. Proc Natl Acad Sci U S A 114:9439–9444. https://doi.org/10.1073/pnas.1707032114

Dong JQ, Varma MV, Wolford A, Ryder T, Di L, Feng B, Terra SG, Sagawa K, Kalgutkar AS (2016) Pharmacokinetics and disposition of the thiouracil derivative PF-06282999, an orally bioavailable, irreversible inactivator of myeloperoxidase enzyme, across animals and humans. Drug Metab Dispos Biol Fate Chem 44:209–219. https://doi.org/10.1124/dmd.115.067868

Eiserich JP, Hristova M, Cross CE, Jones AD, Freeman BA, Halliwell B, van der Vliet A (1998) Formation of nitric oxide-derived inflammatory oxidants by myeloperoxidase in neutrophils. Nature 391:393–397. https://doi.org/10.1038/34923

Fiedler TJ, Davey CA, Fenna RE (2000) X-ray crystal structure and characterization of halide-binding sites of human myeloperoxidase at 1.8 a resolution. J Biol Chem 2000 (275):11964–11971. https://doi.org/10.1074/jbc.275.16.11964

Forbes LV, Furtmüller PG, Khalilova I, Turner R, Obinger C, Kettle AJ (2012) Isoniazid as a substrate and inhibitor of myeloperoxidase: identification of amine adducts and the influence of superoxide dismutase on their formation. Biochem Pharmacol 84:949–960. https://doi.org/10.1016/j.bcp.2012.07.020

Forbes LV, Sjogren T, Auchere F, Jenkins DW, Thong B, Laughton D, Hemsley P, Pairaudeau G, Turner R, Eriksson H, Unitt JF, Kettle AJ (2013) Potent reversible inhibition of myeloperoxidase by aromatic hydroxamates. J Biol Chem 288:36636–36647. https://doi.org/10.1074/jbc.M113.507756

Forghani R, Wojtkiewicz GR, Zhang Y, Seeburg D, Bautz BRM, Pulli B, Milewski AR, Atkinson WL, Iwamoto Y, Zhang ER, Etzrodt M, Rodriguez E, Robbins CS, Swirski FK, Weissleder R, Chen JW (2012) Demyelinating diseases: myeloperoxidase as an imaging biomarker and therapeutic target. Radiology 263:451–460. https://doi.org/10.1148/radiol.12111593

Foye WO, Lemke TL, Williams DA (eds) (2013) Foye's principles of medicinal chemistry, 7th edn. Wolters Kluwer Health/Lippincott Williams & Wilkins, Philadelphia

Furtmüller PG, Burner U, Obinger C (1998) Reaction of myeloperoxidase compound I with chloride, bromide, iodide, and thiocyanate. Biochemistry 37:17923–17930. https://doi.org/10.1021/bi9818772

Furtmüller PG, Obinger C, Hsuanyu Y, Dunford HB (2000) Mechanism of reaction of myeloperoxidase with hydrogen peroxide and chloride ion. Eur J Biochem 267:5858–5864. https://doi.org/10.1046/j.1432-1327.2000.01491.x

Furtmüller PG, Jantschko W, Regelsberger G, Jakopitsch C, Moguilevsky N, Obinger C (2001) A transient kinetic study on the reactivity of recombinant unprocessed monomeric myeloperoxidase. FEBS Lett 503:147–150. https://doi.org/10.1016/s0014-5793(01)02725-9

Furtmüller PG, Arnhold J, Jantschko W, Pichler H, Obinger C (2003) Redox properties of the couples compound I/compound II and compound II/native enzyme of human myeloperoxidase. Biochem Biophys Res Commun 301:551–557. https://doi.org/10.1016/s0006-291x(02)03075-9

Furtmüller PG, Arnhold J, Jantschko W, Zederbauer M, Jakopitsch C, Obinger C (2005) Standard reduction potentials of all couples of the peroxidase cycle of lactoperoxidase. J Inorg Biochem 99:1220–1229. https://doi.org/10.1016/j.jinorgbio.2005.02.021

Furtmüller PG, Zederbauer M, Jantschko W, Helm J, Bogner M, Jakopitsch C, Obinger C (2006) Active site structure and catalytic mechanisms of human peroxidases. Arch Biochem Biophys 445:199–213. https://doi.org/10.1016/j.abb.2005.09.017

Galijasevic S, Abdulhamid I, Abu-Soud HM (2008) Melatonin is a potent inhibitor for myeloperoxidase. Biochemistry 47:2668–2677. https://doi.org/10.1021/bi702016q

Gau J, Furtmüller PG, Obinger C, Prévost M, Van Antwerpen P, Arnhold J, Flemmig J (2016) Flavonoids as promoters of the (pseudo-)halogenating activity of lactoperoxidase and myeloperoxidase. Free Radic Biol Med 97:307–319. https://doi.org/10.1016/j.freeradbiomed.2016.06.026

Green PS, Mendez AJ, Jacob JS, Crowley JR, Growdon W, Hyman BT, Heinecke JW (2004) Neuronal expression of myeloperoxidase is increased in Alzheimer's disease. J Neurochem 90:724–733. https://doi.org/10.1111/j.1471-4159.2004.02527.x

Grishkovskaya I, Paumann-Page M, Tscheliessnig R, Stampler J, Hofbauer S, Soudi M, Sevcnikar B, Oostenbrink C, Furtmüller PG, Djinović-Carugo K, Nauseef WM, Obinger C (2017) Structure of human promyeloperoxidase (proMPO) and the role of the propeptide in processing and maturation. J Biol Chem 292:8244–8261. https://doi.org/10.1074/jbc.M117.775031

Hallingbäck HR, Gabdoulline RR, Wade RC (2006) Comparison of the binding and reactivity of plant and mammalian peroxidases to indole derivatives by computational docking. Biochemistry 45:2940–2950. https://doi.org/10.1021/bi051510e

Hori H, Fenna RE, Kimura S, Ikeda-Saito M (1994) Aromatic substrate molecules bind at the distal heme pocket of myeloperoxidase. J Biol Chem 269:8388–8392

Ihalin R, Loimaranta V, Tenovuo J (2006) Origin, structure, and biological activities of peroxidases in human saliva. Arch Biochem Biophys 445:261–268. https://doi.org/10.1016/j.abb.2005.07.004

Ikeda-Saito M, Shelley DA, Lu L, Booth KS, Caughey WS, Kimura S (1991) Salicylhydroxamic acid inhibits myeloperoxidase activity. J Biol Chem 266:3611–3616

Inghardt T, Johannesson P, Jurva U, Michaëlsson E, Lindstedt-Alstermark E, Tomkinson N, Stonehouse J, Gan L (2016) 1-[2-(aminomethyl)benzyl]-2-thioxo-1,2,3,5-tetrahydro-4h-pyrrolo[3,2-d]pyrimidin-4-ones as inhibitors of myeloperoxidase. WO2016087338A1

Inocencio C, Rivera D, Alcaraz F, Tomás-Barberán FA (2000) Flavonoid content of commercial capers (Capparis spinosa, C. sicula and C. orientalis) produced in mediterranean countries. Eur Food Res Technol 212:70–74. https://doi.org/10.1007/s002170000220

Jantschko W, Furtmüller PG, Allegra M, Livrea MA, Jakopitsch C, Regelsberger G, Obinger C (2002) Redox intermediates of plant and mammalian peroxidases: a comparative transient-kinetic study of their reactivity toward indole derivatives. Arch Biochem Biophys 398:12–22. https://doi.org/10.1006/abbi.2001.2674

Jantschko W, Georg Furtmüller P, Zederbauer M, Lanz M, Jakopitsch C, Obinger C (2003) Direct conversion of ferrous myeloperoxidase to compound II by hydrogen peroxide: an anaerobic stopped-flow study. Biochem Biophys Res Commun 312:292–298. https://doi.org/10.1016/j.bbrc.2003.10.117

Jantschko W, Furtmüller PG, Zederbauer M, Jakopitsch C, Obinger C (2004) Kinetics of oxygen binding to ferrous myeloperoxidase. Arch Biochem Biophys 426:91–97. https://doi.org/10.1016/j.abb.2004.03.019

Jantschko W, Furtmüller PG, Zederbauer M, Neugschwandtner K, Lehner I, Jakopitsch C, Arnhold J, Obinger C (2005) Exploitation of the unusual thermodynamic properties of human myeloperoxidase in inhibitor design. Biochem Pharmacol 69:1149–1157. https://doi.org/10.1016/j.bcp.2005.02.006

Jeelani R, Jahanbakhsh S, Kohan-Ghadr H-R, Thakur M, Khan S, Aldhaheri SR, Yang Z, Andreana P, Morris R, Abu-Soud HM (2017) Mesna (2-mercaptoethane sodium sulfonate) functions as a regulator of myeloperoxidase. Free Radic Biol Med 110:54–62. https://doi.org/10.1016/j.freeradbiomed.2017.05.019

Jiang H, Li X, Tang C (2011) Effect of purple sweet potato flavonoids on metabolism of glucose and lipids in diabetic rats. Zhejiang Xue Xue Bao Yi Xue Ban J Zhejiang Univ Med Sci 40:374–379

Jucaite A, Svenningsson P, Rinne JO, Cselényi Z, Varnäs K, Johnström P, Amini N, Kirjavainen A, Helin S, Minkwitz M, Kugler AR, Posener JA, Budd S, Halldin C, Varrone A, Farde L (2015) Effect of the myeloperoxidase inhibitor AZD3241 on microglia: a PET study in Parkinson's disease. Brain J Neurol 138:2687–2700. https://doi.org/10.1093/brain/awv184

Kettle AJ, Gedye CA, Hampton MB, Winterbourn CC (1995) Inhibition of myeloperoxidase by benzoic acid hydrazides. Biochem J 308:559–563. https://doi.org/10.1042/bj3080559

Kettle AJ, Gedye CA, Winterbourn CC (1997) Mechanism of inactivation of myeloperoxidase by 4-aminobenzoic acid hydrazide. Biochem J 321:503–508. https://doi.org/10.1042/bj3210503

Kim H, Wei Y, Lee JY, Wu Y, Zheng Y, Moskowitz MA, Chen JW (2016) Myeloperoxidase inhibition increases neurogenesis after ischemic stroke. J Pharmacol Exp Ther 359:262–272. https://doi.org/10.1124/jpet.116.235127

Klebanoff SJ (1977) Estrogen binding by leukocytes during phagocytosis. J Exp Med 145:983–998. https://doi.org/10.1084/jem.145.4.983

Klebanoff SJ (2005) Myeloperoxidase: friend and foe. J Leukoc Biol 77:598–625. https://doi.org/10.1189/jlb.1204697

Klebanoff SJ, Green WL (1973) Degradation of thyroid hormones by phagocytosing human leukocytes. J Clin Invest 52:60–72. https://doi.org/10.1172/JCI107174

Klebanoff SJ, Kettle AJ, Rosen H, Winterbourn CC, Nauseef WM (2013) Myeloperoxidase: a front-line defender against phagocytosed microorganisms. J Leukoc Biol 93:185–198. https://doi.org/10.1189/jlb.0712349

Kostyuk VA, Kraemer T, Sies H, Schewe T (2003) Myeloperoxidase/nitrite-mediated lipid peroxidation of low-density lipoprotein as modulated by flavonoids. FEBS Lett 537:146–150. https://doi.org/10.1016/S0014-5793(03)00113-3

La Rocca G, Di Stefano A, Eleuteri E, Anzalone R, Magno F, Corrao S, Loria T, Martorana A, Di Gangi C, Colombo M, Sansone F, Patane F, Farina F, Rinaldi M, Cappello F, Giannuzzi P, Zummo G (2009) Oxidative stress induces myeloperoxidase expression in endocardial endothelial cells from patients with chronic heart failure. Basic Res Cardiol 104:307–320. https://doi.org/10.1007/s00395-008-0761-9

Langley-Evans SC (2000) Antioxidant potential of green and black tea determined using the ferric reducing power (FRAP) assay. Int J Food Sci Nutr 51:181–188. https://doi.org/10.1080/09637480050029683

Lazarević-Pasti T, Leskovac A, Vasić V (2015) Myeloperoxidase inhibitors as potential drugs. Curr Drug Metab 16:168–190. https://doi.org/10.2174/138920021603150812120640

Li Y, Ganesh T, Diebold BA, Zhu Y, McCoy JW, Smith SME, Sun A, Lambeth JD (2015) Thioxo-dihydroquinazolin-one compounds as novel inhibitors of myeloperoxidase. ACS Med Chem Lett 6:1047–1052. https://doi.org/10.1021/acsmedchemlett.5b00287

Lima CAM, Baumann P, Eap CB (2008) Paroxetine plasma concentrations in adult and elderly depressed patients. Rev Psiquiatr Rio Gd Sul 30:13–18. https://doi.org/10.1590/S0101-81082008000100006

Lipinski CA, Lombardo F, Dominy BW, Feeney PJ (2001) Experimental and computational approaches to estimate solubility and permeability in drug discovery and development settings. Adv Drug Deliv Rev 46(3–26):469

Liu C, Desikan R, Ying Z, Gushchina L, Kampfrath T, Deiuliis J, Wang A, Xu X, Zhong J, Rao X, Sun Q, Maiseyeu A, Parthasarathy S, Rajagopalan S (2012) Effects of a novel pharmacologic inhibitor of myeloperoxidase in a mouse atherosclerosis model. PLoS One 7:e50767. https://doi.org/10.1371/journal.pone.0050767

Liu W-Q, Zhang Y-Z, Wu Y, Zhang J-J, Li T-B, Jiang T, Xiong X-M, Luo X-J, Ma Q-L, Peng J (2015) Myeloperoxidase-derived hypochlorous acid promotes ox-LDL-induced senescence of endothelial cells through a mechanism involving β-catenin signaling in hyperlipidemia. Biochem Biophys Res Commun 467:859–865. https://doi.org/10.1016/j.bbrc.2015.10.053

Loke WM, Proudfoot JM, Mckinley AJ, Needs PW, Kroon PA, Hodgson JM, Croft KD (2008) Quercetin and its in vivo metabolites inhibit neutrophil-mediated low-density lipoprotein oxidation. J Agric Food Chem 56:3609–3615. https://doi.org/10.1021/jf8003042

Lundblad RL (2005) Chemical reagents for protein modification, 3rd edn. CRC Press, Boca Raton

Maiocchi SL, Morris JC, Rees MD, Thomas SR (2017) Regulation of the nitric oxide oxidase activity of myeloperoxidase by pharmacological agents. Biochem Pharmacol 135:90–115. https://doi.org/10.1016/j.bcp.2017.03.016

Malik A, Batra JK (2012) Antimicrobial activity of human eosinophil granule proteins: involvement in host defence against pathogens. Crit Rev Microbiol 38:168–181. https://doi.org/10.3109/1040841X.2011.645519

Malle E, Furtmüller PG, Sattler W, Obinger C (2007) Myeloperoxidase: a target for new drug development? Br J Pharmacol 152:838–854. https://doi.org/10.1038/sj.bjp.0707358

Manach C, Scalbert A, Morand C, Rémésy C, Jiménez L (2004) Polyphenols: food sources and bioavailability. Am J Clin Nutr 79:727–747. https://doi.org/10.1093/ajcn/79.5.727

Marquez LA, Huang JT, Dunford HB (1994) Spectral and kinetic studies on the formation of myeloperoxidase compounds I and II: roles of hydrogen peroxide and superoxide. Biochemistry 33:1447–1454. https://doi.org/10.1021/bi00172a022

Nastase M-V, Zeng-Brouwers J, Frey H, Hsieh LT-H, Poluzzi C, Beckmann J, Schroeder N, Pfeilschifter J, Lopez-Mosqueda J, Mersmann J, Ikeda F, Iozzo RV, Dikic I, Schaefer L (2016) An essential role for SHARPIN in the regulation of caspase 1 activity in sepsis. Am J Pathol 186:1206–1220. https://doi.org/10.1016/j.ajpath.2015.12.026

Nauseef WM (2014) Myeloperoxidase in human neutrophil host defence. Cell Microbiol 16:1146–1155. https://doi.org/10.1111/cmi.12312

Nauseef WM (2018) Biosynthesis of human myeloperoxidase. Arch Biochem Biophys 642:1–9. https://doi.org/10.1016/j.abb.2018.02.001

Nicholls SJ (2005) Myeloperoxidase and cardiovascular disease. Arterioscler Thromb Vasc Biol 25:1102–1111. https://doi.org/10.1161/01.ATV.0000163262.83456.6d

Nicholls SJ, Hazen SL (2008) Myeloperoxidase, modified lipoproteins, and atherogenesis. J Lipid Res 50:S346–S351. https://doi.org/10.1194/jlr.R800086-JLR200

Nicolussi A, Auer M, Sevcnikar B, Paumann-Page M, Pfanzagl V, Zámocký M, Hofbauer S, Furtmüller PG, Obinger C (2018) Posttranslational modification of heme in peroxidases – impact on structure and catalysis. Arch Biochem Biophys 643:14–23. https://doi.org/10.1016/j.abb.2018.02.008

Papież MA, Krzyściak W, Wąsik M (2015) Inhibition of myeloperoxidase activity have impact on the formation of DNA double-strand breaks induced by etoposide in HL-60 cell line. Folia Med Cracov 55:43–51

Pattison DI, Davies MJ (2004) Kinetic analysis of the reactions of hypobromous acid with protein components: implications for cellular damage and use of 3-bromotyrosine as a marker of oxidative stress. Biochemistry 43:4799–4809. https://doi.org/10.1021/bi035946a

Pattison DI, Davies MJ (2006) Reactions of myeloperoxidase-derived oxidants with biological substrates: gaining chemical insight into human inflammatory diseases. Curr Med Chem 13:3271–3290. https://doi.org/10.2174/092986706778773095

Paumann-Page M, Katz RS, Bellei M, Schwartz I, Edenhofer E, Sevcnikar B, Soudi M, Hofbauer S, Battistuzzi G, Furtmüller PG, Obinger C (2017) Pre-steady-state kinetics reveal the substrate specificity and mechanism of halide oxidation of truncated human Peroxidasin 1. J Biol Chem 292:4583–4592. https://doi.org/10.1074/jbc.M117.775213

Paumann-Page M, Tscheliessnig R, Sevcnikar B, Katz RS, Schwartz I, Hofbauer S, Pfanzagl V, Furtmüller PG, Obinger C (2020) Monomeric and homotrimeric solution structures of truncated human peroxidasin 1 variants. Biochim Biophys Acta Proteins Proteom 1868(1):140249. https://doi.org/10.1016/j.bbapap.2019.07.002

Poulos TL, Kraut J (1980) The stereochemistry of peroxidase catalysis. J Biol Chem 255:8199–8205

Ramos DR, García MV, Canle LM, Santaballa JA, Furtmüller PG, Obinger C (2008) Myeloperoxidase-catalyzed chlorination: the quest for the active species. J Inorg Biochem 102:1300–1311. https://doi.org/10.1016/j.jinorgbio.2008.01.003

Rasoanaivo P, Wright CW, Willcox ML, Gilbert B (2011) Whole plant extracts versus single compounds for the treatment of malaria: synergy and positive interactions. Malar J 10(Suppl 1): S4. https://doi.org/10.1186/1475-2875-10-S1-S4

Rtibi K, Jabri M-A, Selmi S, Sebai H, Amri M, El-Benna J, Marzouki L (2016) *Ceratonia siliqua* leaves exert a strong ROS-scavenging effect in human neutrophils, inhibit myeloperoxidase in vitro and protect against intestinal fluid and electrolytes secretion in rats. RSC Adv 6:65483–65493. https://doi.org/10.1039/C6RA11297H

Ruf J, Carayon P (2006) Structural and functional aspects of thyroid peroxidase. Arch Biochem Biophys 445:269–277. https://doi.org/10.1016/j.abb.2005.06.023

Ruggeri RB, Buckbinder L, Bagley SW, Carpino PA, Conn EL, Dowling MS, Fernando DP, Jiao W, Kung DW, Orr STM, Qi Y, Rocke BN, Smith A, Warmus JS, Zhang Y, Bowles D, Widlicka DW, Eng H, Ryder T, Sharma R, Wolford A, Okerberg C, Walters K, Maurer TS, Zhang Y, Bonin PD, Spath SN, Xing G, Hepworth D, Ahn K, Kalgutkar AS (2015) Discovery of 2-(6-(5-chloro-2-methoxyphenyl)-4-oxo-2-thioxo-3,4-dihydropyrimidin-1(2 h)-yl)acetamide (pf-06282999): a highly selective mechanism-based myeloperoxidase inhibitor for the treatment of cardiovascular diseases. J Med Chem 58:8513–8528. https://doi.org/10.1021/acs.jmedchem.5b00963

Schewe T, Sies H (2005) Myeloperoxidase-induced lipid peroxidation of LDL in the presence of nitrite. Protection by cocoa flavanols. Biofactors 24:49–58. https://doi.org/10.1002/biof.5520240106

Shao B, Heinecke JW (2009) HDL, lipid peroxidation, and atherosclerosis. J Lipid Res 50:599–601. https://doi.org/10.1194/jlr.E900001-JLR200

Shaw SA, Vokits BP, Dilger AK, Viet A, Clark CG, Abell LM, Locke GA, Duke G, Kopcho LM, Dongre A, Gao J, Krishnakumar A, Jusuf S, Khan J, Spronk SA, Basso MD, Zhao L, Cantor GH, Onorato JM, Wexler RR, Duclos F, Kick EK (2020) Discovery and structure activity relationships of 7-benzyl triazolopyridines as stable, selective, and reversible inhibitors of myeloperoxidase. Bioorg Med Chem 28:115723. https://doi.org/10.1016/j.bmc.2020.115723

Shepherd AM, McNay JL, Ludden TM, Lin MS, Musgrave GE (1981) Plasma concentration and acetylator phenotype determine response to oral hydralazine. Hypertension 3:580–585

Singh AK, Singh N, Sharma S, Singh SB, Kaur P, Bhushan A, Srinivasan A, Singh TP (2008) Crystal structure of lactoperoxidase at 2.4 Å resolution. J Mol Biol 376:1060–1075. https://doi.org/10.1016/j.jmb.2007.12.012

Skaff O, Pattison DI, Davies MJ (2007) Kinetics of hypobromous acid-mediated oxidation of lipid components and antioxidants. Chem Res Toxicol 20:1980–1988. https://doi.org/10.1021/tx7003097

Slimestad R, Fossen T, Vågen IM (2007) Onions: a source of unique dietary flavonoids. J Agric Food Chem 55:10067–10080. https://doi.org/10.1021/jf0712503

Smith GF (2011) Designing drugs to avoid toxicity. Prog Med Chem 50:1–47. https://doi.org/10.1016/B978-0-12-381290-2.00001-X

Soubhye J, Prévost M, Van Antwerpen P, Zouaoui Boudjeltia K, Rousseau A, Furtmüller PG, Obinger C, Vanhaeverbeek M, Ducobu J, Néve J, Gelbcke M, Dufrasne FO (2010) Structure-based design, synthesis, and pharmacological evaluation of 3-(aminoalkyl)-5-fluoroindoles as myeloperoxidase inhibitors. J Med Chem 53:8747–8759. https://doi.org/10.1021/jm1009988

Soubhye J, Aldib I, Elfving B, Gelbcke M, Furtmüller PG, Podrecca M, Conotte R, Colet J-M, Rousseau A, Reye F, Sarakbi A, Vanhaeverbeek M, Kauffmann J-M, Obinger C, Nève J, Prévost M, Zouaoui Boudjeltia K, Dufrasne F, Van Antwerpen P (2013) Design, synthesis, and structure-activity relationship studies of novel 3-alkylindole derivatives as selective and highly potent myeloperoxidase inhibitors. J Med Chem 56:3943–3958. https://doi.org/10.1021/jm4001538

Soubhye J, Aldib I, Prévost M, Elfving B, Gelbcke M, Podrecca M, Conotte R, Colet J-M, Furtmüller PG, Delporte C, Rousseau A, Vanhaeverbeek M, Nève J, Obinger C, Zouaoui-Boudjeltia K, Van Antwerpen P, Dufrasne F (2014) Hybrid molecules inhibiting myeloperoxidase activity and serotonin reuptake: a possible new approach of major depressive disorders with inflammatory syndrome: hybrid MPO and 5-HT reuptake inhibitors. J Pharm Pharmacol 66:1122–1132. https://doi.org/10.1111/jphp.12236

Soubhye J, Aldib I, Delporte C, Prévost M, Dufrasne F, Antwerpen PV (2016a) Myeloperoxidase as a target for the treatment of inflammatory syndromes: mechanisms and structure activity relationships of inhibitors. Curr Med Chem 23:3975–4008

Soubhye J, Meyer F, Furtmüller P, Obinger C, Dufrasne F, Antwerpen PV (2016b) Characterization of chemical features of potent myeloperoxidase inhibitors. Future Med Chem 8:1163–1177. https://doi.org/10.4155/fmc-2016-0031

Soubhye J, Chikh Alard I, Aldib I, Prévost M, Gelbcke M, Tadrent S, Carvalho A, Furtmüller PG, Obinger C, Flemmig J, Meyer F, Rousseau A, Nève J, Mathieu V, Boudjeltia KZ, Dufrasne F, Van Antwerpen P (2017a) Discovery of novel potent reversible and irreversible myeloperoxidase inhibitors using virtual screening procedure. J Med Chem 60:6563–6586. https://doi.org/10.1021/acs.jmedchem.7b00285

Soubhye J, Gelbcke M, Van Antwerpen P, Dufrasne F, Boufadi MY, Nève J, Furtmüller PG, Obinger C, Zouaoui Boudjeltia K, Meyer F (2017b) From dynamic combinatorial chemistry to in vivo evaluation of reversible and irreversible myeloperoxidase inhibitors. ACS Med Chem Lett 8:206–210. https://doi.org/10.1021/acsmedchemlett.6b00417

Soudi M, Zamocky M, Jakopitsch C, Furtmüller PG, Obinger C (2012) Molecular evolution, structure, and function of peroxidasins. Chem Biodivers 9:1776–1793. https://doi.org/10.1002/cbdv.201100438

Soudi M, Paumann-Page M, Delporte C, Pirker KF, Bellei M, Edenhofer E, Stadlmayr G, Battistuzzi G, Boudjeltia KZ, Furtmüller PG, Van Antwerpen P, Obinger C (2015) Multidomain human peroxidasin 1 is a highly glycosylated and stable homotrimeric high spin ferric peroxidase. J Biol Chem 290:10876–10890. https://doi.org/10.1074/jbc.M114.632273

Spalteholz H, Furtmüller PG, Jakopitsch C, Obinger C, Schewe T, Sies H, Arnhold J (2008) Kinetic evidence for rapid oxidation of (−)-epicatechin by human myeloperoxidase. Biochem Biophys Res Commun 371:810–813. https://doi.org/10.1016/j.bbrc.2008.04.139

Stamp LK, Turner R, Khalilova IS, Zhang M, Drake J, Forbes LV, Kettle AJ (2014) Myeloperoxidase and oxidation of uric acid in gout: implications for the clinical consequences of hyperuricaemia. Rheumatology 53:1958–1965. https://doi.org/10.1093/rheumatology/keu218

Stendahl O, Molin L, Dahlgren C (1978) The inhibition of polymorphonuclear leukocyte cytotoxicity by dapsone. A possible mechanism in the treatment of dermatitis herpetiformis. J Clin Invest 62:214–220. https://doi.org/10.1172/JCI109109

Tian R, Ding Y, Peng Y-Y, Lu N (2017) Inhibition of myeloperoxidase- and neutrophil-mediated hypochlorous acid formation in vitro and endothelial cell injury by (−)-epigallocatechin gallate. J Agric Food Chem 65:3198–3203. https://doi.org/10.1021/acs.jafc.7b00631

Tiden A-K, Sjogren T, Svensson M, Bernlind A, Senthilmohan R, Auchere F, Norman H, Markgren P-O, Gustavsson S, Schmidt S, Lundquist S, Forbes LV, Magon NJ, Paton LN, Jameson GNL, Eriksson H, Kettle AJ (2011) 2-thioxanthines are mechanism-based inactivators of myeloperoxidase that block oxidative stress during inflammation. J Biol Chem 286:37578–37589. https://doi.org/10.1074/jbc.M111.266981

Üllen A, Singewald E, Konya V, Fauler G, Reicher H, Nusshold C, Hammer A, Kratky D, Heinemann A, Holzer P, Malle E, Sattler W (2013) Myeloperoxidase-derived oxidants induce blood-brain barrier dysfunction in vitro and in vivo. PLoS One 8:e64034. https://doi.org/10.1371/journal.pone.0064034

Van Antwerpen P, Slomianny MC, Boudjeltia KZ, Delporte C, Faid V, Calay D, Rousseau A, Moguilevsky N, Raes M, Vanhamme L, Furtmüller PG, Obinger C, Vanhaeverbeek M, Nève J, Michalski JC (2010) Glycosylation pattern of mature dimeric leukocyte and recombinant monomeric myeloperoxidase: glycosylation is required for optimal enzymatic activity. J Biol Chem 285:16351–16359. https://doi.org/10.1074/jbc.M109.089748

Winterbourn CC, Kettle AJ (2013) Redox reactions and microbial killing in the neutrophil phagosome. Antioxid Redox Signal 18:642–660. https://doi.org/10.1089/ars.2012.4827

Winterbourn CC, Kettle AJ, Hampton MB (2016) Reactive oxygen species and neutrophil function. Annu Rev Biochem 85:765–792. https://doi.org/10.1146/annurev-biochem-060815-014442

Ximenes VF, Paino IMM, de Faria-Oliveira OMM, da Fonseca LM, Brunetti IL (2005) Indole ring oxidation by activated leukocytes prevents the production of hypochlorous acid. Braz J Med Biol Res 38:1575–1583. https://doi.org/10.1590/S0100-879X2005001100003

Yu G, Liang Y, Huang Z, Jones DW, Pritchard KA, Zhang H (2016) Inhibition of myeloperoxidase oxidant production by N-acetyl lysyltyrosylcysteine amide reduces brain damage in a murine model of stroke. J Neuroinflammation 13:119. https://doi.org/10.1186/s12974-016-0583-x

Zámocký M, Jakopitsch C, Furtmüller PG, Dunand C, Obinger C (2008) The peroxidase-cyclooxygenase superfamily: reconstructed evolution of critical enzymes of the innate immune system. Proteins 71:589–605. https://doi.org/10.1002/prot.21950

Zámocký M, Hofbauer S, Schaffner I, Gasselhuber B, Nicolussi A, Soudi M, Pirker KF, Furtmüller PG, Obinger C (2015) Independent evolution of four heme peroxidase superfamilies. Arch Biochem Biophys 574:108–119. https://doi.org/10.1016/j.abb.2014.12.025

Zederbauer M, Furtmüller PG, Ganster B, Moguilevsky N, Obinger C (2007a) The vinyl-sulfonium bond in human myeloperoxidase: impact on compound I formation and reduction by halides and thiocyanate. Biochem Biophys Res Commun 356(2):450–456. https://doi.org/10.1016/j.bbrc.2007.02.157

Zederbauer M, Furtmüller PG, Brogioni S, Jakopitsch C, Smulevich G, Obinger C (2007b) Heme to protein linkages in mammalian peroxidases: impact on spectroscopic, redox and catalytic properties. Nat Prod Rep 24:571–584. https://doi.org/10.1039/b604178g

Zhang H, Jing X, Shi Y, Xu H, Du J, Guan T, Weihrauch D, Jones DW, Wang W, Gourlay D, Oldham KT, Hillery CA, Pritchard KA (2013) N-acetyl lysyltyrosylcysteine amide inhibits myeloperoxidase, a novel tripeptide inhibitor. J Lipid Res 54:3016–3029. https://doi.org/10.1194/jlr.M038273

Zhang H, Ray A, Miller NM, Hartwig D, Pritchard KA, Dittel BN (2016) Inhibition of myeloperoxidase at the peak of experimental autoimmune encephalomyelitis restores blood-brain barrier integrity and ameliorates disease severity. J Neurochem 136:826–836. https://doi.org/10.1111/jnc.13426

Zhao D, Tang W, Hao Z, Tao J (2015) Identification of flavonoids and expression of flavonoid biosynthetic genes in two coloured tree peony flowers. Biochem Biophys Res Commun 459:450–456. https://doi.org/10.1016/j.bbrc.2015.02.126

Part IV

Stimulating/Substituting ROS

Effects of Mammalian Thioredoxin Reductase Inhibitors

Elias S. J. Arnér

Contents

Abstract

The mammalian thioredoxin system is driven by NADPH through the activities of isoforms of the selenoprotein thioredoxin reductase (TXNRD, TrxR), which in turn help to keep thioredoxins (TXN, Trx) and further downstream targets reduced. Due to a wide range of functions in antioxidant defense, cell proliferation, and redox signaling, strong cellular aberrations are seen upon the targeting of TrxR enzymes by inhibitors. However, such inhibition can nonetheless have rather unexpected consequences. Accumulating data suggest that inhibition of TrxR in normal cells typically yields a paradoxical effect of increased antioxidant defense, with metabolic pathway reprogramming, increased cellular proliferation,

E. S. J. Arnér (✉)
Division of Biochemistry, Department of Medical Biochemistry and Biophysics, Karolinska Institutet, Stockholm, Sweden

Department of Selenoprotein Research, National Institute of Oncology, Budapest, Hungary
e-mail: Elias.Arner@ki.se

H. H. H. W. Schmidt et al. (eds.), *Reactive Oxygen Species*,
Handbook of Experimental Pharmacology 264, https://doi.org/10.1007/164_2020_393

and altered cellular differentiation patterns. Conversely, inhibition of TrxR in cancer cells can yield excessive levels of reactive oxygen species (ROS) resulting in cell death and thus anticancer efficacy. The observed increases in antioxidant capacity upon inhibition of TrxR in normal cells are in part dependent upon activation of the Nrf2 transcription factor, while exaggerated ROS levels in cancer cells can be explained by a non-oncogene addiction of cancer cells to TrxR1 due to their increased endogenous production of ROS. These separate consequences of TrxR inhibition can be utilized therapeutically. Importantly, however, a thorough knowledge of the molecular mechanisms underlying effects triggered by TrxR inhibition is crucial for the understanding of therapy outcomes after use of such inhibitors.

Graphical Abstract

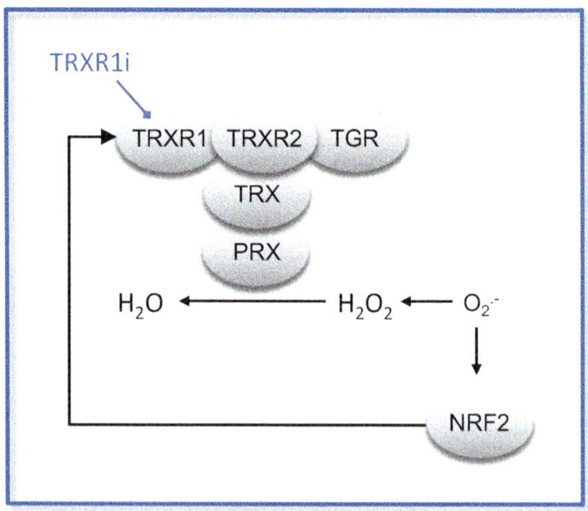

The mammalian thioredoxin system is driven by thioredoxin reductases (TXNRD, TrxR), which keeps thioredoxins (TXN, Trx) and further downstream targets reduced. In normal cells, inhibition of TrxR yields a paradoxical effect of increased antioxidant defense upon activation of the Nrf2 transcription factor. In cancer cells, however, inhibition of TrxR yields excessive reactive oxygen species (ROS) levels resulting in cell death and thus anticancer efficacy, which can be explained by a non-oncogene addiction of cancer cells to TrxR1 due to their increased endogenous production of ROS. These separate consequences of TrxR inhibition can be utilized therapeutically.

Keywords

Reactive oxygen species · Redox signaling · Selenoprotein · Thioredoxin reductase

1 Thioredoxin Reductases and Redox Biology

Redox biology is fundamental to all aspects of life, and altered redox processes are related to several diseases, including aspects of excessive levels of ROS, hypoxia, ischemia-reperfusion injury, and disturbed compartmentalized formation of reactive oxygen species (Forbes et al. 2008; Ryter et al. 2007; Ye et al. 2015). Enzymatically regulated formation of reactive oxygen species, especially H_2O_2, is also essential in several physiologically normal intracellular signaling pathways (Finkel 2000, 2011; Holmstrom and Finkel 2014; Rhee 2006). Different therapies that target the enzymatic systems of redox biology may thereby affect normal physiological events as well as pathways distorted in disease. Most if not all therapies that perturb redox states of cells will be likely to involve, or at least affect, the thioredoxin (Trx) and glutathione (GSH) systems, which are the main mammalian enzyme systems for control of reductive pathways in cells (Arnér 2009; Nordberg and Arnér 2001; Rundlöf and Arnér 2004). Direct drug targeting with inhibition of both of these two redox pathways can have therapeutic effects in cancer treatment (Harris et al. 2015), but simultaneous targeting of both pathways can also result in major unwanted toxicity and severe side effects. Several lines of observations suggest that targeting of Trx reductases (TrxRs) alone may however yield therapeutic efficacy in disease with less severe toxicity to normal cells. This will be discussed here, but first the selenoprotein nature of TrxRs shall be introduced.

Selenium (Se) is an essential trace element for mammals, due to its role as the defining constituent of the 21st amino acid, selenocysteine (Sec), found in selenoproteins (Johansson et al. 2005). The human genome has 25 selenoprotein-encoding genes, mostly encoding enzymes with a single catalytic Sec residue in their active sites (Kryukov et al. 2003). The chemical features of Sec make this amino acid an ideal catalyst for redox reactions, with Sec being much more chemically reactive than its more common sulfur-containing Cys analog (Arnér 2010) and also more resistant to overoxidation (Reich and Hondal 2016). Sec can in many cases be regarded as a "super cysteine," which helps to explain the higher activities of selenoenzymes that are typically seen when compared to their corresponding Sec-to-Cys mutants (Johansson et al. 2005; Reich and Hondal 2016). Some of the mammalian selenoproteins are essential, as illustrated by the early embryonic lethality in mouse knockout models for cytosolic thioredoxin reductase (TrxR1, encoded by *Txnrd1*) (Bondareva et al. 2007), mitochondrial thioredoxin reductase (TrxR2, *Txnrd2*) (Conrad et al. 2004), and glutathione peroxidase 4 (GPx4, *Gpx4*) (Yant et al. 2003). It was also shown that GPx4 protects cells against ferroptosis in a strictly Sec-dependent manner, which may be one of the major functions explaining a need for Sec in this enzyme and for selenoprotein expression overall, at least in certain cell types (Ingold et al. 2017).

Interestingly, cellular TrxR1 status also effectively controls cellular phenotype and differentiation patterns, with its genetic deletion reprogramming metabolism in hepatocytes of mouse liver (Iverson et al. 2013), activating Nrf2 (Cebula et al. 2015) and promoting fibroblasts in culture to undergo adipogenesis (Peng et al. 2016). Such effects of TrxR1 on cellular differentiation relate, among other mechanisms, to modulation of PTP1B signaling linked to tyrosine receptor stimulation (Dagnell et al. 2013b, 2017) and to the direct modulation of redox-sensitive transcription factors such as Nrf2, HIF, and NFκB (Johansson et al. 2017; Kipp et al. 2017). It is clear that TrxR1 can modulate cellular signaling pathways on many different, yet interlinked, levels in cells (Dagnell et al. 2018). A better understanding of those pathways will be important in order to understand and predict the possible outcomes of drug-mediated TrxR inhibition.

1.1 TrxR Genes and Proteins

The mammalian Trx system is an important reductive enzyme system in cells that acts together or in parallel with the glutathione (GSH) system (Arner and Holmgren 2000; Becker et al. 2000; Gromer et al. 2004; Nordberg and Arnér 2001). The Trx system encompasses Trx1 (encoded in human by *TXN*) and several additional Trx-fold enzymes, being kept reduced and thus redox active by the actions of thioredoxin reductases (TrxRs) using NADPH. The Trx-fold proteins can subsequently act to support reductive pathways or modulate redox regulatory systems in a multitude of cellular functions. The human genome encodes three specific TrxR isoenzymes, namely, cytosolic TrxR1 (encoded by *TXNRD1*), mitochondrial TrxR2 (encoded by *TXNRD2*), and testis-specific TGR (encoded by *TXNRD3*), with all three enzymes being selenoproteins (Arner and Holmgren 2000; Arnér 2009; Gromer et al. 2004; Martin 1995; Miranda-Vizuete et al. 2004; Nordberg and Arnér 2001). The differences between these isoforms are discussed further in Sect. 1.2.

Most studies with regard to effects of inhibitors have been performed on TrxR1 or TrxR2, while TGR has been much less studied. However, several pathogenic parasites rely on TGR orthologs, which may be inhibited through drug therapy as a novel form of antiparasitic therapy. This includes targeting of the TGR enzyme in *Schistosoma mansoni* (Kuntz et al. 2007; Lea et al. 2008; Rai et al. 2009; Silvestri et al. 2018; Simeonov et al. 2008), *Schistosoma japonicum* (Huang et al. 2015; Song et al. 2012), *Fasciola gigantica*, *Fasciola hepatica*, and other helminth parasites (Maggioli et al. 2011; Shukla et al. 2018; Williams et al. 2013), the tapeworm *Mesocestoides vogae* (Pasquet et al. 2015), *Taenia crassiceps cysticerci* (Martinez-Gonzalez et al. 2015), *Echinococcus granulosus* (Saiz et al. 2014), and additional cestode and trematode flatworms (Otero et al. 2010; Ross et al. 2012). The rest of this chapter shall however discuss drug targeting of the human forms of TrxR.

1.2 Isoforms and Expression Patterns of Human TrxRs

The human *TXNRD1* gene encodes predominantly cytosolic TrxR1, which is ubiquitously expressed and has Trx1 as its major substrate (Arner and Holmgren 2000; Rundlof and Arner 2004; Rundlof et al. 2004; Sun et al. 2001b). Mitochondrial TrxR2 encoded by *TXNRD2* reduces mitochondrial Trx2 as its main substrate (Lee et al. 1999; Miranda-Vizuete et al. 1999; Rigobello et al. 1998). The *TXNRD3* gene, finally, encodes TGR (thioredoxin glutathione reductase) that has a glutaredoxin (Grx) domain at the N-terminal part of the protein, in addition to its major TrxR module that otherwise is similar in domain structure to that found in TrxR1 and TrxR2. TGR is involved in maturation of sperm cells and mainly expressed in early spermatids (Su et al. 2005; Sun et al. 2001a, 2005).

The *TXNRD1* gene on chromosome 12 (12q23-q24.1) has a complex organization, with numerous transcripts displaying extensive splicing at their 5′-ends, thus producing several different protein isoforms of TrxR1 (Osborne and Tonissen 2001; Rundlof et al. 2000, 2004; Su and Gladyshev 2004; Sun et al. 2001b). One isoform, TXNRD1_v3 (“v3”), has three additional exons encoding a Grx domain, which is expressed in N-terminal fusion to the classical TrxR1 module. This is similar to TGR but v3 has a dithiol active site in contrast to the monothiol site found in TGR (Dammeyer et al. 2008; Rundlof et al. 2004, 2007; Su and Gladyshev 2004). Humans, chimpanzees, and dogs express v3, but mice or rats do not (Su and Gladyshev 2004). The v3 enzyme can be myristoylated and palmitoylated, being targeted to cell membranes where it seems to associate with lipid rafts and trigger formation of filopodia (Cebula et al. 2013; Damdimopoulou et al. 2009; Dammeyer et al. 2008). It is not clear if v3 is also targeted by drugs inhibiting TrxR, but this possibility should not be disregarded. Other major splice variants of TrxR1 are TXNRD1_v1 that is the “classical” form of the enzyme and TXNRD1_v2 (also called TrxR1b) that can be channeled to the nucleus and there interact with transcription factors including the estrogen receptor (Arnér 2009; Damdimopoulos et al. 2004).

The human *TXNRD2* gene is found on chromosome 22 (22q11.21) and mouse *Txnrd2* on chromosome 16. Similarly to *TXNRD1* there is evidence for extensive alternative splicing at the 5′-end of the corresponding transcripts, encoding protein variants with different N-terminal domains (Sun et al. 2001c). Thus, also in the case of TrxR2 there is a chance that drug inhibition of the enzyme also targets several isoforms within the same cells, or in different organs. It should here be noted that not all TrxR isoenzymes are expected to be targeted with the same efficiency upon use of inhibitors, with the final effects both depending upon different affinities for the specific enzymes and upon possible compartmentalization effects. In a side-by-side comparison, it was indeed shown that TrxR1 and TrxR2 differ in their sensitivities to different inhibitors (Rackham et al. 2011) and certain compounds, such as auranofin or isothiocyanates, were shown to target mainly mitochondrial TrxR2 before they inhibit TrxR1 within the cellular context (Brown et al. 2008; Cox et al. 2008).

The human *TXNRD3* gene encoding TGR is located at chromosome 3 (3q21.3), while mouse *Txnrd3* is at chromosome 6. These are yet the least characterized TrxR-encoding genes and also the least characterized TrxR isoenzymes. It should nonetheless be noted that TGR has the same Sec-containing active site motif as the other TrxRs, and it is thus both possible and plausible that also TGR may be targeted upon the use of drugs inhibiting TrxR isoenzymes.

1.3 Catalytic Mechanisms and Propensity for Drug Inhibition of TrxR

All human TrxRs share the same C-terminal -Gly-Cys-Sec-Gly-COOH motif being the proper active site reducing Trx (Arscott et al. 1997; Gladyshev et al. 1996; Lee et al. 2000; Tamura and Stadtman 1996; Zhong et al. 1998, 2000; Zhong and Holmgren 2000). Several crystal structures of Sec-to-Cys substituted mutant enzymes revealed the general domain structure and catalytic mechanism of mammalian TrxRs (Biterova et al. 2005; Eckenroth et al. 2006, 2007a, b; Fritz-Wolf et al. 2007; Sandalova et al. 2001), with a crystal structure of Sec-containing TrxR1 subsequently confirming the proposed formation of a selenenylsulfide at the C-terminus of the oxidized protein (Cheng et al. 2009). Importantly, in the NADPH-reduced enzyme, the selenenylsulfide becomes reduced to a selenolthiol motif, with its highly reactive and nucleophilic Sec residue being fully exposed to solvent and thus serving as a prime target for inhibition by electrophilic compounds (Cheng et al. 2009).

The first part of the reductive half-reaction of TrxR1 utilizes NADPH to reduce an enzyme-bound FAD in one subunit of the dimeric enzyme. The reduced FAD subsequently reduces a disulfide in a -CVNVGC- active site motif present in the same subunit, thus producing a dithiol. This part of the catalytic cycle is similar to that seen in glutathione reductase and other enzymes of the pyridine nucleotide disulfide oxidoreductase family (Williams 1992). However, instead of next reducing a substrate in solution, as with GSSG reduction by glutathione reductase, the C-terminal selenenylsulfide motif in the opposing subunit of TrxR1 is reduced, which may finally reduce substrates of TrxR1 including Trx1 (Cheng et al. 2009). Mammalian TrxRs also reduce several other substrates in addition to Trxs. Two additional and potentially important direct protein substrates of mammalian TrxR1 are glutaredoxin 2 (Johansson et al. 2004) and TRP14 (also called TXNDC17) having several redox signaling roles in cells (Espinosa and Arner 2019; Jeong et al. 2004; Pader et al. 2014; Woo et al. 2004). Another protein substrate of TrxR of potential importance is protein disulfide isomerase (PDI) that, like other ER proteins including CaBP1 and CaBP2 (Erp57), carries Trx domains with active sites that can be reduced by TrxR (Lundström-Ljung et al. 1995). It is interesting that cytosolic TrxR1 somehow reduces ER-resident proteins, which indeed can explain phenomena such as the reductive activation of immunotoxins through PDI being reduced by TrxR (Bellisola et al. 2004) or reduction of the disulfides in misfolded ER proteins being dependent upon TrxR1 (Poet et al. 2017). TrxR1 was also shown to directly reduce the active site of another protein of the Trx family, Trx-like-1 (TXL-1, TXNL-1 or TRP32) (Jimenez et al. 2006), that is a cytosolic protein (Lee et al. 1998) involved in glucose metabolism (Jimenez et al. 2006) and endocytosis (Felberbaum-Corti et al. 2007). Additional protein disulfide substrates for TrxR include Trx isoforms in male germ cells (Jimenez et al. 2002, 2004; Miranda-Vizuete et al. 2004) and the antibacterial peptide NK-lysin (Andersson et al. 1996). Furthermore, TrxR activities are important in controlling the persulfidation states of proteins, including key signaling proteins (Doka et al. 2016, 2020). Again, all of these enzymatic functions may be considered to be inhibited or affected upon the use of TrxR inhibitors.

TrxRs also have non-protein substrates that can play functional roles in a cellular context. This includes reduction of dehydroascorbate (May et al. 1997), lipoic acid (Arnér et al. 1996), cytochrome *c* (Nalvarte et al. 2004), toxoflavin (Gencheva et al. 2018), ubiquinone (Xia et al. 2003), and several other quinone compounds (Cenas et al. 2004). It is not clear if TrxR-mediated reduction of such substrates has a physiological importance, but also these activities will naturally be affected upon TrxR inhibition.

2 Inhibitors of Thioredoxin Reductases

TrxR1 is inhibited by a wide range of different compounds. The relative ease of inhibiting TrxR1 is mainly explained by its exceptionally reactive Sec residue that easily becomes covalently derivatized by many electrophilic inhibitors (Becker et al. 2000; Carvalho et al. 2008; Cebula et al. 2015; Krishnamurthy et al. 2008; Liu et al. 2008a; Prast-Nielsen et al. 2011; Witte et al. 2005). However, TrxR1 is a complex enzyme, and it should not be disregarded that inhibition of the enzyme can be achieved by reversible or irreversible interactions also of other motifs in TrxR1 than its Sec residue. For comprehensive discussions of different classes of TrxR1 inhibitors, see prior reviews on the topic (Arnér 2009; Cai et al. 2012; Cebula et al. 2015; Eriksson et al. 2009; Gromer et al. 2004; Liu et al. 2008a; Rackham et al. 2011; Urig and Becker 2006; Wipf et al. 2004; Zhang et al. 2016, 2018, 2019). Here the different classes of TrxR inhibitors shall not be repeated. Instead, we shall discuss the different cellular consequences of TrxR inhibition and their therapeutic potential.

3 Consequences of Thioredoxin Reductase Inhibition

A large number of compounds that inhibit TrxR1 have anticancer effects, and, moreover, several clinically used anticancer agents are known to inhibit TrxR1 (Arnér 2009; Arnér and Holmgren 2006; Cai et al. 2012; Casini et al. 2008; Chew et al. 2008; Eriksson et al. 2009; Fang et al. 2005; Gromer et al. 2004; Hashemy et al. 2006; Hedstrom et al. 2009; Lincoln et al. 2003; Liu et al. 2008a, b; Lu et al. 2007, 2006; Marzano et al. 2007; Peng et al. 2013; Prast-Nielsen et al. 2010; Prast-Nielsen et al. 2011; Shi et al. 2014; Urig and Becker 2006; Wang et al. 2008; Wipf et al. 2004; Witte et al. 2005). It is not clear, however, whether an efficient anticancer therapy can be developed solely upon TrxR1 inhibition and/or if any specific consequences of TrxR1 targeting can form the basis for a successful anticancer therapy. Some inhibitors of TrxR1 however show clear antitumoral efficacy in mouse models (Stafford et al. 2018; Ye et al. 2017). It is furthermore possible, perhaps even plausible, that TrxR inhibition in normal non-cancerous cells may have therapeutic potentials for use in other diseases than cancer. This will be discussed next.

3.1 Paradoxically Increased Antioxidant Defense

The nuclear factor erythroid-2-related factor 2 (Nrf2) transcription factor activates transcription of several key enzymes supporting cellular antioxidant systems (Copple et al. 2008; Osburn and Kensler 2008; Tong et al. 2006; Zhang 2006). It has been suggested that antitumoral immune system functions require Nrf2 activation (Ghosh et al. 2015; Mougiakakos et al. 2012; Zhang 2006; Zhao et al. 2014) and, interestingly, many inhibitors of TrxR1 also activate Nrf2, indeed suggesting a direct functional link between TrxR1 and Nrf2 (Cebula et al. 2015). A question is whether Nrf2 activation in normal cells can be achieved by drug-mediated TrxR1 inhibition and whether this may have any therapeutic value. It would be possible that such therapy can be used to protect normal cells from damage by excessive ROS levels and indeed also perhaps strengthen the antitumoral immunity. Interestingly, it was, perhaps at first seemingly paradoxically so (Lei et al. 2016), found that TrxR1 inhibition in normal cells becomes highly protective against subsequent oxidative challenges, as a result of a strong Nrf2 activation (Iverson et al. 2013; Locy et al. 2012; Rollins et al. 2010). Such protective effects may help explain how the TrxR1-inhibiting compounds curcumin (Fang et al. 2005; Liu et al. 2008a) or isothiocyanates (Bacon et al. 2007; Brown et al. 2008; Hu et al. 2007; Jakubikova et al. 2006) have chemopreventive effects, provided that Nrf2 has the anticancer preventive capacity that has been proposed (Brigelius-Flohe 2008; Brigelius-Flohe and Banning 2006; Chew et al. 2010; Higgins and Hayes 2011; Hu et al. 2007; Lee et al. 2007; Lu et al. 2006; Poerschke et al. 2012; Surh et al. 2008; Zhang 2006).

3.2 Affected Cell Differentiation Patterns

Mouse embryos lacking TrxR1 die prior to gastrulation and they display a lack of mesoderm formation (Bondareva et al. 2007). Conditionally knocked-out TrxR1 in hepatocytes of mouse liver triggers hyperproliferation, lack of signs of excessive ROS levels, metabolic aberrations, and very strong Nrf2 activation (Prigge et al. 2012a; Rollins et al. 2010; Iverson et al. 2013; Prigge et al. 2017; Suvorova et al. 2009), similar effects as those seen upon drug-mediated inhibition of TrxR1 (Locy et al. 2012), again suggesting that TrxR1 can be linked to control of Nrf2, with activation of Nrf2 upon inhibition or loss of TrxR1 (Cebula et al. 2015; Schmidt 2015). *Txnrd1*-deficient mouse embryonic fibroblasts also display striking features in culture, with an increased cell differentiation, insulin responsiveness, and spontaneous adipogenesis (Peng et al. 2016). Notably, in such cells lacking TrxR1, its major substrate Trx1 is still reduced (Peng et al. 2016), likely through the action of GSH-dependent glutaredoxins (Du et al. 2013). This suggests that any effects of TrxR1 inhibition on cellular phenotypes must not necessarily be due to impaired Trx1 activities. TrxR1-lacking cells nonetheless show increased responses to PDGF in conjunction with exaggerated oxidative inhibition of PTP1B (Dagnell et al. 2013a), again illustrating strong effects of TrxR1 status on cellular signaling pathways. Similar effects may hence be triggered also upon use of TrxR1 inhibitors. In other words, it is possible that inhibition of TrxR1 increases the overall antioxidant capacity of normal cells due to Nrf2 activation, and it may also be possible that a number of immature cell types can become triggered to a propensity for increased differentiation.

3.3 Effects on the Immune System

Antitumoral efficacy of the immune system is an important feature for final eradication of cancer in any form of cancer therapy (Ruffell and Coussens 2015; Shahabi et al. 2015; Vinay et al. 2015). Important in this context is that the Trx system can modulate the effectiveness of the immune system against cancer at least by two different mechanisms. First, activation of Nrf2 seems to be important for antitumoral activities of the immune system (Ghosh et al. 2015; Manda et al. 2015; Mougiakakos et al. 2012; Ruffell and Coussens 2015; Vinay et al. 2015), which may hence be another potentially beneficial consequence of TrxR1 inhibition in cancer therapy. Second, if TrxR1 becomes inhibited in cancer cells, this might increase the secretion from these cells of Trx1 as well as its C-terminally truncated protein Trx80; both of those proteins when present in serum act as co-cytokines and chemokines that may be proposed to attract antitumoral immune cells toward the tumor (Arnér and Holmgren 2000; Arnér and Holmgren 2006; Backman et al. 2007; Hori et al. 1993; Pekkari et al. 2005; Pekkari and Holmgren 2004). It should also be noted that auranofin, a classically used antirheumatic drug, is a very potent inhibitor of TrxR (Cox et al. 2008; Gromer et al. 2002; Marzano et al. 2007; Omata et al. 2006; Rigobello et al. 2005), and although it is not clear if or how TrxR inhibition is part of the antirheumatic efficacy of this gold compound, auranofin is now also being repurposed for use in therapy of cancer and other diseases where TrxR1 inhibition may be viewed as beneficial (Roder and Thomson 2015). It is also of significant interest that TrxR1 targeting yields prevention of STAT3 activation as a secondary downstream effect, which may also contribute to the anticancer efficacy of TrxR1 inhibitors (Busker et al. 2020).

3.4 Anticancer Therapy

It is rather well established that cancer cells have increased endogenous ROS levels (Luo et al. 2009). The activities of their antioxidant systems are thereby also increased, which in turn makes tumor cells more vulnerable to treatments that further enhance their ROS levels (Gorrini et al. 2013; Wondrak 2009). Indeed, Nrf2 is typically highly activated in cancer cells as a means to support their own survival (Brigelius-Flohe and Flohe 2011; Ganan-Gomez et al. 2013; Mitsuishi et al. 2012; Osburn and Kensler 2008; Singh et al. 2008). The expression levels of TrxR1 in turn modulate the cytotoxic profiles of redox active anticancer drugs in cancer cells (Eriksson et al. 2009). It is thus not far-fetched to believe that TrxR1 targeting may be a plausible mechanism of action for anticancer drugs, and the notion that cancer cells have an inherently increased level of ROS that can be targeted for therapy is indeed gaining wide recognition (Galluzzi et al. 2013; Harris et al. 2015; Luo et al. 2009; Manda et al. 2015; Shi et al. 2014; Trachootham et al. 2006, 2009). This property of cancer cells also explains why they typically exhibit high endogenous Nrf2 activities, with increased levels of enzymes in the GSH and Trx systems, as a means of surviving (Brigelius-Flohe and Flohe 2011; Higgins and Hayes 2011; Mitsuishi et al. 2012; Singh et al. 2008; Zhang 2006). It should thus be a natural consequence that inhibition of TrxR in cancer cells should help triggering their cell death, while normal cells should typically survive the loss of TrxR activity (Arnér 2009; Arnér and Holmgren 2006; Chew et al. 2010; Harris et al. 2015; Shi et al. 2014; Trachootham et al. 2009). This may hence be a major principle by which TrxR inhibition can yield anticancer efficacy and reduction of tumor mass. This notion is further corroborated by findings showing that the lack of TrxR1 in cancer cells impairs their capacity to form tumors (Hatfield et al. 2009; Mandal et al. 2010; Yoo et al. 2006, 2007).

An additional effect of drug targeting of TrxR1 in cancer cells, which may contribute to tumor cell death, is the conversion of the enzyme to toxic pro-oxidant redox cycling forms of the protein, named SecTRAPs (*Se*lenium *c*ompromised *t*hioredoxin *r*eductase-derived *a*poptotic *p*roteins) that can further increase ROS levels and thus also help killing cancer cells (Anestål and Arnér 2003; Anestål et al. 2008; Cai et al. 2012; Cebula et al. 2015; Hashemy et al. 2006). These mechanisms of action are also compatible with the activities of novel TrxR1 inhibitors showing anticancer efficacy (Stafford et al. 2018).

As explained above, since compounds that target TrxR1 in cancer cells will likely also induce robust Nrf2 responses in normal cells that, paradoxically, *protect* normal cells from oxidative damage (Iverson et al. 2013; Lei et al. 2016; Locy et al. 2012; Prigge et al. 2012b), this opens the possibility that specific targeting of TrxR1 can have dual effects in anticancer therapy, namely, protection of normal cells with a boost of the immune system by Nrf2 activation on one hand and lethality to cancer cells due to excessive ROS levels on the other.

4 Conclusions

As discussed herein, the outcome of TrxR inhibition will depend upon the cellular context in which the enzyme is inhibited, as well as upon the nature of the isoenzyme (s) or isoform(s) of TrxRs that are being targeted. Notwithstanding the complexity of the profile of TrxR enzymes in cells, the picture emerges that inhibition of these enzymes in normal cells can trigger Nrf2 activation that protects these cells from excessive ROS levels, which in turn boosts the functions of the immune system. In contrast, cancer cells seem to be excessively sensitive to TrxR1 inhibition, and drugs inhibiting the enzyme can thereby have direct anticancer properties. In combination, these consequences of TrxR inhibition suggest that inhibitors of these enzymes are amenable to therapy development for treatment of a number of different diseases, mainly cancer but also diseases where normal cells suffer from injuries due to increased levels of ROS.

Acknowledgments The author wishes to thank his many talented group members and collaborators. Funding from Karolinska Institutet, the Swedish Cancer Society, the Swedish Research Council, the Hungarian National Institute of Oncology, the Hungarian Ministry of Human Capacities, and the Knut and Alice Wallenberg Foundations is also thankfully acknowledged.

References

Andersson M, Holmgren A, Spyrou G (1996) NK-lysin, a disulfide-containing effector peptide of T-lymphocytes, is reduced and inactivated by human thioredoxin reductase. Implication for a protective mechanism against NK-lysin cytotoxicity. J Biol Chem 271:10116–10120

Anestål K, Arnér ESJ (2003) Rapid induction of cell death by selenium-compromised thioredoxin reductase 1 but not by the fully active enzyme containing selenocysteine. J Biol Chem 278:15966–15972

Anestål K, Prast-Nielsen S, Cenas N, Arnér ESJ (2008) Cell death by SecTRAPs – thioredoxin reductase as a prooxidant killer of cells. PLoS One 3:e1846

Arnér ESJ (2009) Focus on mammalian thioredoxin reductases – important selenoproteins with versatile functions. Biochim Biophys Acta 1790:495–526

Arnér ESJ (2010) Selenoproteins-what unique properties can arise with selenocysteine in place of cysteine? Exp Cell Res 316:1296–1303

Arner ES, Holmgren A (2000) Physiological functions of thioredoxin and thioredoxin reductase. Eur J Biochem 267:6102–6109

Arnér ESJ, Holmgren A (2006) The thioredoxin system in cancer. Semin Cancer Biol 16:420–426

Arnér ESJ, Nordberg J, Holmgren A (1996) Efficient reduction of lipoamide and lipoic acid by mammalian thioredoxin reductase. Biochem Biophys Res Commun 225:268–274

Arscott LD, Gromer S, Schirmer RH, Becker K, Williams CH Jr (1997) The mechanism of thioredoxin reductase from human placenta is similar to the mechanisms of lipoamide dehydrogenase and glutathione reductase and is distinct from the mechanism of thioredoxin reductase from *Escherichia coli*. Proc Natl Acad Sci U S A 94:3621–3626

Backman E, Bergh AC, Lagerdahl I, Rydberg B, Sundstrom C, Tobin G, Rosenquist R, Linderholm M, Rosen A (2007) Thioredoxin, produced by stromal cells retrieved from the lymph node microenvironment, rescues chronic lymphocytic leukemia cells from apoptosis in vitro. Haematologica 92:1495–1504

Bacon JR, Plumb GW, Howie AF, Beckett GJ, Wang W, Bao Y (2007) Dual action of sulforaphane in the regulation of thioredoxin reductase and thioredoxin in human HepG2 and Caco-2 cells. J Agric Food Chem 55:1170–1176

Becker K, Gromer S, Schirmer RH, Müller S (2000) Thioredoxin reductase as a pathophysiological factor and drug target. Eur J Biochem 267:6118–6125

Bellisola G, Fracasso G, Ippoliti R, Menestrina G, Rosen A, Solda S, Udali S, Tomazzolli R, Tridente G, Colombatti M (2004) Reductive activation of ricin and ricin A-chain immunotoxins by protein disulfide isomerase and thioredoxin reductase. Biochem Pharmacol 67:1721–1731

Biterova EI, Turanov AA, Gladyshev VN, Barycki JJ (2005) Crystal structures of oxidized and reduced mitochondrial thioredoxin reductase provide molecular details of the reaction mechanism. Proc Natl Acad Sci U S A 102:15018–15023

Bondareva AA, Capecchi MR, Iverson SV, Li Y, Lopez NI, Lucas O, Merrill GF, Prigge JR, Siders AM, Wakamiya M, Wallin SL, Schmidt EE (2007) Effects of thioredoxin reductase-1 deletion on embryogenesis and transcriptome. Free Radic Biol Med 43:911–923

Brigelius-Flohe R (2008) Selenium compounds and selenoproteins in cancer. Chem Biodivers 5:389–395

Brigelius-Flohe R, Banning A (2006) Part of the series: from dietary antioxidants to regulators in cellular signaling and gene regulation. Sulforaphane and selenium, partners in adaptive response and prevention of cancer. Free Radic Res 40:775–787

Brigelius-Flohe R, Flohe L (2011) Basic principles and emerging concepts in the redox control of transcription factors. Antioxid Redox Signal 15:2335–2381

Brown KK, Eriksson SE, Arner ES, Hampton MB (2008) Mitochondrial peroxiredoxin 3 is rapidly oxidized in cells treated with isothiocyanates. Free Radic Biol Med 45:494–502

Busker S, Qian W, Haraldsson M, Espinosa B, Johansson L, Attarha S, Kolosenko I, Liu J, Dagnell M, Grander D, Arner ESJ, Tamm KP, Page BDG (2020) Irreversible TrxR1 inhibitors block STAT3 activity and induce cancer cell death. Sci Adv 6:eaax7945

Cai W, Zhang L, Song Y, Wang B, Zhang B, Cui X, Hu G, Liu Y, Wu J, Fang J (2012) Small molecule inhibitors of mammalian thioredoxin reductase. Free Radic Biol Med 52:257–265

Carvalho CM, Chew EH, Hashemy SI, Lu J, Holmgren A (2008) Inhibition of the human thioredoxin system. A molecular mechanism of mercury toxicity. J Biol Chem 283:11913–11923

Casini A, Gabbiani C, Sorrentino F, Rigobello MP, Bindoli A, Geldbach TJ, Marrone A, Re N, Hartinger CG, Dyson PJ, Messori L (2008) Emerging protein targets for anticancer metallodrugs: inhibition of thioredoxin reductase and cathepsin B by antitumor ruthenium(II)-arene compounds. J Med Chem 51:6773–6781

Cebula M, Moolla N, Capovilla A, Arner ES (2013) The rare TXNRD1_v3 ("v3") splice variant of human thioredoxin reductase 1 protein is targeted to membrane rafts by N-acylation and induces filopodia independently of its redox active site integrity. J Biol Chem 288:10002–10011

Cebula M, Schmidt EE, Arner ES (2015) TrxR1 as a potent regulator of the Nrf2-Keap1 response system. Antioxid Redox Signal 23:823–853

Cenas N, Nivinskas H, Anusevicius Z, Sarlauskas J, Lederer F, Arnér ESJ (2004) Interactions of quinones with thioredoxin reductase – a challenge to the antioxidant role of the mammalian selenoprotein. J Biol Chem 279:2583–2592

Cheng Q, Sandalova T, Lindqvist Y, Arnér ESJ (2009) Crystal structure and catalysis of the selenoprotein thioredoxin reductase 1. J Biol Chem 284:3998–4008

Chew EH, Lu J, Bradshaw TD, Holmgren A (2008) Thioredoxin reductase inhibition by antitumor quinols: a quinol pharmacophore effect correlating to antiproliferative activity. FASEB J 22:2072–2083

Chew EH, Nagle AA, Zhang Y, Scarmagnani S, Palaniappan P, Bradshaw TD, Holmgren A, Westwell AD (2010) Cinnamaldehydes inhibit thioredoxin reductase and induce Nrf2: potential candidates for cancer therapy and chemoprevention. Free Radic Biol Med 48:98–111

Conrad M, Jakupoglu C, Moreno SG, Lippl S, Banjac A, Schneider M, Beck H, Hatzopoulos AK, Just U, Sinowatz F, Schmahl W, Chien KR, Wurst W, Bornkamm GW, Brielmeier M (2004)

Essential role for mitochondrial thioredoxin reductase in hematopoiesis, heart development, and heart function. Mol Cell Biol 24:9414–9423

Copple IM, Goldring CE, Kitteringham NR, Park BK (2008) The Nrf2-Keap1 defence pathway: role in protection against drug-induced toxicity. Toxicology 246:24–33

Cox AG, Brown KK, Arner ES, Hampton MB (2008) The thioredoxin reductase inhibitor auranofin triggers apoptosis through a Bax/Bak-dependent process that involves peroxiredoxin 3 oxidation. Biochem Pharmacol 76:1097–1109

Dagnell M, Frijhoff J, Pader I, Augsten M, Boivin B, Xu J, Mandal PK, Tonks NK, Hellberg C, Conrad M, Arner ES, Ostman A (2013a) Selective activation of oxidized PTP1B by the thioredoxin system modulates PDGF-beta receptor tyrosine kinase signaling. Proc Natl Acad Sci U S A 110:13398–13403

Dagnell M, Frijhoff J, Pader I, Augsten M, Boivin B, Xu J, Mandal PK, Tonks NK, Hellberg C, Conrad M, Arnér ESJ, Östman A (2013b) Selective activation of oxidized PTP1B by the thioredoxin system modulates PDGFβ-receptor tyrosine kinase signaling. Proc Natl Acad Sci U S A 110:13398–13403

Dagnell M, Pace PE, Cheng Q, Frijhoff J, Ostman A, Arner ESJ, Hampton MB, Winterbourn CC (2017) Thioredoxin reductase 1 and NADPH directly protect protein tyrosine phosphatase 1B from inactivation during H2O2 exposure. J Biol Chem 292:14371–14380

Dagnell M, Schmidt EE, Arner ESJ (2018) The A to Z of modulated cell patterning by mammalian thioredoxin reductases. Free Radic Biol Med 115:484–496

Damdimopoulos AE, Miranda-Vizuete A, Treuter E, Gustafsson JÅ, Spyrou G (2004) An alternative splicing variant of the selenoprotein thioredoxin reductase is a modulator of estrogen signaling. J Biol Chem 279:38721–38729

Damdimopoulou PE, Miranda-Vizuete A, Arner ESJ, Gustafsson J-A, Damdimopoulos AE (2009) The human thioredoxin reductase-1 splice variant TXNRD1_v3 is an atypical inducer of cytoplasmic filaments and cell membrane filopodia. BBA-Mol Cell Res 1793:1588–1596

Dammeyer P, Damdimopoulos AE, Nordman T, Jimenez A, Miranda-Vizuete A, Arner ES (2008) Induction of cell membrane protrusions by the N-terminal glutaredoxin domain of a rare splice variant of human thioredoxin reductase 1. J Biol Chem 283:2814–2821

Doka E, Pader I, Biro A, Johansson K, Cheng Q, Ballago K, Prigge JR, Pastor-Flores D, Dick TP, Schmidt EE, Arner ES, Nagy P (2016) A novel persulfide detection method reveals protein persulfide- and polysulfide-reducing functions of thioredoxin and glutathione systems. Sci Adv 2:e1500968

Doka E, Ida T, Dagnell M, Abiko Y, Luong NC, Balog N, Takata T, Espinosa B, Nishimura A, Cheng Q, Funato Y, Miki H, Fukuto JM, Prigge JR, Schmidt EE, Arner ESJ, Kumagai Y, Akaike T, Nagy P (2020) Control of protein function through oxidation and reduction of persulfidated states. Sci Adv 6:eaax8358

Du Y, Zhang H, Zhang X, Lu J, Holmgren A (2013) Thioredoxin 1 is inactivated due to oxidation induced by peroxiredoxin under oxidative stress and reactivated by the glutaredoxin system. J Biol Chem 288:32241–32247

Eckenroth B, Harris K, Turanov AA, Gladyshev VN, Raines RT, Hondal RJ (2006) Semisynthesis and characterization of mammalian thioredoxin reductase. Biochemistry 45:5158–5170

Eckenroth BE, Lacey BM, Lothrop AP, Harris KM, Hondal RJ (2007a) Investigation of the C-terminal redox center of high-Mr thioredoxin reductase by protein engineering and semisynthesis. Biochemistry 46:9472–9483

Eckenroth BE, Rould MA, Hondal RJ, Everse SJ (2007b) Structural and biochemical studies reveal differences in the catalytic mechanisms of mammalian and *Drosophila melanogaster* thioredoxin reductases. Biochemistry 46:4694–4705

Eriksson SE, Prast-Nielsen S, Flaberg E, Szekely L, Arner ES (2009) High levels of thioredoxin reductase 1 modulate drug-specific cytotoxic efficacy. Free Radic Biol Med 47:1661–1671

Espinosa B, Arner ESJ (2019) Thioredoxin-related protein of 14 kDa as a modulator of redox signalling pathways. Br J Pharmacol 176:544–553

Fang J, Lu J, Holmgren A (2005) Thioredoxin reductase is irreversibly modified by curcumin: a novel molecular mechanism for its anticancer activity. J Biol Chem 280:25284–25290

Felberbaum-Corti M, Morel E, Cavalli V, Vilbois F, Gruenberg J (2007) The redox sensor TXNL1 plays a regulatory role in fluid phase endocytosis. PLoS One 2:e1144

Finkel T (2000) Redox-dependent signal transduction. FEBS Lett 476:52–54

Finkel T (2011) Signal transduction by reactive oxygen species. J Cell Biol 194:7–15

Forbes JM, Coughlan MT, Cooper ME (2008) Oxidative stress as a major culprit in kidney disease in diabetes. Diabetes 57:1446–1454

Fritz-Wolf K, Urig S, Becker K (2007) The structure of human thioredoxin reductase 1 provides insights into C-terminal rearrangements during catalysis. J Mol Biol 370:116–127

Galluzzi L, Kepp O, Vander Heiden MG, Kroemer G (2013) Metabolic targets for cancer therapy. Nat Rev Drug Discov 12:829–846

Ganan-Gomez I, Wei Y, Yang H, Boyano-Adanez MC, Garcia-Manero G (2013) Oncogenic functions of the transcription factor Nrf2. Free Radic Biol Med 65:750–764

Gencheva R, Cheng Q, Arner ESJ (2018) Efficient selenocysteine-dependent reduction of toxoflavin by mammalian thioredoxin reductase. Biochim Biophys Acta Gen Subj 1862:2511–2517

Ghosh S, Mukherjee S, Choudhury S, Gupta P, Adhikary A, Baral R, Chattopadhyay S (2015) Reactive oxygen species in the tumor niche triggers altered activation of macrophages and immunosuppression: role of fluoxetine. Cell Signal 27:1398–1412

Gladyshev VN, Jeang K-T, Stadtman TC (1996) Selenocysteine, identified as the penultimate C-terminal residue in human T-cell thioredoxin reductase, corresponds to TGA in the human placental gene. Proc Natl Acad Sci U S A 93:6146–6151

Gorrini C, Harris IS, Mak TW (2013) Modulation of oxidative stress as an anticancer strategy. Nat Rev Drug Discov 12:931–947

Gromer S, Merkle H, Schirmer RH, Becker K (2002) Human placenta thioredoxin reductase: preparation and inhibitor studies. Methods Enzymol 347:382–394

Gromer S, Urig S, Becker K (2004) The thioredoxin system – from science to clinic. Med Res Rev 24:40–89

Harris IS, Treloar AE, Inoue S, Sasaki M, Gorrini C, Lee KC, Yung KY, Brenner D, Knobbe-Thomsen CB, Cox MA, Elia A, Berger T, Cescon DW, Adeoye A, Brustle A, Molyneux SD, Mason JM, Li WY, Yamamoto K, Wakeham A, Berman HK, Khokha R, Done SJ, Kavanagh TJ, Lam CW, Mak TW (2015) Glutathione and thioredoxin antioxidant pathways synergize to drive cancer initiation and progression. Cancer Cell 27:211–222

Hashemy SI, Ungerstedt JS, Zahedi Avval F, Holmgren A (2006) Motexafin gadolinium, a tumor-selective drug targeting thioredoxin reductase and ribonucleotide reductase. J Biol Chem 281:10691–10697

Hatfield DL, Yoo MH, Carlson BA, Gladyshev VN (2009) Selenoproteins that function in cancer prevention and promotion. Biochim Biophys Acta 1790:1541–1545

Hedstrom E, Eriksson S, Zawacka-Pankau J, Arner ES, Selivanova G (2009) p53-dependent inhibition of TrxR1 contributes to the tumor-specific induction of apoptosis by RITA. Cell Cycle 8:3576–3583

Higgins LG, Hayes JD (2011) The cap'n'collar transcription factor Nrf2 mediates both intrinsic resistance to environmental stressors and an adaptive response elicited by chemopreventive agents that determines susceptibility to electrophilic xenobiotics. Chem Biol Interact 192:37–45

Holmstrom KM, Finkel T (2014) Cellular mechanisms and physiological consequences of redox-dependent signalling. Nat Rev Mol Cell Biol 15:411–421

Hori K, Hirashima M, Ueno M, Matsuda M, Waga S, Tsurufuji S, Yodoi J (1993) Regulation of eosinophil migration by adult T cell leukemia-derived factor. J Immunol 151:5624–5630

Hu Y, Urig S, Koncarevic S, Wu X, Fischer M, Rahlfs S, Mersch-Sundermann V, Becker K (2007) Glutathione- and thioredoxin-related enzymes are modulated by sulfur-containing chemopreventive agents. Biol Chem 388:1069–1081

Huang J, Hua W, Li J, Hua Z (2015) Molecular docking to explore the possible binding mode of potential inhibitors of thioredoxin glutathione reductase. Mol Med Rep 12:5787–5795

Ingold I, Berndt C, Schmitt S, Doll S, Poschmann G, Buday K, Roveri A, Peng X, Porto Freitas F, Seibt T, Mehr L, Aichler M, Walch A, Lamp D, Jastroch M, Miyamoto S, Wurst W, Ursini F, Arner ESJ, Fradejas-Villar N, Schweizer U, Zischka H, Friedmann Angeli JP, Conrad M (2017) Selenium utilization by GPX4 is required to prevent hydroperoxide-induced ferroptosis. Cell 172:409–422.

Iverson SV, Eriksson S, Xu J, Prigge JR, Talago EA, Meade TA, Meade ES, Capecchi MR, Arner ES, Schmidt EE (2013) A Txnrd1-dependent metabolic switch alters hepatic lipogenesis, glycogen storage, and detoxification. Free Radic Biol Med 63:369–380

Jakubikova J, Sedlak J, Bod'o J, Bao Y (2006) Effect of isothiocyanates on nuclear accumulation of NF-kappaB, Nrf2, and thioredoxin in caco-2 cells. J Agric Food Chem 54:1656–1662

Jeong W, Yoon HW, Lee SR, Rhee SG (2004) Identification and characterization of TRP14, a thioredoxin-related protein of 14 kDa. New insights into the specificity of thioredoxin function. J Biol Chem 279:3142–3150

Jimenez A, Oko R, Gustafsson JA, Spyrou G, Pelto-Huikko M, Miranda-Vizuete A (2002) Cloning, expression and characterization of mouse spermatid specific thioredoxin-1 gene and protein. Mol Hum Reprod 8:710–718

Jimenez A, Zu W, Rawe VY, Pelto-Huikko M, Flickinger CJ, Sutovsky P, Gustafsson JA, Oko R, Miranda-Vizuete A (2004) Spermatocyte/spermatid-specific thioredoxin-3, a novel Golgi apparatus-associated thioredoxin, is a specific marker of aberrant spermatogenesis. J Biol Chem 279:34971–34982

Jimenez A, Pelto-Huikko M, Gustafsson JA, Miranda-Vizuete A (2006) Characterization of human thioredoxin-like-1: potential involvement in the cellular response against glucose deprivation. FEBS Lett 580:960–967

Johansson C, Lillig CH, Holmgren A (2004) Human mitochondrial glutaredoxin reduces S-glutathionylated proteins with high affinity accepting electrons from either glutathione or thioredoxin reductase. J Biol Chem 279:7537–7543

Johansson L, Gafvelin G, Arnér ESJ (2005) Selenocysteine in proteins – properties and biotechnological use. Biochim Biophys Acta 1726:1–13

Johansson K, Cebula M, Rengby O, Dreij K, Carlstrom KE, Sigmundsson K, Piehl F, Arner ES (2017) Cross talk in HEK293 cells between Nrf2, HIF, and NF-kappaB activities upon challenges with redox therapeutics characterized with single-cell resolution. Antioxid Redox Signal 26:229–246

Kipp AP, Deubel S, Arner ESJ, Johansson K (2017) Time- and cell-resolved dynamics of redox-sensitive Nrf2, HIF and NF-kappaB activities in 3D spheroids enriched for cancer stem cells. Redox Biol 12:403–409

Krishnamurthy D, Karver MR, Fiorillo E, Orru V, Stanford SM, Bottini N, Barrios AM (2008) Gold (I)-mediated inhibition of protein tyrosine phosphatases: a detailed in vitro and cellular study. J Med Chem 51:4790–4795

Kryukov GV, Castellano S, Novoselov SV, Lobanov AV, Zehtab O, Guigo R, Gladyshev VN (2003) Characterization of mammalian selenoproteomes. Science 300:1439–1443

Kuntz AN, Davioud-Charvet E, Sayed AA, Califf LL, Dessolin J, Arner ES, Williams DL (2007) Thioredoxin glutathione reductase from *Schistosoma mansoni*: an essential parasite enzyme and a key drug target. PLoS Med 4:e206

Lea WA, Jadhav A, Rai G, Sayed AA, Cass CL, Inglese J, Williams DL, Austin CP, Simeonov A (2008) A 1,536-well-based kinetic HTS assay for inhibitors of *Schistosoma mansoni* thioredoxin glutathione reductase. Assay Drug Dev Technol 6:551–555

Lee KK, Murakawa M, Takahashi S, Tsubuki S, Kawashima S, Sakamaki K, Yonehara S (1998) Purification, molecular cloning, and characterization of TRP32, a novel thioredoxin-related mammalian protein of 32 kDa. J Biol Chem 273:19160–19166

Lee SR, Kim JR, Kwon KS, Yoon HW, Levine RL, Ginsburg A, Rhee SG (1999) Molecular cloning and characterization of a mitochondrial selenocysteine-containing thioredoxin reductase from rat liver. J Biol Chem 274:4722–4734

Lee SR, Bar-Noy S, Kwon J, Levine RL, Stadtman TC, Rhee SG (2000) Mammalian thioredoxin reductase: oxidation of the C-terminal cysteine/selenocysteine active site forms a thioselenide, and replacement of selenium with sulfur markedly reduces catalytic activity. Proc Natl Acad Sci U S A 97:2521–2526

Lee SB, Cha KH, Selenge D, Solongo A, Nho CW (2007) The chemopreventive effect of taxifolin is exerted through ARE-dependent gene regulation. Biol Pharm Bull 30:1074–1079

Lei XG, Zhu JH, Cheng WH, Bao Y, Ho YS, Reddi AR, Holmgren A, Arner ES (2016) Paradoxical roles of antioxidant enzymes: basic mechanisms and health implications. Physiol Rev 96:307–364

Lincoln DT, Ali Emadi EM, Tonissen KF, Clarke FM (2003) The thioredoxin-thioredoxin reductase system: over-expression in human cancer. Anticancer Res 23:2425–2433

Liu Z, Du ZY, Huang ZS, Lee KS, Gu LQ (2008a) Inhibition of thioredoxin reductase by curcumin analogs. Biosci Biotechnol Biochem 72:2214–2218

Liu Z, Huang S, Li M, Huang Z, Lee KS, Gu L (2008b) Inhibition of thioredoxin reductase by mansonone F analogues: implications for anticancer activity. Chem Biol Interact 177:48–57

Locy ML, Rogers LK, Prigge JR, Schmidt EE, Arner ES, Tipple TE (2012) Thioredoxin reductase inhibition elicits Nrf2-mediated responses in Clara cells: implications for oxidant-induced lung injury. Antioxid Redox Signal 17:1407–1416

Lu J, Papp LV, Fang J, Rodriguez-Nieto S, Zhivotovsky B, Holmgren A (2006) Inhibition of Mammalian thioredoxin reductase by some flavonoids: implications for myricetin and quercetin anticancer activity. Cancer Res 66:4410–4418

Lu J, Chew EH, Holmgren A (2007) Targeting thioredoxin reductase is a basis for cancer therapy by arsenic trioxide. Proc Natl Acad Sci U S A 104:12288–12293

Lundström-Ljung J, Birnbach U, Rupp K, Soling HD, Holmgren A (1995) Two resident ER-proteins, CaBP1 and CaBP2, with thioredoxin domains, are substrates for thioredoxin reductase: comparison with protein disulfide isomerase. FEBS Lett 357:305–308

Luo J, Solimini NL, Elledge SJ (2009) Principles of cancer therapy: oncogene and non-oncogene addiction. Cell 136:823–837

Maggioli G, Silveira F, Martin-Alonso JM, Salinas G, Carmona C, Parra F (2011) A recombinant thioredoxin-glutathione reductase from Fasciola hepatica induces a protective response in rabbits. Exp Parasitol 129:323–330

Manda G, Isvoranu G, Comanescu MV, Manea A, Debelec Butuner B, Korkmaz KS (2015) The redox biology network in cancer pathophysiology and therapeutics. Redox Biol 5:347–357

Mandal PK, Schneider M, Kolle P, Kuhlencordt P, Forster H, Beck H, Bornkamm GW, Conrad M (2010) Loss of thioredoxin reductase 1 renders tumors highly susceptible to pharmacologic glutathione deprivation. Cancer Res 70:9505–9514

Martin JL (1995) Thioredoxin-a fold for all reasons. Structure 3:245–250

Martinez-Gonzalez JJ, Guevara-Flores A, Rendon JL, Arenal IPD (2015) Auranofin-induced oxidative stress causes redistribution of the glutathione pool in Taenia crassiceps cysticerci. Mol Biochem Parasitol 201:16–25

Marzano C, Gandin V, Folda A, Scutari G, Bindoli A, Rigobello MP (2007) Inhibition of thioredoxin reductase by auranofin induces apoptosis in cisplatin-resistant human ovarian cancer cells. Free Radic Biol Med 42:872–881

May JM, Mendiratta S, Hill KE, Burk RF (1997) Reduction of dehydroascorbate to ascorbate by the selenoenzyme thioredoxin reductase. J Biol Chem 272:22607–22610

Miranda-Vizuete A, Damdimopoulos AE, Pedrajas JR, Gustafsson JA, Spyrou G (1999) Human mitochondrial thioredoxin reductase cDNA cloning, expression and genomic organization. Eur J Biochem 261:405–412

Miranda-Vizuete A, Sadek CM, Jimenez A, Krause WJ, Sutovsky P, Oko R (2004) The mammalian testis-specific thioredoxin system. Antioxid Redox Signal 6:25–40

Mitsuishi Y, Motohashi H, Yamamoto M (2012) The Keap1-Nrf2 system in cancers: stress response and anabolic metabolism. Front Oncol 2:200

Mougiakakos D, Okita R, Ando T, Durr C, Gadiot J, Ichikawa J, Zeiser R, Blank C, Johansson CC, Kiessling R (2012) High expression of GCLC is associated with malignant melanoma of low oxidative phenotype and predicts a better prognosis. J Mol Med (Berl) 90:935–944

Nalvarte I, Damdimopoulos AE, Spyrou G (2004) Human mitochondrial thioredoxin reductase reduces cytochrome c and confers resistance to complex III inhibition. Free Radic Biol Med 36:1270–1278

Nordberg J, Arnér ESJ (2001) Reactive oxygen species, antioxidants, and the mammalian thioredoxin system. Free Radic Biol Med 31:1287–1312

Omata Y, Folan M, Shaw M, Messer RL, Lockwood PE, Hobbs D, Bouillaguet S, Sano H, Lewis JB, Wataha JC (2006) Sublethal concentrations of diverse gold compounds inhibit mammalian cytosolic thioredoxin reductase (TrxR1). Toxicol In Vitro 20:882–890

Osborne SA, Tonissen KF (2001) Genomic organisation and alternative splicing of mouse and human thioredoxin reductase 1 genes. BMC Genomics 2:10

Osburn WO, Kensler TW (2008) Nrf2 signaling: an adaptive response pathway for protection against environmental toxic insults. Mutat Res 659:31–39

Otero L, Bonilla M, Protasio AV, Fernandez C, Gladyshev VN, Salinas G (2010) Thioredoxin and glutathione systems differ in parasitic and free-living platyhelminths. BMC Genomics 11:237

Pader I, Sengupta R, Cebula M, Xu J, Lundberg JO, Holmgren A, Johansson K, Arner ES (2014) Thioredoxin-related protein of 14 kDa is an efficient L-cystine reductase and S-denitrosylase. Proc Natl Acad Sci U S A 111:6964–6969

Pasquet V, Bisio H, Lopez GV, Romanelli-Cedrez L, Bonilla M, Saldana J, Salinas G (2015) Inhibition of tapeworm thioredoxin and glutathione pathways by an oxadiazole N-oxide leads to reduced mesocestoides vogae infection burden in mice. Molecules 20:11793–11807

Pekkari K, Holmgren A (2004) Truncated thioredoxin: physiological functions and mechanism. Antioxid Redox Signal 6:53–61

Pekkari K, Goodarzi MT, Scheynius A, Holmgren A, Avila-Carino J (2005) Truncated thioredoxin (Trx80) induces differentiation of human CD14+ monocytes into a novel cell type (TAMs) via activation of the MAP kinases p38, ERK, and JNK. Blood 105:1598–1605

Peng X, Zhang MQ, Conserva F, Hosny G, Selivanova G, Bykov VJ, Arner ES, Wiman KG (2013) APR-246/PRIMA-1MET inhibits thioredoxin reductase 1 and converts the enzyme to a dedicated NADPH oxidase. Cell Death Dis 4:e881

Peng X, Gimenez-Cassina A, Petrus P, Conrad M, Ryden M, Arner ES (2016) Thioredoxin reductase 1 suppresses adipocyte differentiation and insulin responsiveness. Sci Rep 6:28080

Poerschke RL, Franklin MR, Bild AH, Moos PJ (2012) Major differences among chemopreventive organoselenocompounds in the sustained elevation of cytoprotective genes. J Biochem Mol Toxicol 26:344–353

Poet GJ, Oka OB, van Lith M, Cao Z, Robinson PJ, Pringle MA, Arner ES, Bulleid NJ (2017) Cytosolic thioredoxin reductase 1 is required for correct disulfide formation in the ER. EMBO J 36:693–702

Prast-Nielsen S, Cebula M, Pader I, Arner ES (2010) Noble metal targeting of thioredoxin reductase – covalent complexes with thioredoxin and thioredoxin-related protein of 14 kDa triggered by cisplatin. Free Radic Biol Med 49:1765–1778

Prast-Nielsen S, Dexheimer TS, Schultz L, Stafford WC, Cheng Q, Xu J, Jadhav A, Arnér ES, Simeonov A (2011) Inhibition of thioredoxin reductase 1 by porphyrins and other small molecules identified by a high-throughput screening assay. Free Radic Biol Med 50:1114–1123

Prigge JR, Eriksson S, Iverson SV, Meade TA, Capecchi MR, Arner ES, Schmidt EE (2012a) Hepatocyte DNA replication in growing liver requires either glutathione or a single allele of txnrd1. Free Radic Biol Med 52:803–810

Prigge JR, Eriksson S, Iverson SV, Meade TA, Capecchi MR, Arnér ESJ, Schmidt EE (2012b) Hepatocyte DNA replication in growing liver requires either glutathione or a single allele of txnrd1. Free Radic Biol Med 52:803–810

Prigge JR, Coppo L, Martin SS, Ogata F, Miller CG, Bruschwein MD, Orlicky DJ, Shearn CT, Kundert JA, Lytchier J, Herr AE, Mattsson A, Taylor MP, Gustafsson TN, Arner ESJ, Holmgren A, Schmidt EE (2017) Hepatocyte hyperproliferation upon liver-specific co-disruption of thioredoxin-1, thioredoxin reductase-1, and glutathione reductase. Cell Rep 19:2771–2781

Rackham O, Shearwood AM, Thyer R, McNamara E, Davies SM, Callus BA, Miranda-Vizuete A, Berners-Price SJ, Cheng Q, Arnér ES, Filipovska A (2011) Substrate and inhibitor specificities differ between human cytosolic and mitochondrial thioredoxin reductases: implications for development of specific inhibitors. Free Radic Biol Med 50:689–699

Rai G, Sayed AA, Lea WA, Luecke HF, Chakrapani H, Prast-Nielsen S, Jadhav A, Leister W, Shen M, Inglese J, Austin CP, Keefer L, Arner ES, Simeonov A, Maloney DJ, Williams DL, Thomas CJ (2009) Structure mechanism insights and the role of nitric oxide donation guide the development of oxadiazole-2-oxides as therapeutic agents against schistosomiasis. J Med Chem 52:6474–6483

Reich HJ, Hondal RJ (2016) Why nature chose selenium. ACS Chem Biol 11:821–841

Rhee SG (2006) Cell signaling. H2O2, a necessary evil for cell signaling. Science 312:1882–1883

Rigobello MP, Callegaro MT, Barzon E, Benetti M, Bindoli A (1998) Purification of mitochondrial thioredoxin reductase and its involvement in the redox regulation of membrane permeability. Free Radic Biol Med 24:370–376

Rigobello MP, Folda A, Baldoin MC, Scutari G, Bindoli A (2005) Effect of auranofin on the mitochondrial generation of hydrogen peroxide. Role of thioredoxin reductase. Free Radic Res 39:687–695

Roder C, Thomson MJ (2015) Auranofin: repurposing an old drug for a golden new age. Drugs R D 15:13–20

Rollins MF, van der Heide DM, Weisend CM, Kundert JA, Comstock KM, Suvorova ES, Capecchi MR, Merrill GF, Schmidt EE (2010) Hepatocytes lacking thioredoxin reductase 1 have normal replicative potential during development and regeneration. J Cell Sci 123:2402–2412

Ross F, Hernandez P, Porcal W, Lopez GV, Cerecetto H, Gonzalez M, Basika T, Carmona C, Flo M, Maggioli G, Bonilla M, Gladyshev VN, Boiani M, Salinas G (2012) Identification of thioredoxin glutathione reductase inhibitors that kill cestode and trematode parasites. PLoS One 7:e35033

Ruffell B, Coussens LM (2015) Macrophages and therapeutic resistance in cancer. Cancer Cell 27:462–472

Rundlöf A-K, Arnér ESJ (2004) Regulation of the mammalian selenoprotein thioredoxin reductase 1 in relation to cellular phenotype, growth and signaling events. Antioxid Redox Signal 6:41–52

Rundlof AK, Arner ES (2004) Regulation of the mammalian selenoprotein thioredoxin reductase 1 in relation to cellular phenotype, growth, and signaling events. Antioxid Redox Signal 6:41–52

Rundlof AK, Carlsten M, Giacobini MM, Arner ES (2000) Prominent expression of the selenoprotein thioredoxin reductase in the medullary rays of the rat kidney and thioredoxin reductase mRNA variants differing at the 5′ untranslated region. Biochem J 347(Pt 3):661–668

Rundlof AK, Janard M, Miranda-Vizuete A, Arner ES (2004) Evidence for intriguingly complex transcription of human thioredoxin reductase 1. Free Radic Biol Med 36:641–656

Rundlof AK, Fernandes AP, Selenius M, Babic M, Shariatgorji M, Nilsonne G, Ilag LL, Dobra K, Bjornstedt M (2007) Quantification of alternative mRNA species and identification of thioredoxin reductase 1 isoforms in human tumor cells. Differentiation 75:123–132

Ryter SW, Kim HP, Hoetzel A, Park JW, Nakahira K, Wang X, Choi AM (2007) Mechanisms of cell death in oxidative stress. Antioxid Redox Signal 9:49–89

Saiz C, Castillo V, Fontan P, Bonilla M, Salinas G, Rodriguez-Haralambides A, Mahler SG (2014) Discovering Echinococcus granulosus thioredoxin glutathione reductase inhibitors through site-specific dynamic combinatorial chemistry. Mol Divers 18:1–12

Sandalova T, Zhong L, Lindqvist Y, Holmgren A, Schneider G (2001) Three-dimensional structure of a mammalian thioredoxin reductase: implications for mechanism and evolution of a selenocysteine-dependent enzyme. Proc Natl Acad Sci U S A 98:9533–9538

Schmidt EE (2015) Interplay between cytosolic disulfide reductase systems and the Nrf2/Keap1 pathway. Biochem Soc Trans 43:632–638

Shahabi V, Postow MA, Tuck D, Wolchok JD (2015) Immune-priming of the tumor microenvironment by radiotherapy: rationale for combination with immunotherapy to improve anticancer efficacy. Am J Clin Oncol 38:90–97

Shi Y, Nikulenkov F, Zawacka-Pankau J, Li H, Gabdoulline R, Xu J, Eriksson S, Hedstrom E, Issaeva N, Kel A, Arner ES, Selivanova G (2014) ROS-dependent activation of JNK converts p53 into an efficient inhibitor of oncogenes leading to robust apoptosis. Cell Death Differ 21:612–623

Shukla R, Shukla H, Kalita P, Tripathi T (2018) Structural insights into natural compounds as inhibitors of *Fasciola gigantica* thioredoxin glutathione reductase. J Cell Biochem 119:3067–3080

Silvestri I, Lyu H, Fata F, Boumis G, Miele AE, Ardini M, Ippoliti R, Bellelli A, Jadhav A, Lea WA, Simeonov A, Cheng Q, Arner ESJ, Thatcher GRJ, Petukhov PA, Williams DL, Angelucci F (2018) Fragment-based discovery of a regulatory site in thioredoxin glutathione reductase acting as "Doorstop" for NADPH entry. ACS Chem Biol 13:2190–2202

Simeonov A, Jadhav A, Sayed AA, Wang Y, Nelson ME, Thomas CJ, Inglese J, Williams DL, Austin CP (2008) Quantitative high-throughput screen identifies inhibitors of the *Schistosoma mansoni* redox cascade. PLoS Negl Trop Dis 2:e127

Singh A, Boldin-Adamsky S, Thimmulappa RK, Rath SK, Ashush H, Coulter J, Blackford A, Goodman SN, Bunz F, Watson WH, Gabrielson E, Feinstein E, Biswal S (2008) RNAi-mediated silencing of nuclear factor erythroid-2-related factor 2 gene expression in non-small cell lung cancer inhibits tumor growth and increases efficacy of chemotherapy. Cancer Res 68:7975–7984

Song L, Li J, Xie S, Qian C, Wang J, Zhang W, Yin X, Hua Z, Yu C (2012) Thioredoxin glutathione reductase as a novel drug target: evidence from Schistosoma japonicum. PLoS One 7:e31456

Stafford WC, Peng X, Olofsson MH, Zhang X, Luci DK, Lu L, Cheng Q, Tresaugues L, Dexheimer TS, Coussens NP, Augsten M, Ahlzen HM, Orwar O, Ostman A, Stone-Elander S, Maloney DJ, Jadhav A, Simeonov A, Linder S, Arner ESJ (2018) Irreversible inhibition of cytosolic thioredoxin reductase 1 as a mechanistic basis for anticancer therapy. Sci Transl Med 10: eaaf7444

Su D, Gladyshev VN (2004) Alternative splicing involving the thioredoxin reductase module in mammals: a glutaredoxin-containing thioredoxin reductase 1. Biochemistry 43:12177–12188

Su D, Novoselov SV, Sun QA, Moustafa ME, Zhou Y, Oko R, Hatfield DL, Gladyshev VN (2005) Mammalian selenoprotein thioredoxin-glutathione reductase. Roles in disulfide bond formation and sperm maturation. J Biol Chem 280:26491–26498

Sun QA, Kirnarsky L, Sherman S, Gladyshev VN (2001a) Selenoprotein oxidoreductase with specificity for thioredoxin and glutathione systems. Proc Natl Acad Sci U S A 98:3673–3678

Sun QA, Zappacosta F, Factor VM, Wirth PJ, Hatfield DL, Gladyshev VN (2001b) Heterogeneity within animal thioredoxin reductases. Evidence for alternative first exon splicing. J Biol Chem 276:3106–3114

Sun QA, Zappacosta F, Factor VM, Wirth PJ, Hatfield DL, Gladyshev VN (2001c) Heterogeneity within animal thioredoxin reductases. Evidence for alternative first exon splicing. J Biol Chem 276:3106–3114

Sun QA, Su D, Novoselov SV, Carlson BA, Hatfield DL, Gladyshev VN (2005) Reaction mechanism and regulation of mammalian thioredoxin/glutathione reductase. Biochemistry 44:14528–14537

Surh YJ, Kundu JK, Na HK (2008) Nrf2 as a master redox switch in turning on the cellular signaling involved in the induction of cytoprotective genes by some chemopreventive phytochemicals. Planta Med 74:1526–1539

Suvorova ES, Lucas O, Weisend CM, Rollins MF, Merrill GF, Capecchi MR, Schmidt EE (2009) Cytoprotective Nrf2 pathway is induced in chronically txnrd 1-deficient hepatocytes. PLoS One 4:e6158

Tamura T, Stadtman TC (1996) A new selenoprotein from human lung adenocarcinoma cells: purification, properties, and thioredoxin reductase activity. Proc Natl Acad Sci U S A 93:1006–1011

Tong KI, Kobayashi A, Katsuoka F, Yamamoto M (2006) Two-site substrate recognition model for the Keap1-Nrf2 system: a hinge and latch mechanism. Biol Chem 387:1311–1320

Trachootham D, Zhou Y, Zhang H, Demizu Y, Chen Z, Pelicano H, Chiao PJ, Achanta G, Arlinghaus RB, Liu J, Huang P (2006) Selective killing of oncogenically transformed cells through a ROS-mediated mechanism by beta-phenylethyl isothiocyanate. Cancer Cell 10:241–252

Trachootham D, Alexandre J, Huang P (2009) Targeting cancer cells by ROS-mediated mechanisms: a radical therapeutic approach? Nat Rev Drug Discov 8:579–591

Urig S, Becker K (2006) On the potential of thioredoxin reductase inhibitors for cancer therapy. Semin Cancer Biol 16:452–465

Vinay DS, Ryan EP, Pawelec G, Talib WH, Stagg J, Elkord E, Lichtor T, Decker WK, Whelan RL, Kumara HM, Signori E, Honoki K, Georgakilas AG, Amin A, Helferich WG, Boosani CS, Guha G, Ciriolo MR, Chen S, Mohammed SI, Azmi AS, Keith WN, Bilsland A, Bhakta D, Halicka D, Fujii H, Aquilano K, Ashraf SS, Nowsheen S, Yang X, Choi BK, Kwon BS (2015) Immune evasion in cancer: mechanistic basis and therapeutic strategies. Semin Cancer Biol 35 (Suppl):S185–S198

Wang X, Zhang J, Xu T (2008) Thioredoxin reductase inactivation as a pivotal mechanism of ifosfamide in cancer therapy. Eur J Pharmacol 579:66–73

Williams CH Jr (1992) Lipoamide dehydrogenase, glutathione reductase, thioredoxin reductase, and mercuric ion reductase – a family of flavoenzyme transhydrogenases. In: Müller F (ed) Chemistry and biochemistry of flavoenzymes, vol 3. CRC Press, Boca Raton, FL, pp 121–211

Williams DL, Bonilla M, Gladyshev VN, Salinas G (2013) Thioredoxin glutathione reductase-dependent redox networks in platyhelminth parasites. Antioxid Redox Signal 19:735–745

Wipf P, Lynch SM, Birmingham A, Tamayo G, Jimenez A, Campos N, Powis G (2004) Natural product based inhibitors of the thioredoxin-thioredoxin reductase system. Org Biomol Chem 2:1651–1658

Witte AB, Anestål K, Jerremalm E, Ehrsson H, Arnér ESJ (2005) Inhibition of thioredoxin reductase but not of glutathione reductase by the major classes of alkylating and platinum-containing anticancer compounds. Free Radic Biol Med 39:696–703

Wondrak GT (2009) Redox-directed cancer therapeutics: molecular mechanisms and opportunities. Antioxid Redox Signal 11:3013–3069

Woo JR, Kim SJ, Jeong W, Cho YH, Lee SC, Chung YJ, Rhee SG, Ryu SE (2004) Structural basis of cellular redox regulation by human TRP14. J Biol Chem 279:48120–48125

Xia L, Nordman T, Olsson JM, Damdimopoulos A, Björkhem-Bergman L, Nalvarte I, Eriksson LC, Arnér ESJ, Spyrou G, Björnstedt M (2003) The mammalian cytosolic selenoenzyme thioredoxin reductase reduces ubiquinone. A novel mechanism for defense against oxidative stress. J Biol Chem 278:2141–2146

Yant LJ, Ran Q, Rao L, Van Remmen H, Shibatani T, Belter JG, Motta L, Richardson A, Prolla TA (2003) The selenoprotein GPX4 is essential for mouse development and protects from radiation and oxidative damage insults. Free Radic Biol Med 34:496–502

Ye ZW, Zhang J, Townsend DM, Tew KD (2015) Oxidative stress, redox regulation and diseases of cellular differentiation. Biochim Biophys Acta 1850:1607–1621

Ye SF, Li J, Ji SM, Zeng HH, Lu W (2017) Dose-biomarker-response modeling of the anticancer effect of ethaselen in a human non-small cell lung cancer xenograft mouse model. Acta Pharmacol Sin 38:223–232

Yoo MH, Xu XM, Carlson BA, Gladyshev VN, Hatfield DL (2006) Thioredoxin reductase 1 deficiency reverses tumor phenotype and tumorigenicity of lung carcinoma cells. J Biol Chem 281:13005–13008

Yoo MH, Xu XM, Carlson BA, Patterson AD, Gladyshev VN, Hatfield DL (2007) Targeting thioredoxin reductase 1 reduction in cancer cells inhibits self-sufficient growth and DNA replication. PLoS One 2:e1112

Zhang DD (2006) Mechanistic studies of the Nrf2-Keap1 signaling pathway. Drug Metab Rev 38:769–789

Zhang B, Zhang J, Peng S, Liu R, Li X, Hou Y, Han X, Fang J (2016) Thioredoxin reductase inhibitors: a patent review. Expert Opin Ther Pat:1–10

Zhang B, Liu Y, Li X, Xu J, Fang J (2018) Small molecules to target the selenoprotein thioredoxin reductase. Chem Asian J 13:3593–3600

Zhang J, Zhang B, Li X, Han X, Liu R, Fang J (2019) Small molecule inhibitors of mammalian thioredoxin reductase as potential anticancer agents: an update. Med Res Rev 39:5–39

Zhao C, Gillette DD, Li X, Zhang Z, Wen H (2014) Nuclear factor E2-related factor-2 (Nrf2) is required for NLRP3 and AIM2 inflammasome activation. J Biol Chem 289:17020–17029

Zhong L, Holmgren A (2000) Essential role of selenium in the catalytic activities of mammalian thioredoxin reductase revealed by characterization of recombinant enzymes with selenocysteine mutations. J Biol Chem 275:18121–18128

Zhong L, Arnér ESJ, Ljung J, Åslund F, Holmgren A (1998) Rat and calf thioredoxin reductase are homologous to glutathione reductase with a carboxyl-terminal elongation containing a conserved catalytically active penultimate selenocysteine residue. J Biol Chem 273:8581–8591

Zhong L, Arnér ESJ, Holmgren A (2000) Structure and mechanism of mammalian thioredoxin reductase: the active site is a redox-active selenolthiol/selenenylsulfide formed from the conserved cysteine-selenocysteine sequence. Proc Natl Acad Sci U S A 97:5854–5859

Cardiovascular Therapeutic Potential of the Redox Siblings, Nitric Oxide (NO•) and Nitroxyl (HNO), in the Setting of Reactive Oxygen Species Dysregulation

Barbara K. Kemp-Harper, Anida Velagic, Nazareno Paolocci, John D. Horowitz, and Rebecca H. Ritchie

Contents

B. K. Kemp-Harper (✉)
Department of Pharmacology, Cardiovascular Disease Program, Biomedicine Discovery Institute, Monash University, Clayton, VIC, Australia
e-mail: barbara.kemp@monash.edu

A. Velagic
Baker Heart & Diabetes Institute, Melbourne, VIC, Australia

Central Clinical School, Monash University, Melbourne, VIC, Australia

Drug Discovery Biology, Monash Institute of Pharmaceutical Sciences, Monash University, Melbourne, VIC, Australia

N. Paolocci
Division of Cardiology, Department of Medicine, Johns Hopkins School of Medicine, Baltimore, MD, USA

Department of Biomedical Sciences, University of Padova, Padova, Italy

J. D. Horowitz
Basil Hetzel Institute, Queen Elizabeth Hospital, University of Adelaide, Adelaide, SA, Australia

R. H. Ritchie
Department of Pharmacology, Cardiovascular Disease Program, Biomedicine Discovery Institute, Monash University, Clayton, VIC, Australia

Baker Heart & Diabetes Institute, Melbourne, VIC, Australia

Central Clinical School, Monash University, Melbourne, VIC, Australia

Drug Discovery Biology, Monash Institute of Pharmaceutical Sciences, Monash University, Melbourne, VIC, Australia

H. H. H. W. Schmidt et al. (eds.), *Reactive Oxygen Species*,
Handbook of Experimental Pharmacology 264, https://doi.org/10.1007/164_2020_389

Abstract

Reactive oxygen species (ROS) dysregulation is a hallmark of cardiovascular disease, characterised by an imbalance in the synthesis and removal of ROS. ROS such as superoxide ($•O_2^-$), hydrogen peroxide (H_2O_2), hydroxyl (OH•) and peroxynitrite ($ONOO^-$) have a marked impact on cardiovascular function, contributing to the vascular impairment and cardiac dysfunction associated with diseases such as angina, hypertension, diabetes and heart failure. Central to the vascular dysfunction is a reduction in bioavailability and/or physiological effects of vasoprotective nitric oxide (NO•), leading to vasoconstriction, inflammation and vascular remodelling. In a cardiac context, increased ROS generation can also lead to modification of key proteins involved in cardiac contractility. Whilst playing a key role in the pathogenesis of cardiovascular disease, ROS dysregulation also limits the clinical efficacy of current therapies, such as nitrosovasodilators. As such, alternate therapies are sought. This review will discuss the impact of ROS dysregulation on the therapeutic utility of NO• and its redox sibling, nitroxyl (HNO).

Graphical Abstract

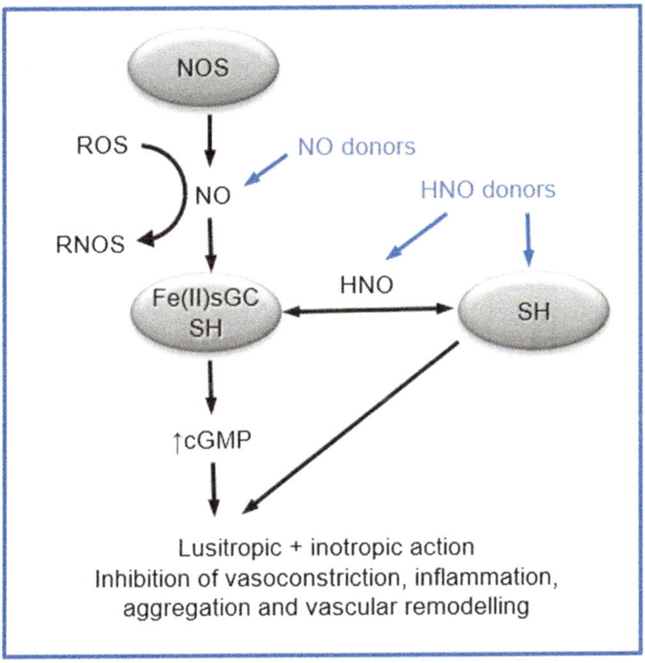

Both nitric oxide (NO) and nitroxyl (HNO) donors signal through soluble guanylyl cyclase (sGC). NO binds to the Fe(II) form of sGC and nitroxyl possibly to both sGC heme and thiol groups. In the vasculature, nitroxyl can also signal through voltage-dependent (K_v) and ATP-sensitive (K_{ATP}) K^+ channels as well as calcitonin gene-related peptide (CGRP). In the heart, HNO directly targets critical thiols to increase myocardial contractility, an effect not seen with NO. The qualitative effects via elevation of cGMP are similar, i.e. lusitropic in the heart and inhibitory on vasoconstriction, inflammation, aggregation and vascular remodelling. Of pathophysiological significance is the fact the efficacy of NO donors is impaired by ROS, e.g. through chemical scavenging of NO, to generate reactive nitrogen oxide species (RNOS), whilst nitroxyl is apparently not.

Keywords

Cardiovascular · Nitric oxide · Nitroxyl · Reactive oxygen species

1 Pharmacology of NO• Versus HNO

HNO, the one-electron reduced and protonated form of NO•, possesses several unique biological and pharmacological properties as compared to its redox sibling that confers therapeutic advantages (Paolocci et al. 2007; Irvine et al. 2008; Bullen et al. 2011a, b). Like NO•, HNO stimulates soluble guanylyl cyclase (sGC), in both the vasculature and myocardium, to increase intracellular cyclic guanosine-3′,5-′-monophosphate (cGMP) and has vasodilatory (Fukuto et al. 1992; Irvine et al. 2003; Bullen et al. 2011a, b; Andrews et al. 2015; Zhu et al. 2015), anti-aggregatory (Bermejo et al. 2005; Chirkov and Horowitz 2007; Bullen et al. 2011a, b), anti-proliferative (Tsihlis et al. 2010) and anti-inflammatory (Andrews et al. 2016) actions. However, given the preference of HNO for interaction with ferric (Fe^{3+}) as opposed to ferrous (Fe^{2+}) heme groups (Miranda et al. 2003), it has been speculated that HNO may preferentially target the oxidised form of sGC (Fe^{3+}-sGC) which predominates in disease conditions associated with ROS dysregulation. To date, however, evidence in support of this concept is lacking (Miller et al. 2009; Zeller et al. 2009). A key distinguishing feature of HNO as compared to NO• is its ability to interact directly with thiols and thiol-containing proteins leading to their oxidation (Fukuto et al. 2013). Specifically, HNO interacts with thiols in a reversible (disulphide formation) or irreversible (sulfinamide formation) manner, whereas NO• first undergoes autoxidation before thiol nitrosation. This evidence is of particular relevance in the heart, where the unique ability of HNO to increase myocardial contractility (an effect not seen with NO•) is due to its interaction with thiols (Paolocci et al. 2001a, b, 2003). Specifically, HNO modulates critical cysteine residues on ryanodine receptors (RyR2) (Cheong et al. 2005) in the sarcoplasmic reticulum (SR), SR Ca^{2+}-ATPase (SERCA2) (Tocchetti et al. 2007) and phospholamban (Sivakumaran et al. 2013; Keceli et al. 2019) to facilitate Ca^{2+} cycling. Moreover, HNO targets cysteine moieties in actin-myosin filaments (Gao et al. 2012) to further enhance myocardial contractility. Therefore, HNO serves as a positive cardiac inotrope, in both healthy and failing hearts (Paolocci et al. 2001a, b, 2003), and has significant clinical potential in the treatment of acute decompensated heart failure (Sabbah et al. 2013; Kemp-Harper et al. 2016) as will be discussed later in this review (see Sect. 5).

2 Relationship Between ROS and NO•/HNO Action

A further distinguishing feature between NO• and HNO is their interaction with ROS, such as superoxide ($•O_2^-$). $•O_2^-$ reacts directly and rapidly ($5–10 \times 10^9\ M^{-1}\ s^{-1}$) with NO• (Gryglewski et al. 1986; Beckman and Koppenol 1996), leading to the generation of the powerful oxidant, peroxynitrite ($ONOO^-$). $ONOO^-$ further exacerbates the oxidative environment, uncoupling endothelial NO synthase (eNOS) (Kuzkaya et al. 2003), leading to $•O_2^-$ generation and reduced NO• synthesis. $ONOO^-$ can also attenuate NO• signalling by oxidising sGC and rendering it unresponsive to NO• (Evgenov et al. 2006; Stasch et al. 2006). Thus, the direct interaction of NO• with ROS limits the bioavailability and efficacy of endogenous NO• (Paolocci et al. 2001a, b) and underlies much of the pathology of cardiovascular disease. Indeed, in the vasculature, a loss in vasoprotective NO• augments vasospasm, promotes immune cell infiltration, stimulates vascular smooth muscle cell proliferation and promotes platelet aggregation (Ritchie et al. 2017). Conversely in the myocardium, a deficiency in endogenous NO• can impact diastolic function, with a reduction in cardiac relaxation and compliance, together with increased cardiac hypertrophy and fibrosis (Ritchie et al. 2017). Besides, the clinical utility of NO•-based therapeutics (e.g. nitrosovasodilators) is impacted by increased ROS generation such that tolerance, pseudo-tolerance and NO• resistance can arise (see Sect. 5).

In direct contrast to NO•, HNO is, for the most part, resistant to scavenging by ROS (Miranda et al. 2002; Leo et al. 2012). The interaction between $\bullet O_2^-$ and HNO is unfavourable and as such oxidants are not generated in the presence of HNO and ROS dysregulation. Importantly, this opens up the potential for HNO donors in cardiovascular disease settings whereby the efficacy of HNO may be preserved yet that of NO• donors compromised. Indeed, preclinical studies in cardiovascular disease models (e.g. hypercholesterolemia, hypertension, diabetes, heart failure) (Paolocci et al. 2003; Bullen et al. 2011a, b; Leo et al. 2012; Wynne et al. 2012; Irvine et al. 2013a, b; Tare et al. 2017; Qin et al. 2020) have provided evidence for preserved efficacy of HNO donors (e.g. Angeli's salt; isopropylamine NONOate, IPA/NO) in the face of impaired NO• signalling, together with resistance to tolerance development with continued use (Irvine et al. 2011, 2007; Andrews et al. 2015). Moreover, in the clinical setting, the anti-aggregatory actions of HNO are maintained in patients with coronary artery disease, yet those to NO• are impaired (Dautov et al. 2013).

Whilst we have addressed the impact of ROS upon NO• and HNO, it is also relevant to consider the ability of these redox siblings themselves to limit ROS production. Within the vasculature and myocardium, the NADPH oxidase (Nox) family of ROS-generating enzymes (e.g. Nox1, 2, 4 and 5) are key contributors to ROS dysregulation (Selemidis et al. 2008; Thomas et al. 2008; Drummond et al. 2011). NO• donors (e.g. DETA/NONOate) inhibit Nox2 oxidase activity in human endothelial cells by S-nitrosylation of the organiser subunit, p47 phox (Selemidis et al. 2007). Similarly, we have shown that the HNO donors, Angeli's salt and IPA/NO, directly target vascular Nox2 oxidase to suppress $\bullet O_2^-$ generation in the cerebral vasculature, potentially via interaction with reactive cysteine thiols within p47 phox (Miller et al. 2013). These actions of HNO are rapid (3 min) as compared to NO• (6 h) and may reflect the direct vs. indirect interaction of HNO and NO• with thiols, respectively (Fukuto et al. 2013). Also, HNO may serve as a one-electron reductant (donating its hydrogen atom) and can stimulate the activity of the antioxidant protein, heme oxygenase-1 (Mondoro et al. 2001).

Collectively, HNO possesses a suite of properties which are amenable to use in the treatment of cardiovascular diseases associated with ROS dysregulation including (1) resistance to scavenging by $\bullet O_2^-$, (2) an ability to limit $\bullet O_2^-$ generation, (3) resistance to tolerance development, (4) direct interaction with thiols, (5) vasoprotective actions (vasodilator, anti-aggregatory, anti-proliferative and anti-inflammatory) and (6) cardioprotective actions (positive cardiac inotrope and lusitrope, anti-hypertrophic) (Sabbah et al. 2013; Cao et al. 2015; Bullen et al. 2011a, b). The impact of ROS dysregulation and the therapeutic utility of HNO versus NO• in vascular and cardiac diseases will be explored.

3 Vascular Actions of NO• Versus HNO: Impact of ROS Dysregulation

3.1 Angina: Impact of NO• Versus HNO

Patients with stable angina pectoris experience chest pain, mainly under exertion, a condition underpinned by coronary artery disease and marked vascular ROS generation. The presence of atherosclerotic lesions within the coronary vasculature compromises coronary blood flow reserve and oxygen supply and is associated with a loss of vasoprotective endothelial-derived NO• (Cox et al. 1989; Schachinger et al. 2000; Tousoulis et al. 2014) and increased ROS generation (Azumi et al. 2002; Guzik et al. 2006). Organic nitrates, such as glyceryl trinitrate (GTN) or isosorbide dinitrate (ISDN), donate NO• and have been mainstay treatments for this condition for >100 years, leading to an improvement in coronary blood flow, unloading of the heart (via peripheral vasodilation) and improved myocardial oxygen matching. Nitrates also inhibit platelet aggregation, beneficial actions in the maintenance of perfusion to the ischemic myocardium. However, the susceptibility of organic nitrates to tolerance development with continued use (Daiber et al. 2008; Munzel et al. 2013) necessitates the implementation of nitrate-free periods, placing patients with angina at risk of a cardiovascular event. Tolerance development arises, to a large extent, as a consequence of the impaired generation of NO• from organic nitrates and is associated with ROS dysregulation (Fayers et al. 2003) (see Sect. 5).

Moreover, organic nitrates and other NO•-based therapies are subject to the phenomenon of NO• resistance, whereby their vasodilatory and anti-aggregatory efficacy is impaired, independent of prior exposure (Chirkov and Horowitz 2007). Such a loss in anti-platelet and vasodilatory efficacy of nitrosovasodilators is a significant limitation for this drug class in the treatment of acute cardiovascular emergencies, such as acute myocardial infarction, transient myocardial ischaemia and acute decompensated heart failure. Importantly, ROS are a key contributor to this loss in NO• responsiveness (NO• resistance), with $\bullet O_2^-$ scavenging NO• and promoting the oxidation of its target, sGC (Ritchie et al. 2017) (see Sect. 5).

HNO donors offer a clear advantage over traditional organic nitrates in the treatment of stable angina pectoris. Thus we have shown in preclinical studies that (1) the vasodilatory and anti-aggregatory actions of HNO are preserved in hypercholesterolemia in the presence of increased $\bullet O_2^-$ generation (Bullen et al. 2011a, b), (2) HNO donors are not susceptible to the development of tolerance either in vitro (Irvine et al. 2007; Andrews et al. 2015) or in vivo (Irvine et al. 2011) and (3) HNO donors do not cause endothelial dysfunction (Irvine et al. 2011). Moreover, our studies in human isolated arteries have shown that HNO donors do not induce cross-tolerance to GTN (Andrews et al. 2015) and vice versa. In addition, in patients with coronary artery disease, we have shown that the HNO donor IPA/NO partially circumvents NO• resistance (Dautov et al. 2013). Thus, the ability of IPA/NO to increase platelet cGMP and inhibit platelet aggregation was found to be substantially higher than that of the NO• donor sodium nitroprusside. Such favourable properties of HNO are due to a large extent to its resistance to scavenging by $\bullet O_2^-$ (Miranda et al. 2002; Leo et al. 2012). These findings suggest that HNO donors may be of use in the treatment of angina, either as a stand-alone therapy or co-administered with GTN to allow the use of lower concentrations of nitrates, thereby minimising the potential for tolerance development. We also speculate that HNO donors may have clinical utility in the treatment of acute cardiovascular emergencies (Velagic et al. 2020).

3.2 Hypertension: Impact of NO• Versus HNO

Hypertension (average systolic blood pressure $\geq$ 140 mmHg, average diastolic blood pressure $>$ 90 mmHg) is a major risk factor for heart attack or stroke (Messerli et al. 2007). It is associated with endothelial dysfunction, increased vascular tone, vascular inflammation and remodelling (Touyz 2004; Iadecola and Davisson 2008; Dinh et al. 2014). Elevated ROS generation is a major contributor to the pathogenesis of the disease (Lee and Griendling 2008) with the NADPH oxidase family of enzymes, particularly Nox1 and Nox2, playing a key role (Wingler et al. 2001; Wind et al. 2010; Drummond et al. 2011). ROS generation has a significant impact on the NO-sGC-cGMP signalling pathway in hypertension (Schachinger et al. 2000), leading to impaired endogenous NO• synthesis and/or effect (endothelial dysfunction) (Schulz et al. 2011), partially via sGC oxidation, which renders the enzyme unresponsive to NO• (Ruetten et al. 1999; Kloss et al. 2000; Zalba et al. 2000). Such dysfunction is particularly evident in aged, hypertensive rodents. Whilst nitrosovasodilators are potent vasodilators and may compensate for the hypertension-associated reduction in endogenous NO• bioavailability, they themselves are susceptible to scavenging by $\bullet O_2^-$ and may have reduced efficacy due to sGC dysfunction. Their clinical utility is further limited by tolerance development, systemic hypotension and reflex tachycardia. Therefore, NO•-based therapeutics, such as sodium nitroprusside, are only used in the acute management of hypertensive crises.

HNO donors, on the other hand, may offer an alternative therapeutic strategy. We and others have shown that HNO is a potent vasodilator, lowering blood pressure in vivo (De Witt et al. 2001; Paolocci et al. 2003; Irvine et al. 2013a, b) and causing relaxation in isolated rodent (Irvine et al. 2003; Favaloro and Kemp-Harper 2007; Bullen et al. 2011a, b; Wynne et al. 2012) and human arteries (Andrews et al. 2015). In the vasculature, HNO signals predominantly via the sGC-cGMP pathway but unlike NO• can also target voltage-dependent (K_v) (Irvine et al. 2003; Favaloro and Kemp-Harper 2009) and ATP-sensitive (K_{ATP}) K^+ channels and calcitonin gene-related peptide (CGRP) (Favaloro and Kemp-Harper 2007) to evoke vasorelaxation. Interestingly, at high concentrations, HNO can oxidise critical thiols on sGC, inhibiting its activity, a potential mechanism to limit excessive vasodilatation to HNO donors (Miller et al. 2009). Importantly the vasorelaxant efficacy of HNO is preserved in hypertension, with vasorelaxant responses to the HNO donors Angeli's salt and IPA/NO maintained in isolated aorta from angiotensin II-infused mice (Wynne et al. 2012) and spontaneously hypertensive rats (SHR) (Irvine et al.

2013a, b) and the blood pressure-lowering actions preserved in conscious SHR (Irvine et al. 2013a, b). Together with increased ROS generation, hypertension is also associated with thiol depletion. Given HNO does not interact with $\bullet O_2^-$ (Miranda et al. 2002; Leo et al. 2012) and is scavenged by thiols (Wink et al. 1998), preserved efficacy of HNO in the context of hypertension is supported.

3.3 Diabetes: Impact of NO• Versus HNO

Diabetes is associated with a plethora of cardiovascular complications, including coronary artery disease, peripheral vascular disease and heart failure (Almourani et al. 2019). Central to the disease pathology is hyperglycaemia-induced ROS generation (Ritchie and Abel 2020). Thus in the diabetic vasculature, increased activity of NADPH oxidase (Nox1, Nox2) is evident (Drummond et al. 2011; Antonopoulos et al. 2015) together with mitochondrial dysfunction and ROS generation (Nishikawa et al. 2000) and elevated thioredoxin-interacting protein (TXNIP; negative regulator of antioxidant thioredoxin) (Schulze et al. 2004). These pathways work in concert to promote diabetes-associated vascular ROS dysregulation and have a significant impact on NO•-sGC/cGMP signalling.

Specifically, hyperglycaemia can cause a reduction in endogenous NO• generation via decreased synthesis and oxidation (via $ONOO^-$ generation) of the eNOS cofactor tetrahydrobiopterin (BH_4) (Xu et al. 2007) leading to impaired endothelium-dependent vasodilation (Heitzer et al. 2000; Okon et al. 2005). Such endothelial dysfunction may also reflect a loss in NO• responsiveness per se (NO• resistance) in the diabetic vasculature. This concept is supported by the finding that in patients with type 2 diabetes, vasorelaxation to SNP is attenuated in isolated mammary arteries (Okon et al. 2005) and brachial-artery flow-mediated vasodilation to SNP impaired (Williams et al. 1996; van Etten et al. 2002; Shemyakin et al. 2012). Indeed, a reduction in vascular sGC expression has been reported in experimental models of diabetes (Silva et al. 2015) and oxidation of sGC evident in patients with type 2 diabetes (Stasch et al. 2006). Importantly, NO• resistance extends beyond the diabetic vasculature with resistance to the anti-aggregatory actions of NO• also apparent in patients with type 2 diabetes (Anderson et al. 2005).

NO• resistance has significant implications when treating diabetic patients during a cardiovascular emergency (e.g. acute myocardial infarct, transient myocardial ischaemia). In such a situation, immediate vasodilatory and anti-aggregatory actions are required. However, due to the reduced responsiveness of the diabetic vasculature and platelets to NO•, NO•-based therapies are precluded. As suggested above, HNO donors may be of use in the face of NO• resistance. In the setting of diabetes and associated ROS dysregulation, we and others have shown that the vasoprotective actions of HNO are maintained. Indeed, in the diabetic rat vasculature endothelium-dependent relaxation mediated by NO• is impaired, but that mediated by HNO is preserved (Leo et al. 2012; Kahlberg et al. 2016; Tare et al. 2017). Furthermore, vasorelaxation to HNO donors is preserved in the resistance vessels (Tare et al. 2017) and coronary vasculature (Qin et al. 2020) of diabetic animals. With an ability to cause venous (unload the heart) and arterial (improve coronary blood flow) dilation, coupled with anti-aggregatory actions, HNO has the potential to mitigate an ischemic event (Pagliaro et al. 2003).

4 Cardiac Actions of NO• Versus HNO: Impact of ROS Dysregulation

4.1 Acute Impact of NO• Versus HNO on Cardiac Function

The ability of a cardioactive agent to acutely enhance cardiac function, both cardiac relaxation and compliance (diastolic function) and contractility (systolic function), is a favourable therapeutic property in several cardiac pathologies. The NO redox siblings NO• and HNO exhibit such traits, to varying degrees. Of the spectrum of cardiac actions described for HNO in particular, its ability to enhance cardiac contractile function has attracted considerable research attention. Paolocci and colleagues provided the first evidence of a direct HNO-mediated inotropic effect at the level of individual cardiomyocytes (Tocchetti et al. 2007), in which Angeli's salt elicited prompt (<1 s) concentration-dependent enhancement of cardiomyocyte shortening in adult mouse cardiomyocytes in vitro. This HNO enhancement of contractile function has been confirmed by several subsequent reports, for both prototypical and next-generation HNO donors, across cardiomyocytes (Kohr et al. 2010) (Yong et al. 2010; Sivakumaran et al. 2013; Roof et al. 2017) and myocardial preparations in vitro (Dai et al. 2007; El-Armouche et al. 2010; Gao et al. 2012; Sabbah et al. 2013; Chin et al. 2014, 2016; Roof et al. 2017; Qin et al. 2020), as well as in large animal models in vivo (Paolocci et al. 2001a, b, 2003; Sabbah et al. 2013; Hartman et al. 2018). Whilst current evidence suggests that HNO donors enhance cardiac output in humans, their potential to acutely and specifically improve myocardial contractile function (to the extent that this has now been demonstrated in small and large animal models) remains to be investigated (Tita et al. 2017; Cowart et al. 2019), as noted by a recent editorial (Parissis et al. 2017). Contributing mechanisms to the HNO donor-mediated enhanced contractile function include heightened sensitivity of cardiomyocyte myofilaments to calcium and increased calcium release from, and reuptake by, the sarcoplasmic reticulum (Cheong et al. 2005; Dai et al. 2007; Tocchetti et al. 2007; Froehlich et al. 2008; Lancel et al. 2009; Kohr et al. 2010; Ding et al. 2011; Gao et al. 2012; Sivakumaran et al. 2013; Hartman et al. 2018; Keceli et al. 2019). These positive inotropic actions of HNO are largely considered attributed to the reactivity of HNO with key thiol residues (e.g. on cysteines in calcium-handling and myofilament proteins, including ryanodine receptors, SERCA2a and phospholamban), with a potential smaller contribution from sGC-mediated actions (Kemp-Harper et al. 2016). The precise nature by which HNO interacts with these specific thiols to elicit these actions, and what properties discriminate these thiols from other thiols, remains, however, to be resolved (Fukuto 2019).

In contrast, evidence of potential positive inotropic actions of NO• in the literature is harder to find, and this likely reflects the relative lack of reactivity of NO• with critical thiol residues capable of enhancing intracellular calcium flux and/or myofilament calcium sensitivity, as previously reviewed (Irvine et al. 2008; Ritchie et al. 2017). Our own efforts have suggested that the pure NO• donor DEA-NO can augment LV systolic and developed pressures, as well as LV + dP/dt, in the rodent myocardium in vitro, but only at relatively high doses that likely are secondary to changes in haemodynamic load (Chin et al. 2014). Similar to its positive inotropic actions, HNO elicits robust relaxation responses in the myocardium (Chin et al. 2014, 2016; Roof et al. 2017; Hartman et al. 2018). There is evidence that NO• also enhances cardiac relaxation in this context (Carroll et al. 1986; Paulus et al. 1994; Chin et al. 2014).

The specific impact of ROS dysregulation on the positive inotropic and lusitropic actions of HNO and NO• has mostly been inferred from their relative changes in potency in disease states associated with high ROS levels and from the reported impact of ROS, such as $\bullet O_2^-$, on NO sibling bioavailability (outlined above). Given that HNO is impervious to reactivity with $\bullet O_2^-$, whereas NO• is highly susceptible to it (Irvine et al. 2008, 2013a, b), the positive inotropic and lusitropic actions of HNO are more likely to be preserved than those of NO• under elevated oxidative stress; this is however yet to be specifically interrogated. For example, Paolocci and others have shown that HNO donors maintain these actions of enhanced contractile function and relaxation in animal models of heart failure (Paolocci et al. 2003; Sabbah et al. 2013; Roof et al. 2017; Hartman et al. 2018), but concomitant readouts of ROS generation were not obtained. Similarly, a recent Phase II clinical trial of the HNO donor BMS-986231 (previously known as CXL-1427) suggests improved cardiac index and pulmonary capillary wedge pressure in patients with chronic heart failure (Tita et al. 2017), where a high level of ROS is likely evident (Ritchie et al. 2009, 2017). The specific impact of ROS dysregulation on the acute inotropic and lusitropic actions of the two redox siblings may extend beyond their differential susceptibility to reactivity with ROS. As recently reviewed, myofilament proteins themselves are targets for oxidative post-translational modifications (Cuello et al. 2018), particularly at cysteine residues. Hence future studies examining the impact of concomitant elevated ROS levels (resulting from either acute exposure to pro-oxidants such as hydrogen peroxide or pyrogallol or to, e.g., sustained heart failure, ischaemic heart disease, diabetes or atherosclerosis, etc.) on the acute inotropic and lusitropic responses to HNO and NO• (together with the simultaneous impact on levels of both ROS and ROS-mediated post-translational modifications) are clearly warranted.

4.2 Impact of NO• Versus HNO on Cardiac Remodelling

Increased left ventricular (LV) mass is a key characteristic of heart failure, which represents a combination of LV cardiomyocyte hypertrophy and cardiac fibrosis (Rudolph et al. 2009). These structural manifestations precede (and contribute to the development of) impairments in LV contractility and relaxation. Of the various cell types of which the heart is comprised, cardiomyocytes and cardiac fibroblasts (and associated extracellular matrix proteins) are the primary culprits of increased LV mass (Hannan et al. 2003; Ritchie et al. 2009). Given that increased LV mass remains a significant independent risk factor for cardiovascular morbidity and mortality, as first revealed by the Framingham Heart Study (Dunn et al. 1990; Ho et al. 1993), effective therapeutic means to limit this is favourable. The NO redox siblings NO• and HNO both exhibit the ability to impact on cardiac remodelling, to variable degrees, exploiting cGMP to achieve this (Ritchie et al. 2009).

Both exogenous (through use of NO• donors) and endogenous NO• exhibit robust anti-hypertrophic actions in neonatal and adult rodent cardiomyocytes in vitro (Calderone et al. 1998; Ritchie et al. 1998; Wollert and Drexler 2002; Irvine et al. 2013a, b). These anti-hypertrophic NO• actions are also seen in the intact heart. Direct evidence supporting this comes from acute ex vivo perfusion of isolated hearts with an NO• donor (Rosenkranz et al. 2002). Similar in vivo anti-hypertrophic NO• actions are suggested by the ability of the NOS substrate L-arginine to limit L-NAME-induced LV hypertrophy (Paulis et al. 2008), in addition to the impact of genetic manipulation of nNOS and eNOS on cardiac morphology (Barouch et al. 2002; Loyer et al. 2008). Further in humans, it has been suggested that impaired endogenous NO• generation is a contributor to increased LV mass (Sverdlov et al. 2011). Concerning the other predominant component of LV remodelling, anti-fibrotic NO• actions have been reported in both primary cardiac fibroblasts in vitro (Calderone et al. 1998) and inferred in vivo in the context of NOS3 deficiency (Ichinose et al. 2004), as has been reviewed (Ritchie et al. 2009, 2017; Farah et al. 2018). Given the high (and rapid) susceptibility of NO• to reactivity with ROS such as $\bullet O_2^-$ (as described above) and resultant $ONOO^-$ generation (Munzel et al. 2005), together with the combined phenomena of NO• resistance and nitrate tolerance (Horowitz 2004; Chirkov and Horowitz 2007; Ritchie et al. 2017), the utility of NO• donors as a therapeutic approach to blunt LV remodelling in the clinic is limited.

In direct contrast to NO•, its redox sibling HNO is mostly impervious to reactivity with ROS and hence offers potential therapeutic promise for targeting cardiac remodelling, even in the context of concomitant ROS dysregulation (Ritchie et al. 2009, 2017; Kemp-Harper et al. 2016). Our own efforts provided the first evidence of anti-hypertrophic HNO actions, taking advantage of the prototypical HNO donors Angeli's salt and IPA-NO, in neonatal rat cardiomyocytes (Lin et al. 2012; Irvine et al. 2013a, b). These actions were dependent on sGC/cGMP signalling and were associated with suppression of NADPH oxidase activity and expression. Further, the ability of an HNO donor to stimulate sGC activity was impervious to prior exposure of the purified enzyme to the oxidant pyrogallol (whereas the ability of an NO• donor in this context was impaired). Taking these in vitro, relatively short-term studies (48 h) into the intact heart over the longer term in vivo, daily administration of the HNO donor 1-nitrosocyclo hexyl acetate (1-NCA) over the final 4 weeks of diabetes limited cardiac remodelling (including cardiomyocyte hypertrophy and pro-fibrotic connective tissue growth factor expression) in vivo in mice, actions associated with a reduction in myocardial NADPH oxidase (Cao et al. 2015). Comparable evidence for anti-hypertrophic and/or anti-fibrotic NO• actions, in the face of enhanced ROS generation, in this in vivo context are however lacking.

4.3 Ischaemia-Reperfusion (I-R) Injury: Impact of NO• Versus HNO

Myocardial I-R impairs both cardiomyocyte viability and LV function in vitro and in vivo Elevated ROS are not only evident in the acute phase of I-R-induced LV dysfunction (Goh et al. 2007) but also during the evolving heart failure that develops over the longer-term post-myocardial infarction (Hill and Singal 1996). I-R injury reflects a combination of damage to the coronary vasculature that perfuses the myocardium and to the cardiomyocytes responsible for maintaining contractile function. Given that both NO• and HNO elicit robust vasoprotective dilator actions, limit cardiac remodelling and can acutely enhance cardiac function, we now consider their respective potential protective actions in the context of I-R.

Both redox siblings have been shown to limit infarct size and cardiomyocyte injury resulting from I-R insults (via direct actions specifically at the cardiomyocyte as well as secondary to their vasodilator properties), as we have recently reviewed (Ritchie et al. 2017). A shared contributing mechanism of these cardioprotective actions of both HNO and NO• is sGC/cGMP/PKG, although both may elicit a component of cardiomyocyte-sparing actions in I-R injury that are independent of this pathway (Garreffa et al. 2006; Phillips et al. 2009; Sun et al. 2013; Tullio et al. 2017), with reports suggesting roles for S-nitrosothiols (in the case of NO•) and calcitonin gene-related peptide (specific for HNO) in the literature. Secondly, we consider their respective abilities to acutely enhance coronary vasodilation, both in normal physiology and in the context of I-R. Whilst the HNO donor Angeli's salt and the NO• donor DEA/NO elicit comparable dose-dependent vasodilation responses in the coronary vasculature (Chin et al. 2016), following I-R injury, the vasodilator response to the HNO donor was preserved, whilst that of the NO• donor was markedly impaired. These observations reflect the relative sensitivity of an NO• donor (but not an HNO donor) to vascular ROS dysregulation (Leo et al. 2012). These preserved vasoprotective actions are in contrast to the impact of I-R injury on responsiveness to both redox siblings. Under normoxic conditions, the HNO donor was superior to the NO• donor with respect to acute dose-dependent inotropic and lusitropic responses; neither nitrogen oxide donor elicited a robust acute myocardial response following I-R injury (Chin et al. 2016). However, despite this impact of I-R injury on the inotropic responsiveness to acute exposure to Angeli's salt (~2 min), administration of an HNO donor for the full duration of post-ischaemic reperfusion (~30 min) offers cardioprotective recovery of LV contractile function, with reduced ventricular fibrillation (Pagliaro et al. 2003; Chin et al. 2016). NO• reactivity with ROS and sGC oxidation (secondary to I-R-induced ROS dysregulation) are the likely potential contributing mechanisms to loss of vasodilator responsiveness to NO• in this context.

4.4 Chronic Heart Failure: Potential Impact of NO• Versus HNO

In recent years, next-generation HNO donors have been developed and have progressed into clinical trials, mainly focussing on acute decompensated heart failure as the therapeutic indication (as reviewed (Kemp-Harper et al. 2016)). Drug development in this field focussed in particular on avoiding the generation of by-products simultaneously with the desired HNO (as is seen with both Angeli's salt and IPA-NO) whilst considering aqueous solubility, stability at physiological pH and half-lives of HNO donors. In the context of heart failure, the ability of HNO donors to acutely enhance inotropic function has been well-documented (discussed above). The ability of NO redox siblings to limit heart failure with chronic administration over the longer term has, however, received considerably less research attention. This fact is perhaps less surprising in the case of NO• donors, given their susceptibility to reactivity with ROS, as well as to nitrate tolerance and NO• resistance, but such consideration for the case of HNO donors in the management of chronic heart failure is clearly warranted. As detailed above, HNO donors are in contrast impervious to reactivity with ROS, exhibit anti-hypertrophic and anti-fibrotic actions and elicit robust enhancement of LV function and vasodilator effects. As we have reviewed, both heart failure with reduced ejection fraction (HFrEF) and heart failure with preserved ejection fraction (known as HFpEF) are aetiologies mostly without effective therapeutic options, conditions in which longer-acting HNO donors offer promise (Kemp-Harper et al. 2016). The HNO donor 1-NCA has been shown to limit diabetes-induced diastolic dysfunction in mice when administered over a 4-week period (Cao et al. 2015), but its physical properties limit its utility in the clinic (existing as a blue oil).

5 Current Clinical Status of NO• Donors in the Context of ROS Dysregulation and Cardiovascular Disease

NO• donors can be classified as:

1. Agents which release NO• via enzymatic cleavage (the organic nitrates)
2. Agents which release NO• via a non-enzymatic process (sodium nitroprusside, molsidomine)

Additionally, inorganic nitrates and nitrite may generate NO•: the process involves a complex process of enzymatic conversion of nitrate to nitrite (typically by oral bacteria) followed by systemic reduction of nitrite to NO• (Amdahl et al. 2019). The organic nitrates are sometimes complexed with molecules exerting other effects. For example, nicorandil, a prophylactic anti-anginal agent, consists of an organic nitrate moiety coupled with a potassium channel opening agent. However, in practice, many of the effects of nicorandil suggest that effects of NO• release may predominate (Rajaratnam et al. 1999).

In general, organic nitrates are utilised for the prophylaxis of angina pectoris and play a secondary role in the management of systolic heart failure (Elkayam et al. 2004). However, there is evidence that the combination of ISDN with hydralazine may increase survival in patients with systolic heart failure (e.g. African Americans) (Taylor et al. 2004). Furthermore, short-acting preparations of both GTN and ISDN are used for the acute relief of angina symptoms, whilst intravenously infused organic nitrates are useful in the management of unstable angina pectoris and have accessory roles in the emergency management of acute myocardial infarction (Pasupathy et al. 2017). Sodium nitroprusside is utilised mainly for the acute management of hypertensive crises.

The phenomenon of *nitrate tolerance*, or progressive attenuation of organic nitrate effect during long-term continuous therapy, has represented a major concern regarding the clinical efficacy of organic nitrates (Munzel et al. 2014). Apart from the strategy of ensuring that organic nitrates are utilised with a "nitrate-free interval" incorporated into the treatment regimen, no means for limiting nitrate tolerance have thus far been identified.

As regards the mechanisms underlying the development of tolerance, there is substantial evidence that, at least in the clinical setting, tolerance is organic nitrate-selective and is associated with impaired cleavage of NO• from organic nitrate molecules (Sage et al. 2000). It is clear that several enzymes are involved in this process of organic nitrate cleavage and concomitant NO• release, but the strongest evidence implicates aldehyde dehydrogenase type 2 (ALDH2), which is inactivated in the presence of nitrate tolerance (Daiber and Munzel 2015).

Many recent studies have raised the possibility that the development of nitrate tolerance is associated with dysregulation of ROS. For example, this can be demonstrated in the context of the continuous administration of large doses of organic nitrates to healthy subjects (Gori et al. 2001; Daiber and Munzel 2015). The implications of association of incremental ROS generation with continuous administration of nitrates include the potential for nitrate tolerance to impact on responses to endogenous as well as exogenous NO• and for the long-term administration of organic nitrates to be counterproductive: indeed, some, but not all, clinical data support this postulate (Sage et al. 2000). Similarly, some, but not all, clinical studies have demonstrated the presence of cross-tolerance between organic nitrates and NO•.

The problem of *pseudo-tolerance* to organic nitrates is essentially a rebound phenomenon that occurs when organic nitrate therapy is suddenly withdrawn, for example, by an abrupt cessation of infusions or of a controlled delivery source of NO• release (Ferratini 1994). Nitrate withdrawal angina was first observed in the contest of the munitions industry but remains a clinical problem today. The rebound reflects nitrate-induced release of catecholamines in many patients, and the algebraic consequences on the vascular tone of sudden diminution of NO-induced vasodilator effect, combined with continued catecholamine-related increases in vasomotor tone. Thus, strategies involving the use of "nitrate-free periods" are not clinically ideal.

Finally, nitrates, as well as endogenous NO•, are subject to the phenomenon of *NO• resistance,* which presents as de novo, rather than acquired, impairment of haemodynamic and anti-aggregatory responses (Chirkov and Horowitz 2007). NO• resistance has now been described in a large number of conditions associated with ROS dysregulation, including stable and unstable angina pectoris; congestive heart failure; diabetes, especially with concomitant hyperglycaemia; aortic stenosis; coronary artery spasm; and polycystic ovarian syndrome. As such, NO• resistance may contribute to the progression and worsening of these conditions and attenuates the effectiveness of acute organic nitrate therapy to treat crises such as acute ischaemia or heart failure. A number of agents have been suggested as possible means for limiting NO• resistance clinically: these include ACE inhibitors in the context of heart failure (Willoughby et al. 2012), the "metabolic" anti-ischaemic agent perhexiline in the case of severe myocardial ischaemia (Willoughby et al. 2002), hydralazine in chronic heart failure and intravenously infused N-acetylcysteine in acute ischaemia and heart failure (Velagic et al. 2020). Rapid correction of hyperglycaemia via insulin infusion also markedly potentiates responses to NO• (Worthley et al. 2007).

6 Current Clinical Status of HNO Donors in the Context of ROS Dysregulation and Cardiovascular Disease

The seminal finding by Paolocci and colleagues that HNO serves as a positive cardiac inotrope in the setting of heart failure (Paolocci et al. 2001a, b, 2003), together with its spectrum of vaso- and cardioprotective actions, catapulted interest in developing HNO donors for the treatment of acute decompensated heart failure (ADHF). In patients with ADHF, pre-existing heart failure is exacerbated together with a varying degree of left ventricular systolic dysfunction. ROS dysregulation is evident in these patients with elevated plasma levels of uric acid (Bishu et al. 2012). Key to the treatment of ADHF is an improvement in arterial oxygenation, and loop diuretics are used for this purpose (Cotter et al. 1998). However, this approach does not address the underlying cause of ADHF development, and alternative therapies are sought. Such approaches include positive cardiac inotropes (e.g. dobutamine, levosimendan) to enhance cardiac contractility (Follath et al. 2002; Hsiao and Greenberg 2016), yet currently available agents only provide short-term improvement. There is also interest in the use of vasodilator therapies in this disease setting, yet the underlying ROS generation and resistance of some heart failure patients to organic nitrates (Armstrong et al. 1980) limit the use of NO•-based therapies. With an ability to enhance cardiac contractility and unload the heart (venous dilation), resistance to scavenging by $\bullet O_2^-$, ability to circumvent NO• resistance and lack of tolerance development, HNO donors offer an exciting option in the treatment of ADHF.

Towards this goal, novel synthetic HNO donors have recently been developed. The prototypical HNO donors, Angeli's salt (Hughes and Cammack 1999; Demoncheaux et al. 2003; DuMond and King 2011), IPA/NO (Miranda et al. 2005; Shoman et al. 2011), Piloty's acid (Pino and Feelisch 1994; Zamora et al. 1995) and acyloxy nitroso compounds (e.g. 1-NCA) (Sha et al. 2006) used in

Table 1 HNO donors and their properties

HNO donor	Structure	Properties
Angeli's salt	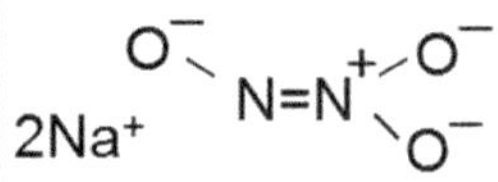	• Dissociates at physiological pH and temperature to yield HNO • $t_{1/2} \sim 2.5$ min • NO$^{\bullet}$ donor at pH < 4 and high concentrations (>10 μM) • Co-releases nitrite
Piloty's acid	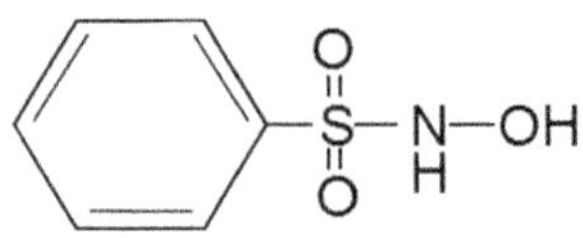	• Decomposes to release HNO • Rate of HNO release dependent upon pH • $t_{1/2} = 36$ h • HNO donor only above physiological pH • NO$^{\bullet}$ donor at physiological pH • Co-releases benzensulfinate
Isopropylamine NONOate (IPA-NO)	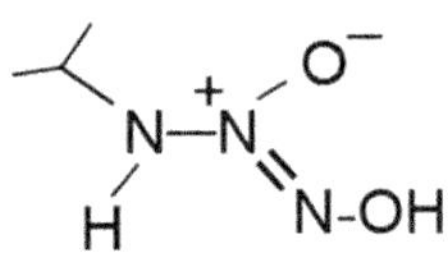	• Dissociates at physiological pH and temperature to yield HNO • $t_{1/2} \sim 2.3$ min • Donates NO$^{\bullet}$ at pH < 7 • No co-release of nitrite • Nitrosamine by-product
1-Nitrosocyclo hexyl acetate (1-NCA)	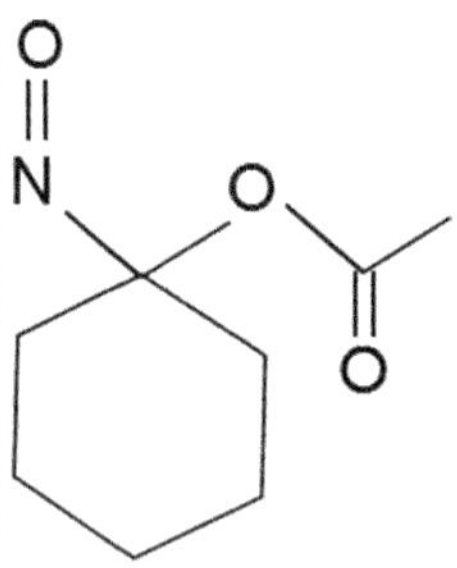	• Undergoes hydrolysis to yield HNO • Rate of HNO release dependent upon pH • $t_{1/2} = >13$ h at neutral pH • Co-releases nitrite and NO$^{\bullet}$
CXL-1020	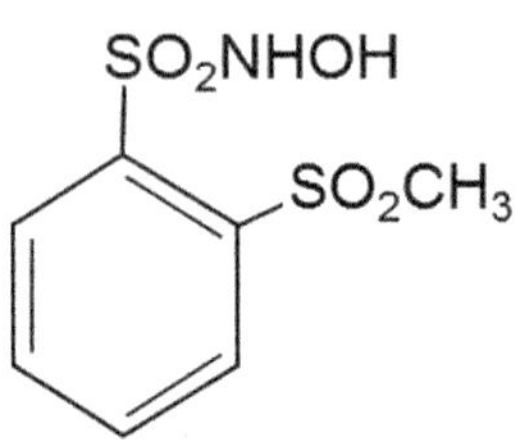	• Spontaneously decomposes at physiological pH to yield HNO • $t_{1/2} = 2–3$ min • Inert organic by-product, CXL-1051
CXL-1036	Not disclosed	• 40% oral bioavailability • $t_{1/2} = 30$ min
BMS-986231 (CXL-1427)	Not disclosed	• Decomposes at physiological pH • $t_{1/2} = 0.7–2.5$ h

preclinical studies, were not suitable for clinical use given their short-half lives, highly alkaline vehicles and active by-products (Table 1). A series of next-generation pure HNO donors have been synthesised and include compounds such as CXL-1020 (Sabbah et al. 2013), CXL-1036 (Kemp-Harper et al. 2016) and

BMS-986231 (formerly known as CXL-1427) (Cowart et al. 2019) (Table 1). CXL-1020, 2-methylsulfonyl benzene N-hydroxy sulphonamide, spontaneously decomposes under physiological pH to generate HNO (half-life 2 min) and an inert organic by-product, CXL-1051 (Sabbah et al. 2013). CXL-1020 has efficacy in patients with ADHF, with a 6 h intravenous infusion leading to a reduction in pulmonary capillary wedge pressure and increase in cardiac and stroke volume index (Sabbah et al. 2013). No serious adverse effects were noted with only a modest decrease in systemic vascular resistance and no increase in heart rate. These findings were promising and highlighted that HNO donors, unlike legacy inotropes, were not associated with adverse effects such as tachycardia and arrhythmias. However, longer periods of infusion of CXL-1020 (24–48 h, i.v.) were associated with irritation at the site of infusion (Arcaro et al. 2014; Mebazaa et al. 2015); therefore the focus turned to new-generation HNO donors without such limitations.

BMS-986231 (half-life, 40–144 min) is a key candidate and has a favourable safety profile, with a recent Phase I clinical trial in healthy volunteers demonstrating BMS-986231 (24 h and 48 h intravenous infusion) to be well tolerated (Cowart et al. 2019) up to a dose of 10 μg/kg/min, with headaches the most frequently reported side effect (common for vasodilator therapy). Importantly, haemodynamic efficacy was evident with BMS-986231 lowering systolic and diastolic blood pressure and increasing cardiac index, in a well-tolerated and dose-dependent manner. A Phase IIa clinical trial in patients with decompensated heart failure reported that BMS-986231 (6 h intravenous infusion) caused dose-dependent (5, 7 and 12 μg/kg/min) reductions in pulmonary capillary wedge pressure together with increases in stroke volume and cardiac index and a reduction in total peripheral resistance (Tita et al. 2017). The clinical utility of BMS-986231 is being further evaluated via an ongoing multicentre, randomised, double-blind, placebo-controlled clinical trial, the StandUP-AHF study (Study Assessing Nitroxyl Donor Upon Presentation with Acute Heart Failure) (Felker et al. 2019). Collectively, clinical studies to date have demonstrated efficacy, and favourable safety profiles, of new-generation HNO donors in the setting of heart failure, where ROS dysregulation is evident. We eagerly await the outcome of the StandUP AHF study to fully evaluate the therapeutic potential of HNO donors in ADHF.

7 Concluding Remarks

ROS dysregulation associated with cardiovascular disease markedly limits the therapeutic utility of NO•-based drugs, contributing to NO• resistance and tolerance development. The unique pharmacological properties of HNO, encompassing direct thiol interaction leading to positive cardiac inotropic and lusitropic effects, vasodilator and anti-aggregatory capacity, resistance to scavenging by $\bullet O_2^-$, circumvention of NO• resistance and the ability to rapidly limit $\bullet O_2^-$ generation, confer therapeutic advantages and preserved efficacy in the face of increased ROS levels. Recent clinical studies suggest that HNO donors will continue to emerge as a new pharmacotherapy for the management of acute heart failure scenarios and we speculate they may also be of use in the treatment of acute cardiovascular emergencies. Moreover, the aforementioned cardio- and vaso-protective actions of HNO, coupled with anti-remodelling and anti-inflammatory properties and lack of tolerance development, suggest HNO donors may have value in the long-term treatment of other cardiovascular pathologies such as chronic heart failure, angina, diabetes and hypertension.

Acknowledgements **Grants** This work was supported in part by both the National Health and Medical Research Council (NHMRC) of Australia, including APP1045140 (to RHR, JDH and BKK) and APP1059960 (to RHR), and the Victorian Government's Operational Infrastructure Support Program. AV is supported by an Australian Government Research Training Program Scholarship and a Monash Graduate Excellence Scholarship (MGES).

Disclosures The authors declare that there are no conflicts of interest.

References

Almourani R, Chinnakotla B, Patel R, Kurukulasuriya LR, Sowers J (2019) Diabetes and cardiovascular disease: an update. Curr Diab Rep 19(12):161

Amdahl MB, DeMartino AW, Gladwin MT (2019) Inorganic nitrite bioactivation and role in physiological signaling and therapeutics. Biol Chem 401(1):201–211

Anderson RA, Ellis GR, Evans LM, Morris K, Chirkov YY, Horowitz JD, Jackson SK, Rees A, Lewis MJ, Frenneaux MP (2005) Platelet nitrate responsiveness in fasting and postprandial type 2 diabetes. Diab Vasc Dis Res 2(2):88–93

Andrews KL, Lumsden NG, Farry J, Jefferis AM, Kemp-Harper BK, Chin-Dusting JP (2015) Nitroxyl: a vasodilator of human vessels that is not susceptible to tolerance. Clin Sci (Lond) 129 (2):179–187

Andrews KL, Sampson AK, Irvine JC, Shihata WA, Michell DL, Lumsden NG, Lim C, Huet O, Drummond GR, Kemp-Harper BK, Chin-Dusting JP (2016) Nitroxyl (HNO) reduces endothelial and monocyte activation and promotes M2 macrophage polarization. Clin Sci (Lond) 130 (18):1629–1640

Antonopoulos AS, Margaritis M, Coutinho P, Shirodaria C, Psarros C, Herdman L, Sanna F, De Silva R, Petrou M, Sayeed R, Krasopoulos G, Lee R, Digby J, Reilly S, Bakogiannis C, Tousoulis D, Kessler B, Casadei B, Channon KM, Antoniades C (2015) Adiponectin as a link between type 2 diabetes and vascular NADPH oxidase activity in the human arterial wall: the regulatory role of perivascular adipose tissue. Diabetes 64(6):2207–2219

Arcaro A, Lembo G, Tocchetti CG (2014) Nitroxyl (HNO) for treatment of acute heart failure. Curr Heart Fail Rep 11(3):227–235

Armstrong PW, Armstrong JA, Marks GS (1980) Pharmacokinetic-hemodynamic studies of intravenous nitroglycerin in congestive cardiac failure. Circulation 62(1):160–166

Azumi H, Inoue N, Ohashi Y, Terashima M, Mori T, Fujita H, Awano K, Kobayashi K, Maeda K, Hata K, Shinke T, Kobayashi S, Hirata K, Kawashima S, Itabe H, Hayashi Y, Imajoh-Ohmi S, Itoh H, Yokoyama M (2002) Superoxide generation in directional coronary atherectomy specimens of patients with angina pectoris: important role of NAD(P)H oxidase. Arterioscler Thromb Vasc Biol 22(11):1838–1844

Barouch LA, Harrison RW, Skaf MW, Rosas GO, Cappola TP, Kobeissi ZA, Hobai IA, Lemmon CA, Burnett AL, O'Rourke B, Rodriguez ER, Huang PL, Lima JA, Berkowitz DE, Hare JM (2002) Nitric oxide regulates the heart by spatial confinement of nitric oxide synthase isoforms. Nature 416(6878):337–339

Beckman JS, Koppenol WH (1996) Nitric oxide, superoxide, and peroxynitrite: the good, the bad, and ugly. Am J Phys 271(5 Pt 1):C1424–C1437

Bermejo E, Saenz DA, Alberto F, Rosenstein RE, Bari SE, Lazzari MA (2005) Effect of nitroxyl on human platelets function. Thromb Haemost 94(3):578–584

Bishu K, Deswal A, Chen HH, LeWinter MM, Lewis GD, Semigran MJ, Borlaug BA, McNulty S, Hernandez AF, Braunwald E, Redfield MM (2012) Biomarkers in acutely decompensated heart failure with preserved or reduced ejection fraction. Am Heart J 164(5):763–770.e763

Bullen ML, Miller AA, Andrews KL, Irvine JC, Ritchie RH, Sobey CG, Kemp-Harper BK (2011a) Nitroxyl (HNO) as a vasoprotective signaling molecule. Antioxid Redox Signal 14 (9):1675–1686

Bullen ML, Miller AA, Dharmarajah J, Drummond GR, Sobey CG, Kemp-Harper BK (2011b) Vasorelaxant and antiaggregatory actions of the nitroxyl donor isopropylamine NONOate are maintained in hypercholesterolemia. Am J Physiol Heart Circ Physiol 301(4):H1405–H1414

Calderone A, Thaik CM, Takahashi N, Chang DL, Colucci WS (1998) Nitric oxide, atrial natriuretic peptide, and cyclic GMP inhibit the growth-promoting effects of norepinephrine in cardiac myocytes and fibroblasts. J Clin Invest 101(4):812–818

Cao N, Wong YG, Rosli S, Kiriazis H, Huynh K, Qin C, Du XJ, Kemp-Harper BK, Ritchie RH (2015) Chronic administration of the nitroxyl donor 1-nitrosocyclo hexyl acetate limits left ventricular diastolic dysfunction in a mouse model of diabetes mellitus in vivo. Circ Heart Fail 8 (3):572–581

Carroll JD, Lang RM, Neumann AL, Borow KM, Rajfer SI (1986) The differential effects of positive inotropic and vasodilator therapy on diastolic properties in patients with congestive cardiomyopathy. Circulation 74(4):815–825

Cheong E, Tumbev V, Abramson J, Salama G, Stoyanovsky DA (2005) Nitroxyl triggers Ca2+ release from skeletal and cardiac sarcoplasmic reticulum by oxidizing ryanodine receptors. Cell Calcium 37(1):87–96

Chin KY, Qin C, Cao N, Kemp-Harper BK, Woodman OL, Ritchie RH (2014) The concomitant coronary vasodilator and positive inotropic actions of the nitroxyl donor Angeli's salt in the intact rat heart: contribution of soluble guanylyl cyclase-dependent and -independent mechanisms. Br J Pharmacol 171(7):1722–1734

Chin KY, Michel L, Qin CX, Cao N, Woodman OL, Ritchie RH (2016) The HNO donor Angeli's salt offers potential haemodynamic advantages over NO or dobutamine in ischaemia-reperfusion injury in the rat heart ex vivo. Pharmacol Res 104:165–175

Chirkov YY, Horowitz JD (2007) Impaired tissue responsiveness to organic nitrates and nitric oxide: a new therapeutic frontier? Pharmacol Ther 116(2):287–305

Cotter G, Metzkor E, Kaluski E, Faigenberg Z, Miller R, Simovitz A, Shaham O, Marghitay D, Koren M, Blatt A, Moshkovitz Y, Zaidenstein R, Golik A (1998) Randomised trial of high-dose isosorbide dinitrate plus low-dose furosemide versus high-dose furosemide plus low-dose isosorbide dinitrate in severe pulmonary oedema. Lancet 351(9100):389–393

Cowart D, Venuti RP, Lynch K, Guptill JT, Noveck RJ, Foo SY (2019) A phase 1 randomized study of single intravenous infusions of the novel nitroxyl donor BMS-986231 in healthy volunteers. J Clin Pharmacol 59(5):717–730

Cox DA, Vita JA, Treasure CB, Fish RD, Alexander RW, Ganz P, Selwyn AP (1989) Atherosclerosis impairs flow-mediated dilation of coronary arteries in humans. Circulation 80(3):458–465

Cuello F, Wittig I, Lorenz K, Eaton P (2018) Oxidation of cardiac myofilament proteins: priming for dysfunction? Mol Asp Med 63:47–58

Dai T, Tian Y, Tocchetti CG, Katori T, Murphy AM, Kass DA, Paolocci N, Gao WD (2007) Nitroxyl increases force development in rat cardiac muscle. J Physiol 580(Pt.3):951–960

Daiber A, Munzel T (2015) Organic nitrate therapy, nitrate tolerance, and nitrate-induced endothelial dysfunction: emphasis on redox biology and oxidative stress. Antioxid Redox Signal 23 (11):899–942

Daiber A, Wenzel P, Oelze M, Munzel T (2008) New insights into bioactivation of organic nitrates, nitrate tolerance and cross-tolerance. Clin Res Cardiol 97(1):12–20

Dautov RF, Ngo DT, Licari G, Liu S, Sverdlov AL, Ritchie RH, Kemp-Harper BK, Horowitz JD, Chirkov YY (2013) The nitric oxide redox sibling nitroxyl partially circumvents impairment of platelet nitric oxide responsiveness. Nitric Oxide 35:72–78

De Witt BJ, Marrone JR, Kaye AD, Keefer LK, Kadowitz PJ (2001) Comparison of responses to novel nitric oxide donors in the feline pulmonary vascular bed. Eur J Pharmacol 430 (2–3):311–315

Demoncheaux EA, Foster PJ, Borland CD, Smith AP, Higenbottam TW, Davies MB (2003) Determination of trace concentrations of dissolved nitric oxide in a biological buffer. Analyst 128(10):1281–1285

Ding W, Li Z, Shen X, Martin J, King SB, Sivakumaran V, Paolocci N, Gao WD (2011) Reversal of isoflurane-induced depression of myocardial contraction by nitroxyl via myofilament sensitization to Ca2+. J Pharmacol Exp Ther 339(3):825–831

Dinh QN, Drummond GR, Sobey CG, Chrissobolis S (2014) Roles of inflammation, oxidative stress, and vascular dysfunction in hypertension. Biomed Res Int 2014:406960

Drummond GR, Selemidis S, Griendling KK, Sobey CG (2011) Combating oxidative stress in vascular disease: NADPH oxidases as therapeutic targets. Nat Rev Drug Discov 10(6):453–471

DuMond JF, King SB (2011) The chemistry of nitroxyl-releasing compounds. Antioxid Redox Signal 14(9):1637–1648

Dunn FG, McLenachan J, Isles CG, Brown I, Dargie HJ, Lever AF, Lorimer AR, Murray GD, Pringle SD, Robertson JW (1990) Left ventricular hypertrophy and mortality in hypertension: an analysis of data from the Glasgow blood pressure clinic. J Hypertens 8(8):775–782

El-Armouche A, Wahab A, Wittkopper K, Schulze T, Bottcher F, Pohlmann L, King SB, DuMond JF, Gerloff C, Boger RH, Eschenhagen T, Carrier L, Donzelli S (2010) The new HNO donor, 1-nitrosocyclohexyl acetate, increases contractile force in normal and beta-adrenergically desensitized ventricular myocytes. Biochem Biophys Res Commun 402(2):340–344

Elkayam U, Bitar F, Akhter MW, Khan S, Patrus S, Derakhshani M (2004) Intravenous nitroglycerin in the treatment of decompensated heart failure: potential benefits and limitations. J Cardiovasc Pharmacol Ther 9(4):227–241

Evgenov OV, Pacher P, Schmidt PM, Hasko G, Schmidt HH, Stasch JP (2006) NO-independent stimulators and activators of soluble guanylate cyclase: discovery and therapeutic potential. Nat Rev Drug Discov 5(9):755–768

Farah C, Michel LYM, Balligand JL (2018) Nitric oxide signalling in cardiovascular health and disease. Nat Rev Cardiol 15(5):292–316

Favaloro JL, Kemp-Harper BK (2007) The nitroxyl anion (HNO) is a potent dilator of rat coronary vasculature. Cardiovasc Res 73(3):587–596

Favaloro JL, Kemp-Harper BK (2009) Redox variants of NO (NO{middle dot} and HNO) elicit vasorelaxation of resistance arteries via distinct mechanisms. Am J Physiol Heart Circ Physiol 296(5):H1274–H1280

Fayers KE, Cummings MH, Shaw KM, Laight DW (2003) Nitrate tolerance and the links with endothelial dysfunction and oxidative stress. Br J Clin Pharmacol 56(6):620–628

Felker GM, Borentain M, Cleland JG, DeSouza MM, Kessler PD, O'Connor CM, Seiffert D, Teerlink JR, Voors AA, McMurray JJV (2019) Rationale and design for the development of a novel nitroxyl donor in patients with acute heart failure. Eur J Heart Fail 21(8):1022–1031

Ferratini M (1994) Risk of rebound phenomenon during nitrate withdrawal. Int J Cardiol 45 (2):89–96

Follath F, Cleland JG, Just H, Papp JG, Scholz H, Peuhkurinen K, Harjola VP, Mitrovic V, Abdalla M, Sandell EP, Lehtonen L, Steering C, S. Investigators of the Levosimendan Infusion versus Dobutamine (2002) Efficacy and safety of intravenous levosimendan compared with dobutamine in severe low-output heart failure (the LIDO study): a randomised double-blind trial. Lancet 360(9328):196–202

Froehlich JP, Mahaney JE, Keceli G, Pavlos CM, Goldstein R, Redwood AJ, Sumbilla C, Lee DI, Tocchetti CG, Kass DA, Paolocci N, Toscano JP (2008) Phospholamban thiols play a central

role in activation of the cardiac muscle sarcoplasmic reticulum calcium pump by nitroxyl. Biochemistry 47(50):13150–13152
Fukuto JM (2019) A recent history of nitroxyl chemistry, pharmacology and therapeutic potential. Br J Pharmacol 176(2):135–146
Fukuto JM, Chiang K, Hszieh R, Wong P, Chaudhuri G (1992) The pharmacological activity of nitroxyl: a potent vasodilator with activity similar to nitric oxide and/or endothelium-derived relaxing factor. J Pharmacol Exp Ther 263(2):546–551
Fukuto JM, Cisneros CJ, Kinkade RL (2013) A comparison of the chemistry associated with the biological signaling and actions of nitroxyl (HNO) and nitric oxide (NO). J Inorg Biochem 118:201–208
Gao WD, Murray CI, Tian Y, Zhong X, DuMond JF, Shen X, Stanley BA, Foster DB, Wink DA, King SB, Van Eyk JE, Paolocci N (2012) Nitroxyl-mediated disulfide bond formation between cardiac myofilament cysteines enhances contractile function. Circ Res 111(8):1002–1011
Garreffa AM, Woodman OL, Cao AH, Ritchie RH (2006) Sodium nitroprusside protects adult rat cardiac myocytes from cellular injury induced by simulated ischemia: role for a non-cGMP-dependent mechanism of nitric oxide protection. J Cardiovasc Pharmacol 47(1):1–8
Goh SS, Woodman OL, Pepe S, Cao AH, Qin C, Ritchie RH (2007) The red wine antioxidant resveratrol prevents cardiomyocyte injury following ischemia-reperfusion via multiple sites and mechanisms. Antioxid Redox Signal 9(1):101–113
Gori T, Mak SS, Kelly S, Parker JD (2001) Evidence supporting abnormalities in nitric oxide synthase function induced by nitroglycerin in humans. J Am Coll Cardiol 38(4):1096–1101
Gryglewski RJ, Palmer RM, Moncada S (1986) Superoxide anion is involved in the breakdown of endothelium-derived vascular relaxing factor. Nature 320(6061):454–456
Guzik TJ, Sadowski J, Guzik B, Jopek A, Kapelak B, Przybylowski P, Wierzbicki K, Korbut R, Harrison DG, Channon KM (2006) Coronary artery superoxide production and nox isoform expression in human coronary artery disease. Arterioscler Thromb Vasc Biol 26(2):333–339
Hannan RD, Jenkins A, Jenkins AK, Brandenburger Y (2003) Cardiac hypertrophy: a matter of translation. Clin Exp Pharmacol Physiol 30(8):517–527
Hartman JC, Del Rio CL, Reardon JE, Zhang K, Sabbah HN (2018) Intravenous infusion of the novel HNO donor BMS-986231 is associated with beneficial inotropic, lusitropic, and vasodilatory properties in 2 canine models of heart failure. JACC Basic Trans Sci 3(5):625–638
Heitzer T, Krohn K, Albers S, Meinertz T (2000) Tetrahydrobiopterin improves endothelium-dependent vasodilation by increasing nitric oxide activity in patients with type II diabetes mellitus. Diabetologia 43(11):1435–1438
Hill MF, Singal PK (1996) Antioxidant and oxidative stress changes during heart failure subsequent to myocardial infarction in rats. Am J Pathol 148(1):291–300
Ho KK, Pinsky JL, Kannel WB, Levy D (1993) The epidemiology of heart failure: the Framingham study. J Am Coll Cardiol 22(4 Suppl A):6A–13A
Horowitz JD (2004) Tolerance induction during therapy with long-acting nitrates: how extensive is the "collateral damage"? Cardiovasc Drugs Ther 18(1):11–12
Hsiao R, Greenberg B (2016) Contemporary treatment of acute heart failure. Prog Cardiovasc Dis 58(4):367–378
Hughes MN, Cammack R (1999) Synthesis, chemistry, and applications of nitroxyl ion releasers sodium trioxodinitrate or Angeli's salt and Piloty's acid. Methods Enzymol 301:279–287
Iadecola C, Davisson RL (2008) Hypertension and cerebrovascular dysfunction. Cell Metab 7 (6):476–484
Ichinose F, Bloch KD, Wu JC, Hataishi R, Aretz HT, Picard MH, Scherrer-Crosbie M (2004) Pressure overload-induced LV hypertrophy and dysfunction in mice are exacerbated by congenital NOS3 deficiency. Am J Physiol Heart Circ Physiol 286(3):H1070–H1075
Irvine JC, Favaloro JL, Kemp-Harper BK (2003) NO- activates soluble guanylate cyclase and Kv channels to vasodilate resistance arteries. Hypertension 41(6):1301–1307
Irvine JC, Favaloro JL, Widdop RE, Kemp-Harper BK (2007) Nitroxyl anion donor, Angeli's salt, does not develop tolerance in rat isolated aortae. Hypertension 49(4):885–892

Irvine JC, Ritchie RH, Favaloro JL, Andrews KL, Widdop RE, Kemp-Harper BK (2008) Nitroxyl (HNO): the Cinderella of the nitric oxide story. Trends Pharmacol Sci 29(12):601–608

Irvine JC, Kemp-Harper BK, Widdop RE (2011) Chronic administration of the HNO donor Angeli's salt does not lead to tolerance, cross-tolerance, or endothelial dysfunction: comparison with GTN and DEA/NO. Antioxid Redox Signal 14(9):1615–1624

Irvine JC, Cao N, Gossain S, Alexander AE, Love JE, Qin C, Horowitz JD, Kemp-Harper BK, Ritchie RH (2013a) HNO/cGMP-dependent antihypertrophic actions of isopropylamine-NONOate in neonatal rat cardiomyocytes: potential therapeutic advantages of HNO over NO. Am J Physiol Heart Circ Physiol 305(3):H365–H377

Irvine JC, Ravi RM, Kemp-Harper BK, Widdop RE (2013b) Nitroxyl donors retain their depressor effects in hypertension. Am J Physiol Heart Circ Physiol 305(6):H939–H945

Kahlberg N, Qin CX, Anthonisz J, Jap E, Ng HH, Jelinic M, Parry LJ, Kemp-Harper BK, Ritchie RH, Leo CH (2016) Adverse vascular remodelling is more sensitive than endothelial dysfunction to hyperglycaemia in diabetic rat mesenteric arteries. Pharmacol Res 111:325–335

Keceli G, Majumdar A, Thorpe CN, Jun S, Tocchetti CG, Lee DI, Mahaney JE, Paolocci N, Toscano JP (2019) Nitroxyl (HNO) targets phospholamban cysteines 41 and 46 to enhance cardiac function. J Gen Physiol 151(6):758–770

Kemp-Harper BK, Horowitz JD, Ritchie RH (2016) Therapeutic potential of nitroxyl (HNO) donors in the management of acute decompensated heart failure. Drugs 76(14):1337–1348

Kloss S, Bouloumie A, Mulsch A (2000) Aging and chronic hypertension decrease expression of rat aortic soluble guanylyl cyclase. Hypertension 35(1 Pt 1):43–47

Kohr MJ, Kaludercic N, Tocchetti CG, Dong Gao W, Kass DA, Janssen PM, Paolocci N, Ziolo MT (2010) Nitroxyl enhances myocyte Ca2+ transients by exclusively targeting SR Ca2+-cycling. Front Biosci (Elite Ed) 2:614–626

Kuzkaya N, Weissmann N, Harrison DG, Dikalov S (2003) Interactions of peroxynitrite, tetrahydrobiopterin, ascorbic acid, and thiols: implications for uncoupling endothelial nitric-oxide synthase. J Biol Chem 278(25):22546–22554

Lancel S, Zhang J, Evangelista A, Trucillo MP, Tong X, Siwik DA, Cohen RA, Colucci WS (2009) Nitroxyl activates SERCA in cardiac myocytes via glutathiolation of cysteine 674. Circ Res 104 (6):720–723

Lee MY, Griendling KK (2008) Redox signaling, vascular function, and hypertension. Antioxid Redox Signal 10(6):1045–1059

Leo CH, Joshi A, Hart JL, Woodman OL (2012) Endothelium-dependent nitroxyl-mediated relaxation is resistant to superoxide anion scavenging and preserved in diabetic rat aorta. Pharmacol Res 66(5):383–391

Lin EQ, Irvine JC, Cao AH, Alexander AE, Love JE, Patel R, McMullen JR, Kaye DM, Kemp-Harper BK, Ritchie RH (2012) Nitroxyl (HNO) stimulates soluble guanylyl cyclase to suppress cardiomyocyte hypertrophy and superoxide generation. PLoS One 7(4):e34892

Loyer X, Gomez AM, Milliez P, Fernandez-Velasco M, Vangheluwe P, Vinet L, Charue D, Vaudin E, Zhang W, Sainte-Marie Y, Robidel E, Marty I, Mayer B, Jaisser F, Mercadier JJ, Richard S, Shah AM, Benitah JP, Samuel JL, Heymes C (2008) Cardiomyocyte overexpression of neuronal nitric oxide synthase delays transition toward heart failure in response to pressure overload by preserving calcium cycling. Circulation 117(25):3187–3198

Mebazaa A, Longrois D, Metra M, Mueller C, Richards AM, Roessig L, Seronde MF, Sato N, Stockbridge NL, Gattis Stough W, Alonso A, Cody RJ, Cook Bruns N, Gheorghiade M, Holzmeister J, Laribi S, Zannad F (2015) Agents with vasodilator properties in acute heart failure: how to design successful trials. Eur J Heart Fail 17(7):652–664

Messerli FH, Williams B, Ritz E (2007) Essential hypertension. Lancet 370(9587):591–603

Miller TW, Cherney MM, Lee AJ, Francoleon NE, Farmer PJ, King SB, Hobbs AJ, Miranda KM, Burstyn JN, Fukuto JM (2009) The effects of nitroxyl (HNO) on soluble guanylate cyclase activity: interactions at ferrous heme and cysteine thiols. J Biol Chem 284(33):21788–21796

Miller AA, Maxwell KF, Chrissobolis S, Bullen ML, Ku JM, Michael De Silva T, Selemidis S, Hooker EU, Drummond GR, Sobey CG, Kemp-Harper BK (2013) Nitroxyl (HNO) suppresses vascular Nox2 oxidase activity. Free Radic Biol Med 60:264–271

Miranda KM, Yamada K, Espey MG, Thomas DD, DeGraff W, Mitchell JB, Krishna MC, Colton CA, Wink DA (2002) Further evidence for distinct reactive intermediates from nitroxyl and peroxynitrite: effects of buffer composition on the chemistry of Angeli's salt and synthetic peroxynitrite. Arch Biochem Biophys 401(2):134–144

Miranda KM, Nims RW, Thomas DD, Espey MG, Citrin D, Bartberger MD, Paolocci N, Fukuto JM, Feelisch M, Wink DA (2003) Comparison of the reactivity of nitric oxide and nitroxyl with heme proteins. A chemical discussion of the differential biological effects of these redox related products of NOS. J Inorg Biochem 93(1–2):52–60

Miranda KM, Nagasawa HT, Toscano JP (2005) Donors of HNO. Curr Top Med Chem 5 (7):649–664

Mondoro TH, Ryan BB, Hrinczenko BW, Schechter AN, Vostal JG, Alayash AI (2001) Biological action of nitric oxide donor compounds on platelets from patients with sickle cell disease. Br J Haematol 112(4):1048–1054

Munzel T, Daiber A, Ullrich V, Mulsch A (2005) Vascular consequences of endothelial nitric oxide synthase uncoupling for the activity and expression of the soluble guanylyl cyclase and the cGMP-dependent protein kinase. Arterioscler Thromb Vasc Biol 25(8):1551–1557

Munzel T, Daiber A, Gori T (2013) More answers to the still unresolved question of nitrate tolerance. Eur Heart J 34(34):2666–2673

Munzel T, Steven S, Daiber A (2014) Organic nitrates: update on mechanisms underlying vasodilation, tolerance and endothelial dysfunction. Vasc Pharmacol 63(3):105–113

Nishikawa T, Edelstein D, Du XL, Yamagishi S, Matsumura T, Kaneda Y, Yorek MA, Beebe D, Oates PJ, Hammes HP, Giardino I, Brownlee M (2000) Normalizing mitochondrial superoxide production blocks three pathways of hyperglycaemic damage. Nature 404(6779):787–790

Okon EB, Chung AW, Rauniyar P, Padilla E, Tejerina T, McManus BM, Luo H, van Breemen C (2005) Compromised arterial function in human type 2 diabetic patients. Diabetes 54 (8):2415–2423

Pagliaro P, Mancardi D, Rastaldo R, Penna C, Gattullo D, Miranda KM, Feelisch M, Wink DA, Kass DA, Paolocci N (2003) Nitroxyl affords thiol-sensitive myocardial protective effects akin to early preconditioning. Free Radic Biol Med 34(1):33–43

Paolocci N, Biondi R, Bettini M, Lee CI, Berlowitz CO, Rossi R, Xia Y, Ambrosio G, L'Abbate A, Kass DA, Zweier JL (2001a) Oxygen radical-mediated reduction in basal and agonist-evoked NO release in isolated rat heart. J Mol Cell Cardiol 33(4):671–679

Paolocci N, Saavedra WF, Miranda KM, Martignani C, Isoda T, Hare JM, Espey MG, Fukuto JM, Feelisch M, Wink DA, Kass DA (2001b) Nitroxyl anion exerts redox-sensitive positive cardiac inotropy in vivo by calcitonin gene-related peptide signaling. Proc Natl Acad Sci U S A 98 (18):10463–10468

Paolocci N, Katori T, Champion HC, John MES, Miranda KM, Fukuto JM, Wink DA, Kass DA (2003) Positive inotropic and lusitropic effects of HNO/NO- in failing hearts: independence from beta-adrenergic signaling. Proc Natl Acad Sci U S A 100(9):5537–5542

Paolocci N, Jackson MI, Lopez BE, Miranda K, Tocchetti CG, Wink DA, Hobbs AJ, Fukuto JM (2007) The pharmacology of nitroxyl (HNO) and its therapeutic potential: not just the Janus face of NO. Pharmacol Ther 113(2):442–458

Parissis J, Bistola V, Ikonomidis I, Triposkiadis F (2017) Nitroxyl donors for acute heart failure: promising newcomers. Eur J Heart Fail 19(10):1333–1334

Pasupathy S, Tavella R, Grover S, Raman B, Procter NEK, Du YT, Mahadavan G, Stafford I, Heresztyn T, Holmes A, Zeitz C, Arstall M, Selvanayagam J, Horowitz JD, Beltrame JF (2017) Early use of N-acetylcysteine with nitrate therapy in patients undergoing primary percutaneous coronary intervention for ST-segment-elevation myocardial infarction reduces myocardial infarct size (the NACIAM trial [N-acetylcysteine in acute myocardial infarction]). Circulation 136(10):894–903

Paulis L, Matuskova J, Adamcova M, Pelouch V, Simko J, Krajcirovicova K, Potacova A, Hulin I, Janega P, Pechanova O, Simko F (2008) Regression of left ventricular hypertrophy and aortic remodelling in NO-deficient hypertensive rats: effect of L-arginine and spironolactone. Acta Physiol (Oxf) 194(1):45–55

Paulus WJ, Vantrimpont PJ, Shah AM (1994) Acute effects of nitric oxide on left ventricular relaxation and diastolic distensibility in humans. Assessment by bicoronary sodium nitroprusside infusion. Circulation 89(5):2070–2078

Phillips L, Toledo AH, Lopez-Neblina F, Anaya-Prado R, Toledo-Pereyra LH (2009) Nitric oxide mechanism of protection in ischemia and reperfusion injury. J Investig Surg 22(1):46–55

Pino RZ, Feelisch M (1994) Bioassay discrimination between nitric oxide (NO.) and nitroxyl (NO-) using L-cysteine. Biochem Biophys Res Commun 201(1):54–62

Qin CX, Anthonisz J, Leo CH, Kahlberg N, Velagic A, Li M, Jap E, Woodman OL, Parry LJ, Horowitz JD, Kemp-Harper BK, Ritchie RH (2020) Nitric oxide resistance, induced in the myocardium by diabetes, is circumvented by the nitric oxide redox sibling, nitroxyl. Antioxid Redox Signal 32(1):60–77

Rajaratnam R, Brieger DB, Hawkins R, Freedman SB (1999) Attenuation of anti-ischemic efficacy during chronic therapy with nicorandil in patients with stable angina pectoris. Am J Cardiol 83 (7):1120–1124, A1129

Ritchie RH, Abel ED (2020) Basic mechanisms of diabetic heart disease. Circ Res 126 (11):1501–1525

Ritchie RH, Schiebinger RJ, LaPointe MC, Marsh JD (1998) Angiotensin II-induced hypertrophy of adult rat cardiomyocytes is blocked by nitric oxide. Am J Phys 275(4 Pt 2):H1370–H1374

Ritchie RH, Irvine JC, Rosenkranz AC, Patel R, Wendt IR, Horowitz JD, Kemp-Harper BK (2009) Exploiting cGMP-based therapies for the prevention of left ventricular hypertrophy: NO* and beyond. Pharmacol Ther 124(3):279–300

Ritchie RH, Drummond GR, Sobey CG, De Silva TM, Kemp-Harper BK (2017) The opposing roles of NO and oxidative stress in cardiovascular disease. Pharmacol Res 116:57–69

Roof SR, Ueyama Y, Mazhari R, Hamlin RL, Hartman JC, Ziolo MT, Reardon JE, Del Rio CL (2017) CXL-1020, a novel nitroxyl (HNO) prodrug, is more effective than milrinone in models of diastolic dysfunction-A cardiovascular therapeutic: an efficacy and safety study in the rat. Front Physiol 8:894

Rosenkranz AC, Hood SG, Woods RL, Dusting GJ, Ritchie RH (2002) Acute antihypertrophic actions of bradykinin in the rat heart: importance of cyclic GMP. Hypertension 40(4):498–503

Rudolph A, Abdel-Aty H, Bohl S, Boye P, Zagrosek A, Dietz R, Schulz-Menger J (2009) Noninvasive detection of fibrosis applying contrast-enhanced cardiac magnetic resonance in different forms of left ventricular hypertrophy relation to remodeling. J Am Coll Cardiol 53 (3):284–291

Ruetten H, Zabel U, Linz W, Schmidt HH (1999) Downregulation of soluble guanylyl cyclase in young and aging spontaneously hypertensive rats. Circ Res 85(6):534–541

Sabbah HN, Tocchetti CG, Wang M, Daya S, Gupta RC, Tunin RS, Mazhari R, Takimoto E, Paolocci N, Cowart D, Colucci WS, Kass DA (2013) Nitroxyl (HNO): a novel approach for the acute treatment of heart failure. Circ Heart Fail 6(6):1250–1258

Sage PR, de la Lande IS, Stafford I, Bennett CL, Phillipov G, Stubberfield J, Horowitz JD (2000) Nitroglycerin tolerance in human vessels: evidence for impaired nitroglycerin bioconversion. Circulation 102(23):2810–2815

Schachinger V, Britten MB, Zeiher AM (2000) Prognostic impact of coronary vasodilator dysfunction on adverse long-term outcome of coronary heart disease. Circulation 101(16):1899–1906

Schulz E, Gori T, Munzel T (2011) Oxidative stress and endothelial dysfunction in hypertension. Hypertens Res 34(6):665–673

Schulze PC, Yoshioka J, Takahashi T, He Z, King GL, Lee RT (2004) Hyperglycemia promotes oxidative stress through inhibition of thioredoxin function by thioredoxin-interacting protein. J Biol Chem 279(29):30369–30374

Selemidis S, Dusting GJ, Peshavariya H, Kemp-Harper BK, Drummond GR (2007) Nitric oxide suppresses NADPH oxidase-dependent superoxide production by S-nitrosylation in human endothelial cells. Cardiovasc Res 75(2):349–358

Selemidis S, Sobey CG, Wingler K, Schmidt HH, Drummond GR (2008) NADPH oxidases in the vasculature: molecular features, roles in disease and pharmacological inhibition. Pharmacol Ther 120(3):254–291

Sha X, Isbell TS, Patel RP, Day CS, King SB (2006) Hydrolysis of acyloxy nitroso compounds yields nitroxyl (HNO). J Am Chem Soc 128(30):9687–9692

Shemyakin A, Kovamees O, Rafnsson A, Bohm F, Svenarud P, Settergren M, Jung C, Pernow J (2012) Arginase inhibition improves endothelial function in patients with coronary artery disease and type 2 diabetes mellitus. Circulation 126(25):2943–2950

Shoman ME, DuMond JF, Isbell TS, Crawford JH, Brandon A, Honovar J, Vitturi DA, White CR, Patel RP, King SB (2011) Acyloxy nitroso compounds as nitroxyl (HNO) donors: kinetics, reactions with thiols, and vasodilation properties. J Med Chem 54(4):1059–1070

Silva MA, Bruder-Nascimento T, Cau SB, Lopes RA, Mestriner FL, Fais RS, Touyz RM, Tostes RC (2015) Spironolactone treatment attenuates vascular dysfunction in type 2 diabetic mice by decreasing oxidative stress and restoring NO/GC signaling. Front Physiol 6:269

Sivakumaran V, Stanley BA, Tocchetti CG, Ballin JD, Caceres V, Zhou L, Keceli G, Rainer PP, Lee DI, Huke S, Ziolo MT, Kranias EG, Toscano JP, Wilson GM, O'Rourke B, Kass DA, Mahaney JE, Paolocci N (2013) HNO enhances SERCA2a activity and cardiomyocyte function by promoting redox-dependent phospholamban oligomerization. Antioxid Redox Signal 19 (11):1185–1197

Stasch JP, Schmidt PM, Nedvetsky PI, Nedvetskaya TY, Arun Kumar HS, Meurer S, Deile M, Taye A, Knorr A, Lapp H, Muller H, Turgay Y, Rothkegel C, Tersteegen A, Kemp-Harper B, Muller-Esterl W, Schmidt HH (2006) Targeting the heme-oxidized nitric oxide receptor for selective vasodilatation of diseased blood vessels. J Clin Invest 116(9):2552–2561

Sun J, Aponte AM, Kohr MJ, Tong G, Steenbergen C, Murphy E (2013) Essential role of nitric oxide in acute ischemic preconditioning: S-nitros(yl)ation versus sGC/cGMP/PKG signaling? Free Radic Biol Med 54:105–112

Sverdlov AL, Ngo DT, Nightingale AK, Rajendran S, Mishra K, Heresztyn T, Ritchie RH, Marwick TH, Frenneaux MP, Horowitz JD (2011) The endogenous NOS inhibitor asymmetric dimethylarginine (ADMA) predicts LV mass independent of afterload. Nitric Oxide 25 (1):41–46

Tare M, Kalidindi RS, Bubb KJ, Parkington HC, Boon WM, Li X, Sobey CG, Drummond GR, Ritchie RH, Kemp-Harper BK (2017) Vasoactive actions of nitroxyl (HNO) are preserved in resistance arteries in diabetes. Naunyn Schmiedeberg's Arch Pharmacol 390:397–408

Taylor AL, Ziesche S, Yancy C, Carson P, D'Agostino R Jr, Ferdinand K, Taylor M, Adams K, Sabolinski M, Worcel M, Cohn JN, I. African-American Heart Failure Trial (2004) Combination of isosorbide dinitrate and hydralazine in blacks with heart failure. N Engl J Med 351 (20):2049–2057

Thomas SR, Witting PK, Drummond GR (2008) Redox control of endothelial function and dysfunction: molecular mechanisms and therapeutic opportunities. Antioxid Redox Signal 10 (10):1713–1765

Tita C, Gilbert EM, Van Bakel AB, Grzybowski J, Haas GJ, Jarrah M, Dunlap SH, Gottlieb SS, Klapholz M, Patel PC, Pfister R, Seidler T, Shah KB, Zielinski T, Venuti RP, Cowart D, Foo SY, Vishnevsky A, Mitrovic V (2017) A phase 2a dose-escalation study of the safety, tolerability, pharmacokinetics and haemodynamic effects of BMS-986231 in hospitalized patients with heart failure with reduced ejection fraction. Eur J Heart Fail 19(10):1321–1332

Tocchetti CG, Wang W, Froehlich JP, Huke S, Aon MA, Wilson GM, Di Benedetto G, O'Rourke B, Gao WD, Wink DA, Toscano JP, Zaccolo M, Bers DM, Valdivia HH, Cheng H, Kass DA, Paolocci N (2007) Nitroxyl improves cellular heart function by directly enhancing cardiac sarcoplasmic reticulum Ca2+ cycling. Circ Res 100(1):96–104

Tousoulis D, Simopoulou C, Papageorgiou N, Oikonomou E, Hatzis G, Siasos G, Tsiamis E, Stefanadis C (2014) Endothelial dysfunction in conduit arteries and in microcirculation. Novel therapeutic approaches. Pharmacol Ther 144(3):253–267

Touyz RM (2004) Reactive oxygen species, vascular oxidative stress, and redox signaling in hypertension: what is the clinical significance? Hypertension 44(3):248–252

Tsihlis ND, Murar J, Kapadia MR, Ahanchi SS, Oustwani CS, Saavedra JE, Keefer LK, Kibbe MR (2010) Isopropylamine NONOate (IPA/NO) moderates neointimal hyperplasia following vascular injury. J Vasc Surg 51(5):1248–1259

Tullio F, Penna C, Cabiale K, Femmino S, Galloni M, Pagliaro P (2017) Cardioprotective effects of calcitonin gene-related peptide in isolated rat heart and in H9c2 cells via redox signaling. Biomed Pharmacother 90:194–202

van Etten RW, de Koning EJ, Verhaar MC, Gaillard CA, Rabelink TJ (2002) Impaired NO-dependent vasodilation in patients with type II (non-insulin-dependent) diabetes mellitus is restored by acute administration of folate. Diabetologia 45(7):1004–1010

Velagic A, Qin C, Woodman OL, Horowitz JD, Ritchie RH, Kemp-Harper BK (2020) Nitroxyl: a novel strategy to circumvent diabetes associated impairments in nitric oxide signaling. Front Pharmacol 11:727

Williams SB, Cusco JA, Roddy MA, Johnstone MT, Creager MA (1996) Impaired nitric oxide-mediated vasodilation in patients with non-insulin-dependent diabetes mellitus. J Am Coll Cardiol 27(3):567–574

Willoughby SR, Stewart S, Chirkov YY, Kennedy JA, Holmes AS, Horowitz JD (2002) Beneficial clinical effects of perhexiline in patients with stable angina pectoris and acute coronary syndromes are associated with potentiation of platelet responsiveness to nitric oxide. Eur Heart J 23(24):1946–1954

Willoughby SR, Rajendran S, Chan WP, Procter N, Leslie S, Liberts EA, Heresztyn T, Chirkov YY, Horowitz JD (2012) Ramipril sensitizes platelets to nitric oxide: implications for therapy in high-risk patients. J Am Coll Cardiol 60(10):887–894

Wind S, Beuerlein K, Armitage ME, Taye A, Kumar AH, Janowitz D, Neff C, Shah AM, Wingler K, Schmidt HH (2010) Oxidative stress and endothelial dysfunction in aortas of aged spontaneously hypertensive rats by NOX1/2 is reversed by NADPH oxidase inhibition. Hypertension 56(3):490–497

Wingler K, Wunsch S, Kreutz R, Rothermund L, Paul M, Schmidt HH (2001) Upregulation of the vascular NAD(P)H-oxidase isoforms Nox1 and Nox4 by the renin-angiotensin system in vitro and in vivo. Free Radic Biol Med 31(11):1456–1464

Wink DA, Feelisch M, Fukuto J, Chistodoulou D, Jourd'heuil D, Grisham MB, Vodovotz Y, Cook JA, Krishna M, DeGraff WG, Kim S, Gamson J, Mitchell JB (1998) The cytotoxicity of nitroxyl: possible implications for the pathophysiological role of NO. Arch Biochem Biophys 351(1):66–74

Wollert KC, Drexler H (2002) Regulation of cardiac remodeling by nitric oxide: focus on cardiac myocyte hypertrophy and apoptosis. Heart Fail Rev 7(4):317–325

Worthley MI, Holmes AS, Willoughby SR, Kucia AM, Heresztyn T, Stewart S, Chirkov YY, Zeitz CJ, Horowitz JD (2007) The deleterious effects of hyperglycemia on platelet function in diabetic patients with acute coronary syndromes mediation by superoxide production, resolution with intensive insulin administration. J Am Coll Cardiol 49(3):304–310

Wynne BM, Labazi H, Tostes RC, Webb RC (2012) Aorta from angiotensin II hypertensive mice exhibit preserved nitroxyl anion mediated relaxation responses. Pharmacol Res 65(1):41–47

Xu J, Wu Y, Song P, Zhang M, Wang S, Zou MH (2007) Proteasome-dependent degradation of guanosine 5′-triphosphate cyclohydrolase I causes tetrahydrobiopterin deficiency in diabetes mellitus. Circulation 116(8):944–953

Yong QC, Hu LF, Wang S, Huang D, Bian JS (2010) Hydrogen sulfide interacts with nitric oxide in the heart: possible involvement of nitroxyl. Cardiovasc Res 88(3):482–491

Zalba G, Beaumont FJ, San Jose G, Fortuno A, Fortuno MA, Etayo JC, Diez J (2000) Vascular NADH/NADPH oxidase is involved in enhanced superoxide production in spontaneously hypertensive rats. Hypertension 35(5):1055–1061

Zamora R, Grzesiok A, Weber H, Feelisch M (1995) Oxidative release of nitric oxide accounts for guanylyl cyclase stimulating, vasodilator and anti-platelet activity of Piloty's acid: a comparison with Angeli's salt. Biochem J 312(Pt 2):333–339

Zeller A, Wenzl MV, Beretta M, Stessel H, Russwurm M, Koesling D, Schmidt K, Mayer B (2009) Mechanisms underlying activation of soluble guanylate cyclase by the nitroxyl donor Angeli's salt. Mol Pharmacol 76(5):1115–1122

Zhu G, Groneberg D, Sikka G, Hori D, Ranek MJ, Nakamura T, Takimoto E, Paolocci N, Berkowitz DE, Friebe A, Kass DA (2015) Soluble guanylate cyclase is required for systemic vasodilation but not positive inotropy induced by nitroxyl in the mouse. Hypertension 65 (2):385–392

Tetrahydrobiopterin and Nitric Oxide Synthase Recouplers

Keith M. Channon

Contents

K. M. Channon (✉)
BHF Field Marshal Earl Alexander Professor of Cardiovascular Medicine, University of Oxford and Oxford University Hospitals, Oxford, UK
e-mail: keith.channon@cardiov.ox.ac.uk

H. H. H. W. Schmidt et al. (eds.), *Reactive Oxygen Species*,
Handbook of Experimental Pharmacology 264, https://doi.org/10.1007/164_2020_390

Graphical Abstract

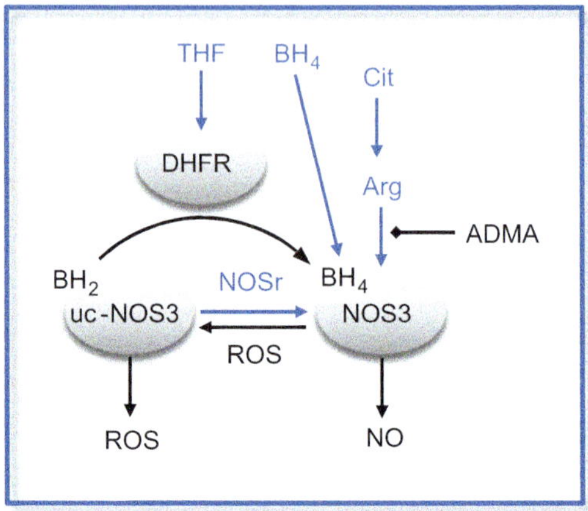

Generation of nitric oxide (NO) by the nitric oxide synthase (NOS) enzymes plays multiple signalling roles in every organ system, with crucial roles in the cardiovascular system, mediated by endothelial nitric oxide synthase (eNOS, encoded by *NOS3*) and neuronal nitric oxide synthase (nNOS, encoded by *NOS1*) in regulation of blood pressure, flow, oxygen delivery and cardiac function. Loss of normal NO-mediated functions in cardiovascular disease state is associated with changes in nitroso-redox signalling that are not dependent solely upon altered NO generation, but increased generation of reactive oxygen species (ROS). The NOS enzymes can also generate ROS, in a catalytic mode whereby the generation of NO from L-arginine is 'uncoupled' from the reduction of molecular oxygen. NOS uncoupling is determined by several factors, including the availability and oxidation state of the required NOS cofactor, tetrahydrobiopterin (BH4). The duality of NOS functions as enzymes that generate both NO and ROS under different regulatory states has emerged as an important pathophysiologic mechanism, and is a potential therapeutic target, via agents that can maintain or restore NOS coupling, for example via effects on BH4 availability.

Keywords

Cardiovascular disease · Endothelium · Nitric oxide · Reactive oxygen species · Tetrahydrobiopterin

1 Importance of Nitric Oxide and Nitroso-Redox Signalling

Generation of nitric oxide (NO) by the nitric oxide synthase (NOS) enzymes plays multiple signalling roles in every organ system, with crucial roles in neurotransmission, immune function, cardiovascular regulation and energy metabolism. In the cardiovascular system, endothelial nitric oxide synthase (eNOS, encoded by *NOS3*) and neuronal nitric oxide synthase (nNOS, encoded by *NOS1*) regulate blood pressure, blood flow, oxygen delivery by haemoglobin, cardiac contractility and relaxation (Forstermann and Sessa 2012). Cardiovascular disease states, particularly those associated with deranged endothelial function, including hypertension, diabetes, high cholesterol and other risk factors associated with atherosclerosis, are all characterised by impairment or loss of normal endothelial-derived NO production and/or NO bioactivity. In particular, the hallmark of endothelial-derived NO, that of endothelium-dependent vasorelaxation, is impaired in diseased blood vessels or in patients or animal models with vascular disease states. Impaired endothelial NO bioactivity contributes to cardiovascular pathology such as hypertension, atherosclerotic disease and pulmonary arterial hypertension, suggesting that restoration or augmentation of NO generation and/or NO bioactivity is a rational therapeutic strategy. However, in cardiovascular disease states, impairment of NO generation or bioactivity is typically associated with changes in nitroso-redox signalling that are not solely dependent upon altered NO generation. Specifically, increased generation of reactive oxygen species (ROS) leads to activation of other nitroso-redox-dependent signalling pathways, particularly those mediated by ROS such as superoxide, hydrogen peroxide and the consequent disturbances of cellular redox state leading to changes in cellular glutathione and mitochondrial function. Several different sources of cellular ROS contribute to disordered nitroso-redox signalling in cardiovascular disease states, including mitochondrial ROS regeneration, the NADPH oxidases and xanthine oxidoreductase. However, the NOS enzymes can also generate ROS, and the duality of NOS functions as enzymes that generate both NO and ROS under different regulatory states has emerged as an important pathophysiologic mechanism and is a potential therapeutic target (Channon 2004).

1.1 Nitric Oxide Synthase Coupling and Uncoupling

The nitric oxide synthases generate nitric oxide by the five-electron oxidation of the guanidino nitrogen of L-arginine. The oxidation occurs at the heme active site of NOS, via two successive mono-oxygenation reactions involving an N-hydroxyl-L-arginine intermediate. Electron flow occurs from cellular NADPH, via flavin groups (FAD and FMN) in the C-terminal reductase domain, which is activated by binding of calmodulin, causing a transformational switch enabling electron flow to the heme group where oxidation of L-arginine by molecular oxygen generates nitric oxide and L-citrulline.

The NOS enzymes are homodimeric, co-ordinated by a zinc centre, and require the co-factor tetrahydrobiopterin (BH4), located in the oxygenase domain close to the heme active site, that is required for the oxidation of L-arginine. Specifically, the BH4 molecule activates oxygen at the heme site by donating an electron, which is then recycled without consumption of BH4. The co-ordinated electron flow from the flavins in the reductase domain, via BH4 and the sequential oxidation of L-arginine with the generation of NO, is termed "NOS coupling". However, the nitric oxide synthases may also function in an "uncoupled" state, whereby electron transfer is no longer tightly coupled to the oxidation of the substrate L-arginine. Under these conditions, reduction of molecular oxygen occurs independent of L-arginine oxidation and results in the generation of superoxide rather than nitric oxide. Thus, nitric oxide synthases may be a source of ROS in cardiovascular disease states, where NOS uncoupling leads to associated loss of NO and increased ROS. Indeed, loss of the BH4 co-factor appears to be a critical switch that causes NOS uncoupling, among other factors (see below). As a redox active co-factor that is readily oxidised, BH4 itself is susceptible to oxidation, forming dihydrobiopterin (BH2). Generation of BH2 from BH4 increases NOS uncoupling, since BH2 can competitively bind NOS in place of BH4, but is unable to support electron transfer, resulting in a further increase in NOS uncoupling.

The complex regulation of the nitric oxide synthases and their functions as either coupled enzymes that generate NO or uncoupled enzymes that generate ROS confers the function of subcellular "redox hubs" that both sense and effect nitroso-redox functions. The coupling vs. uncoupling of the nitric oxide synthase is modulated by a number of important factors.

1.2 Availability of L-Arginine Substrate and Presence of Methylarginine Inhibitors

The availability of the substrate L-arginine is a critical factor in maintaining NOS coupling and NO generation. Under physiological conditions, it is presumed that intracellular L-arginine levels exceed the limiting concentration for NO production, determined by the Km of NOS. This has been estimated at approximately 100 μM L-arginine, based on experimental studies in cultured cells and reconstituted purified enzyme systems. However, the availability of the substrate L-arginine within the cell is likely regulated by microdomain compartmentalisation, by the distribution and activity of the L-arginine Y+ cationic transporter and by "competing" enzyme systems that also use L-arginine as a sub-substrate. In particular, the arginase enzymes metabolise L-arginine to urea and are regulated in cardiovascular disease states. Arginine methyltransferase generates asymmetric dimethylarginine (ADMA) from L-arginine. ADMA is also derived from methylation of L-arginine residues in cellular proteins and is a circulating plasma biomarker of cardiovascular risk. In addition to loss of L-arginine as a substrate for NO generation, increased ADMA levels, and those of other substituted arginines, directly inhibit NOS by competition for L-arginine binding at the active site. Furthermore, ADMA appears to specifically induce NOS uncoupling, blocking NO generation and augmenting ROS production (Antoniades et al. 2009). Thus, strategies to augment cellular or subcellular L-arginine availability and/or reduce the production of endogenous methylarginines, including ADMA, may be an approach to augment NO production and by promoting NOS coupling and limiting NOS uncoupling.

1.3 NOS Post-translational Modification

The NOS enzymes are post-translationally modified with multiple modifications at multiple sites, and this is an important aspect of NOS regulation. Phosphorylation of endothelial NOS at serine 1,177 by AKT1 (and potentially other kinases including AMPK, CaMKII and PKA) activates the enzyme to agonist or fluid shear stress. Phosphorylation of eNOS at threonine 495 reduces enzyme activity, with phosphorylation by PKC and de-phosphorylation by protein phosphatase 1 (PP1). In addition to activation/inhibition of NO synthases, phosphorylation of threonine 495, which may be related to activation by calcium calmodulin, appears to have an additional specific role in promoting eNOS uncoupling.

Post-translational modification also regulates subcellular trafficking and compartmentalisation of eNOS. eNOS is localised to plasma membrane caveolin and the cytoplasmic aspect of the Golgi apparatus, mediated by myristoylation at glycine 2 and palmitoylation at cysteine 15 and cysteine 26. A protein-protein interaction between eNOS and caveolin 1 (or, for nNOS, caveolin 3) also co-localises eNOS with caveolin and maintains eNOS in an inactive state. Following stimulation, the binding of the calcium-calmodulin complex breaks the caveolin-eNOS inhibitory interaction, translocating eNOS into the cytoplasm where it binds chaperones such as Hsp90 and AKT1, allowing phosphorylation and activation. The physiological activation of eNOS is critically dependent on this subcellular localisation and trafficking, since interventions to mis-traffic eNOS results in loss of activation.

eNOS itself can be nitrosylated, by nitric oxide, at specific cysteine residues. In particular, the critical cysteine residues 94 and 99, required for co-ordinating the zinc cluster that is necessary for homodimerisation, can be modified either by nitrosylation or by nitration, by peroxynitrite, and this can lead to eNOS uncoupling.

An important post-translational modification that both "senses" changes in cellular redox signalling and can cause eNOS uncoupling is the post-translational S-glutathionylation of eNOS at two conserved cysteines in the eNOS reductase domain (cysteine 869 and cysteine 908). S-glutathionylation at these residues leads to loss of NO production and increased eNOS-dependent superoxide generation and appears to be an important mechanism in pathophysiologic states such as ischemia-reperfusion. Importantly, uncoupled eNOS mediated by S-glutathionylation leads to oxidation of BH4 to BH2, which in turn induces eNOS uncoupling due to loss of BH4. Conversely, induction of eNOS uncoupling by loss of BH4 also leads to S-glutathionylation of the enzyme, demonstrating the mutuality of critical mechanisms that induce eNOS uncoupling.

2 BH4-Dependent eNOS Uncoupling as a Therapeutic Target

The potential to target BH4 as a means of restoring or maintaining eNOS coupling in disease states was suggested by initial observations revealing that BH4 levels are significantly reduced, associated with uncoupling of eNOS, in endothelial cells, in animal model systems and in human blood vessels associated with a range of disease states including diabetes (Guzik et al. 2002), hypercholesterolemia, smoking, hypertension and atherosclerosis. Loss of BH4 in these disease states is typically associated with a relative increase in the oxidised product, dihydrobiopterin (BH2), which is unable to support eNOS coupling, and may compete with BH4, preventing its binding with eNOS. In particular, oxidation of BH4 by peroxynitrite, generated from the interaction of nitric oxide and superoxide, appears to be a potent cause for BH4 loss in vascular diseases.

The observation that loss of BH4, associated with eNOS uncoupling, is a consistent feature of vascular disease states led to many studies testing the effects of pharmacologic BH4 supplementation. Many studies in cell culture, in animal models and in patients demonstrated that administration of supplementary BH4 could have salutary effects on endothelial function, haemodynamic dysregulation and vascular diseases including atherosclerosis. Nevertheless, the validity of BH4 supplementation as a therapy that specifically targets eNOS uncoupling remained presumptive, with many studies questioning whether high-dose pharmacologic supplementation of BH4 exerted effects on endothelial function through other mechanisms, including ROS scavenging or other NOS-independent redox actions. Indeed, such effects of BH4 have been recently described, suggesting that restoration of eNOS coupling is not the only potential therapeutic target for BH4.

Strong mechanistic evidence supporting a causative role for BH4 in eNOS coupling and cardiovascular disease pathogeneses has come from genetically targeted mouse models. An endothelial cell-specific transgenic mouse overexpressing the rate-limiting enzyme in BH4 synthesis (GTP cyclohydrolase 1, GTPCH) resulted in augmentation of endothelial cell BH4 levels, but without alteration in plasma or systemic BH4 levels, thus enabling the question of selective endothelial cell BH4 effects to be addressed, without the confounding effects of high-level systemic BH4 supplementation (Alp et al. 2003). Transgenic augmentation of endothelial cell BH4 synthases rescued the endothelial dysfunction state characteristic of diabetes, inhibited atherosclerotic plaque formation when crossed onto the $ApoE^{-/-}$ mouse, improved vascular remodelling and endothelial cell repopulation after vascular injury and limited the functional and remodelling response to hypoxia-induced pulmonary hypertension. Furthermore, the paradoxical increase in atherosclerosis caused by transgenic overexpression of eNOS results in eNOS uncoupling due to inadequate levels of BH4. Transgenic restoration of BH4 levels rescued the uncoupled eNOS resulting from transgenic eNOS overexpression and in the $ApoE^{-/-}$ mouse normalised the paradoxical increase in atherosclerosis provoked by transgenic eNOS overexpression without a corresponding increase in BH4. The requirement for endothelial cell BH4 has recently been demonstrated by the generation of an endothelial specific conditional knockout of GTPCH, resulting in endothelial cells that are selectively BH4 deficient in vivo. This results in mild hypertension and a switch in eNOS-mediated vasodilatation from an NO-mediated response to vasorelaxation responses mediated by hydrogen peroxide, generated by uncoupled eNOS (Chuaiphichai et al. 2014). These mice with endothelial cell-specific deficiency in BH4 are more susceptible to early initiation of atherosclerosis and to development of aortic aneurysms following angiotensin II infusion.

Thus, many observational studies, experimental medicine studies in humans and genetically targeted mouse models strongly suggest that BH4-mediated eNOS coupling is a rational and tractable therapeutic target. Nevertheless, harnessing the therapeutic potential of BH4-dependant eNOS coupling remains largely unrealised.

2.1 Clinical Trials of Oral BH4 in Cardiovascular Disease

Some clinical trials of oral BH4 have tested the ability to improve endothelial function in humans. In a study in 22 patients with hypercholesterolemia, with healthy volunteers as controls, subjects were randomised to 4 weeks of oral BH4 (400 mg twice daily) for placebo (Cosentino et al. 2008). Apart from hypercholesterolemia (LDL cholesterol greater than 4.5 mM), the patients were all healthy, mean age 57 years without hypertension, diabetes or smoking. Treatment with BH4 for 4 weeks resulted in an improvement in endothelial function as measured by venous occlusion plethysmography forearm blood flow and reduced plasma levels of 8-F2 isoprostane, a marker of systemically elevated ROS levels. BH4 treatment increased circulating plasma BH4 levels from approximately 5uM to 40uM. Levels of the oxidised biopterin, BH2, were not reported in this study.

In a double-blind randomised controlled trial, oral BH4 (700 mg per day or 400 mg per day) or placebo was administered in 49 patients with coronary artery disease scheduled for elective coronary artery bypass surgery (Cunnington et al. 2012). Treatment was administered for 2–6 weeks prior to surgery, enabling both non-invasive assessment of vascular function and the analysis of blood and vascular tissue specimens obtained at the time of surgery, following either BH4 treatment or placebo. BH4 treatment resulted in an approximate threefold increase in circulating plasma BH4, but this was accompanied with a proportionately similar increase in plasma levels of the oxidised biopterin, BH2. There was no significant improvement in endothelial function, measured non-invasively either by MR imaging of brachial artery flow-mediated dilatation or by organ chamber studies of ACh-dependent vasorelaxations in isolated samples of blood vessels obtained at the time of surgery. Most importantly, analyses of biopterin levels in vascular tissue samples revealed a modest increase in tissue BH4 levels in samples of saphenous vein, but an equivalent increase in BH2 levels. In samples of internal mammary artery, where baseline BH4 levels were somewhat higher, there was no significant increase in arterial tissue BH4 levels following oral BH4 treatment. Further studies showed that BH4 is rapidly oxidised to BH2 when incubated in either blood or plasma and that supplementation of human vascular tissue with BH4, following BH4 incubation, requires co-incubation with the antioxidant dithioerythritol.

Taken together, these clinical trials of oral BH4 therapy as a strategy to recouple eNOS suggest that BH4 is likely to be ineffective in patients with existing cardiovascular disease states, due to BH4 oxidation in the circulation and/or the vascular wall. This leads to minimal augmentation of endothelial BH4 levels and a proportionate increase in BH2 levels, which does not rescue uncoupled eNOS.

2.2 Augmenting BH4 Bioavailability by Co-administration of Antioxidants and Folates

2.2.1 Ascorbic Acid

The observation that administration of BH4 as in vivo as a supplementation strategy typically leads to BH4 oxidation had led to the evaluation of other agents that may reduce BH4 loss by oxidation and hence improve the bioavailability of BH4 to support NOS coupling.

Substantial evidence supports a role for ascorbic acid (vitamin C) in vitro, ascorbic acid increases the availability of BH4 and applied to cultured cells leading to increased NO synthesis, and this is consistent with the known short half-life of BH4 in aqueous media, in the absence of a reducing agent, when it is rapidly oxidised to BH2. Specifically, BH4 appears particularly susceptible to oxidation by peroxynitrite, leading to formation of the trihydro-biopterin radical. The trihydro-biopterin radical can be reduced back to BH4 by ascorbate, whereas thiol reductants are less efficient in regenerating BH4. Correspondingly, in vivo studies with ascorbic acid administration increase vascular tissue levels of BH4 and eNOS activity in experimental models of endothelial dysfunction and ischemia-reperfusion. However, there are no formal clinical trials of combined BH4 and ascorbic acid treatment in either healthy subjects or patients with cardiovascular disease; this remains an important consideration for strategies to recouple eNOS.

In addition to effects on BH4 oxidation, or regeneration of BH4, ascorbic acid may also augment NO bioactivity by direct effects on eNOS function. In cultured cells, ascorbic acid treatment leads to a rapid increase in serine 1,177 phosphorylation and a decrease in threonine 495 phosphorylation, an effect meditated by protein phosphatase 2A and AMPK. Ascorbic acid may also act with other antioxidants such as N-acetyl-cysteine and with increased L-arginine levels, to improve NOS coupling.

2.2.2 Folates

A close relationship between BH4 and folate metabolism has long been recognised, in part because folates and pterins share structural and biochemical similarities. Indeed, folate co-factors are referred to as "conjugated pterins", and in plants and bacteria, folate and pterin biochemistry is closely linked. In mammals, the shared enzymology is via dihydrofolate reductase (DHFR), whereby generation of the reduced folate, tetrahydrofolate (THF), is also a key enzyme in the "salvage" of BH2 to regenerate BH4. The specific role of DHFR is discussed below.

Initial clinical studies showed that 5-methyl tetrahydrofolate (5MTHF) administration could improve endothelial function in patients with familial hypercholesterolaemia and that 5MTHF increased the activity of recombinant eNOS in vitro, with a reduction in superoxide generation. Some studies have suggested that 5MTHF combined directly to the active site of eNOS and can "substitute" for BH4. However, most other studies suggest that folates are more likely to act through other mechanisms that reduce intracellular ROS generation, including the known effects of folate supplementation on reducing homocysteine levels.

More direct evidence for the effects of 5MTHF in patients with cardiovascular disease was obtained by studies of the effects of IV infusion of 5MTHF immediately prior to bypass surgery in patients with coronary artery disease and from observing the direct effects of 5MTHF incubation on human vessels in vitro (Antoniades et al. 2006). 5MTHF infusion improved NO-mediated endothelial function and reduced ROS generation. There was no evidence for direct scavenging of superoxide by 5MTHF nor of lowering plasma homocysteine levels. However, 5MTHF was identified as a scavenger of peroxynitrite, leading to an increase in vascular tissue BH4 and an improvement in eNOS uncoupling. Furthermore, there was a significant correlation between the relative level of BH4 (in relation to oxidised biopterin species) and superoxide generation from uncoupled eNOS that was shifted by 5MTHF towards higher BH4 levels and greater eNOS coupling.

Thus, administration of reduced folates such as 5MTHF, or co-administration with BH4, is a rational strategy to improve eNOS coupling, but remains to be tested in a formal clinical trial in patients.

2.3 Activation of BH4 Synthesis by Targeting GTPCH Activity

The de novo synthesis of BH4 from GTP is synthesised by a three-step pathway, of which the first and rate-limiting step is catalysed by GTP cyclohydrolase 1 (GTPCH, encoded by GCH1). GTPCH is a requirement for cellular BH4 biosynthesis. Global deletion of *Gch1* in mice is embryonically lethal (Douglas et al. 2015). The level of *Gch1* expression, and GTPCH protein, correlates with cellular BH4 levels across a broad range. Targeted deletion of *Gch1* in specific cell types in vivo renders the cells deficient of BH4, even if plasma BH4 levels remain the normal range, indicating that, at physiologic levels, plasma BH4 is not able to rescue cellular BH4 deficiency, even though pharmacologic BH4 supplementation is able to increase tissue BH4 levels. Thus, modulating GTPCH levels and/or activity is a promising therapeutic strategy to target NOS uncoupling, when BH4 levels are limiting.

GTPCH is a homodecamer of two identical homopentamers. Levels of GTPCH protein and enzymatic activity are very high in hepatocytes (where BH4 acts as a co-factor for phenylalanine hydroxylase), and GTPCH expression is induced in inflammatory cells following cytokine stimulation. However, loss of GTPCH protein is a feature of disease states characterised by increased ROS production. Loss of GTPCH protein is mediated by proteasomal degradation. For example, in conditions of hyperglycaemia or in diabetes, GTPCH is ubiquitinated leading to proteasome-dependent degradation. Furthermore, peroxynitrite can oxidise the GTPCH zinc cluster co-ordinated by cysteine 141, leading to loss of GTPCH activity and further increase in the rate of GTPCH ubiquitination. Proteasome inhibition is able to reduce GTPCH degradation. Mitochondrial mediated ROS production, for example, stimulated by ADMA, can also stimulate proteasome-dependent degradation of

GTPCH. Risk factors related to cardiovascular disease, such as cigarette smoking, induce eNOS uncoupling in endothelial cells that is associated with increased degradation of GTPCH (29524646). Thus, drugs that reduce or protect GTPCH from proteasomal degradation would be expected to maintain or increase BH4 levels in cardiovascular disease states and prevent eNOS uncoupling. Some evidence suggests that established drug treatments such as metformin, via activation of AMPK, not only activate eNOS by serine 1,177 phosphorylation but can also maintain eNOS coupling by prevention of GTPCH degradation; furthermore, AMPK activation inhibits the Nox NADPH oxidases which are an important source of ROS leading to eNOS uncoupling.

2.3.1 GTPCH Post-translational Modification and Protein-Protein Interactions

GTPCH is regulated by important auto-inhibitory domain functions and by interactions with GTPCH regulatory feedback protein (GFRP) with an allosteric activation of the inhibitory GTPCH/GRFP interaction by phenylalanine, a mechanism most relevant in hepatocytes. However, the GTPCH/GRFP complex also appears to be important in other cell types. For example, administration of phenylalanine in vivo leads to a sustained elevation in vascular BH4 levels, suggesting that the GTPCH/GRFP interaction is also important in endothelial cells and could be a therapeutic target for activating GTPCH and BH4 synthases (Hussein et al. 2015). Important further insights to GTPCH activity regulation have come from the discovery that GTPCH may be phosphorylated at a specific serine residue (serine 81) and that phosphorylation at this site increases GTPCH activity and BH4 production, which in endothelial cells is stimulated by laminar shear stress. Phosphorylation of serine 81 inhibits GFRP binding and inhibition, with de-association of the GTPCH/GRFP complex allowing GTPCH phosphorylation in endothelial cells. Maintaining GTPCH phosphorylation and downregulating GFRP can prevent eNOS uncoupling due to loss of laminar shear stress. Accordingly, phosphorylation of GTPCH and/or modulation of the GTPCH/GRFP interaction is a rational therapeutic target to activate BH4 synthases and prevent eNOS uncoupling endothelial cells. Indeed, a novel mechanism for high-throughput screening of compounds that target the GTPCH/GRFP interaction, using a FRET-based high-throughput assay, was used to identify molecules that could potentially increase GTPCH activity and BH4 levels (Li et al. 2011).

2.3.2 Statins

Statin drugs have been found to have salutary effects on eNOS expression, on eNOS activity and on BH4 synthesis, in endothelial cells and in blood vessels. Statin treatment leads to rapid improvement in BH4 availability in patients with coronary artery disease, mediated by increased GTPCH activity, and occurring more rapidly and in independently of any effect of statins on lowering of plasma cholesterol. These effects of statins are dependent on Rac1 signalling and may be mediated either by inhibition of the Nox NADPH oxidases or by another mechanism directly affecting GTPCH expression and/or activity (Antoniades et al. 2011). In endothelial cells, fluvastatin increases *GCH1* mRNA, and statins may also exert effects via microRNA (miR) expression. For example, miR 133a expression in endothelial cells is inhibited by statins, which led to inhibition of the deleterious effects of miR 133a on GTPCH protein and BH4 levels, and the resulting impairment of endothelial function in mice with either high cholesterol or diabetes. Thus, statin drugs or novel agents identified from the "pleiotropic" effects of statins (i.e. independent of LDL cholesterol lowering) have the potential to increase BH4 and eNOS coupling in the endothelium.

2.4 Recycling BH4 from BH2 by Dihydrofolate Reductase (DHFR)

Dihydrofolate reductase (DHFR) is a key enzyme in folate biochemistry, which reduces dihydrofolic acid to tetrahydrofolic acid, using NADPH as electron donor. Tetrahydrofolic acid plays key roles as a precursor of co-factors in one carbon transfer reactions (e.g. synthesis of purines, thymidylic acid and some amino acids). These precursors are required for nucleic acid synthesis and proliferation. Accordingly, agents that inhibit bacterial DHFR (e.g. trimethoprim) are used as antibiotics, whilst methotrexate, which inhibits mammalian DHFR, is widely used as a chemotherapy and anti-inflammatory drug, by inhibition of cancer or immune cell proliferation.

In addition to its classical role in production of DHF to MTF, DHFR can also catalyse the reduction of BH2 to BH4, using NADPH as the electron donor. This reaction is slower than the reduction of DHF to THF, but is inhibited by methotrexate, and is a critical contributor to maintenance of BH4 levels in endothelial cells (Crabtree et al. 2009). In addition, DHFR is required for reduction of folic acid (a commonly used synthetic supplement) to dihydrofolate, which in turn requires reduction to THF.

The importance of endothelial cell DHFR for eNOS activity and NO bioavailability was identified by knocking down DHFR in cultured endothelial cells and by observing the effects of angiotensin II that led to hydrogen peroxide-mediated downregulation of DHFR activity and resultant loss of BH4. The hydrogen peroxide-mediated reduction in DHFR activity appeared to be related to an initial burst of ROS generation leading to eNOS uncoupling. Subsequent studies in endothelial cells confirmed a requirement for DHFR-mediated BH4 recycling from BH2 for maintaining eNOS coupling, by either DHFR knockdown or inhibition using methotrexate. Loss of DHFR function decreased the BH4:BH2 ratio, and the effect on eNOS uncoupling was more marked under conditions of low BH4 synthesis (i.e. reduced GTPCH expression) (Crabtree and Channon 2011).

Human endothelial cells appear to be more dependent on DHFR, since BH4 supplementation of human endothelial cells tends to accumulate BH2 rather than

BH4. In human endothelial cells, the activity of human DHFR appears kinetically limiting for maintenance of BH4 levels, with inhibition of BH2 reduction by folic acid, presumably due to substrate competition. These initial findings identified endothelial cell DHFR as an important factor in both regulating BH4 availability in endothelial cells and also mediating the potential therapeutic effects of BH4 supplementation. Combination of BH4 with folates, such as folic acid, might be expected to have a detrimental effect due to the requirement and competition for DHFR conversion to DHF and then MTHF, whereas fully reduced folates such as THF derivatives (e.g. 5MTHF) might be expected to both stabilise BH4 and increase BH4 recycling from BH2. Whereas folic acid appeared to inhibit the activity of DHFR for BH4 recycling, studies in hypoxic pulmonary hypertension suggested that folic acid can act to upregulate DHFR protein levels in the hypoxic endothelium. This effect of folic acid was also seen in ApoE$^{-/-}$ mice where abdominal aortic aneurysm was induced by angiotensin II infusion. Oral folic acid increased the expression and activity of DHFR in endothelial cells and led to restoration of eNOS coupling.

A novel aspect of DHFR regulation was revealed by the observation that DHFR is polyubiquitinated and undergoes proteasomal degradation in endothelial cells, an effect inhibited by eNOS-derived NO. Proteasomal degradation of DHFR was mediated by S-nitrosylation of DHFR cysteine 7 that in turn regulates endothelial BH4 levels and eNOS coupling.

The role and regulation of DHFR in endothelial cell BH4 availability and eNOS coupling may be particularly important under conditions whereby de novo synthesis BH4, via GTPCH, is limiting (e.g. ROS-mediated proteasomal degradation of GTPCH), when cellular BH4 availability becomes more dependent upon recycling of BH4 from BH2 rather than de novo synthesis. However, a number of important questions remain. The S-nitrosylation of DHFR leading to increased activity would expect to be more important under conditions of eNOS coupling (i.e. with greater NO bioavailability) rather than eNOS uncoupling. Furthermore, the effects of folic acid on DHFR activity and protein levels appear to be different between mouse and human cells. Finally, the cellular transport and uptake of BH4 versus BH2 may be regulated by, and/or dependent upon, the ability of DHFR to reduce BH2 to BH4. Nevertheless, DHFR remains an important target to increase BH4 availability and restore eNOS coupling in cardiovascular disease states (Fig. 1).

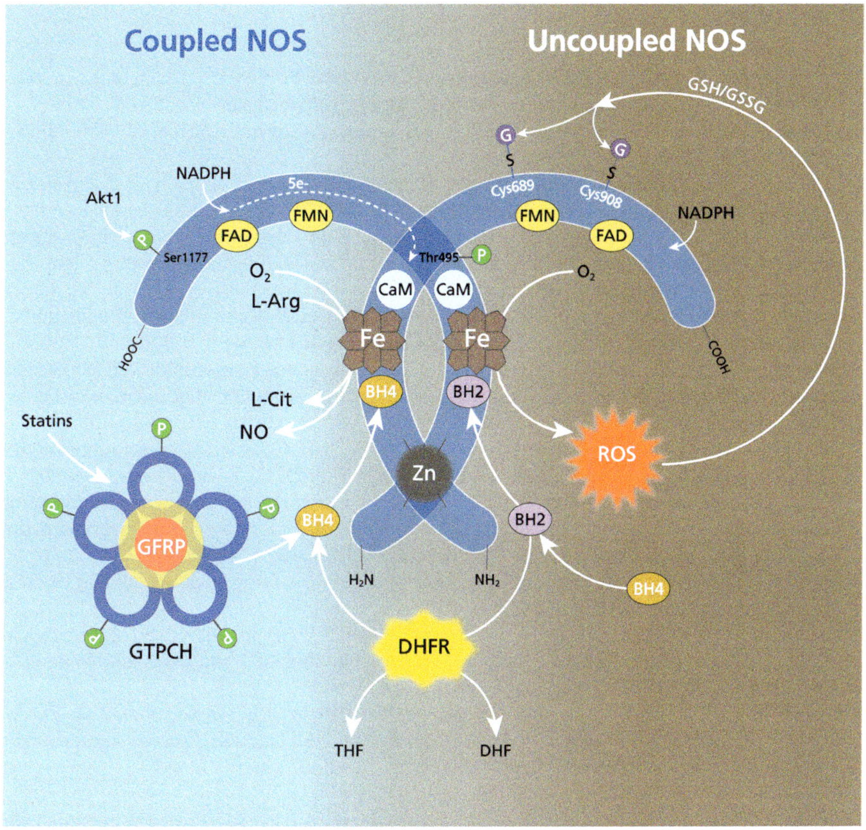

Fig. 1 Regulation of NOS coupling vs. uncoupling. In the *coupled* state (left side of diagram), nitric oxide synthase (NOS) is activated by Akt1-dependent phosphorylation at serine 1,177 and by binding of calcium calmodulin (CaM). Enzyme activation enables electron (e^-) flow from NADPH via the flavin domain (containing FAD and FMN) to the heme active site on the complementary subunit of the homodimer, co-ordinated by the zinc centre (Zn). Nitric oxide (NO) is generated by the oxidation of L-arginine (L-Arg) by molecular oxygen (O_2), via a hydroxy-arginine intermediate, generating L-citrulline (L-Cit). Tetrahydrobiopterin (BH4) is required for electron transfer during L-arginine oxidation and for stabilising the NOS homodimer. BH4 is synthesised from GTP via the rate-limiting enzyme GTP cyclohydrolase 1 (GTPCH) which is regulated by phosphorylation and induced by drugs including statins. Levels of BH4 may also be increased or stabilised by antioxidants such as vitamin C by reduced folates and by delivery in liposomes. In the *uncoupled* state (right side of diagram), flavin-mediated electron flow from NADPH results in reduction of molecular oxygen at the active site, but is not coupled to L-arginine oxidation, resulting in generation of reactive oxygen species (ROS). Activation is inhibited, and uncoupling promoted, by phosphorylation of threonine (Thr) 495. Loss of BH4 and/or binding of the oxidised form dihydrobiopterin (BH2) to the active site promotes uncoupling, which is also promoted by asymmetric dimethylarginine (ADMA) and by post-translational S-glutathionylation of cysteine residues 689 and 908. S-glutathionylation is promoted through increased RS generation by increased cellular oxidised to reduced glutathione ratio (GSSG/GSH). Loss of BH4 via oxidation to BH2 can be regenerated by the enzyme dihydrofolate reductase (DHFR)

References

Alp NJ, Mussa S, Khoo J, Cai S, Guzik T, Jefferson A, Goh N, Rockett KA, Channon KM (2003) Tetrahydrobiopterin-dependent preservation of nitric oxide-mediated endothelial function in diabetes by targeted transgenic GTP-cyclohydrolase I overexpression. J Clin Invest 112:725–735

Antoniades C, Shirodaria C, Warrick N, Cai S, de Bono J, Lee J, Leeson P, Neubauer S, Ratnatunga C, Pillai R, Refsum H, Channon KM (2006) 5-methyltetrahydrofolate rapidly improves endothelial function and decreases superoxide production in human vessels: effects on vascular tetrahydrobiopterin availability and endothelial nitric oxide synthase coupling. Circulation 114:1193–1201

Antoniades C, Shirodaria C, Leeson P, Antonopoulos A, Warrick N, Van-Assche T, Cunnington C, Tousoulis D, Pillai R, Ratnatunga C, Stefanadis C, Channon KM (2009) Association of plasma asymmetrical dimethylarginine (ADMA) with elevated vascular superoxide production and endothelial nitric oxide synthase uncoupling: implications for endothelial function in human atherosclerosis. Eur Heart J 30:1142–1150

Antoniades C, Bakogiannis C, Leeson P, Guzik TJ, Zhang MH, Tousoulis D, Antonopoulos AS, Demosthenous M, Marinou K, Hale A, Paschalis A, Psarros C, Triantafyllou C, Bendall J, Casadei B, Stefanadis C, Channon KM (2011) Rapid, direct effects of statin treatment on arterial redox state and nitric oxide bioavailability in human atherosclerosis via tetrahydrobiopterin-mediated endothelial nitric oxide synthase coupling. Circulation 124:335–345

Channon KM (2004) Tetrahydrobiopterin: regulator of endothelial nitric oxide synthase in vascular disease. Trends Cardiovasc Med 14:323–327

Chuaiphichai S, McNeill E, Douglas G, Crabtree MJ, Bendall JK, Hale AB, Alp NJ, Channon KM (2014) Cell-autonomous role of endothelial GTP cyclohydrolase 1 and tetrahydrobiopterin in blood pressure regulation. Hypertension 64:530–540

Cosentino F, Hurlimann D, Delli Gatti C, Chenevard R, Blau N, Alp NJ, Channon KM, Eto M, Lerch P, Enseleit F, Ruschitzka F, Volpe M, Luscher TF, Noll G (2008) Chronic treatment with tetrahydrobiopterin reverses endothelial dysfunction and oxidative stress in hypercholesterolaemia. Heart 94:487–492

Crabtree MJ, Channon KM (2011) Synthesis and recycling of tetrahydrobiopterin in endothelial function and vascular disease. Nitric Oxide 25:81–88

Crabtree MJ, Tatham AL, Hale AB, Alp NJ, Channon KM (2009) Critical role for tetrahydrobiopterin recycling by dihydrofolate reductase in regulation of endothelial nitric-oxide synthase coupling: relative importance of the de novo biopterin synthesis versus salvage pathways. J Biol Chem 284:28128–28136

Cunnington C, Van Assche T, Shirodaria C, Kylintireas I, Lindsay AC, Lee JM, Antoniades C, Margaritis M, Lee R, Cerrato R, Crabtree MJ, Francis JM, Sayeed R, Ratnatunga C, Pillai R, Choudhury RP, Neubauer S, Channon KM (2012) Systemic and vascular oxidation limits the efficacy of oral tetrahydrobiopterin treatment in patients with coronary artery disease. Circulation 125:1356–1366

Douglas G, Hale AB, Crabtree MJ, Ryan BJ, Hansler A, Watschinger K, Gross SS, Lygate CA, Alp NJ, Channon KM (2015) A requirement for Gch1 and tetrahydrobiopterin in embryonic development. Dev Biol 399:129–138

Forstermann U, Sessa WC (2012) Nitric oxide synthases: regulation and function. Eur Heart J 33:829–837, 837a-837d

Guzik TJ, Mussa S, Gastaldi D, Sadowski J, Ratnatunga C, Pillai R, Channon KM (2002) Mechanisms of increased vascular superoxide production in human diabetes mellitus: role of NAD(P)H oxidase and endothelial nitric oxide synthase. Circulation 105:1656–1662

Hussein D, Starr A, Heikal L, McNeill E, Channon KM, Brown PR, Sutton BJ, McDonnell JM, Nandi M (2015) Validating the GTP-cyclohydrolase 1-feedback regulatory complex as a therapeutic target using biophysical and in vivo approaches. Br J Pharmacol 172:4146–4157

Li L, Du Y, Chen W, Fu H, Harrison DG (2011) A novel high-throughput screening assay for discovery of molecules that increase cellular tetrahydrobiopterin. J Biomol Screen 16:836–844

Part V

Repairing ROS Damage

Soluble Guanylate Cyclase Stimulators and Activators

Peter Sandner, Daniel P. Zimmer, G. Todd Milne, Markus Follmann, Adrian Hobbs, and Johannes-Peter Stasch

Contents

The original version of this chapter was revised. A correction to this chapter is available at https://doi.org/10.1007/164_2019_249.

P. Sandner (✉)
Bayer AG, Pharmaceuticals R&D, Pharma Research Center, Wuppertal, Germany

Department of Pharmacology, Hannover Medical School, Hannover, Germany
e-mail: peter.sandner@bayer.com

D. P. Zimmer · G. T. Milne
Ironwood Pharmaceuticals, Cambridge, MA, USA

M. Follmann
Bayer AG, Pharmaceuticals R&D, Pharma Research Center, Wuppertal, Germany

A. Hobbs
Barts and the London School of Medicine and Dentistry QMUL, London, UK

J.-P. Stasch
Bayer AG, Pharmaceuticals R&D, Pharma Research Center, Wuppertal, Germany

Institute of Pharmacy, University Halle-Wittenberg, Halle, Germany

H. H. H. W. Schmidt et al. (eds.), *Reactive Oxygen Species*,
Handbook of Experimental Pharmacology 264, https://doi.org/10.1007/164_2018_197

Abstract

When Furchgott, Murad, and Ignarro were honored with the Nobel prize for the identification of nitric oxide (NO) in 1998, the therapeutic implications of this discovery could not be fully anticipated. This was due to the fact that available therapeutics like NO donors did not allow a constant and long-lasting cyclic guanylyl monophosphate (cGMP) stimulation and had a narrow therapeutic window. Now, 20 years later, the stimulator of soluble guanylate cyclase (sGC), riociguat, is on the market and is the only drug approved for the treatment of two forms of pulmonary hypertension (PAH/CTEPH), and a variety of other sGC stimulators and sGC activators are in preclinical and clinical development for additional indications. The discovery of sGC stimulators and sGC activators is a milestone in the field of NO/sGC/cGMP pharmacology. The sGC stimulators and sGC activators bind directly to reduced, heme-containing and oxidized, heme-free sGC, respectively, which results in an increase in cGMP production. The action of sGC stimulators at the heme-containing enzyme is independent of NO but is enhanced in the presence of NO whereas the sGC activators interact with the heme-free form of sGC. These highly innovative pharmacological principles of sGC stimulation and activation seem to have a very broad therapeutic potential. Therefore, in both academia and industry, intensive research and development efforts have been undertaken to fully exploit the therapeutic benefit of these new compound classes. Here we summarize the discovery of sGC stimulators and sGC activators and the current developments in both compound classes, including the mode of action, the chemical structures, and the genesis of the terminology and nomenclature. In addition, preclinical studies exploring multiple aspects of their in vitro, ex vivo, and in vivo pharmacology are reviewed, providing an overview of multiple potential applications. Finally, the clinical developments, investigating the treatment potential of these compounds in various diseases like heart failure, diabetic kidney disease, fibrotic diseases, and hypertension, are reported. In summary, sGC stimulators and sGC activators have a unique mode of action with a broad treatment potential in cardiovascular diseases and beyond.

Graphical Abstract

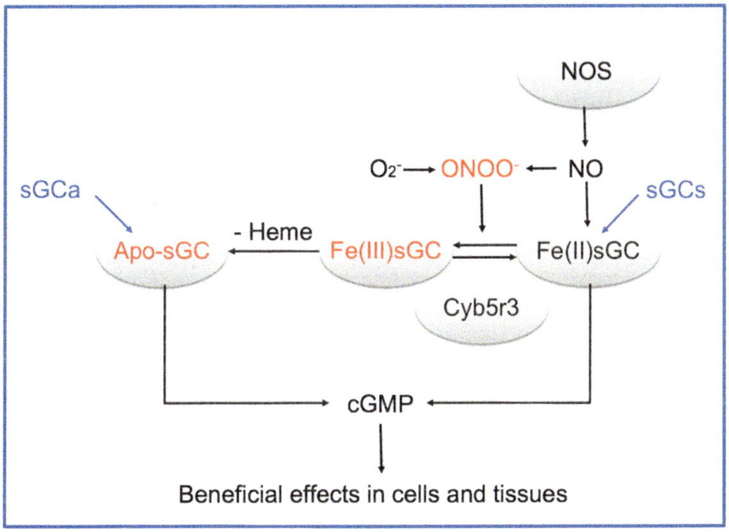

Keywords
cGMP · Cyclic guanosine monophosphate · Nitric oxide · sGC · sGC activator · sGC stimulator · Soluble guanylyl cyclase

1 Heme-Containing and Heme-Free sGC: Structure, Function, and Regulation

The second messenger cyclic guanosine monophosphate (cGMP) is generated by the heterodimeric α/β-heme protein soluble guanylate cyclase (sGC) upon activation by its endogenous ligand nitric oxide (NO) (Derbyshire and Marletta 2012). The β sGC subunit carries the N-heme-nitric oxide binding domain (H-NOX). Since the H-NOX domain binds NO, this enzyme is also known as the NO-sensitive guanylyl cyclase. NO-dependent sGC stimulation triggers formation of cGMP and promotes vasodilation and inhibits smooth muscle proliferation, leukocyte recruitment, platelet aggregation, and vascular remodeling through a number of downstream targets such as protein kinases, cyclic nucleotide-gated channels, and phosphodiesterases, making the NO/sGC/cGMP signaling a central vasoprotective signaling pathway (Lucas et al. 2000; Feil et al. 2003; Lundberg et al. 2015).

sGC is a key signal-transduction enzyme in the cardiovascular system, and many cardiovascular diseases, such as hypertension, pulmonary hypertension, heart failure, chronic kidney disease, and erectile dysfunction, are associated with dysfunction of the NO/sGC/cGMP-signaling pathway (Kemp-Harper and Feil 2008; Schulz et al. 2008; Stasch et al. 2011; Klinger and Kadowitz 2017). NO/sGC/cGMP signaling can be impaired in a variety of ways: increased ROS production by NADPH-oxidases and uncoupled NO-synthases, scavenging of NO via the reaction of NO and O^{2-} to form peroxynitrite, and oxidation of sGC to its NO-insensitive Fe^{3+} state and subsequent loss of the NO binding site on the prosthetic heme group (Stasch and Hobbs 2009; Stasch et al. 2011; Pan et al. 2016). Oxidative stress ultimately results in a reduced bioavailability of NO. The heme-free form of sGC is unresponsive to NO and prone to ubiquitin-mediated degradation (Stasch et al. 2006; Meurer et al. 2009; Hoffmann et al. 2009). In addition, sGC transcription and the stability of sGC mRNA are also affected by oxidative stress (Sharina and Martin 2017). Oxidative stress is associated with several cardiovascular diseases and is characterized by increased formation of reactive oxygen species (ROS) (Ritchie et al. 2017).

There is growing evidence supporting the relationship between genetic variants in the NO/sGC/cGMP pathway, and the prevalence and progression of cardiovascular, pulmonary, and renal diseases (Leineweber et al. 2017). Importantly, genetic alterations of the GUCY1A3 gene, which encodes the α1 subunit of the sGC, are associated with coronary artery disease as well as Moyamoya disease, achalasia, and hypertension (Erdmann et al. 2013; Kessler et al. 2017; Wallace et al. 2016). Moreover, associations with other mechanisms of sGC regulation have been described, such as membrane association and binding to the chaperone CCTη or heat shock protein 90 (HSP90) (Erdmann et al. 2013; Ghosh and Stuehr 2017).

Two distinct compound classes capable of activating sGC in an NO-independent manner were discovered at Bayer, the so-called sGC stimulators and sGC activators (Stasch and Hobbs 2009; Schmidt et al. 2009). Both classes of compounds directly bind to sGC and are allosteric modulators of guanylyl cyclase activity (Fig. 1).

sGC stimulators have a dual mode of action: they directly stimulate the native form of the enzyme independently of NO and they are also able to sensitize sGC to low levels of NO by stabilizing NO–sGC binding (Stasch et al. 2001; Stacy et al. 2018). The binding site of sGC stimulators has been a long-standing question that was recently addressed with a set of experiments incorporating photo-affinity crosslinking with LC-MS/MS and NMR approaches. Results from these experiments indicate that sGC stimulators likely bind near a previously identified tunnel of possible importance for NO escape from the heme pocket within the H-NOX domain of the β1 subunit. A potential mechanism of action of sGC stimulators involves the occlusion of a tunnel release by stimulator binding, thus leading to an observed higher affinity of NO to the ferrous heme-moiety (Wales et al. 2018; Winter et al. 2011). Maintaining sGC heme in the ferrous state is essential for sGC/cGMP signaling via NO and sGC stimulators. The ferrous heme group is non-covalently bound to the ß1 subunit of sGC via the proximal heme ligand H105 and the heme-binding motif Y-x-S-x-R, provided by ßTyr135, ßSer137, and ßArg139 (Schmidt et al. 2004).

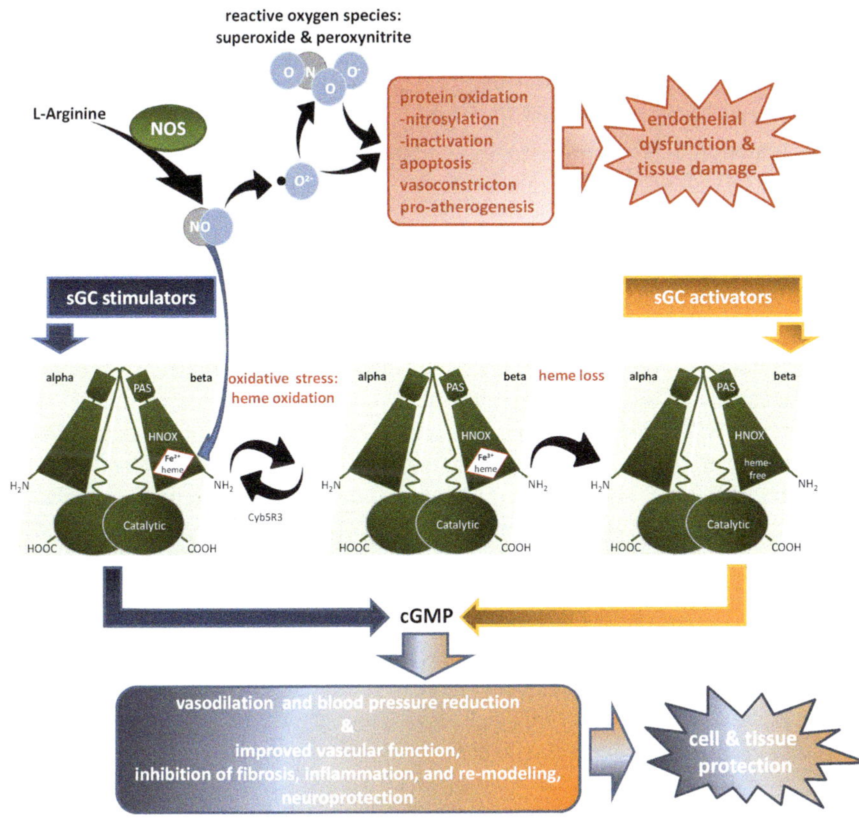

Fig. 1 Schematic representation of the sGC structure and the importance of heme-containing, native sGC and heme-free, dysfunctional form of sGC and its redox equilibrium. Oxidative stress is a risk factor for several cardiovascular diseases and is associated with increased formation of superoxide radicals, which react with NO to form the strong oxidant, peroxynitrite which is known to oxidize and inactivate many biomolecules, culminating in tissue damage. In particular, peroxynitrite oxidizes sGC, resulting in oxidized sGC and loss of the heme group which makes the enzyme unresponsive to NO. Balancing these effects, Cyb5R3, a heme iron reductase ubiquitously expressed in vascular smooth muscle cells, reduces the sGC heme iron and thereby resensitizes sGC to NO. The sGC stimulators and sGC activators are targeting the NO/sGC/cGMP pathway by stimulating the heme-containing sGC and the heme-free sGC, respectively, which is triggering the formation of cGMP which mediates the beneficial effects in cells and tissues

In contrast to sGC stimulators, sGC activators bind to the unoccupied heme-binding domain, thereby mimicking NO-bound heme, and activate the pathologically changed, heme-free, NO-unresponsive form of sGC (Stasch et al. 2015). Importantly, in isolated cells, in ex vivo blood vessels, and in vivo, sGC activators such as cinaciguat had greater pharmacological activity under pathophysiological and oxidative stress conditions compared to sGC stimulators (Stasch et al. 2001; Thoonen et al. 2015). This therapeutic principle may preferentially dilate diseased versus normal blood vessels and therefore have far-reaching implications for the clinical use of sGC activators as unique diagnostic tools and innovative vascular therapy (Armitage et al. 2009; Stasch et al. 2006; Gladwin 2006).

Recently, heme-deficient sGC mice have been generated by a gene replacement approach; the codon for His105 is replaced by the codon for Phe105 (Thoonen et al. 2015). These mice represent a unique experimental platform to distinguish between heme-dependent and heme-independent effects of sGC as well as sGC stimulators and sGC activators. Furthermore, the in vivo relevance of heme-free, dysfunctional sGC could be investigated for the first time. The phenotype of these heme-deficient sGC mice is affected. Blood pressure was higher in these mice than in wild type (WT) mice. The heme-deficient sGC mice also showed gastrointestinal (GI) tract abnormalities, growth retardation, and a reduced life span. Importantly, the ability of aortic rings to relax in response to NO was completely abolished in aortas taken from heme-deficient sGC mice. In contrast, the sGC activator cinaciguat relaxed precontracted aortas from heme-deficient sGC mice at a lower concentration than required to relax those from WT mice. Consistent with the in vitro findings, in vivo NO effects were also abolished in heme-deficient sGC mice, and cinaciguat decreased blood pressure to a greater extent in heme-deficient sGC mice than in WT mice. This indicates the presence of a heme-free or dysfunctional sGC pool in vivo, and shows that it can be reactivated by sGC activators to overcome the pathophysiology of a disrupted NO/sGC/cGMP signaling pathway. Diseases associated with NO resistance would appear to be ideally suited for therapies directed at restoring redox homeostasis, sGC activity, and NO sensitivity.

There is a growing appreciation for the role of redox state in modulating NO/sGC/cGMP signaling. Data from research using stimulator and activator families of sGC agonists have provided support for the thesis that sGC bioactivity is redox regulated. Both agonism of native, heme-bound sGC by stimulators and of heme-free sGC by activators leads to increased formation of cGMP, which exerts multifaceted cellular and tissue effects (Stasch et al. 2011, 2015). However, oxidative stress shifts intracellular levels of native sGC toward the oxidized, dysfunctional, heme-free form that is unresponsive to both endogenous and exogenous NO (Evgenov et al. 2006; Munzel et al. 2007). This concept of NO resistance provides the rationale for sGC activators that bind to the unoccupied sGC heme binding site, thereby favoring the active enzyme state. In addition, assuming a sensitive balance between heme-free, oxidized, and heme-containing sGC in cells and tissues, it is proposed that sGC activators by virtue of very low Kd values are capable of shifting this equilibrium towards the heme-free sGC (Stasch et al. 2002; Schmidt et al. 2003; Kollau et al. 2018). While sGC undergoes proteasomal degradation once its heme is oxidized, this process is prevented when agents such as sGC activators bind the sGC heme binding site (Meurer et al. 2009; Hoffmann et al. 2009).

In cardiovascular disease, the protective NO/sGC/cGMP signaling pathway is impaired due to a decreased pool of NO-sensitive heme-containing sGC accompanied by a concomitant increase in NO-insensitive heme-free sGC. However, no direct method exists to detect cellular heme-free sGC other than its activation by sGC activators (Stasch et al. 2006; Gladwin 2006). Fluorescence dequenching, based on the interaction of the optical active prosthetic heme group and the attached biarsenical fluorophor FlAsH, was used to detect changes in cellular sGC heme status (Hoffmann et al. 2011). Loss of the prosthetic group by oxidative stress was corroborated by an observed decrease in NO-induced sGC activity, reduced sGC protein levels, and an increased effect of sGC activators. The applicability of this approach based on the cellular expression of an engineered sGC variant is limited to recombinant expression systems. Nevertheless, it allows monitoring of sGC's redox regulation in living cells and future enhancements might be able to extend this approach to in vivo conditions.

While the oxidation of heme sGC under pathophysiological conditions and its association with enhanced sGC activation by sGC activators under these conditions are well documented, most of the hypothesized relationships between the function of ferrochelatase in heme biosynthesis and sGC regulation remain to be investigated (Patel et al. 2017). Mitochondrial heme biosynthesis is an important factor in controlling the expression and function of sGC and systems influencing superoxide generation and actions. Modulation of heme biosynthesis by ferrochelatase inhibition with *N*-methyl protoporphyrin IX promoted sGC depletion, superoxide elevation, and attenuation of relaxation to NO donors (Alruwaili et al. 2017). These studies suggest that disruption of heme biosynthesis resulting in a loss of cGMP production may serve as a contributing mechanism to the progression of cardiovascular disease.

Recently, a further important step in the enzymatic process that modulates sGC redox state and cGMP signaling has been discovered (Rahaman et al. 2017). Nicotinamide adenine dinucleotide (NADH) cytochrome b5 reductase 3 (Cyb5R3), a heme iron reductase ubiquitously expressed in vascular smooth muscle cells, sensitizes sGC to NO by reducing the sGC heme iron and thereby controls cGMP production and blood vessel dilation (Fig. 1). Consequently, Cyb5R3 expression and activity may also influence responses to therapeutics that activate and stimulate sGC (Rahaman et al. 2017).

2 NO Donors and Phosphodiesterase 5 (PDE5) Inhibitors as cGMP Increasing Drugs

Given the substantial disease relevance of impaired NO/sGC/cGMP signaling, it is no surprise that modulators that target the NO/sGC/cGMP signaling cascade other than sGC stimulators and activators have been successfully employed as pharmacological interventions. Drugs acting on this pathway are useful for treating a variety of diseases. Although having distinct limitations, drugs have been successfully developed that act at the top of the NO/sGC/cGMP pathway to increase NO bioavailability (nitrates and NO donors) and that prolong signaling by stabilizing cGMP (PDE5 inhibitors).

2.1 NO Donors

In the nineteenth century, a long time before the discovery of NO and cGMP signaling, amylnitrate and nitroglycerine were known to be beneficial for the treatment of patients with angina pectoris (Brunton 1867; Murrell 1879). In fact, Alfred Nobel, who suffered from angina pectoris, was treated with nitroglycerine (glyceryltrinitrate GTN). Within the last almost 150 years of using NO donors, medicinal chemists synthesized a variety of NO-liberating drugs and organic nitrates that have been approved for angina pectoris: isosorbide mono and dinitrate (ISDN, ISMN), sodium nitroprusside (SNP), but also molsidomin in order to increase half-life. These NO donors liberate NO enzymatically or nonenzymatically and potently relax coronary blood vessels. Despite these intensive efforts and broad application, the main disadvantages of NO donors were only partly resolved. Nitrates still have a small therapeutic range and lead to tachyphylaxis. In addition, released NO reacts with ROS such as superoxide anions to produce peroxynitrite, which can cause tissue damage. Thus, stable and NO-independent stimulation of cGMP production could have major therapeutic advantages over NO donors.

2.2 PDE5 Inhibitors

Levels of cGMP can be increased by the use of phosphodiesterase type 5 inhibitors (PDE5i), which inhibit degradation of cGMP and were introduced into medical therapy for the treatment of erectile dysfunction (ED). The first compound approved for ED treatment was sildenafil (Viagra™) in 1998 followed by vardenafil (Levitra™) and tadalafil (Cialis™) in 2003. In 2007 and 2009, sildenafil and tadalafil were also approved for the treatment of pulmonary arterial hypertension (PAH) as Revatio™ and Adcirca™, respectively, followed by an additional approval of tadalafil for the treatment of symptomatic benign prostatic hyperplasia (BPH) in 2011. These different applications show the broad treatment potential of cGMP-enhancing drugs. Despite these advances, a substantial number of ED patients (estimated at 30–50% of all patients) do not sufficiently respond to PDE5i (Shabsigh 2004; Bruzziches et al. 2013). In addition, some pulmonary hypertension patients do not adequately respond to PDE5i therapy (Oudiz et al. 2011; Shapiro et al. 2012; Hoeper et al. 2017a, b). This nonresponse to PDE5i therapy could be mechanistically explained by the mode of action of PDE5i, which inhibit only cGMP degradation. Importantly, there are multiple phosphodiesterases that degrade cGMP and are differentially expressed with in cells and highly compartmentalized (Fischmeister et al. 2006). As a consequence, the pharmacology of PDE5i is limited to tissues that express PDE5 and where PDE5 represents the primary mechanism of cGMP metabolism. Upon blockade of PDE5, other cGMP-metabolizing PDEs may compensate (Stasch et al. 2011). In addition, the efficacy of PDE5i may be substantially limited under conditions of very low endogenous NO production, resulting in low intracellular cGMP production. Decoupling of the NOS/NO/cGMP signaling cascade and low NO/cGMP production has been shown in ED patients (Bivalacqua et al. 2003) and also in patients with PAH and heart failure where endothelial dysfunction leads to impaired NO synthesis (Breitenstein et al. 2017). The low endogenous NO production could be due to aging (Garbán et al. 1995), but also metabolic syndrome, dyslipidemia (Mulhall et al. 2006), and obesity with and without hypogonadism (Gurbuz et al. 2008). Diabetes has also been associated with impaired NO production (Cartledge et al. 2001; Musicki and Burnett 2007).

The therapeutic success of NO-donors and PDE5 inhibitors validates the key pharmacological role of the NO/sGC/cGMP pathway and the broad therapeutic utility of targeting this pathway. However, distinct limitations with regard to tolerance, tissue expression, and robustness of pharmacological response underscore the opportunity for agents such as sGC stimulators and sGC activators that specifically target this pathway.

3 Nomenclature of sGC Stimulators and sGC Activators and INN Names

The evolution of terminology to define sGC stimulators and activators has followed closely the elucidation of, and distinction between, the mechanisms of action of these two series of sGC agonists. The name "sGC stimulator" was first coined by Bayer scientists who used a high-throughput approach to build upon earlier successes in the development of small molecules that directly triggered enzyme activity in a NO-independent fashion, yet also synergized with NO (e.g., YC-1, CFM-1571) (Ko et al. 1994; Selwood et al. 2001). Whereas compounds that did not promote enzyme turnover in synergy with NO, but rather triggered cGMP synthesis in oxidized or heme-deficient protein, were called "sGC activators" (e.g., BAY 58-2667, HMR1766). Indeed, the term "sGC activator" was arguably first described by Abbott researchers who reported a series of ortho-substituted sulfanyl-cinnamic acid (aminoalkyl) amides, highlighted by A-350619 (Miller et al. 2003), that were structurally dissimilar to YC-1, but utilized a similar mode of action (i.e., acting independently of NO and requiring the presence of a reduced heme moiety). Thus, despite identifying what are now termed "stimulators" they named the compounds GC "activators." In many ways, this exemplifies the flawed designation since stimulator and activator are used synonymously in scientific English to describe a mechanism of agonism at an enzyme, transcription factor, or receptor. Yet the differentiation between these two chemical classes is key to understanding not only the pharmacology but also novel (patho) physiological roles of the NO-sensitive cyclase. Specifically, it is believed that enzyme oxidation might occur during, and contribute to, cardiovascular disease by leading to heme loss and NO insensitivity (Stasch et al. 2006); moreover, administration of "sGC activators" may proffer a means to target diseased vessels or organs. This slightly ambiguous terminology was in many respects improved upon by several groups, including scientists at Merck (Bittner et al. 2009), who created the labels "heme-dependent" and "heme-independent" sGC activators (HDA and HIA, respectively). This approach defined the disparate mechanisms of action more clearly, but muddied the waters by using the term "activator" for both sets of compounds, implying that each might work on oxidized/heme-free enzyme according to the original Bayer nomenclature.

The terminology to describe these novel classes of guanylyl cyclase ligands is further complicated by a recent revision of the enzyme nomenclature surrounding the target. Despite the ingrained use of sGC "stimulators" and "activators," and their formal approval as drug classes (i.e., name in approved drug labels or NDA/), the nomenclature surrounding their target protein, NO-sensitive guanylyl cyclase, has recently been updated (Alexander et al. 2015). Whilst these enzymes were originally termed soluble guanylyl cyclases, it has become clear that the term "soluble" is a misnomer, and these proteins are often associated with the cytoplasmic membrane, seemingly via interaction with chaperone proteins such as Hsp70 and Hsp90 (Balashova et al. 2005; Venema et al. 2003). The nomenclature of the NO-sensitive guanylyl cyclase isoforms have recently been modified to align with that of the homologous membrane-spanning proteins (e.g., GC-A, GC-B, GC-C; those that act as cognate receptors for the natriuretic peptide and guanylin family of hormones, or play a role in sensory perception (Kuhn 2016). Specifically, the ubiquitous NO-sensitive guanylyl cyclase comprising an α_1 and β_1 subunit is now referred to as GC-1, and the more tissue-specific (e.g., CNS, kidney, placenta) $\alpha_2\beta_1$ heterodimer is now termed GC-2 (Alexander et al. 2015). In addition, there has been considerable debate as to the correct chemical transformation catalyzed by the cGMP-synthesizing cyclase family; guanylate or guanylyl. Original discussions in the mid-1970s between scientists involved in the characterization of both cGMP- and cAMP-synthesizing enzymes resulted in the terms guanylate and adenylate begin adopted, if anything for ease of pronunciation rather than biochemical precision. However, from a chemical perspective the accurate nomenclature is unequivocally guanylyl, rather than guanylate (based on equivalent reactions of, for example, acetyl and acetate), since the α-oxygen of GTP leaves with the diphosphate group (Walseth et al. 1981) concomitant with reaction of the α-phosphorus with the ribose hydroxyl to cyclize GMP. Regardless of the new terminology, the identifiers "sGC stimulators" and "sGC activators" will persist to distinguish these family of molecules with distinct mechanisms of action, and their respective drug classes.

4 Discovery of sGC Stimulators

In 1994, scientists at Bayer started a screening campaign for substances that could induce an increase in NO synthesis and thereby stimulate sGC in porcine endothelial cells (Evgenov et al. 2006; Stasch and Hobbs 2009). These studies involved measurement of cGMP levels by radioimmunoassay, leading to the unexpected discovery of

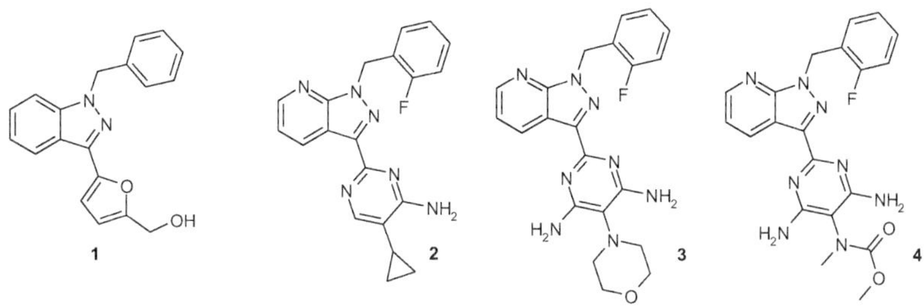

YC-1	BAY 41-2272	BAY 41-8543	riociguat
cGMP formation MEC = 100 μM Rabbit aorta IC_{50} = 10 μM	MEC = 0.03 μM IC_{50} = 0.30 μM	MEC = 0.03 μM IC_{50} = 0.10 μM	MEC = 0.03 μM IC_{50} = 0.12 μM

Scheme 1 sGC stimulators YC-1 (**1**), BAY 41-2272 (**2**), BAY 41-8543 (**3**), and riociguat (**4**)

NO-independent sGC stimulators. At the same time, researchers at the National Taiwan University Taipei and Yung Shin Pharmaceuticals, Taiwan reported that a benzyl indazole compound named YC-1 (1) (Scheme 1) inhibited platelet aggregation via stimulation of cGMP synthesis. YC-1 (1) was subsequently characterized as a direct NO-independent, but heme-dependent, sGC stimulator. It stimulated isolated sGC by a factor of 30× to 40× at 100 μM, and showed a strong synergistic effect when combined with NO-releasing compounds and a loss of stimulation after oxidation or removal of the prosthetic heme moiety of sGC. YC-1 (1) exhibited a promising profile in various pharmacological studies. However, in addition to its relatively weak sGC stimulating potency, it revealed a poor pharmacokinetic profile and a lack of specificity as it was found to inhibit phosphodiesterases and to modulate many cGMP-independent effects. Therefore, further optimization of potency, pharmacokinetic properties, and specificity was required to realize the full therapeutic potential of this novel class of drugs.

Based on these initial results, extensive structure–activity relationship (SAR) studies were performed at Bayer to systematically optimize the structure of YC-1. The in vitro potency of the compounds was assessed by two different methods, a cGMP formation assay in sGC-overexpressing Chinese hamster ovary (CHO) cells and a functional assay based on the inhibition of phenylephrine-induced contraction of rabbit aortic rings. A first breakthrough in terms of improved potency resulted from the replacement of the benzyl indazole moiety of YC-1 by a (2-fluorobenzyl)pyrazolopyridine moiety and, even more importantly, the exchange of the (hydroxymethyl)furan portion for a 5-substituted 4-aminopyrimidine or 4,6-diaminopyrimidine group. Small molecule X-ray structures revealed a coplanar arrangement of this biaryl system, which is apparently important for achieving high potency. The 5-cyclopropyl-4-aminopyrimidine derivative BAY 41-2272 (2) (Scheme 1) showed a greatly improved sGC stimulating potency, with an IC_{50} of 0.3 μM for the contraction of rabbit aortic rings (YC-1, IC_{50} = 10 μM), and a minimum effective concentration (MEC) of 0.03 μM for cGMP formation in CHO cells (YC-1, MEC = 10 μM). In contrast to YC-1, BAY 41-2272 is a highly specific sGC stimulator and no relevant inhibition of phosphodiesterases was observed. Whereas the 1-(2-fluorobenzyl)-1H-pyrazolo[3,4-b]pyridine part of this new lead series turned out to be essential for potent sGC stimulating activity, the pyrimidine moiety allowed for broad variations. Further studies led to the 4,6-diamino-5-morpholino analogue BAY 41-8543 (3), displaying threefold higher potency in the phenylephrine-induced contraction of rabbit aorta ($IC_{50} = 0.10\ \mu M$).

BAY 41-2272 and BAY 41-8543, however, displayed low metabolic stability and low oral bioavailability in rats, and BAY 41-2272 showed a strong inhibition as well as induction of metabolizing cytochrome P450 (CYP) enzymes. While these properties precluded further development, both compounds were used as tool compounds to study this novel class of drugs in numerous pharmacological experiments, resulting in more than 200 publications from various research groups around the world. Metabolite identification studies of BAY 41-2272 and BAY 41-8543 revealed an oxidative metabolism at the cyclopropyl and morpholino substituent, respectively. It was further demonstrated that, in contrast to compounds with small, lipophilic substituents at the pyrimidine C5-position (BAY 41-2272), derivatives with larger, more polar 5-substituents displayed no relevant CYP inhibition (BAY 41-8543). Continuous efforts to introduce other polar, potentially more stable substituents at the pyrimidine C5-position culminated in the identification of the *N*,*O*-dimethylcarbamate **4** (Scheme 1), BAY 63-2521 [International Nonproprietary Name (INN): riociguat]. Riociguat showed no relevant CYP interaction and a superior pharmacokinetic profile, including good oral bioavailability across different species.

In vitro, riociguat stimulated purified, recombinant sGC up to 73-fold, from 0.1 to 100 μM, and showed the typical profile of sGC stimulators: strong synergistic enzyme activation when combined with NO-releasing agents and crucial dependency on the presence of the reduced prosthetic heme moiety.

In conscious, spontaneously hypertensive rats, oral administration of riociguat resulted in a long-lasting and dose-dependent blood pressure decrease (Mittendorf et al. 2009). Importantly, and in contrast to nitrates, the effect is preserved over several weeks or when the rats are rendered nitrate tolerant. Riociguat was also investigated in different animal models of pulmonary hypertension (PH), including mice subjected to chronic hypoxia and rats injected subcutaneously with monocrotaline. In these experimental models, riociguat improved pulmonary hemodynamics and prevented, and even partially reversed, features of adverse structural remodeling such as right ventricular hypertrophy and muscularization of small pulmonary arteries (Stasch et al. 2011). Based on its combined profile of excellent potency, specificity, efficacy, and safety, riociguat was selected as a drug development candidate for the treatment of different forms of pulmonary hypertension (PH).

Riociguat was the first sGC stimulator to successfully transition from animal experiments to controlled clinical studies in patients. In randomized, double-blind, placebo-controlled Phase III trials in patients with the PH subforms, pulmonary arterial hypertension (PAH), and chronic thromboembolic PH (CTEPH), riociguat met the primary endpoint in exercise capacity (6-min-walking-distance, 6MWD). Riociguat showed a significant improvement in the 6MWD versus the placebo (+36 m, PAH; +46 m, CTEPH). Additionally, improvements were observed across

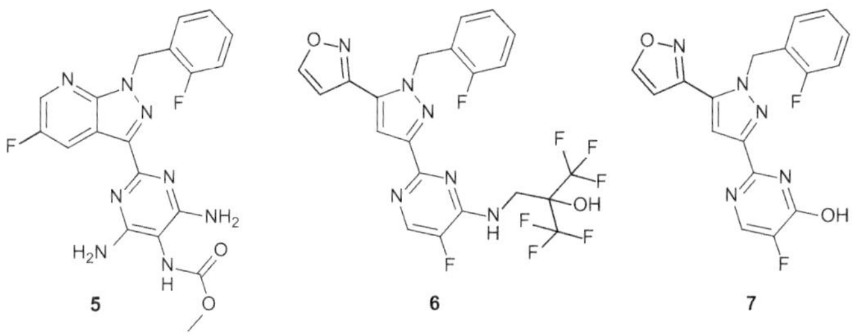

5 vericiguat

6 praliciguat

7 IWP-051

Scheme 2 sGC stimulators vericiguat (**5**), Praliciguat (**6**) and IWP-051 (**7**)

secondary endpoints, including pulmonary hemodynamics, functional class, and time to clinical worsening. Riociguat (Adempas^{TM}) is the first drug that has demonstrated efficacy in two life-threatening PH indications: CTEPH and PAH, and it is the only drug approved for CTEPH.

4.1 Activities Towards Next-Generation sGC Stimulators

Based on an increasing knowledge associated with this mode of action, the promising pharmacological effects of sGC stimulators and the clinical success of riociguat, several companies have pursued programs to further explore the structure–activity relationships (SAR) of the bis-heterocyclic pyrimidino-substituted pyrazolopyridines (Bayer, Pfizer, Merck) or to identify new lead series of sGC stimulators (Astellas, Bayer, Ironwood). From these efforts, three more sGC stimulators have made a successful transition to clinical studies: vericiguat (BAY 1021189) (**5**, Scheme 2) currently in phase 3 trials for heart failure with reduced ejection fraction (HFrEF), praliciguat (IW-1973; 6, scheme 2) currently in phase 2 trials for diabetic nephropathy and heart failure with preserved ejection fraction (HFpEF), and olinciguat (IW-1701) recently completed a phase 2a study in achalasia and currently in a phase 2 trial for sickle cell disease.

Vericiguat resulted from an optimization approach to identify orally bioavailable sGC stimulators with a longer duration of action than riociguat, in order to support a profile allowing for a once-daily oral dosing, and less oxidative metabolism in order to reduce drug interaction potential. Riociguat has a moderate half-life in different animal species and this pharmacokinetic profile translated into a three times daily dosing regimen in patients (Frey et al. 2017). The strategy was to further optimize the metabolic stability of riociguat mainly catalyzed by CYP1A1, and also by CYP3A4, CYP3A5, and CYP2J2 and hence reduce blood clearance to achieve a longer half-life. In these studies, vericiguat exhibited the best overall pharmacokinetic profile, with a low clearance and long half-life in rats and dogs after intravenous dosing, as well as high oral bioavailability (Follmann et al. 2017). In addition, vericiguat (**5**) had no inhibitory effects on major CYP isoforms (1A2, 2A6, 2B6, 2C8, 2C9, 2C19, 2D6, and

3A4), as indicated by IC_{50} values of >50 μM (Follmann et al. 2017). After thorough preclinical DMPK studies, vericiguat was selected as a clinical candidate and proved to have a pharmacokinetic profile in humans suitable for once-daily dosing.

Additional in vivo studies in animal models of hypertension, heart failure, and kidney disease have revealed dose-dependent antifibrotic and organ-protective properties in line with the sGC stimulator mode of action. Vericiguat is currently being investigated in a phase 3 clinical trial in patients with HFrEF (NCT02861534, Armstrong et al. 2017) and in a phase 2 clinical trial in patients with HFpEF (NCT03547583).

Researchers at Ironwood Pharmaceuticals have discovered several novel sGC stimulators and advanced three compounds into development (Buys et al. 2018). The medicinal chemistry effort that led to the bis-heteroaryl pyrazole IWP-051(7, scheme 2), a pharmacodynamically active compound with low clearance and a long half-life in rats, has been described (Nakai et al. 2016). Ironwood has advanced two other sGC stimulators, praliciguat and olinciguat, into clinical studies, and a third compound, IW-6463, that readily crosses the blood–brain barrier, is under preclinical evaluation for the potential treatment of CNS diseases.

Praliciguat is an sGC stimulator with a long half-life in preclinical species, extensive tissue distribution, and mainly hepatic clearance (Tobin et al. 2018). Its pharmacokinetic half-life in humans is consistent with QD dosing (Hanrahan et al. 2018). Praliciguat has completed Phase 1 studies in healthy subjects and Phase 2a exploratory studies in patients with type 2 diabetes and a history of hypertension (NCT02906579, NCT03091920) Praliciguat is currently under investigation for treatment of diabetic nephropathy (NCT03217591) and heart failure with preserved ejection fraction (NCT03254485).

Olinciguat is an-sGC stimulator that has completed phase 1 studies as well as a Phase 2a study in patients with achalasia (NCT02931565). Olinciguat is also under investigation for the treatment of sickle cell disease (NCT03285178). In clinical studies in healthy adults, olinciguat demonstrated a long half-life and low peak-to-trough plasma ratio with QD dosing (Mittleman et al. 2017).

5 Discovery of sGC Activators

Following the discovery of the NO-independent, heme-dependent sGC stimulators, scientists at Bayer performed a high-throughput screen (HTS) in 1997 with the goal of identifying additional sGC stimulator leads. For this effort, the cGMP formation assay in sGC-overexpressing CHO cells was utilized. Surprisingly, a compound with an unprecedented and distinct dicarboxylic acid motif (6) was identified as a potent agonist of sGC (Scheme 3). After further mechanistic in vitro studies, it was established that this compound behaved in a completely different manner to the sGC stimulators, stimulating sGC in an NO- as well as heme-independent fashion. Thus, the novel pharmacological class was designated an sGC activator to clearly differentiate this new behavior and mode of action from sGC stimulators.

This serendipitous discovery provided a tool to further explore redox regulation of sGC and its role in the pathogenesis of several cardiovascular disorders. More specifically, this offered the prime opportunity to design drugs for selective binding

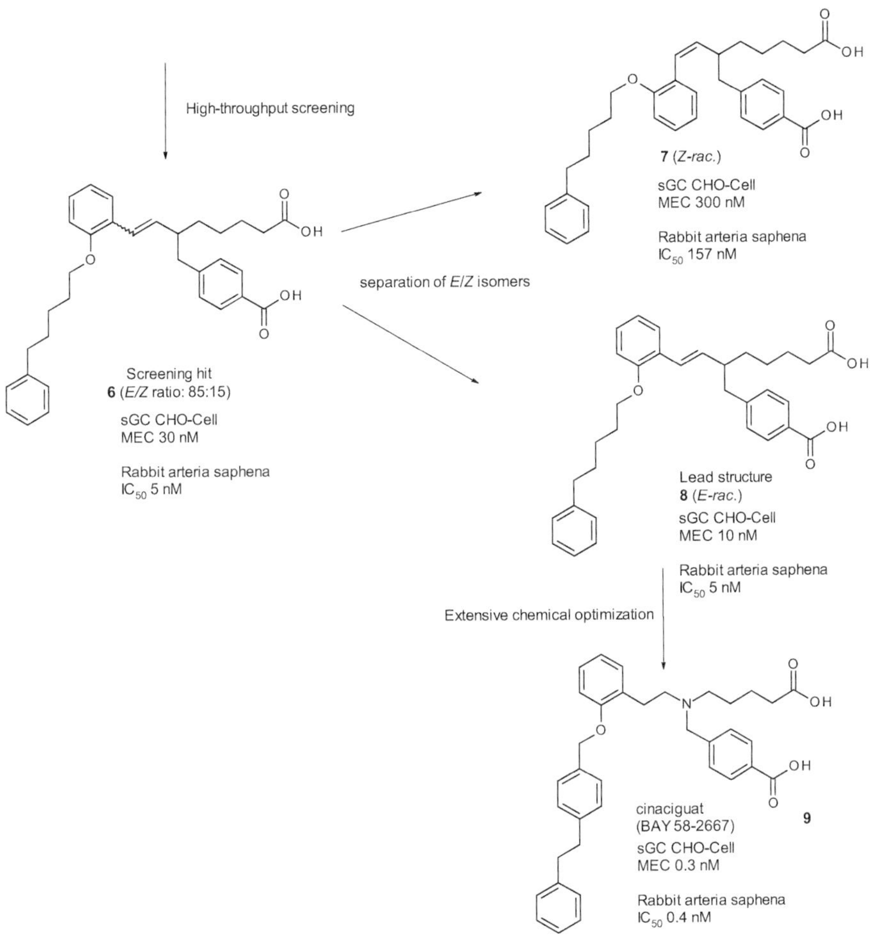

Scheme 3 Discovery of sGC activators by HTS and evolution towards cinaciguat (9)

to the oxidized, heme-free sGC generated by the influence of oxidative stress causally involved in many cardiovascular diseases.

Screening hit (6) presented as a racemic 85:15E/Z-mixture. After separation, the racemic E-isomer (8) turned out to be 30-fold more potent than the racemic Z-isomer (7). Subsequent separation of enantiomers revealed that the R,E-isomer of 8 is 70-fold more potent than the corresponding S-enantiomer. Moreover, lead structure 8 also showed promising in vitro potency on isolated recombinant sGC and relaxation of precontracted rabbit arteria saphena rings. Based on these initial results, an extensive lead optimization program was initiated with the goal of identifying a candidate suitable for intravenous dosing. The exchange of the central allylic moiety

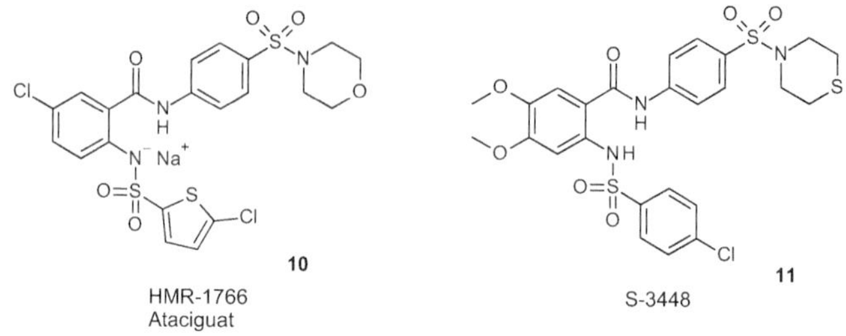

10 **11**

Scheme 4 sGC activators reported by Hoechst Marion Roussel: ataciguat (10) and S-3448 (11)

for an ethylamino linkage and modification of the phenylpentyl side chain resulted in the discovery of clinical candidate BAY 58-2667 (9) (INN: cinaciguat).

The pharmacological efficacy profile of cinaciguat was explored in various in vivo models of myocardial infarction, chronic renal failure, arterial and pulmonary hypertension, and chronic heart failure. In a canine model of congestive heart failure (HF), intravenous administration of cinaciguat resulted in dose-dependent reductions in cardiac preload and afterload, and a concomitant increase in cardiac output and renal blood flow without further neurohumoral activation (Boerrigter et al. 2007).

In 2001, researchers from Hoechst Marion Roussel disclosed anthranilic acid derivatives that represented a novel structural class of compounds also reported to activate the oxidized and/or heme-free form of sGC. The best-described examples are HMR 1766 (10) (INN: ataciguat) and S-3448 (11) (Scheme 4).

The inhibition of phenylephrine-induced contraction of rat aortic rings by both compounds is only moderate to weak; however, the pharmacological efficacy of ataciguat and S-3448 was demonstrated in various in vivo models of atherosclerosis and peripheral arterial occlusive disease. Chronic treatment of streptozotocin diabetic rats with ataciguat improved endothelial function and normalized platelet activation. Additionally, reduced atherosclerosis and improved endothelium-dependent vasorelaxation were observed in ApoE$^{-/-}$ mice treated with ataciguat. Stage II peripheral arterial occlusive disease is mainly characterized by exercise-induced muscle fatigue. Ataciguat improved ischemia-induced muscle fatigue in Zucker Diabetic Fatty (ZDF) rats with unilateral hind-limb ischemia as an experimental model of peripheral arterial occlusive disease.

5.1 Activities Towards Second-Generation sGC Activators

The search for novel sGC activators has become an increasingly competitive field. Several approaches have been reported in the recent patent literature. Interestingly, all second-generation sGC activators contain a monocarboxylic acid moiety.

In 2009, Merck (12) and GlaxoSmithKline (GSK) (13) disclosed very similar sGC activators incorporating an identical 5-(trifluoromethyl)pyrazole-4-carboxylic acid moiety attached to a pyridine scaffold (Scheme 5). Even more recently,

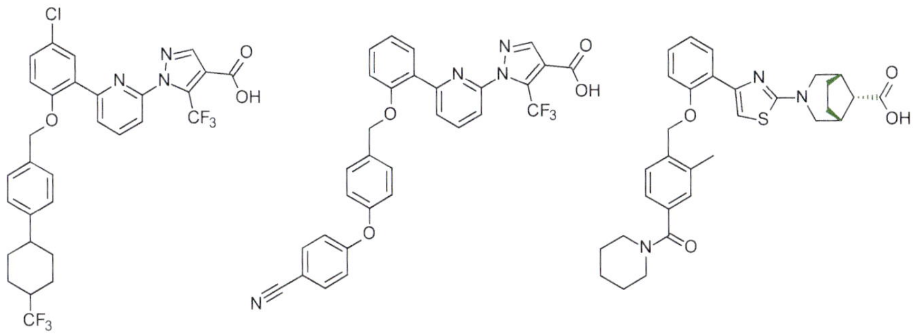

12 13 14

Scheme 5 sGC activators described by Merck, GSK, and Boehringer Ingelheim

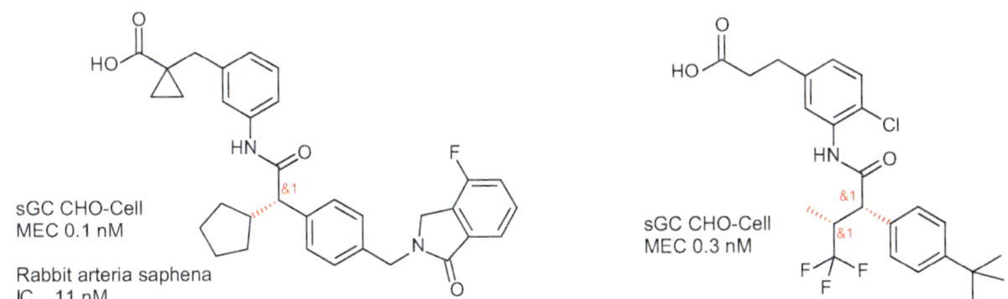

15 16

Scheme 6 3-Phenylpropionic acid sGC activators disclosed by Bayer

Boehringer Ingelheim disclosed BI 703704 (14) displaying in vivo activity in a ZSF1 rat model of type 2 diabetes mellitus (T2DM)–induced nephropathy.

Bayer has also disclosed monocarboxylic acids with novel structural features, highlighting branched 3-phenylpropionic acid derivatives, as exemplified by compound 15 (Scheme 6). With the aim of improving the DMPK profile of these compounds, lower molecular weight 3-phenylpropionic acid congeners have been prepared, as exemplified by compound 16 (Follmann et al. 2013). Bayer is currently developing three new generation sGC activators in phase 1 clinical development for pulmonary hypertension (PH), acute respiratory distress syndrome (ARDS) and chronic kidney disease (CKD).

6 Therapeutic Applications of sGC Stimulators and sGC Activators: A Preclinical Perspective

cGMP is a universal second messenger that regulates the function of many cell types, including smooth muscle cells, cardiomyocytes, fibroblasts, adipocytes, and neurons. Whereas the downstream signaling pathway remains to be fully elucidated, it is abundantly evident that cGMP is critical for the maintenance of cellular and organ homeostasis, and that NO/sGC/cGMP dysfunction is linked to the pathogenesis of numerous diseases. This ubiquitous signaling pathway is the pharmacologic target of sGC stimulators and sGC activators. These pharmacologic agents have been intensively profiled in vitro, ex vivo, and in vivo in mechanistic and disease-relevant animal models to better understand their mode of action and to search for new therapeutic applications. Preclinical and clinical studies have revealed that sGC agonists affect contractility and proliferation of smooth muscle, reduce inflammation and fibrosis, positively impact metabolic risk factors (including weight gain, glucose, and cholesterol), and affect neuronal health and function.

The sGC stimulator riociguat is approved for the treatment of different forms of pulmonary hypertension. However, sGC stimulators and sGC activators have shown beneficial effects in animal models of a variety of other disease conditions. At the cGMP conference held in Bamberg, Germany in June 2017, world experts discussed not only novel targets for cGMP, but also new therapeutic applications of sGC modulators (Friebe et al. 2017). Although the details are beyond the scope of this chapter, some current and future lines of potential therapeutic applications are summarized below.

6.1 Cardiovascular Diseases: Pulmonary Hypertension, Arterial Hypertension, and Heart Failure

NO/sGC/cGMP signaling plays a central role in the cardiovascular system and thus is an obvious area of therapeutic interest, particularly with regard to pulmonary hypertension (PH), systemic hypertension, and heart failure. In 2013, riociguat (BAY 63-2521) was approved as first-in-class sGC stimulator for the treatment of pulmonary arterial hypertension (PAH) and chronic thromboembolic pulmonary hypertension (CTEPH) (Humbert and Ghofrani 2016; Hoeper 2015). In preclinical models of pulmonary hypertension, including hypoxia models and a monocrotaline model, the sGC stimulator BAY 41-2272 and the sGC activator cinaciguat reduced pulmonary hypertension, right ventricular hypertrophy, and lung vascular remodeling in a chronic hypoxia model of pulmonary hypertension, and both compounds reversed hemodynamic and structural changes in a rat monocrotaline model of severe pulmonary hypertension (Dumitrascu et al. 2006). In the Su5416/hypoxia model of pulmonary hypertension, riociguat decreased RV hypertrophy, increased cardiac output, and decreased total pulmonary resistance (Lang et al. 2012). In addition, riociguat reduced PH, pulmonary vascular remodeling and improved right ventricular function in a mouse TAC model of Group 2 PH, but did not improve left ventricular function or hypertrophy (Pradhan et al. 2016). Subsequently, there have been more than 30 preclinical publications demonstrating the effect of sGC stimulators including riociguat in cardiopulmonary diseases and pulmonary hypertension (Stasch and Evgenov 2013). These preclinical results anticipated clinical findings of reduction in PVR and NT-proBNP in PH patients, as well as improvement in exercise tolerance, which provided the basis for regulatory approvals of riociguat for treatment of PAH and CTEPH (Ghofrani et al. 2013a, b).

Activation of NO/sGC/cGMP signaling causes vascular smooth muscle relaxation and vasorelaxation. Consistent with a hypertensive phenotype in mouse sGC knockout models (Friebe et al. 2007; Buys et al. 2008), human genetic variants in the NO/sGC/cGMP pathway, including variants of sGC, have been associated with elevated blood pressure and increased cardiovascular disease risk (International Consortium for Blood Pressure Genome-Wide Association 2011). Although there are more than six classes of drugs that are used to treat hypertension, many patients do not achieve blood pressures below the guideline-recommended levels (Whelton et al. 2017; Pimenta and Calhoun 2016) of 130/80 mmHg. Of the drugs used to treat hypertension, only sodium nitroprusside targets the NO/sGC/cGMP pathway, but, due to tachyphylaxis, its use is limited to acute treatment of hypertensive crisis. It is expected that sGC stimulators and activators, which dose-dependently reduce blood pressure in animal models of hypertension (Mittendorf et al. 2009; Geschka et al. 2011; Tobin et al. 2018), would lower blood pressure in patients who are not at goal despite treatment with current standard of care. sGC stimulators in particular may provide potent blood pressure reduction in refractory patients with salt-sensitive hypertension, which is associated with endothelial dysfunction. Indeed, sGC stimulators have shown potent, dose-dependent blood pressure reduction in the Dahl salt-sensitive model of hypertension (Geschka et al. 2011; Tobin et al. 2018).

Chronic heart failure constitutes a major health problem worldwide. Pharmacological therapies, targeting the renin-angiotensin system or sympathetic nervous system, have limited efficacy (Lewis et al. 2017). Drugs targeting the cGMP pathway, including isosorbide dinitrate/hydralazine (BiDil) and the angiotensin receptor neprilysin inhibitor sacubitril combined with valsartan (Entresto), have proven effective in treatment of heart failure. In recent years, heart failure has been categorized into heart failure with reduced ejection fraction (HFrEF) and heart failure with preserved ejection fraction (HFpEF). The prevalence of HFpEF is now nearly equal to the prevalence of HFrEF. Although HFpEF has been characterized as a heterogenous condition and has been particularly refractory to pharmacologic treatment, an evolving understanding of the disease suggests that microvascular inflammation resulting in cardiac and systemic endothelial dysfunction may be a common underlying pathophysiology, highlighting the potential pharmacologic utility of sGC stimulators (Paulus and Tschope 2013). Nitrates have a long history of use in angina as well as acute and chronic heart failure, but their utility may be limited by development of tolerance and propensity to form highly damaging peroxynitrites. The rationale for treating both HFrEF and HFpEF with sGC agonists is strong and was recently reviewed (Breitenstein et al. 2017). sGC is expressed in critical cardiovascular tissues including vascular smooth muscle, the heart, and the kidney, and sGC $\alpha1^{-/-}$ knockout mice have impaired ventricular relaxation and reduced cardiac output (Irvine et al. 2012). The sGC activator cinaciguat reduced cardiac hypertrophy and improved systolic and diastolic function in a diabetic cardiomyopathy model, and prevented cardiac hypertrophy in a pressure-overload model (Mátyás et al. 2015). Similarly, in Dahl salt-sensitive rats, in the angiotensin II – pressure-overload models, or in post-myocardial infarction models of heart failure, treatment with sGC stimulators has consistently shown beneficial effects on cardiac function, remodeling, and fibrosis, as well as on levels of the ventricular stress hormone NT-proBNP (Masuyama et al. 2006; Masuyama et al. 2009; Methner et al. 2013; Geschka et al. 2011). Finally, NO/sGC/cGMP signaling plays an important role in regulating inflammation, which is believed to be a common underlying pathophysiology in HFpEF. In a double transgenic rat (dTGR) model of HFpEF, treatment with the sGC stimulator BAY 41-8543 dramatically improved survival rate, reduced cardiac fibrosis, macrophage infiltration, and gap junction remodeling. The expression of dysregulated cardiac genes associated with fibrosis, inflammation, apoptosis, oxidative stress, and ion channel function was restored in treated dTGR in the direction of healthy controls. Treatment reduced systemic blood pressure levels and improved endothelium-dependent vasorelaxation of resistance vessels. Further comprehensive in vivo phenotyping showed an improved diastolic cardiac function, improved hemodynamics, and also less susceptibility to ventricular arrhythmias. Thus, sGC stimulation was highly effective in improving several HFpEF facets in this animal model, underscoring its potential value for patients (Wilck et al. 2018). Moreover, in mouse studies, it has been shown that treatment with NO and sGC stimulators reduced P-selectin expression and leukocyte recruitment (Ahluwalia et al. 2004; Tchernychev et al. 2017), indicating potential to reduce microvascular inflammation.

6.2 Kidney Diseases

The potential benefits of cGMP increase in the kidney and of sGC stimulators and activators in renal disease were recently reviewed (Stasch et al. 2015; Krishnan et al. 2018). Chronic kidney disease, defined as reduced eGFR and/or increased urinary albumin, is associated with high morbidity and mortality, and treatment options are limited. NO/sGC/cGMP signaling is involved in vital renal functions including regulation of renal blood flow and glomerular hemodynamics as well as water and salt transport in the tubular system (Kone 1997). Indeed, advanced nephropathy is associated with progressive decline in levels of NO (Prabhakar 2004). sGC stimulators and activators have been tested in several models of kidney diseases including hypertensive nephropathy, unilateral ureteral obstruction, diabetic nephropathy, and acute glomerular nephritis. Treatment with both stimulators and activators has resulted in reduced proteinuria and/or albuminuria; decreased renal glomerulosclerosis, fibrosis and markers of fibrosis, and improved podocyte health (reviewed in Stasch et al. 2015). Data from in vitro studies in renal fibroblasts suggest cGMP-mediated suppression of TGFβ/P-smad3 signaling may contribute to antifibrotic effects in the kidney (Schinner et al. 2017). In human renal proximal tubular cells in vitro, treatment with sGC stimulators reduced TGFβ-induced apoptosis and TNFα-induced increases in MCP-1 (Liu et al. 2016). In addition, NO/sGC/cGMP stimulation mediates renal protection through systemic effects on the cardiovascular system, which, in health, may suppress tubular degeneration and subsequent renal fibrosis and hypertrophy.

6.3 Fibrotic Diseases (Lung Fibrosis and Systemic Sclerosis)

There is accumulating evidence that cGMP elevation can have an antifibrotic effect via directly targeting fibroblasts and myofibroblasts (Sandner et al. 2017; Sandner and Stasch 2017). Understanding these effects extends the mode of action of cGMP beyond vasodilation and may provide the basis for completely new applications of cGMP-enhancing drugs. As noted above, in models of hypertension, cardiomyopathy, and chronic kidney disease, treatment with sGC agonists has been associated with antifibrotic effects on target organs. In recent years, antifibrotic effects of sGC modulators have been explored in non-hypertensive models of fibrosis in organs outside of the cardiovascular system, including the lung and the skin. In a bleomycin mouse model of pulmonary fibrosis, riociguat treatment reduced pulmonary hypertension, right ventricular hypertrophy, inflammation, and pulmonary fibrosis (Evgenov et al. 2011). The sGC stimulator BAY 41-2272 halted development of skin fibrosis in bleomycin and Tsk-1 mouse models as demonstrated by reduced dermal thickening, myofibroblast number, and hydroxyproline content (Beyer et al. 2012). In many of these studies, sGC agonists reduced fibrosis at doses that did not affect blood pressure or heart rate, suggesting that the antifibrotic effects are independent from the hemodynamic effects and may be due to direct effects of cGMP on fibrotic processes. The antifibrotic mechanism has been explored in vitro in fibroblasts from lung, skin, kidney, liver, and heart (Vettel et al. 2014; Beyer et al. 2015; Hewitson et al. 2004; Hall et al. 2018; Lambers et al. 2014). Increasing cGMP signaling in these cellular studies suppressed TGFβ-mediated increases in collagen and ECM production, inhibited fibroblast-myofibroblast differentiation, and/or reduced fibroblast proliferation.

6.4 Liver Diseases

Chronic liver diseases such as hepatitis and alcoholic and nonalcoholic liver disease can lead to cirrhosis. Liver cirrhosis is characterized by extensive fibrotic scarring of the liver, which is associated with impaired hepatic function and leads to complications such as portal hypertension and esophageal varices. sGC agonists have shown promising effects in preclinical models of liver fibrosis (Knorr et al. 2008; Hall et al. 2017). The sGC stimulator riociguat was shown to reduce liver fibrosis and portal pressure in cirrhotic rats (Schwabl et al. 2018) and mechanistic studies suggest that sGC modulators may inhibit fibrotic differentiation of hepatic stellate cells (Xiao et al. 2015; Hall et al. 2017). Nonalcoholic steatohepatitis (NASH) is a liver disease with characteristics of steatosis, inflammation, and fibrosis, and is a growing health concern globally. Patients with NASH are at risk for developing cirrhosis, and also have elevated cardiovascular event risk. There are no approved treatments for NASH, and most of the treatments that are in clinical development address only one aspect of the NASH pathophysiology (i.e., steatosis, inflammation, or fibrosis). sGC agonists have the potential to impact all three aspects of NASH pathophysiology: inflammation [discussed above (Ahluwalia et al. 2004)], fibrosis, and steatosis (see Sect. 6.5 below) (Hoffmann et al. 2015); indeed, in an experimental NASH model, the sGC stimulator praliciguat affected all three aspects (Flores-Costa et al. 2017).

6.5 Metabolic Disease

Elevated plasma glucose, excess visceral fat, abnormal cholesterol or triglyceride levels, and high blood pressure are components of the metabolic syndrome. When these factors occur together, they increase an individual's risk of developing heart disease, stroke, and diabetes. Treatment with the sGC stimulator BAY 41-8543 improved metabolic measures (weight gain, fat mass, diabetic phenotype) in a mouse diet-induced obesity (DIO) model (Hoffmann et al. 2015). In a similar DIO model, treatment with the sGC stimulator praliciguat improved glucose tolerance and insulin sensitivity and lower triglycerides (Schwartzkopf et al. 2018) Furthermore, olinciguat and praliciguat reduced fasting glucose in the ZSF1 model of diabetic nephropathy (Profy et al. 2017; Masferrer et al. 2016), and the sGC stimulator praliciguat reduced hepatic steatosis in an experimental NASH model (Flores-Costa et al. 2017). The promising metabolic effects suggest evaluation of sGC stimulators in individuals with metabolic diseases including obesity, diabetes, hyperlipidemia, and metabolic syndrome, and NASH may be warranted. One mechanism for the metabolic effects of sGC agonism may involve increased lipid uptake into brown adipose and increased whole-body energy expenditure (Hoffmann et al. 2015). This area of research warrants further exploration and may have broad relevance to treatment of metabolic disease and associated comorbidities.

6.6 Central and Peripheral Nervous System Disorders

The importance of cGMP in neuronal and sensory signaling, cognitive function, and brain health has gained greater appreciation in recent years. Both the ubiquitous NO-sensitive guanylyl cyclase comprising $\alpha 1$ and $\beta 1$ subunits now referred to as GC-1, and the more tissue-specific (e.g., CNS, kidney, placenta) $\alpha 2 \beta 1$ heterodimer now termed GC-2 are expressed in the brain (Mergia et al. 2003; Ibarra et al. 2001), and cGMP has been shown to mediate memory formation and LTP (Bollen et al. 2014). In addition, NO and sGC regulate local blood flow in the CNS in response to neuronal activity through a process known as functional hyperemia (Faraco and Iadecola 2013). Vascular dysfunction may underlie forms of dementia and Alzheimer's disease as systemic hypertension is a leading risk factor for these diseases. There is also a growing interest in the role that neuroinflammation may play in the deterioration of brain health and cognitive function. Pharmacologic data with PDE5i suggest that cGMP signaling may suppress neuroinflammation (Agusti et al. 2017; Christina Alves et al. 2015; Raffaella et al. 2016). Drugs affecting the NO/sGC/cGMP signaling pathway may address multiple aspects of the pathophysiology of dementia. Inhibitors of PDE9, a cGMP-specific phosphodiesterase, have shown promising results in preclinical models of learning and memory (van der Staay et al. 2008). However, a PDE9 inhibitor did not improve cognition in the clinic (Schwam et al. 2014). Relative to PDE9 inhibitors, sGC agonists, and particularly sGC stimulators, which enhance neuronal *and vascular* NO signaling, may have the potential to address the broader constellation of deficiencies in dementia. CNS-targeted sGC agonists have not been available for clinical investigation. IW-6463 is a novel sGC stimulator that penetrates the blood–brain barrier being evaluated for potential use in CNS diseases.

6.7 Gastrointestinal Motility Disorders

NO released from nitrergic neurons in the GI tract is an important regulator of GI smooth muscle relaxation and motility (Groneberg et al. 2016). Mice lacking sGC develop fatal GI obstruction (Friebe et al. 2007). There is strong evidence that dysfunctional nitrergic signaling is involved in GI motility disorders such as achalasia, gastroparesis, slow transit constipation, and Hirschsprung's disease. sGC is found in several cell types in the GI tract, including smooth muscle cells, interstitial cells of Cajal (ICC), and fibroblast-like cells; smooth muscle and ICC-specific sGC knockouts have increased understanding of the roles of sGC in each cell type in the regulation of intestinal peristalsis (Groneberg et al. 2016).

Achalasia is a swallowing disorder in which the lower esophageal sphincter (LES) remains in a contracted state, limiting passage of food from the esophagus into the stomach. Achalasia has been associated with loss of NO signaling neurons in the LES (Hoshino et al. 2013), mice deficient in neuronal nitric oxide synthase (nNOS) develop LES hypertension (Sivarao et al. 2001), and individuals with a rare homozygous loss of sGC mutation develop achalasia (Herve et al. 2014). NO donors and PDE5 inhibitors have reduced LES pressure in patients with achalasia (Patel et al. 2015; Bortolotti et al. 2000; Eherer et al. 2002). The sGC stimulator olinciguat was shown to relax human LES ex vivo (Zimmer et al. 2017), and a Phase 2a exploratory study was recently completed.

6.8 Hematologic (Sickle Cell Disease)

Sickle cell disease (SCD) is an inherited blood disorder resulting from an allele of the hemoglobin beta gene that results in sickling of red blood cells (Ingram 1956). Individuals with SCD can develop a number of complications including anemia, acute chest syndrome, pulmonary hypertension, fatigue, and vaso-occlusive crisis, which is characterized by extreme pain. Sickle cell disease is associated with endothelial and NO dysfunction (Nahavandi et al. 2002) resulting from increased circulating levels of free hemoglobin (an NO scavenger), arginase (which degrades the nitric oxide synthase substrate arginine), and ADMA (a nitric oxide synthase inhibitor). A main clinical feature of SCD is unpredictable and recurrent severe pain associated with sickle-cell-mediated small vessel vaso-occlusion, which may be triggered or potentiated by vascular dysfunction and inflammation. Effective therapies targeting the SCD symptoms and quality of life including the cause of vaso-occlusive pain are needed.

Hydroxyurea is approved as a chronic use drug treatment for SCD. Although aspects of its mechanism of action are not completely understood, hydroxyurea may prevent red blood cell sickling by increasing fetal hemoglobin expression via NO release (Cokic et al. 2003). Red blood cells appear to have a functional NO/sGC/cGMP signaling pathway; furthermore, red blood cells from patients with endothelial dysfunction (associated with coronary artery disease) are responsive to NO as well as both sGC stimulators and sGC activators (Cortese-Krott et al. 2018). Stimulation of the NO/sGC/cGMP pathway has been shown to decrease vascular inflammation in vivo. As noted above, the sGC stimulator BAY 41-2272 and NO reduced leukocyte rolling and adhesion in an eNOS deficient mouse model (Ahluwalia et al. 2004), and the sGC stimulator olinciguat reduced makers of vascular inflammation and increased leukocyte rolling and velocity in a TNFα mouse model (Tchernychev et al. 2017). Chronic oral administration of the sGC activator cinaciguat improved endothelial function and reversed pulmonary hypertension and cardiac remodeling in a mouse model of SCD without affecting systemic blood pressure (Potoka et al. 2018). In a humanized SCD mouse model of TNFα-induced acute vaso-occlusion, BAY 73-6691, a PDE9 inhibitor, reduced leukocyte recruitment and red blood cell–leukocyte interactions, and improved leukocyte rolling and adhesion (Almeida et al. 2012). Finally, vasodilation mediated by sGC agonists is expected to increase blood flow in small vessels, preventing vaso-occlusion. In summary, sGC agonists could affect blood flow, vascular inflammation, and red blood cell sickling thereby preventing multiple complications of SCD.

6.9 Ocular Diseases

Glaucoma is a progressive optic neuropathy and a leading cause of blindness worldwide. Ocular pressure is a risk factor for development of primary open angle glaucoma, the most prevalent form of glaucoma. Mice deficient in sGC exhibit ophthalmic pathology resembling glaucoma, including increased intraocular pressure, optic neuropathy, and retinal vascular dysfunction. Additionally, human candidate gene studies revealed that a variant in the locus encoding genes for GC1 are associated with one form of primary open angle glaucoma (Buys et al. 2013). Furthermore, several studies have demonstrated that NO, cGMP, and sGC modulators may reduce intraocular pressure through regulation of aqueous humor outflow from the anterior chamber through the trabecular meshwork and Schlemm's canal (Kotikoski et al. 2003; Ge et al. 2016). Emerging data suggest that modulators of cGMP availability may also prevent optic nerve damage, independent of effects on ocular pressure. However, very recently a sGC activator from Novartis (MGV354) was profiled preclinically and clinically in Glaucoma. Despite promising preclinical results in animal models in which MGV354 significantly lowered intraocular pressure (Prasanna et al. 2018), MGV354 failed in the phase 1/2 clinical trial (Stacy et al. 2018).

6.10 Preclinical Summary

In summary, the multifaceted pharmacology of sGC modulators with effects on vascular function, inflammation, fibrosis, neuronal health and signaling, and metabolism affords the opportunity to positively impact a variety of pathologic conditions and organ systems. The availability of sGC stimulators and sGC activators enables the investigation of these unique mechanisms in both the preclinical and clinical settings.

7 Clinical Developments of sGC Stimulators and sGC Activators

Given the broad impact of the NO/sGC/cGMP pathway on regulation of cell, tissue, and body function, it is not surprising that there are many clinical trials, both completed and ongoing, investigating the treatment potential of sGC stimulators and sGC activators. Clinical trials with early sGC agonists that are no longer in clinical development are listed here for reference and completeness (Table 1).

Based on a broad preclinical profiling of riociguat in animal models of pulmonary hypertension in which a significant reduction of pulmonary artery pressure could be demonstrated, riociguat was developed for the treatment of pulmonary hypertension. Riociguat showed efficacy in two Phase 3 trials in pulmonary hypertension patients, namely, the PATENT trial in pulmonary hypertension patients, WHO Group 1 (pulmonary arterial hypertension – PAH) and the CHEST trial in WHO Group 4 (chronic thromboembolic pulmonary hypertension – CTEPH) (Ghofrani et al. 2013a, b; Frey et al. 2017). Treatment effects of riociguat were sustained for at least 2 years in the long-term Phase 3 extension studies PATENT-2 and CHEST-2 trials (Rubin et al. 2015; Simonneau et al. 2016; Ghofrani et al. 2016). A broad range of completed and ongoing Phase 3b, Phase 4, and investigator-initiated clinical trials with riociguat have also been performed and cannot be covered fully in the scope of this review. Importantly, riociguat was also tested in patients with pulmonary arterial hypertension with insufficient response to PDE5 inhibitors. This single-arm, open-label, uncontrolled study (RESPITE, NCT0200762) indicated that replacing PDE5i with riociguat may be a feasible and effective treatment strategy in these patients (Hoeper et al. 2017a, b). A randomized controlled, open-label multicenter international study is currently ongoing to confirm the results (REPLACE, NCT02891850). In addition, in the MOTION (NCT 02191137) study, an open-label Phase 4 program, treatment-naïve pulmonary arterial hypertension patients were studied for patient-reported outcome using three different quality of life instruments. The results showed a positive impact of riociguat treatment on patient-reported quality of life.

Beyond pulmonary hypertension there are also clinical trials ongoing to treat systemic, arterial hypertension with sGC stimulators. Vasorelaxation is a prominent effect of sGC stimulators at higher dose levels. Since sGC stimulators actively augment cGMP and downstream vasodilation rather than block vasoconstriction,

Table 1 Lists a selection of Phase 2 and 3 trials with sGC stimulators and sGC activators from the public ClinicalTrials.gov database (https://clinicaltrials.gov/)

sGC stimulators	Indication	Phase	NCT number	Study name	Status	
Riociguat (BAY 63-2521)	PAH	3	NCT00810693	PATENT	Completed	Approved for PAH in 2013
	CTEPH	3	NCT00855465	CHEST	Completed	Approved for CTEPH in 2013
	PAH	3	NCT00863681	PATENT-2	Completed	
	CTEPH	3	NCT00910429	CHEST_2	Completed	
	PAH children	3	NCT02562235	PATENT CHILD	Ongoing	
	PAH	3	NCT02007629	RESPITE	Completed	
	PH-LVD	2	NCT01065454	LEPHT	Completed	
	PH-IIPs	2	NCT02138825	RISE IIP	Terminated	
	dcSSc[a]	2	NCT02283762	RISE SSc	Completed	
	CF	2	NCT02170025		Terminated	
	SCD	2	NCT02633397		Ongoing	
Nelociguat (BAY 60-4552)	ED	2	NCT01168817		Completed	
Vericiguat (BAY 102-1189)	HFrEF	2	NCT01951625	SOCRATES-REDUCED	Completed	
	HFpEF	2	NCT01951638	SOCRATES-PRESERVED	Completed	
	HFrEF[a]	3	NCT02861534	VICTORIA	Ongoing	
	HFpEF[a]	2	NCT03547583	VITALITY-HFpEF	Ongoing	
Olinciguat (IW-1701)	Achalasia	2	NCT02931565		Completed	
	SCD	2	NCT03285178	STRONG SCD	Ongoing	
Praliciguat (IW-1973)	T2D and HTN	2	NCT03091920		Completed	
	T2D and HTN	2	NCT02906579		Completed	
	HFpEF	2	NCT03254485	CAPACITY-HFpEF	Ongoing	
	Diabetic Nephropathy	2	NCT03217591		Ongoing	

This table does not include Phase 1 clinical studies but does include development programs that are completed and terminated to reflect the full range of potential indications

[a]Bayer/MSD codevelopment; BAY 102-1189 = MK-1242

sGC stimulators might provide advantages over classical antihypertensive therapies. Bayer investigated a new chemical class of long-acting sGC stimulators in Phase 1 studies for the treatment of difficult to treat hypertension patients. Ironwood recently completed two Phase 2 studies with praliciguat in patients with T2DM and a history of hypertension on stable regimens of anti-glycemic and anti-hypertensive agents (Hanrahan et al. 2018a, b). Consistent with preclinical observations, treatment with praliciguat led to reductions in blood pressure and improvement in metabolic parameters including fasting plasma glucose and cholesterol levels in this patient population. These studies confirmed the pharmacokinetic profile of praliciguat supporting once-daily dosing and broad tissue distribution and set the stage for ongoing studies of this compound in patients with diabetic nephropathy and HFpEF discussed below.

The NO/sGC/cGMP signaling pathway plays a pivotal role in the regulation of the cardiovascular system, and sGC stimulators and sGC activators have the potential for broad impact on the treatment of cardiovascular diseases. cGMP increase by these compounds may result in systemic improvements driven by the vascular effects but could also have direct effects in cardiac or renal tissues improving heart and kidney function. In addition, there are hints from preclinical and clinical studies on metabolic effects and effects on adipose tissues. For these reasons, clinical trials, both completed and ongoing, have investigated the effects of sGC stimulators in chronic heart failure. Two Phase 2 studies, the so-called SOCRATES trials (SOluble guanylate Cyclase stimulatoR in heArT failure), were conducted in chronic heart failure patients with reduced and preserved ejection fraction, SOCRATES-REDUCED and SOCRATES-PRESERVED, respectively (Pieske et al. 2014). In the SOCRATES-REDUCED study (NCT01951625), the exploratory analysis suggested a dose-dependent reduction of NT-proBNP and a trend for reduction of CV deaths and HF hospitalizations (Gheorghiade et al. 2015). In the SOCRATES-PRESERVED study (NCT01951638), no significant effect on NT-proBNP was observed but there was an improvement in quality of life scores (Pieske et al. 2017; Filippatos et al. 2017). Currently a Phase 3 confirmatory trial with vericiguat in HFrEF, the so-called VICTORIA trial (VerICiguaT Global Study in Subjects With Heart Failure With Reduced Ejection Fraction, NCT02861534) is ongoing (Armstrong et al. 2017). More recently, Ironwood initiated the Phase 2 CAPACITY-HFpEF trial (NCT03254485) with praliciguat and Bayer together with MSD started a Phase 2 VITALITY-HFpEF trial (NCT03547583) with vericiguat. Both studies are evaluating the potential benefit of sGC stimulators in treating heart failure patients with preserved ejection fraction. Building on the preclinical data supporting positive renal effects of sGC stimulators (Stasch et al. 2015; Tobin et al. 2018), Praliciguat is also being studied in patients with diabetic nephropathy (NCT03217591).

In addition to the indications focusing on cardiopulmonary, cardiovascular, and heart and kidney diseases there are other indications under investigation in clinical trials. Based on preclinical profiling, several proof of concept and Phase 2 trials have been or are being conducted to explore the potential beneficial effects in patients. To explore the potential antifibrotic effects that have been observed in preclinical models of lung and skin fibrosis, Phase 2b trials were initiated. Riociguat was investigated in patients with symptomatic PH, associated with idiopathic interstitial pneumonias including idiopathic pulmonary fibrosis (RISE-IIP, NCT02138825) (Nathan et al. 2017). Moreover, the effects of riociguat on skin fibrosis in SSc patients (RISE-SSc, NCT2283762) are currently studied. The RISE-IIP study was terminated prematurely due to an unfavorable risk versus benefit ratio in these patients (Nathan et al. 2017). The RISE-SSc study is ongoing with recruiting finished and data expected in 2018. Based on the various modes of actions and expression of sGC in different tissues and organs, smaller studies have been conducted or are underway exploring effects on rare diseases. The effects of the sGC stimulator olinciguat were evaluated in a recently completed exploratory study in patients with achalasia (NCT02931565). Olinciguat and riociguat are also in Phase 2 trials in sickle cell disease patients (NCT03285178 and NCT02633397, respectively). In addition to these more advanced clinical programs, there are still other sGC stimulators in preclinical development that might increase the number of compounds available for the benefit of patients (Friebe et al. 2017).

7.1 sGC Activators

Compared to sGC stimulators, the development pipeline of sGC activators is relatively limited and less advanced. There are no sGC activators in late stage development or approved to date. Multiple sGC activator projects have been terminated in Phase 2. sGC activators have not been explored in chronic heart failure; however, cinaciguat (BAY 58-2667) has been characterized in acute heart failure. Based on promising preclinical results, a Phase 2 study in patients with acute decompensated heart failure (ADHF) was initiated. Continuous intravenous infusion of cinaciguat was well tolerated and resulted in an improvement of cardiopulmonary hemodynamics. The subsequent clinical Phase 2b program studied the effects of cinaciguat in three randomized, double-blind, placebo-controlled studies in ADHF patients; however, the clinical development of cinaciguat was terminated prematurely because of hypotensive events without clear benefit (Breitenstein et al. 2017). The oral sGC activator ataciguat (HMR 1766) was investigated in Phase 2 studies for the treatment of peripheral arterial occlusive disease (PAD) and neuropathic pain. These projects were terminated due at least in part to the long-lasting blood pressure-lowering effects of these compounds in the absence of clear therapeutic benefit. In addition, the understanding of diseases with increased oxidative stress burden as the mode of action of sGC activators is still incompletely understood. Bayer recently reported three sGC activators in Phase 1 with the intention of potentially treating chronic kidney disease pulmonary hypertension and acute respiratory distress syndrome (ARDS). In addition, Boehringer-Ingelheim (BI) recently reported the early development of an sGC activator for chronic kidney disease (Friebe et al. 2017). It will become a very interesting topic in the future, how these compound class of sGC activators – acting on the heme-free sGC – could be differentiated from sGC stimulators. Especially in patients with diseases accompanied by increased oxidative stress burden, this could broaden the treatment potential or increase efficacy.

Acknowledgments The authors would like to thank Christian Meier, Kelly Lewis, and Shalini Murali at Bayer and Jennifer Chickering, Albert Profy, Emmanuel Buys, Joon Jung, Paul Renhowe, Yueh-tyng Chien, Regina Graul, and Chris Winrow at Ironwood for contributions to and critical reading of the manuscript.

Conflict of Interest GTM and DPZ are employees of Ironwood Pharmaceuticals, and MF, PS, and JPS are employees of Bayer AG Pharmaceuticals.

References

Agusti A, Hernandez-Rabaza V, Balzano T, Taoro-Gonzalez L, Ibanez-Grau A, Cabrera-Pastor A, Fustero S, Llansola M, Montoliu C, Felipo V (2017) Sildenafil reduces neuroinflammation in cerebellum, restores GABAergic tone, and improves motor in-coordination in rats with hepatic encephalopathy. CNS Neurosci Ther 23(5):386–394. https://doi.org/10.1111/cns.12688

Ahluwalia A, Foster P, Scotland RS, McLean PG, Mathur A, Perretti M, Moncada S, Hobbs AJ (2004) Antiinflammatory activity of soluble guanylate cyclase: cGMP-dependent down-regulation of P-selectin expression and leukocyte recruitment. Proc Natl Acad Sci U S A 101 (5):1386–1391. https://doi.org/10.1073/pnas.0304264101

Alexander SP, Fabbro D, Kelly E, Marrion N, Peters JA, Benson HE, Faccenda E, Pawson AJ, Sharman JL, Southan C, Davies JA, Collaborators CGTP (2015) The concise guide to PHARMACOLOGY 2015/16: enzymes. Br J Pharmacol 172(24):6024–6109. https://doi.org/10.1111/bph.13354

Almeida CB, Scheiermann C, Jang JE, Prophete C, Costa FF, Conran N, Frenette PS (2012) Hydroxyurea and a cGMP-amplifying agent have immediate benefits on acute vaso-occlusive events in sickle cell disease mice. Blood 120(14):2879–2888. https://doi.org/10.1182/blood-2012-02-409524

Alruwaili N, Sun D, Wolin MS (2017) Modulation of heme biosynthesis by ferrochelatase inhibition controls soluble guanylate cyclase expression and superoxide production in bovine coronary arteries. FASEB J 31(1_Supp):1080–1015

Armitage ME, Wingler K, Schmidt HH, La M (2009) Translating the oxidative stress hypothesis into the clinic: NOX versus NOS. J Mol Med 87(11):1071–1076. https://doi.org/10.1007/s00109-009-0544-2

Armstrong PW, Roessig L, Patel MJ, Anstrom KJ, Butler J, Voors AA, Lam CSP, Ponikowski P, Temple T, Pieske B, Ezekowitz J, Hernandez AF, Koglin J, O'Connor CM (2017) A multicenter, randomized, double-blind, placebo-controlled trial of the efficacy and safety of the oral soluble guanylate cyclase stimulator: the VICTORIA trial. JACC Heart Fail 6(2):96–104. https://doi.org/10.1016/j.jchf.2017.08.013

Balashova N, Chang FJ, Lamothe M, Sun Q, Beuve A (2005) Characterization of a novel type of endogenous activator of soluble guanylyl cyclase. J Biol Chem 280(3):2186–2196. https://doi.org/10.1074/jbc.M411545200

Beyer C, Reich N, Schindler SC, Akhmetshina A, Dees C, Tomcik M, Hirth-Dietrich C, von Degenfeld G, Sandner P, Distler O, Schett G, Distler JH (2012) Stimulation of soluble guanylate cyclase reduces experimental dermal fibrosis. Ann Rheum Dis 71(6):1019–1026. https://doi.org/10.1136/annrheumdis-2011-200862

Beyer C, Zenzmaier C, Palumbo-Zerr K, Mancuso R, Distler A, Dees C, Zerr P, Huang J, Maier C, Pachowsky ML, Friebe A, Sandner P, Distler O, Schett G, Berger P, Distler JH (2015) Stimulation of the soluble guanylate cyclase (sGC) inhibits fibrosis by blocking non-canonical TGFbeta signalling. Ann Rheum Dis 74(7):1408–1416. https://doi.org/10.1136/annrheumdis-2013-204508

Bittner AR, Sinz CJ, Chang J, Kim RM, Mirc JW, Parmee ER, Tan Q (2009) Soluble guanylate cyclase activators. WIPO, Geneva

Bivalacqua TJ, Usta MF, Champion HC, Kadowitz PJ (2003 Nov-Dec) Hellstrom WJ (2003) Endothelial dysfunction in erectile dysfunction: role of the endothelium in erectile physiology and disease. J Androl 24(6 Suppl):S17–S37

Boerrigter G, Costello-Boerrigter LC, Cataliotti A, Lapp H, Stasch JP, Burnett JC Jr (2007) Targeting heme-oxidized soluble guanylate cyclase in experimental heart failure. Hypertension 49(5):1128–1133

Bollen E, Puzzo D, Rutten K, Privitera L, De Vry J, Vanmierlo T, Kenis G, Palmeri A, D'Hooge R, Balschun D, Steinbusch HM, Blokland A, Prickaerts J (2014) Improved long-term memory via enhancing cGMP-PKG signaling requires cAMP-PKA signaling. Neuropsychopharmacology 39(11):2497–2505. https://doi.org/10.1038/npp.2014.106

Bortolotti M, Mari C, Lopilato C, Porrazzo G, Miglioli M (2000) Effects of sildenafil on esophageal motility of patients with idiopathic achalasia. Gastroenterology 118(2):253–257

Breitenstein S, Roessig L, Sandner P, Lewis KS (2017) Novel sGC stimulators and sGC activators for the treatment of heart failure. Handb Exp Pharmacol 243:225–247. https://doi.org/10.1007/164_2016_100

Brunton TL (1867) On the use of nitrite of amyl in angina pectoris. Lancet 90(2290):97–98

Bruzziches R, Francomano D, Gareri P, Lenzi A, Aversa A (2013) An update on pharmacological treatment of erectile dysfunction with phosphodiesterase type 5 inhibitors. Expert Opin Pharmacother 14(10):1333–1344. https://doi.org/10.1517/14656566.2013.799665

Buys ES, Ko YC, Alt C, Hayton SR, Jones A, Tainsh LT, Ren R, Giani A, Clerte M, Abernathy E, Tainsh RE, Oh DJ, Malhotra R, Arora P, de Waard N, Yu B, Turcotte R, Nathan D, Scherrer-Crosbie M, Loomis SJ, Kang JH, Lin CP, Gong H, Rhee DJ, Brouckaert P, Wiggs JL, Gregory MS, Pasquale LR, Bloch KD, Ksander BR (2013) Soluble guanylate cyclase alpha1-deficient mice: a novel murine model for primary open angle glaucoma. PLoS One 8(3):e60156. https://doi.org/10.1371/journal.pone.0060156

Buys ES, Sips P, Vermeersch P, Raher MJ, Rogge E, Ichinose F, Dewerchin M, Bloch KD, Janssens S, Brouckaert P (2008) Gender-specific hypertension and responsiveness to nitric oxide in sGCalpha1 knockout mice. Cardiovasc Res 79(1):179–186. https://doi.org/10.1093/cvr/cvn068

Buys ES, Zimmer DP, Chickering J, Graul R, Chien YT, Profy A, Hadcock JR, Masferrer JL, Milne GT (2018) Discovery and development of next generation sGC stimulators with diverse multidimensional pharmacology and broad therapeutic potential. Nitric Oxide 78:72–80

Cartledge JJ, Eardley I, Morrison JF (2001) Nitric oxide-mediated corpus cavernosal smooth muscle relaxation is impaired in ageing and diabetes. BJU Int 87(4):394–401

Christina Alves P, Peixoto CA, Nunes AK, Garcia-Osta A, Ana Karolina Santana N, Ana G-O (2015) Phosphodiesterase-5 inhibitors: action on the signaling pathways of neuroinflammation, neurodegeneration, and cognition. Mediators Inflamm 2015:940207. https://doi.org/10.1155/2015/940207

Cokic VP, Smith RD, Beleslin-Cokic BB, Njoroge JM, Miller JL, Gladwin MT, Schechter AN (2003) Hydroxyurea induces fetal hemoglobin by the nitric oxide-dependent activation of soluble guanylyl cyclase. J Clin Invest 111(2):231–239. https://doi.org/10.1172/JCI16672

Cortese-Krott MM, Mergia E, Kramer CM, Lückstädt W, Yang J, Wolff G, Panknin C, Bracht T, Sitek B, Pernow J, Stasch JP, Feelisch M, Koesling D, Kelm M (2018) Identification of a soluble guanylate cyclase in RBCs: preserved activity in patients with coronary artery disease. Redox Biol 14:328–337. https://doi.org/10.1016/j.redox.2017.08.020

Derbyshire ER, Marletta MA (2012) Structure and regulation of soluble guanylate cyclase. Annu Rev Biochem 81:533–559. https://doi.org/10.1146/annurev-biochem-050410-100030

Dumitrascu R, Weissmann N, Ghofrani HA, Dony E, Beuerlein K, Schmidt H, Stasch JP, Gnoth MJ, Seeger W, Grimminger F, Schermuly RT (2006) Activation of soluble guanylate cyclase reverses experimental pulmonary hypertension and vascular remodeling. Circulation 113(2):286–295. https://doi.org/10.1161/CIRCULATIONAHA.105.581405

Eherer AJ, Schwetz I, Hammer HF, Petnehazy T, Scheidl SJ, Weber K, Krejs GJ (2002) Effect of sildenafil on oesophageal motor function in healthy subjects and patients with oesophageal motor disorders. Gut 50(6):758–764

Erdmann J, Stark K, Esslinger UB, Rumpf PM, Koesling D, de Wit C, Kaiser FJ, Braunholz D, Medack A, Fischer M, Zimmermann ME, Tennstedt S, Graf E, Eck S, Aherrahrou Z, Nahrstaedt J, Willenborg C, Bruse P, Brænne I, Nöthen MM, Hofmann P, Braund PS, Mergia E, Reinhard W, Burgdorf C, Schreiber S, Balmforth AJ, Hall AS, Bertram L, Steinhagen-Thiessen E, Li SC, März W, Reilly M, Kathiresan S, McPherson R, Walter U, Ott J, Samani NJ, Strom TM, Meitinger T, Hengstenberg C, Schunkert H, CARDIoGRAM (2013) Dysfunctional nitric oxide signalling increases risk of myocardial infarction. Nature 504 (7480):432–436. https://doi.org/10.1038/nature12722

Evgenov OV, Pacher P, Schmidt PM, Hasko G, Schmidt HHHW, Stasch J-P (2006) NO-independent stimulators and activators of soluble guanylate cyclase: discovery and therapeutic potential. Nat Rev Drug Discov 5(9):755–768. https://doi.org/10.1038/nrd2038

Evgenov OV, Zou L, Zhang M, Mino-Kenudson M, Mark EJ, Buys ES, Raher MJ, Li Y, Feng Y, Jones RC, Stasch J-P, Chao W (2011) Nitric oxide-independent stimulation of soluble guanylate cyclase attenuates pulmonary fibrosis. BMC Pharmacol 11(1):O9. https://doi.org/10.1186/1471-2210-11-s1-o9

Faraco G, Iadecola C (2013) Hypertension: a harbinger of stroke and dementia. Hypertension 62 (5):810–817. https://doi.org/10.1161/HYPERTENSIONAHA.113.01063

Feil R, Lohmann SM, de Jonge H, Walter U, Hofmann F (2003) Cyclic GMP-dependent protein kinases and the cardiovascular system: insights from genetically modified mice. Circ Res 93 (10):907–916

Filippatos G, Maggioni AP, Lam CSP, Pieske-Kraigher E, Butler J, Spertus J, Ponikowski P, Shah SJ, Solomon SD, Scalise AV, Mueller K, Roessig L, Bamber L, Gheorghiade M, Pieske B (2017) Patient-reported outcomes in the SOluble guanylate Cyclase stimulatoR in heArT failurE patientS with PRESERVED ejection fraction (SOCRATES-PRESERVED) study. Eur J Heart Fail 19(6):782–791. https://doi.org/10.1002/ejhf.800

Fischmeister R, Castro LR, Abi-Gerges A, Rochais F, Jurevicius J, Leroy J, Vandecasteele G (2006) Compartmentation of cyclic nucleotide signaling in the heart: the role of cyclic nucleotide phosphodiesterases. Circ Res 99(8):816–828

Flores-Costa R, Alcaraz-Quiles J, Titos E, López-Vicario C, Casulleras M, Duran-Güell M, Rius B, Diaz A, Hall K, Shea C, Sarno R, Masferrer JL, Claria J (2017) The soluble guanylate cyclase stimulator IW-1973 prevents inflammation and fibrosis in experimental non-alcoholic steatohepatitis. Br J Pharmacol 175(6):953–967. https://doi.org/10.1111/bph.14137

Follmann M, Griebenow N, Hahn MG, Hartung I, Mais F-J, Mittendorf J, Schaefer M, Schirok H, Stasch J-P, Stoll F, Straub A (2013) The chemistry and biology of soluble guanylate cyclase stimulators and activators. Angew Chem Int Ed 52:9442–9462

Follmann M, Ackerstaff J, Redlich G, Wunder F, Lang D, Kern A, Fey P, Griebenow N, Kroh W, Becker-Pelster EM, Kretschmer A, Geiss V, Li V, Straub A, Mittendorf J, Jautelat R, Schirok H, Schlemmer KH, Lustig K, Gerisch M, Knorr A, Tinel H, Mondritzki T, Trübel H, Sandner P, Stasch JP (2017) Discovery of the soluble guanylate cyclase stimulator vericiguat (BAY 1021189) for the treatment of chronic heart failure. J Med Chem 60(12):5146–5161

Frey R, Becker C, Saleh S, Unger S, van der Mey D, Mück W (2017) Clinical pharmacokinetic and pharmacodynamic profile of riociguat. Clin Pharmacokinet 57(6):647–661. https://doi.org/10.1007/s40262-017-0604-7

Friebe A, Mergia E, Dangel O, Lange A, Koesling D (2007) Fatal gastrointestinal obstruction and hypertension in mice lacking nitric oxide-sensitive guanylyl cyclase. Proc Natl Acad Sci U S A 104(18):7699–7704. https://doi.org/10.1073/pnas.0609778104

Friebe A, Sandner P, Schmidtko A (2017) Meeting report of the 8th International Conference on cGMP "cGMP: generators, effectors and therapeutic implications" at Bamberg, Germany from June 23rd to 25th 2017. Naunyn Schmiedebergs Arch Pharmacol 390(12):1177–1188. https://doi.org/10.1007/s00210-017-1429-5

Garbán H, Vernet D, Freedman A, Rajfer J, González-Cadavid N (1995) Effect of aging on nitric oxide-mediated penile erection in rats. Am J Physiol 268(1. Pt 2):H467–H475

Ge P, Navarro ID, Kessler MM, Bernier SG, Perl NR, Sarno R, Masferrer J, Hannig G, Stamer WD (2016) The soluble guanylate cyclase stimulator iwp-953 increases conventional outflow facility in mouse eyes. Invest Ophthalmol Vis Sci 57(3):1317–1326. https://doi.org/10.1167/iovs.15-18958

Geschka S, Kretschmer A, Sharkovska Y, Evgenov OV, Lawrenz B, Hucke A, Hocher B, Stasch J (2011) Soluble guanylate cyclase stimulation prevents fibrotic tissue remodeling and improves survival in salt-sensitive Dahl rats. PLoS One 6(7):e21853. https://doi.org/10.1371/journal.pone.0021853

Gheorghiade M, Greene SJ, Butler J, Filippatos G, Lam CS, Maggioni AP, Ponikowski P, Shah SJ, Solomon SD, Kraigher-Krainer E, Samano ET, Müller K, Roessig L, Pieske B, SOCRATES-REDUCED Investigators and Coordinators (2015) Effect of Vericiguat, a soluble guanylate cyclase stimulator, on natriuretic peptide levels in patients with worsening chronic heart failure and reduced ejection fraction: the SOCRATES-REDUCED randomized trial. JAMA 314 (21):2251–2262. https://doi.org/10.1001/jama.2015.15734

Ghofrani HA, D'Armini AM, Grimminger F, Hoeper MM, Jansa P, Kim NH, Mayer E, Simonneau G, Wilkins MR, Fritsch A, Neuser D, Weimann G, Wang C, Group C-S (2013a) Riociguat for the treatment of chronic thromboembolic pulmonary hypertension. N Engl J Med 369(4):319–329. https://doi.org/10.1056/NEJMoa1209657

Ghofrani HA, Galie N, Grimminger F, Grunig E, Humbert M, Jing ZC, Keogh AM, Langleben D, Kilama MO, Fritsch A, Neuser D, Rubin LJ (2013b) Riociguat for the treatment of pulmonary arterial hypertension. N Engl J Med 369(4):330–340. https://doi.org/10.1056/NEJMoa1209655

Ghofrani HA, Grimminger F, Grünig E, Huang Y, Jansa P, Jing ZC, Kilpatrick D, Langleben D, Rosenkranz S, Menezes F, Fritsch A, Nikkho S, Humbert M (2016) Predictors of long-term outcomes in patients treated with riociguat for pulmonary arterial hypertension: data from the PATENT-2 open-label, randomised, long-term extension trial. Lancet Respir Med 4 (5):361–371. https://doi.org/10.1016/S2213-2600(16)30019-4

Ghosh A, Stuehr DJ (2017) Regulation of sGC via hsp90, cellular heme, sGC agonists, and NO: new pathways and clinical perspectives. Antioxid Redox Signal 26(4):182–190. https://doi.org/10.1089/ars.2016.6690

Gladwin MT (2006) Deconstructing endothelial dysfunction: soluble guanylyl cyclase oxidation and the NO resistance syndrome. J Clin Invest 116(9):2330–2332. https://doi.org/10.1172/JCI29807

Groneberg D, Voussen B, Friebe A (2016) Integrative control of gastrointestinal motility by nitric oxide. Curr Med Chem 23(24):2715–2735

Gurbuz N, Mammadov E, Usta MF (2008) Hypogonadism and erectile dysfunction: an overview. Asian J Androl 10(1):36–43

Hall K, Jacobson S, Zhang P, Liu G, Sarno R, Catanzano V, Bernier S, Currie M, Masferrer J (2017) Inhibition of fibrosis and inflammation by a soluble guanylate cyclase stimulator in models of liver disease. Paper presented at The Liver Meeting, Washington, DC

Hall K, Bernier S, Jacobson S, Liu G, Sarno R, Catanzano V, Sheppeck J, Hadcock J, Currie M, Masferrer J (2018) Stimulation of soluble guanylate cyclase inhibited fibrosis and inflammation in human liver microtissues and in an animal model of liver disease. J Hepatol 68:S397. https://doi.org/10.1016/S0168-8278(18)31030-4

Hanrahan JP, Wakefield JD, Wilson PJ, Mihova M, Chickering JG, Ruff D, Hall M, Milne TM, Currie MG, Profy AT (2018) A randomized, placebo-controlled, multiple-ascending-dose study to assess the safety, tolerability, pharmacokinetics, and pharmacodynamics of the soluble guanylate cyclase stimulator praliciguat in healthy subjects. Clin Pharmacol Drug Dev. https://doi.org/10.1002/cpdd.627. [Epub ahead of print]

Hanrahan JP, Wakefield JD, Wilson PJ, Miller P, Chickering J, Morrow L, Hall ML, Currie M, Milne GT, Profy AT (2018a) Fourteen-day study of praliciguat, a soluble guanylate cyclase stimulator, in patients with diabetes and hypertension. Diabetes 67(Supplement 1):74-OR. https://doi.org/10.2337/db18-74-OR

Hanrahan JP, Wakefield JD, Wilson PJ, Zimmer DP, Mihova M, Chickering J, Ruff D, Hall ML, Currie M, Milne GT, Profy AT (2018b) Rapid dose escalation study of praliciguat, a soluble guanylate cyclase stimulator, in patients with diabetes and hypertension. Diabetes 67(Supplement 1):1207-P. https://doi.org/10.2337/db18-1207-P

Herve D, Philippi A, Belbouab R, Zerah M, Chabrier S, Collardeau-Frachon S, Bergametti F, Essongue A, Berrou E, Krivosic V, Sainte-Rose C, Houdart E, Adam F, Billiemaz K, Lebret M, Roman S, Passemard S, Boulday G, Delaforge A, Guey S, Dray X, Chabriat H, Brouckaert P, Bryckaert M, Tournier-Lasserve E (2014) Loss of alpha1beta1 soluble guanylate cyclase, the major nitric oxide receptor, leads to moyamoya and achalasia. Am J Hum Genet 94(3):385–394. https://doi.org/10.1016/j.ajhg.2014.01.018

Hewitson TD, Martic M, Darby IA, Kelynack KJ, Bisucci T, Tait MG, Becker GJ (2004) Intracellular cyclic nucleotide analogues inhibit in vitro mitogenesis and activation of fibroblasts derived from obstructed rat kidneys. Nephron Exp Nephrol 96(2):e59–e66. https://doi.org/10.1159/000076405

Hoeper MM (2015) Pharmacological therapy for patients with chronic thromboembolic pulmonary hypertension. Eur Respir Rev 24(136):272–282

Hoeper MM, Klinger JR, Benza RL, Simonneau G, Langleben D, Naeije R, Corris PA (2017a) Rationale and study design of RESPITE: an open-label, phase 3b study of riociguat in patients with pulmonary arterial hypertension who demonstrate an insufficient response to treatment with phosphodiesterase-5 inhibitors. Respir Med 122(Suppl 1):S18–S22. https://doi.org/10.1016/j.rmed.2016.11.001

Hoeper MM, Simonneau G, Corris PA, Ghofrani HA, Klinger JR, Langleben D, Naeije R, Jansa P, Rosenkranz S, Scelsi L, Grünig E, Vizza CD, Chang M, Colorado P, Meier C, Busse D, Benza RL (2017b) RESPITE: switching to riociguat in pulmonary arterial hypertension patients with inadequate response to phosphodiesterase-5 inhibitors. Eur Respir J 50(3):1602425. https://doi.org/10.1183/13993003.02425-2016

Hoffmann LS, Etzrodt J, Willkomm L, Sanyal A, Scheja L, Fischer AW, Stasch JP, Bloch W, Friebe A, Heeren J, Pfeifer A (2015) Stimulation of soluble guanylyl cyclase protects against obesity by recruiting brown adipose tissue. Nat Commun 6:7235. https://doi.org/10.1038/ncomms8235

Hoffmann LS, Schmidt PM, Keim Y, Hoffmann C, Schmidt HH, Stasch JP (2011) Fluorescence dequenching makes haem-free soluble guanylate cyclase detectable in living cells. PLoS One 6(8):e23596. https://doi.org/10.1371/journal.pone.0023596

Hoffmann LS, Schmidt PM, Keim Y, Schaefer S, Schmidt HH, Stasch JP (2009) Distinct molecular requirements for activation or stabilization of soluble guanylyl cyclase upon haem oxidation-induced degradation. Br J Pharmacol 157(5):781–795. https://doi.org/10.1111/j.1476-5381.2009.00263.x

Hoshino M, Omura N, Yano F, Tsuboi K, Kashiwagi H, Yanaga K (2013) Immunohistochemical study of the muscularis externa of the esophagus in achalasia patients. Dis Esophagus 26(1):14–21. https://doi.org/10.1111/j.1442-2050.2011.01318.x

Humbert M, Ghofrani HA (2016) The molecular targets of approved treatments for pulmonary arterial hypertension. Thorax 71(1):73–83. https://doi.org/10.1136/thoraxjnl-2015-207170

Ibarra C, Nedvetsky PI, Gerlach M, Riederer P, Schmidt HH (2001) Regional and age-dependent expression of the nitric oxide receptor, soluble guanylyl cyclase, in the human brain. Brain Res 907(1–2):54–60

Ingram VM (1956) A specific chemical difference between the globins of normal human and sickle-cell anaemia haemoglobin. Nature 178(4537):792–794

International Consortium for Blood Pressure Genome-Wide Association S (2011) Genetic variants in novel pathways influence blood pressure and cardiovascular disease risk. Nature 478(7367):103–109. https://doi.org/10.1038/nature10405

Irvine JC, Ganthavee V, Love JE, Alexander AE, Horowitz JD, Stasch JP, Kemp-Harper BK, Ritchie RH (2012) The soluble guanylyl cyclase activator bay 58-2667 selectively limits

cardiomyocyte hypertrophy. PLoS One 7(11):e44481. https://doi.org/10.1371/journal.pone.0044481
Kemp-Harper B, Feil R (2008) Meeting report: cGMP matters. Sci Signal 1(9):pe12. https://doi.org/10.1126/stke.19pe12
Kessler T, Wobst J, Wolf B, Eckhold J, Vilne B, Hollstein R, von Ameln S, Dang TA, Sager HB, Moritz Rumpf P, Aherrahrou R, Kastrati A, Björkegren JLM, Erdmann J, Lusis AJ, Civelek M, Kaiser FJ, Schunkert H (2017) Functional characterization of the GUCY1A3 coronary artery disease risk locus. Circulation 136(5):476–489. https://doi.org/10.1161/CIRCULATIONAHA.116.024152
Klinger JR, Kadowitz PJ (2017) The nitric oxide pathway in pulmonary vascular disease. Am J Cardiol 120(8S):S71–S79. https://doi.org/10.1016/j.amjcard.2017.06.012
Knorr A, Hirth-Dietrich C, Alonso-Alija C, Harter M, Hahn M, Keim Y, Wunder F, Stasch JP (2008) Nitric oxide-independent activation of soluble guanylate cyclase by BAY 60-2770 in experimental liver fibrosis. Arzneimittelforschung 58(2):71–80. https://doi.org/10.1055/s-0031-1296471
Ko FN, Wu CC, Kuo SC, Lee FY, Teng CM (1994) YC-1, a novel activator of platelet guanylate cyclase. Blood 84(12):4226–4233
Kollau A, Opelt M, Wölkart G, Gorren ACF, Russwurm M, Koesling D, Mayer B, Schrammel A (2018) Irreversible activation and stabilization of soluble guanylate cyclase by the protoporphyrin IX mimetic cinaciguat. Mol Pharmacol 93(2):73–78. https://doi.org/10.1124/mol.117.109918
Kone BC (1997) Nitric oxide in renal health and disease. Am J Kidney Dis 30(3):311–333
Kotikoski H, Vapaatalo H, Oksala O (2003) Nitric oxide and cyclic GMP enhance aqueous humor outflow facility in rabbits. Curr Eye Res 26(2):119–123
Krishnan SM, Kraehling JR, Eitner F, Bénardeau A, Sandner P (2018) The Impact of the nitric oxide (NO)/soluble guanylyl cyclase (sGC) signaling cascade on kidney health and disease: a preclinical perspective. Int J Mol Sci 19(6):1712. https://doi.org/10.3390/ijms19061712
Kuhn M (2016) Molecular physiology of membrane guanylyl cyclase receptors. Physiol Rev 96 (2):751–804. https://doi.org/10.1152/physrev.00022.2015
Lambers C, Roth M, Hofbauer E, Petkov V, Block LH (2014) Anti-remodeling potencies of the soluble guanylate cyclase activator BAY 41-2272 in human lung fibroblasts. Eur Respir J 44 (Suppl 58):3423
Lang M, Kojonazarov B, Tian X, Kalymbetov A, Weissmann N, Grimminger F, Kretschmer A, Stasch JP, Seeger W, Ghofrani HA, Schermuly RT, Baktybek K, Xia T, Anuar K, Norbert W, Friedrich G, Axel K, Johannes-Peter S, Werner S, Hossein Ardeschir G, Ralph Theo S (2012) The soluble guanylate cyclase stimulator Riociguat ameliorates pulmonary hypertension induced by hypoxia and SU5416 in rats. PLoS One 7(8):e43433. https://doi.org/10.1371/journal.pone.0043433
Leineweber K, Moosmang S, Paulson D (2017) Genetics of NO deficiency. Am J Cardiol 120(8S): S80–S88. https://doi.org/10.1016/j.amjcard.2017.06.013
Lewis KS, Butler J, Bauersachs J, Sandner P (2017) The three-decade long journey in heart failure drug development. Handb Exp Pharmacol 243:1–14. https://doi.org/10.1007/164_2016_101
Liu G, Shea C, Ranganath S, Im GY, Sheppeck JE, Masferrer JL (2016) The sGC stimulator IWP-121 inhibits renal inflammation and fibrosis in human RTPC and Dahl-ss rat model. Paper presented at the Keystone symposia fibrosis: from basic mechanisms to targeted therapies, Keystone, CO, Feb
Lucas KA, Pitari GM, Kazerounian S, Ruiz-Stewart I, Park J, Schulz S, Chepenik KP, Waldman SA (2000) Guanylyl cyclases and signaling by cyclic GMP. Pharmacol Rev 52(3):375–414
Lundberg JO, Gladwin MT, Weitzberg E (2015) Strategies to increase nitric oxide signalling in cardiovascular disease. Nat Rev Drug Discov 14(9):623–641. https://doi.org/10.1038/nrd4623
Masferrer JL, Shea C, Lonie E, Liu G, Profy A, Milne GT, Currie MG (2016) Novel sGC stimulator IW-1701 prevents the progression of diabetic nephropathy when administered in combination with Enalapril in the ZSF1 rat model. Paper presented at the American Society of Nephrology Kidney Week, Chicago, IL, Nov 15–20

Masuyama H, Tsuruda T, Kato J, Imamura T, Asada Y, Stasch JP, Kitamura K, Eto T (2006) Soluble guanylate cyclase stimulation on cardiovascular remodeling in angiotensin II-induced hypertensive rats. Hypertension 48(5):972–978. https://doi.org/10.1161/01.HYP.0000241087.12492.47

Masuyama H, Tsuruda T, Sekita Y, Hatakeyama K, Imamura T, Kato J, Asada Y, Stasch JP, Kitamura K (2009) Pressure-independent effects of pharmacological stimulation of soluble guanylate cyclase on fibrosis in pressure-overloaded rat heart. Hypertens Res 32(7):597–603. https://doi.org/10.1038/hr.2009.64

Mátyás C, Németh BT, Oláh A, Hidi L, Birtalan E, Kellermayer D, Ruppert M, Korkmaz-Icöz S, Kökény G, Horváth EM, Szabó G, Merkely B, Radovits T (2015) The soluble guanylate cyclase activator cinaciguat prevents cardiac dysfunction in a rat model of type-1 diabetes mellitus. Cardiovasc Diabetol 14:145. https://doi.org/10.1186/s12933-015-0309-x

Mergia E, Russwurm M, Zoidl G, Koesling D (2003) Major occurrence of the new alpha2beta1 isoform of NO-sensitive guanylyl cyclase in brain. Cell Signal 15(2):189–195

Methner C, Buonincontri G, Hu CH, Vujic A, Kretschmer A, Sawiak S, Carpenter A, Stasch JP, Krieg T (2013) Riociguat reduces infarct size and post-infarct heart failure in mouse hearts: insights from MRI/PET imaging. PLoS One 8(12):e83910. https://doi.org/10.1371/journal.pone.0083910

Meurer S, Pioch S, Pabst T, Opitz N, Schmidt PM, Beckhaus T, Wagner K, Matt S, Gegenbauer K, Geschka S, Karas M, Stasch JP, Schmidt HH, Müller-Esterl W (2009) Nitric oxide-independent vasodilator rescues heme-oxidized soluble guanylate cyclase from proteasomal degradation. Circ Res 105(1):33–41. https://doi.org/10.1161/CIRCRESAHA

Miller LN, Nakane M, Hsieh GC, Chang R, Kolasa T, Moreland RB, Brioni JD (2003) A-350619: a novel activator of soluble guanylyl cyclase. Life Sci 72(9):1015–1025

Mittleman RS, Wilson P, Sykes K, Mihova M, Chickering JG, Ruff D, Hall M, Milne TG, Currie MG, Chien Y (2017) Multiple-ascending-dose study of the soluble guanylate cyclase stimulator, IW-1701, in healthy subjects. Blood 130(Suppl 1):3533. http://www.bloodjournal.org/content/130/Suppl_1/3533. Accessed 07 Dec 2018

Mittendorf J, Weigand S, Alonso-Alija C, Bischoff E, Feurer A, Gerisch M, Kern A, Knorr A, Lang D, Muenter K, Radtke M, Schirok H, Schlemmer KH, Stahl E, Straub A, Wunder F, Stasch JP (2009) Discovery of riociguat (BAY 63-2521): a potent, oral stimulator of soluble guanylate cyclase for the treatment of pulmonary hypertension. ChemMedChem 4(5):853–865. https://doi.org/10.1002/cmdc.200900014

Mulhall J, Teloken P, Brock G, Kim E (2006) Obesity, dyslipidemias and erectile dysfunction: a report of a subcommittee of the sexual medicine society of North America. J Sex Med 3(5):778–786. https://doi.org/10.1111/j.1743-6109.2006.00286.x

Munzel T, Genth-Zotz S, Hink U (2007) Targeting heme-oxidized soluble guanylate cyclase. solution for all cardiorenal problems in heart failure? Hypertension 49(5):974–976

Murrell W (1879) Nitro-glycerin as a remedy for angina pectoris. Lancet 113(2890):80–81 ff

Musicki B, Burnett AL (2007) Endothelial dysfunction in diabetic erectile dysfunction. Int J Impot Res 19(2):129–138

Nahavandi M, Tavakkoli F, Wyche MQ, Perlin E, Winter WP, Castro O (2002) Nitric oxide and cyclic GMP levels in sickle cell patients receiving hydroxyurea. Br J Haematol 119(3):855–857

Nakai T, Perl NR, Barden TC, Carvalho A, Fretzen A, Germano P, Im GY, Jin H, Kim C, Lee TW, Long K, Moore J, Rohde JM, Sarno R, Segal C, Solberg EO, Tobin J, Zimmer DP, Currie MG (2016) Discovery of IWP-051, a novel orally bioavailable sGC stimulator with once-daily dosing potential in humans. ACS Med Chem Lett 7(5):465–469. https://doi.org/10.1021/acsmedchemlett.5b00479

Nathan S, Behr J, Collard HR, Cottin V, Hoeper MM, Martinez F, Corte T, Keogh A, Leuchte H, Mogulkoc N, Ulrich S, Wuyts W, Malcolm S, Shah S, Yao M, Wells A (2017) RISE-IIP: Riociguat for the treatment of pulmonary hypertension associated with idiopathic interstitial pneumonia. Eur Respir J 50:OA1985. https://doi.org/10.1183/1393003.congress-2017.OA1985

Oudiz R, Shapiro S, Torres F, Feldman J, Frost A, Allard M, Blair C, Gillies H (2011) ATHENA-1: hemodynamic improvements following the addition of ambrisentan to background PDE5i therapy in patients with pulmonary arterial hypertension. Chest 140:905A (4_MeetingAbstracts)

Pan J, Zhang X, Yuan H, Xu Q, Zhang H, Zhou Y, Huang ZX, Tan X (2016) The molecular mechanism of heme loss from oxidized soluble guanylate cyclase induced by conformational change. Biochim Biophys Acta 1864(5):488–500. https://doi.org/10.1016/j.bbapap.2016.02.012

Patel D, Lakhkar A, Wolin MS (2017) Redox mechanisms influencing cGMP signaling in pulmonary vascular physiology and pathophysiology. Adv Exp Med Biol 967:227–240. https://doi.org/10.1007/978-3-319-63245-2_13

Patel DA, Kim HP, Zifodya JS, Vaezi MF (2015) Idiopathic (primary) achalasia: a review. Orphanet J Rare Dis 10:89. https://doi.org/10.1186/s13023-015-0302-1

Paulus WJ, Tschope C (2013) A novel paradigm for heart failure with preserved ejection fraction: comorbidities drive myocardial dysfunction and remodeling through coronary microvascular endothelial inflammation. J Am Coll Cardiol 62(4):263–271. https://doi.org/10.1016/j.jacc.2013.02.092

Pieske B, Butler J, Filippatos G, Lam C, Maggioni AP, Ponikowski P, Shah S, Solomon S, Kraigher-Krainer E, Samano ET, Scalise AV, Müller K, Roessig L, Gheorghiade M, SOCRATES Investigators and Coordinators (2014) Rationale and design of the SOluble guanylate Cyclase stimulatoR in heArT failurE Studies (SOCRATES). Eur J Heart Fail 16 (9):1026–1038. https://doi.org/10.1002/ejhf.135

Pieske B, Maggioni AP, Lam CSP, Pieske-Kraigher E, Filippatos G, Butler J, Ponikowski P, Shah SJ, Solomon SD, Scalise AV, Mueller K, Roessig L, Gheorghiade M (2017) Vericiguat in patients with worsening chronic heart failure and preserved ejection fraction: results of the SOluble guanylate Cyclase stimulatoR in heArTfailurE patientS with PRESERVED EF (SOCRATES-PRESERVED) study. Eur Heart J 38(15):1119–1127. https://doi.org/10.1002/ejhf.135

Pimenta E, Calhoun DA (2016) Drug development for hypertension: do we need another antihypertensive agent for resistant hypertension? Curr Hypertens Rep 18(4):25. https://doi.org/10.1007/s11906-016-0634-9

Potoka KP, Wood KC, Baust JJ, Bueno M, Hahn S, Vanderpool RR, Bachman T, Mallampalli GM, Hwedieh DO, Schrott V, Bullock GC, Becker-Pelster EM, Stampfuss J, Mather I, Stasch JP, Truebel H, Sandner P, Mora AL, Straub AC, Gladwin MT (2018) Nitric oxide-independent activation of soluble guanylate cyclase improves vascular function and reverses cardiac remodeling in sickle cell disease. Am J Respir Cell Mol Biol 58(5):636–647

Prabhakar SS (2004) Role of nitric oxide in diabetic nephropathy. Semin Nephrol 24(4):333–344

Pradhan K, Sydykov A, Tian X, Mamazhakypov A, Neupane B, Luitel H, Weissmann N, Seeger W, Grimminger F, Kretschmer A, Stasch JP, Ghofrani HA, Schermuly RT (2016) Soluble guanylate cyclase stimulator riociguat and phosphodiesterase 5 inhibitor sildenafil ameliorate pulmonary hypertension due to left heart disease in mice. Int J Cardiol 216:85–91. https://doi.org/10.1016/j.ijcard.2016.04.098

Prasanna G, Ferrara L, Adams C, Ehara T, Li B, Yang L, Xiang C, Ng CTH, Kim S, Towler C, Topley T, McAllister C, Ghosh M, Newton R, Stacy R, Rice DS, Mogi MA (2018) novel selective soluble guanylate cyclase activator, MGV354, lowers intraocular pressure in preclinical models, following topical ocular dosing. Invest Ophthalmol Vis Sci 59(5):1704–1716. https://doi.org/10.1167/iovs.18-23772

Profy AT, Shea C, Lonie E, Liu G, Milne GT, Currie MG, Masferrer J (2017) IW-1973, a soluble guanylate cyclase stimulator, inhibits progression of diabetic nephropathy in the ZSF1 rat model. Paper presented at the American Diabetes Association 77th scientific sessions, San Diego, CA, June 9–13

Raffaella M, Moretti R, Leger PL, Besson VC, Csaba Z, Pansiot J, Di Criscio L, Gentili A, Titomanlio L, Bonnin P, Baud O, Charriaut-Marlangue C, Pierre-Louis L, Valérie CB, Zsolt C, Julien P, Di Lorena C, Andrea G, Luigi T, Philippe B, Olivier B, Christiane C-M

(2016) Sildenafil, a cyclic GMP phosphodiesterase inhibitor, induces microglial modulation after focal ischemia in the neonatal mouse brain. J Neuroinflammation 13(1):95. https://doi.org/10.1186/s12974-016-0560-4

Rahaman MM, Nguyen AT, Miller MP, Hahn SA, Sparacino-Watkins C, Jobbagy S, Carew NT, Cantu-Medellin N, Wood KC, Baty CJ, Schopfer FJ, Kelley EE, Gladwin MT, Martin E, Straub AC (2017) Cytochrome b5 reductase 3 modulates soluble guanylate cyclase redox state and cGMP signaling. Circ Res 121(2):137–148. https://doi.org/10.1161/CIRCRESAHA.117.310705

Ritchie RH, Drummond GR, Sobey CG, De Silva TM, Kemp-Harper BK (2017) The opposing roles of NO and oxidative stress in cardiovascular disease. Pharmacol Res 116:57–69. https://doi.org/10.1016/j.phrs.2016.12.017

Rubin LJ, Galiè N, Grimminger F, Grünig E, Humbert M, Jing ZC, Keogh A, Langleben D, Fritsch A, Menezes F, Davie N, Ghofrani HA (2015) Riociguat for the treatment of pulmonary arterial hypertension: a long-term extension study (PATENT-2). Eur Respir J 45(5):1303–1313. https://doi.org/10.1183/09031936.00090614

Sandner P, Berger P, Zenzmaier C (2017) The potential of sGC modulators for the treatment of age-related fibrosis: a mini-review. Gerontology 63(3):216–227. https://doi.org/10.1159/000450946

Sandner P, Stasch JP (2017) Anti-fibrotic effects of soluble guanylate cyclase stimulators and activators: a review of the preclinical evidence. Respir Med 122(Suppl 1):S1–S9. https://doi.org/10.1016/j.rmed.2016.08.022

Schinner E, Wetzl V, Schramm A, Kees F, Sandner P, Stasch JP, Hofmann F, Schlossmann J (2017) Inhibition of the TGFbeta signalling pathway by cGMP and cGMP-dependent kinase I in renal fibrosis. FEBS Open Bio 7(4):550–561. https://doi.org/10.1002/2211-5463.12202

Schmidt HH, Schmidt PM, Stasch JP (2009) NO- and haem-independent soluble guanylate cyclase activators. Handb Exp Pharmacol 191:309–339. https://doi.org/10.1007/978-3-540-68964-5_14

Schmidt P, Schramm M, Schröder H, Stasch JP (2003) Receptor binding assay for nitric oxide- and heme-independent activators of soluble guanylate cyclase. Anal Biochem 314(1):162–165

Schmidt PM, Schramm M, Schröder H, Wunder F, Stasch JP (2004) Identification of residues crucially involved in the binding of the heme moiety of soluble guanylate cyclase. J Biol Chem 279(4):3025–3032

Schulz E, Jansen T, Wenzel P, Daiber A, Munzel T (2008) Nitric oxide, tetrahydrobiopterin, oxidative stress, and endothelial dysfunction in hypertension. Antioxid Redox Signal 10(6):1115–1126. https://doi.org/10.1089/ars.2007.1989

Schwabl P, Brusilovskaya K, Supper P, Bauer D, Königshofer P, Riedl F, Hayden H, Fuchs CD, Stift J, Oberhuber G, Aschauer S, Bonderman D, Gnad T, Pfeifer A, Uschner FE, Trebicka J, Rohr-Udilova N, Podesser BK, Peck-Radosavljevic M, Trauner M, Reiberger T (2018) The soluble guanylate cyclase stimulator riociguat reduces fibrogenesis and portal pressure in cirrhotic rats. Sci Rep 8(1):9372. https://doi.org/10.1038/s41598-018-27656-y

Schwam EM, Nicholas T, Chew R, Billing CB, Davidson W, Ambrose D, Altstiel LD (2014) A multicenter, double-blind, placebo-controlled trial of the PDE9A inhibitor, PF-04447943, in Alzheimer's disease. Curr Alzheimer Res 11(5):413–421

Schwartzkopf CD, Hadcock J, Jones JE, Currie M, Milne GT, Masferrer J (2018) Praliciguat, a clinical-stage sGC stimulator, improved glucose tolerance and insulin sensitivity and lowered triglycerides in a mouse diet-induced obesity model. Diabetes 67(Supplement 1):1886-P. https://doi.org/10.2337/db18-1886-P

Selwood DL, Brummell DG, Budworth J, Burtin GE, Campbell RO, Chana SS, Charles IG, Fernandez PA, Glen RC, Goggin MC, Hobbs AJ, Kling MR, Liu Q, Madge DJ, Meillerais S, Powell KL, Reynolds K, Spacey GD, Stables JN, Tatlock MA, Wheeler KA, Wishart G,

Woo CK (2001) Synthesis and biological evaluation of novel pyrazoles and indazoles as activators of the nitric oxide receptor, soluble guanylate cyclase. J Med Chem 44:78–93
Shabsigh R (2004) Therapy of ED: PDE-5 inhibitors. Endocrine 23(2–3):135–141
Shapiro S, Gillies H, Allard M, Blair C, Oudiz RJ (2012) ATHENA-1: long term clinical improvements following the addition of ambrisentan to background PDE5i therapy in patients with pulmonary arterial hypertension. J Heart Lung Transplant 31(4):S28eS29
Sharina IG, Martin E (2017) The role of reactive oxygen and nitrogen species in the expression and splicing of nitric oxide receptor. Antioxid Redox Signal 26(3):122–136. https://doi.org/10.1089/ars.2016.6687
Simonneau G, D'Armini AM, Ghofrani HA, Grimminger F, Jansa P, Kim NH, Mayer E, Pulido T, Wang C, Colorado P, Fritsch A, Meier C, Nikkho S, Hoeper MM (2016) Predictors of long-term outcomes in patients treated with riociguat for chronic thromboembolic pulmonary hypertension: data from the CHEST-2 open-label, randomised, long-term extension trial. Lancet Respir Med 4(5):372–380. https://doi.org/10.1016/S2213-2600(16)30022-4
Sivarao DV, Mashimo HL, Thatte HS, Goyal RK (2001) Lower esophageal sphincter is achalasic in nNOS(-/-) and hypotensive in W/W(v) mutant mice. Gastroenterology 121(1):34–42
Stacy R, Huttner K, Watts J, Peace J, Wirta D, Walters T, Sall K, Seaman J, Ni X, Prasanna G, Mogi M, Adams C, Yan JH, Wald M, He Y, Newton R, Kolega R, Grosskreutz C (2018) A Randomized, controlled phase I/II study to evaluate the safety and efficacy of MGV354 for ocular hypertension or glaucoma. Am J Ophthalmol 192:113–123. https://doi.org/10.1016/j.ajo.2018.05.015. Epub 2018 May 24
Stasch JP, Becker EM, Alonso-Alija C, Apeler H, Dembowsky K, Feurer A, Gerzer R, Minuth T, Perzborn E, Pleiss U, Schröder H, Schroeder W, Stahl E, Steinke W, Straub A, Schramm M (2001) NO-independent regulatory site on soluble guanylate cyclase. Nature 410(6825):212–215
Stasch JP, Evgenov OV (2013) Soluble guanylate cyclase stimulators in pulmonary hypertension. Handb Exp Pharmacol 218:279–313. https://doi.org/10.1007/978-3-642-38664-0_12
Stasch J-P, Hobbs AJ (2009) NO-independent, haem-dependent soluble guanylate cyclase stimulators. Handb Exp Pharmacol 191:277–308. https://doi.org/10.1007/978-3-540-68964-5_13
Stasch JP, Pacher P, Evgenov OV (2011) Soluble guanylate cyclase as an emerging therapeutic target in cardiopulmonary disease. Circulation 123(20):2263–2273. https://doi.org/10.1161/CIRCULATIONAHA.110.981738
Stasch JP, Schlossmann J, Hocher B (2015) Renal effects of soluble guanylate cyclase stimulators and activators: a review of the preclinical evidence. Curr Opin Pharmacol 21:95–104. https://doi.org/10.1016/j.coph.2014.12.014
Stasch JP, Schmidt P, Alonso-Alija C, Apeler H, Dembowsky K, Haerter M, Heil M, Minuth T, Perzborn E, Pleiss U, Schramm M, Schroeder W, Schröder H, Stahl E, Steinke W, Wunder F (2002) NO- and haem-independent activation of soluble guanylyl cyclase: molecular basis and cardiovascular implications of a new pharmacological principle. Br J Pharmacol 136 (5):773–783
Stasch JP, Schmidt PM, Nedvetsky PI, Nedvetskaya TY, Hs AK, Meurer S, Deile M, Taye A, Knorr A, Lapp H, Muller H, Turgay Y, Rothkegel C, Tersteegen A, Kemp-Harper B, Muller-Esterl W, Schmidt HH (2006) Targeting the heme-oxidized nitric oxide receptor for selective vasodilatation of diseased blood vessels. J Clin Invest 116:2552–2561
Tchernychev BT, Feil S, Germano P, Warren W, Lonie E, Feil R, Milne GT, Hadcock J, Chien Y-T, Currie MG, Graul R (2017) The clinical-stage sGC stimulator IW-1701 prevents increase of plasma biomarkers of intravascular inflammation and suppresses leukocyte-endothelial interactions in TNFalpha-treated mice. Paper presented at the American Society of Hematology – 59th Annual Meeting, Atlanta, GA

Thoonen R, Cauwels A, Decaluwe K, Geschka S, Tainsh RE, Delanghe J, Hochepied T, De Cauwer L, Rogge E, Voet S, Sips P, Karas RH, Bloch KD, Vuylsteke M, Stasch JP, Van de Voorde J, Buys ES, Brouckaert P (2015) Cardiovascular and pharmacological implications of haem-deficient NO-unresponsive soluble guanylate cyclase knock-in mice. Nat Commun 6:8482. https://doi.org/10.1038/ncomms9482

Tobin JV, Zimmer DP, Shea C, Germano P, Bernier SG, Liu G, Long K, Miyashiro J, Ranganath S, Jacobson S, Tang K, Im GJ, Sheppeck J, Moore JD, Sykes K, Wakefield J, Sarno R, Banijamali AR, Profy AT, Milne GT, Currie MG, Masferrer JL (2018) Pharmacological characterization of IW-1973, a novel soluble guanylate cyclase stimulator with extensive tissue distribution, anti-hypertensive, anti-inflammatory, and anti-fibrotic effects in preclinical models of disease. J Pharmacol Exp Ther 365:664–675. https://doi.org/10.1124/jpet.117.247429

van der Staay FJ, Rutten K, Bärfacker L, Devry J, Erb C, Heckroth H, Karthaus D, Tersteegen A, van Kampen M, Blokland A, Prickaerts J, Reymann KG, Schröder UH, Hendrix M (2008) The novel selective PDE9 inhibitor BAY 73-6691 improves learning and memory in rodents. Neuropharmacology 55(5):908–918. https://doi.org/10.1016/j.neuropharm.2008.07.005

Venema RC, Venema VJ, Ju H, Harris MB, Snead C, Jilling T, Dimitropoulou C, Maragoudakis ME, Catravas JD (2003) Novel complexes of guanylate cyclase with heat shock protein 90 and nitric oxide synthase. Am J Physiol Heart Circ Physiol 285(2):H669–H678

Vettel C, Lammle S, Ewens S, Cervirgen C, Emons J, Ongherth A, Dewenter M, Lindner D, Westermann D, Nikolaev VO, Lutz S, Zimmermann WH, El-Armouche A (2014) PDE2-mediated cAMP hydrolysis accelerates cardiac fibroblast to myofibroblast conversion and is antagonized by exogenous activation of cGMP signaling pathways. Am J Physiol Heart Circ Physiol 306(8):H1246–H1252. https://doi.org/10.1152/ajpheart.00852.2013

Wales JA, Chen CY, Breci L, Weichsel A, Bernier SG, Sheppeck JE 2nd, Solinga R, Nakai T, Renhowe PA, Jung J, Montfort WR (2018) Discovery of stimulator binding to a conserved pocket in the heme domain of soluble guanylyl cyclase. J Biol Chem 293(5):1850–1864. https://doi.org/10.1074/jbc.RA117.000457

Wallace S, Guo DC, Regalado E, Mellor-Crummey L, Bamshad M, Nickerson DA, Dauser R, Hanchard N, Marom R, Martin E, Berka V, Sharina I, Ganesan V, Saunders D, Morris SA, Milewicz DM (2016) Disrupted nitric oxide signaling due to GUCY1A3 mutations increases risk for moyamoya disease, achalasia and hypertension. Clin Genet 90(4):351–360. https://doi.org/10.1111/cge.12739

Walseth TF, Graff G, Krick TP, Goldberg ND (1981) The fate of 18O in guanosine monophosphate during enzymic transformations leading to guanosine 3′,5′-monophosphate generation. J Biol Chem 256:2176–2179

Whelton PK, Carey RM, Aronow WS, Casey DE Jr, Collins KJ, Dennison Himmelfarb C, DePalma SM, Gidding S, Jamerson KA, Jones DW, MacLaughlin EJ, Muntner P, Ovbiagele B, Smith SC Jr, Spencer CC, Stafford RS, Taler SJ, Thomas RJ, Williams KA Sr, Williamson JD, Wright JT Jr (2017) 2017 ACC/AHA/AAPA/ABC/ACPM/AGS/APhA/ASH/ASPC/NMA/PCNA guideline for the prevention, detection, evaluation, and management of high blood pressure in adults: executive summary: a report of the American College of Cardiology/American Heart Association Task Force on clinical practice guidelines. Hypertension 138(17):e426–e483. https://doi.org/10.1161/HYP.0000000000000066

Wilck N, Markó L, Balogh A, Kräker K, Herse F, Bartolomaeus H, Szijártó IA, Reichhart N, Strauß O, Heuser A, Brockschnieder D, Kretschmer A, Lesche R, Stasch JP, Sandner P, Luft FC, Müller DN, Dechend R, Haase N (2018) Nitric oxide sensitive guanylyl cyclase stimulation improves experimental heart failure in rats with preserved ejection fraction. JCI Insight 3(4):96006

Winter MB, Herzik MA Jr, Kuriyan J, Marletta MA (2011) Tunnels modulate ligand flux in a heme nitric oxide/oxygen binding (H-NOX) domain. Proc Natl Acad Sci U S A 108(43):E881–E889. https://doi.org/10.1073/pnas.1114038108

Xiao J, Jin C, Liu Z, Guo S, Zhang X, Zhou X, Wu X (2015) The design, synthesis, and biological evaluation of novel YC-1 derivatives as potent anti-hepatic fibrosis agents. Org Biomol Chem 13(26):7257–7264. https://doi.org/10.1039/c5ob00710k

Zimmer DP, Silva IA, Chien Y-T, Milne GT, Currie M (2017) The soluble guanylate cyclase stimulator IW-1701 enhances nitric oxide-mediated relaxation of human lower esophageal sphincter *ex vivo*. Gastroenterology 152(5):S699. https://doi.org/10.1016/S0016-5085(17)32443-5

Inhibitors of Advanced Glycation End Product (AGE) Formation and Accumulation

Karly C. Sourris, Anna Watson, and Karin Jandeleit-Dahm

Contents

Abstract

A range of chemically different compounds are known to inhibit the formation and accumulation of advanced glycation end products (AGEs) or disrupt associated signalling pathways. There is evidence that some of these agents can provide

K. C. Sourris · A. Watson · K. Jandeleit-Dahm (✉)
Department of Diabetes, Central Clinical School, Monash University, Melbourne, VIC, Australia
e-mail: karin.jandeleit-dahm@monash.edu

H. H. H. W. Schmidt et al. (eds.), *Reactive Oxygen Species*,
Handbook of Experimental Pharmacology 264, https://doi.org/10.1007/164_2020_391

end-organ protection in chronic diseases including diabetes. Whilst this group of therapeutics are structurally and functionally different and have a range of mechanisms of action, they ultimately reduce the deleterious actions and the tissue burden of advanced glycation end products. To date it remains unclear if this is due to the reduction in tissue AGE levels per se or the modulation of downstream signal pathways. Some of these agents either stimulate antioxidant defence or reduce the formation of reactive oxygen species (ROS), modify lipid profiles and inhibit inflammation. A number of existing treatments for glucose lowering, hypertension and hyperlipidaemia are also known to reduce AGE formation as a by-product of their action. Targeted AGE formation inhibitors or AGE cross-link breakers have been developed and have shown beneficial effects in animal models of diabetic complications as well as other chronic conditions. However, only a few of these agents have progressed to clinical development. The failure of clinical translation highlights the importance of further investigation of the advanced glycation pathway, the diverse actions of agents which interfere with AGE formation, cross-linking or AGE receptor activation and their effect on the development and progression of chronic diseases including diabetic complications.

Graphical Abstract

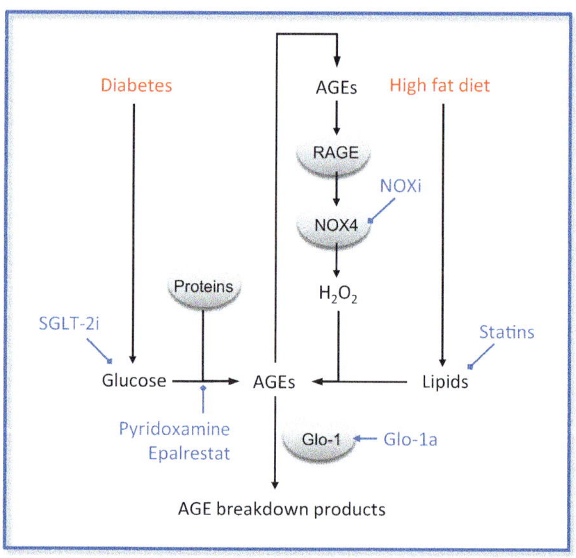

Advanced glycation end products (AGEs) are (1) proteins or lipids that become glycated as a result of exposure to sugars or (2) non-proteinaceous oxidised lipids. They are implicated in ageing and the development, or worsening, of many degenerative diseases, such as diabetes, atherosclerosis, chronic kidney and Alzheimer's disease. Several antihypertensive and antidiabetic agents and statins also indirectly lower AGEs. Direct AGE inhibitors currently investigated include pyridoxamine and epalrestat, the inhibition of the formation of reactive dicarbonyls such as methylglyoxal as an important precursor of AGEs via increased activation of the detoxifying enzyme Glo-1 and inhibitors of NOX-derived ROS to reduce the AGE/RAGE signalling.

Keywords

Advanced glycation end products · Diabetes · RAGE · Reactive oxygen species · Receptors · Signalling · Treatments

1 What Are AGEs?

Advanced glycation end products are formed as a result of non-enzymatic biochemical reactions called the Maillard reaction which includes the reaction of glucose with amino residues on proteins and lipids (Lutgers et al. 2006; Mulder et al. 2006) (Fig. 1). Early glycation products such as methylglyoxal can be transformed into more stable AGEs, which may be irreversibly cross-linked with proteins or DNA and may subsequently alter the organ function and structure. AGEs are a heterogeneous and complex group of modifications (Maillard 1912).

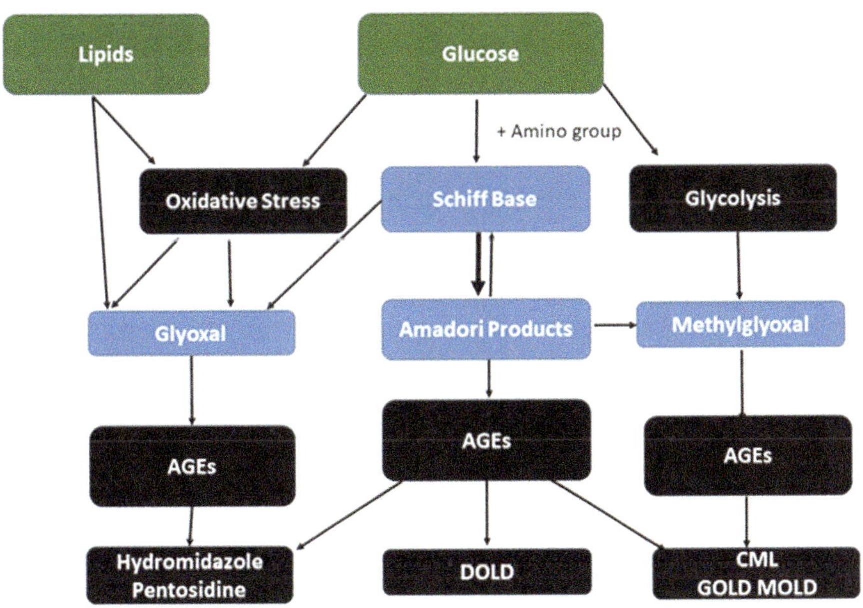

Fig. 1 Schematic representation of AGE formation and different types of AGEs. *AGE* advanced glycation end product, *CML* N_ε-carboxymethyl-lysine, *DOLD* 3-deoxyglucosone-derived lysine dimer, *GOLD* glyoxal-derived lysine dimer, *MOLD* methylglyoxal-derived lysine dimer

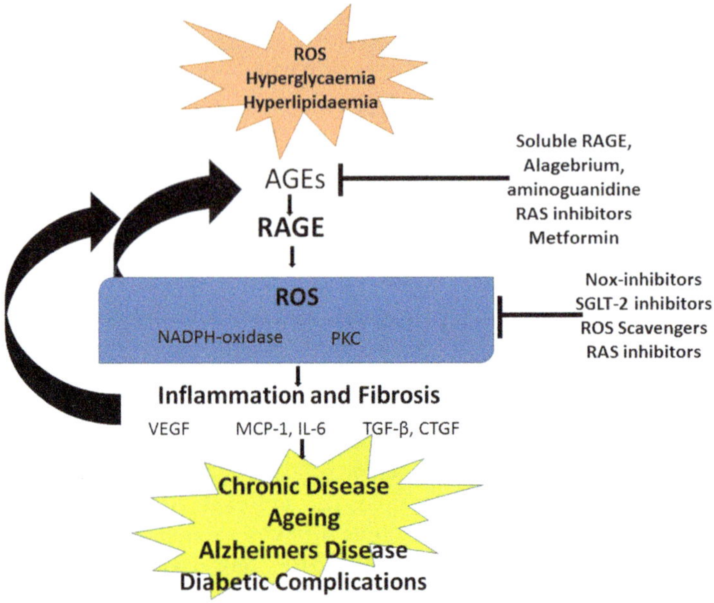

Fig. 2 Bidirectional actions of AGEs and ROS and downstream signalling

Increased formation of AGEs occurs in stages of increased or dysregulated reactive oxygen species (ROS) formation, such a hyperglycaemia, hyperlipidaemia and inflammation all of which are common characteristics of diabetic complications (Forbes and Cooper 2013; Sourris and Forbes 2009) (Fig. 2).

AGEs accumulate in tissues including in the vasculature, kidney and heart resulting in changes to organ structure and function (Soro-Paavonen et al. 2008). In the vasculature, AGE accumulation is associated with endothelial dysfunction and vascular stiffness and ultimately can contribute to the development of atherosclerosis. Within the kidney they lead to alterations in filtration and increased matrix accumulation ultimately leading to kidney fibrosis (Watson et al. 2012). In the heart, the accumulation of AGEs is associated with increased diastolic stiffness and dysfunction and can lead to cardiac fibrosis. Studies in type 1 diabetic patients show that AGE accumulation may predict the severity of micro- and macrovascular complications. Specifically, serum AGE levels are significantly elevated with the progression from normo- to microalbuminuria and subsequently with the development of overt nephropathy (Sell et al. 1992). In addition, skin collagen-associated AGE concentrations correlate with the severity of microvascular complications in patients with long-standing type 1 diabetes (Monnier et al. 1999) and with carotid intimal thickening (Nathan et al. 2003).

In type 2 diabetic patients, ischemic heart disease and hypertension correlate with circulating AGE levels, suggesting that they may be potential biomarkers of diabetic cardiovascular risk (Sugiyama et al. 1998). It has been demonstrated that serum AGE levels are not completely associated with glycaemic control, as assessed by

HbA1c in the clinical setting (Monnier et al. 1999; Steffes et al. 2003). This may explain the progression of diabetic complications in some patients with relatively good glycaemic control. The Diabetes Control and Complications Trial (DCCT) study demonstrated that AGE levels were a better predictor of progression to complications than HbA_{1C}. Differences in AGEs accounted for over a third of the variance (Monnier et al. 1999). Thus other factors, such as ROS, may contribute to AGEs production and accumulation in patients with adequate glycaemic control (Baynes and Thorpe 1999; Babaei-Jadidi et al. 2003).

It is yet to be determined which AGE modifications are the most pathogenic in disease. AGE cross-linked moieties, such as pentosidine, have intrinsic fluorescence. Tissue and plasma fluorescence can then be used as indicator AGE modifications with increased fluorescence within the kidney (Soulis-Liparota et al. 1995; Watson et al. 2011), retina (Stitt et al. 2002), skin (Lutgers et al. 2006; Genuth et al. 2005) as well as other organs affected by diabetic microvascular disease (Soulis et al. 1997a). This has been shown to increase with diabetes progression. Changes in renal and hepatic function are also linked to increases in tissue fluorescence, reflecting the role these organs play in clearing AGEs from the body (Makita et al. 1994). Furthermore, in type 1 and type 2 diabetic patients, circulating levels of fluorescent AGEs correlate with complications (Miura et al. 2003; Kalousova et al. 2006).

Other AGEs, such as N-carboxymethyllysine (CML), are not cross-linked and do not fluoresce however they have are elevated in the serum of type 1 diabetic patients and in diabetic rodent models (Watson et al. 2011; Makita et al. 1991). Type 2 diabetic patients also demonstrated increased levels in circulating CML (Wautier et al. 2003) and the AGE precursor dicarbonyl methylglyoxal (Kilhovd et al. 2003). Elevations in CML levels have been associated with the presence of microvascular complications, including retinopathy and nephropathy (Beisswenger et al. 1995).

2 Measurement of AGEs

Several AGEs and their intermediates have been identified and can be measured in serum and urine including CML, pentosidine and the dicarbonyl precursor, methylglyoxal. Circulating levels of AGEs have been measured in diabetic patients as well as in experimental models of diabetes. Our group has demonstrated that the measurement of total AGEs within the serum of diabetic patients does not necessarily correlate with progressive diabetic renal dysfunction (Coughlan and Forbes 2011). By contrast, we have found that circulating high-molecular-weight AGEs in type 2 diabetic patients correlate with the decline in renal dysfunction (Penfold et al. 2010). Furthermore, Coughlan et al. demonstrated that urinary levels of AGEs are strongly associated with progressive renal decline in both type 1 and type 2 diabetics and may be used as a biomarker (Coughlan and Forbes 2011). This has been verified by others who have shown that circulating AGEs are associated with progression of other diabetic complications (Makita et al. 1992).

2.1 Accumulation of AGEs in the Skin

The cross-linking of AGEs in the skin is associated with the generation of autofluorescence which may be measured in skin by a non-invasive AGE reader. The accumulation of low-molecular AGEs within the skin was shown to be higher in patients with diabetes correlating with glomerular filtration rate (GFR); those with a higher GFR were found to have a lower fluorescence (Thomas et al. 2005a). Hartog et al. found that diabetic patients on dialysis exhibited higher levels of skin autofluorescence when compared to a control group (Hartog et al. 2005). More recently it has been demonstrated that skin autofluorescence in a healthy population is directly associated with age, smoking, waist circumference and diet (Kellow et al. 2018). In disease states associated with increased ROS production, there is a significant increase in skin AGE autofluorescence.

3 AGE Receptors

In addition to the direct deleterious effects of AGE accumulation in tissues, there are receptor-mediated effects. Vascular, renal, neuronal and haematopoietic cells are all known to express receptors for AGEs (Goldin et al. 2006). The receptors for AGE are important modulators of the deleterious effects of these compounds. Receptors for AGEs are generally identified as either inflammatory (the receptor for AGEs (RAGE), AGE-R2) or clearance type receptors (AGE-R1, AGE-R3, CD36, Scr-II, FEEL-1 and FEEL-2) (Sourris and Forbes 2009; Alikhani et al. 2005; Schrijvers et al. 2004; Singh et al. 2001; Vlassara 1997; Vlassara and Bucala 1996).

3.1 Receptor for Advanced-Glycation End Products (RAGE)

The receptor for AGEs (RAGE) is a member of the immunoglobulin superfamily, expressed on the surface of monocytes, proximal tubular cells (Morcos et al. 2002), neurons, macrophages, and glomerular epithelial cells (podocytes) (Wendt et al. 2003a), mesangial, endothelial, smooth muscle, and fibroblast cells (Schmidt et al. 1994a; Wautier and Guillausseau 2001; Bierhaus et al. 2005a; Koulis et al. 2015). RAGE is a multiligand receptor which is capable of binding to a number of ligands other than AGEs including amyloid A, s100A8-9, amyloid-β-peptides, calgranulins, and amphoterin (HMGB1) (Yan et al. 1997; Bierhaus et al. 2005a; Yan et al. 2006). Its major physiological role is thought to be in host–pathogen defence (Bierhaus et al. 2005a).

The RAGE gene is located on chromosome 6 adjacent to the HLA locus in both human and mouse (Wautier and Guillausseau 2001; Bierhaus et al. 2005a) and its transcription is known to be both constitutive and inducible. RAGE is expressed during embryogenesis whilst it is generally down-regulated in adult life in most tissues (Bierhaus et al. 2005a). In chronic diseases such as Alzheimers, ageing, and diabetes, RAGE is known to be elevated (Gao et al. 2008; Son et al. 2017).

The RAGE protein consists of three immunoglobulin-like regions, one v-domain and two c-domains, in addition to transmembrane and cytoplasmic regions (Neeper et al. 1992; Schmidt et al. 1994b). There are a number of isoforms of RAGE, which lack either the cytoplasmic or extracellular domains. These include soluble RAGE (sRAGE), thought to be the result of proteolytic shedding of RAGE from the cell surface (Humpert et al. 2007). Soluble RAGE binds AGEs with a high affinity and has been considered as a decoy receptor for AGEs (Schlueter et al. 2003). Endothelial cells are known to secrete an isoform of RAGE (es-RAGE), which is a c-terminal splice variant of RAGE and lacks a trans-membrane and effector domain. Finally, NT-RAGE lacks an amino terminus; however, its function is still unclear (Yonemura and Tsukita 1999; Bierhaus et al. 2005a, b; Bohlender et al. 2005). Diabetic mice genetically manipulated to over-express RAGE have significant glomerulosclerosis (Yamamoto et al. 2001; Inagi et al. 2006). By contrast, it has been shown that RAGE knockout (KO) mice have less vascular and renal injury with diabetes (Myint et al. 2006; Soro-Paavonen et al. 2008; Coughlan et al. 2009; Sourris et al. 2010; Watson et al. 2012; Koulis et al. 2014). sRAGE treatment significantly attenuated diabetes-associated atherosclerosis development in animal models (Bucciarelli et al. 2002).

3.2 AGE-Clearance Receptors (AGE-R1, AGE-2 and AGE-R3)

The AGE receptor complex comprises of AGE-R1, AGE-R2 and AGE-R3 (also known as galectin-3) and is known to be central in the clearance of AGEs. The interaction amongst these receptors is thought to drive the degradation of AGE-modified molecules into smaller fragments for clearance by the kidney (Forbes and Cooper 2013).

AGE-R1 (OST-48, 48 kDa), as a member of the oligosaccharyl-transferase protein family, is a type 1 integral membrane protein (Yang et al. 1991; Vlassara 2001). AGE-R1 was the first AGE receptor cell surface clearance receptor identified and is anchored within the endoplasmic reticulum where it is thought to be a stabilising molecule for the oligosaccharyltransferase (OST) complex and thus is also referred to as OST-48 (Yang et al. 1991; Vlassara 2001). There is reduced renal and white blood cell expression of AGE-R1 in diabetes, and it has been suggested that AGE-R1 could be a potential target for therapy. Mice which have a transgenic overexpression of AGE-R1 do not develop diabetic nephropathy (DN) (Liu et al. 2005). Increasing the levels of AGE-R1 in renal cells produces a concomitant down regulation of RAGE; thus increasing AGE-R1 expression would likely provide additional benefits. This is yet to be clarified as we have recently demonstrated that the overexpression of AGE-R1 is associated with increased accumulation of liver AGEs leading to a concomitant increase in hepatic injury (Zhuang et al. 2017).

AGE-R2 (80K-H, 90 kDa) is a tyrosine-phosphorylated protein which is located within the plasma membrane. Whilst initially thought to act as a substrate for protein kinase C (PKC), more recently it was found to be part of the intracellular signalling pathway for fibroblast growth factor (FGF) receptor (Stitt et al. 1999, 2002; Makita et al. 1991; Schrijvers et al. 2004; Vlassara 2001; Yang et al. 1994). AGE-R2 is thought to contribute to the early stages of AGE signal transduction (Yang et al. 1991; Stitt et al. 1999).

AGE-R3 (also known as galectin-3, 32 kDa) is expressed in the nucleus, cytoplasm and cell surface of eosinophils, mast cells, the epithelium of the gastrointestinal and respiratory tracts, macrophages, renal cells, sensory neurons as well as in aortic endothelial cells. It has also been demonstrated in atherosclerotic plaques (Iacobini et al. 2005; Wada and Yagihashi 2005; Watson et al. 2014). It has been found to bind carbohydrates, laminin and IgE molecules. Cellular functions of AGE-R3 have been found to include apoptosis, inflammation and tumour growth (Vlassara 2001; Wada and Yagihashi 2005; Vlassara 1995; Kikuchi et al. 2004, 2005; Nangia-Makker et al. 2007). The inflammatory role of AGE-R3 has been extensively investigated. It has also been shown to have immunoregulatory potential and attracts eosinophils when expressed in T lymphocytes (Matsumoto et al. 1998). Increased expression of AGE-R3 in endothelial, pancreatic and melanoma cancer cells was found to promote proliferation and survival, demonstrating anti-apoptotic effects (Jiang et al. 2008; Johnson et al. 2007; Prieto et al. 2006). In addition, we have previously demonstrated that AGE-R3 levels correlate with diabetes-associated atherosclerosis in RAGE-deficient mice (Watson et al. 2014).

3.3 Scavenger Receptors (CD36, FEEL-1 and FEEL-2)

CD36 is an 88 kDa glycosylated protein which binds various molecules including fatty acids, collagen and oxidised LDL (oxLDL) (Febbraio et al. 2001; Nicholson et al. 2001). It is expressed on the cell surface of both macrophages and adipocytes (Kuniyasu et al. 2003). CD36's major pathophysiological functions include scavenging of oxLDL in macrophages and fatty acid transport in a number of cell types including adipocytes (Horiuchi et al. 2005). AGEs can also bind to CD36 with a high affinity leading to subsequent receptor-mediated endocytosis (Ohgami et al. 2001a, b, 2003).

Fasciclin, EGF-like, laminin-type EGF-like and link domain-containing scavenger receptors 1 and 2 (FEEL-1 and FEEL-2) bind AGE-modified proteins. As with other AGE receptors, FEEL-1 and FEEL-2 are multiligand receptors which endocytose bacteria, modified LDL as well as AGEs (Prevo et al. 2004). At the mRNA level, both receptors have been detected in lymph nodes and spleen; however, cell surface expression has only been identified for FEEL-1 on CD14-positive mononuclear cells (Horiuchi et al. 1996, 2005; Tamura et al. 2003).

3.4 Downstream Signalling of AGEs (Fig. 2)

There has been a growing body of evidence suggesting that AGEs activate PKC-mediated signal transduction leading to diabetic complications (Inoguchi et al. 1992; Xia et al. 1994). PKC has 11 different isoforms, many of which have been shown to be involved in diabetic complications (Fig. 2).

Most therapeutic approaches targeting advanced glycation also have direct or indirect effects on PKC. The attenuation of PKC-α phosphorylation and reduced translocation with alagebrium (ALA) has been shown in both in vivo models of DN and in vitro studies (Thallas-Bonke et al. 2004). It remains to be determined if this action of ALA on PKC-α phosphorylation partly explains its renoprotective actions. Modulation of PKC activity within the diabetic kidney has also been shown by vitamin B derivatives (Babaei-Jadidi et al. 2003; Hammes et al. 2003). AT1 receptor antagonists also attenuate diabetes-induced PKCε activity increases within the diabetic heart (Malhotra et al. 1997). Both aminoguanidine and ACE inhibitors prevent diabetes-associated increases in PKC-β activation in renal glomeruli (Osicka et al. 2000). Modulation of PKC has been demonstrated in vascular endothelial cells with aspirin (Dragomir et al. 2004) and the insulin sensitising agent metformin (Isoda et al. 2006). More recently, NOX-4 signalling has also been linked to PKC actions. The genetic deletion of NOX-4 in experimental models demonstrated a reduction in PKC expression and signalling within the diabetic kidney (Thallas-Bonke et al. 2014).

NF-κB is a transcription factor composed of two subunits, the most common of which are the p50 and p65 subunits (Barnes and Larin 1997). The active p65 subunit is central in the activation of numerous genes including, adhesion molecules, cytokines and many other inflammatory and proliferative proteins implicated in the process of chronic disease and diabetic complications (Barnes and Larin 1997; Bierhaus et al. 2001). NF-κB is activated by a range of stimuli including glucose (Pieper and Riazulhaq 1997) and ROS (Nishikawa et al. 2000). AGEs are also involved in activation of NF-κB mostly via a RAGE-dependent pathway leading to its translocation to the nucleus where it induces transcription of target genes such as IL-6 and TNF-α (Yan et al. 1994).

Pyrrolidine dithiocarbamate (PDTC) is a NF-κB inhibitor which has been used in both diabetic (Lee et al. 2004; Liu et al. 1999a, b; Rangan et al. 1999) and nondiabetic animal models of renal disease where it was found to be renoprotective (Rangan et al. 1999), although its toxicity does not allow for direct translation to the clinical setting. The importance of NF-κB in the pathogenesis of early renal macrophage infiltration in experimental diabetes has been shown to be modulated by interruption of the RAS (Lee et al. 2004; Liu et al. 2006). Diabetes-induced increases in NF-κB activation have been shown to be attenuated by several interventions including metformin (Isoda et al. 2006), aspirin (Zheng and Guan 2007), vitamin B derivatives (Hammes et al. 2003), carnosine (Odashima et al. 2006) and thiazolidinediones (Marx et al. 2004). It remains unclear if effects on NF-κB are required for the actions of AGE-lowering therapies in chronic diseases.

There is a growing body of evidence about the central role of inflammation in the development and progression of chronic disease including DC. In particular, monocyte chemoattractant protein (MCP-1), transforming growth factor-β1 (TGF-β1), connective tissue growth factor (cTGF) and vascular endothelial growth factor (VEGF) have all been implicated in inflammation and end-organ damage.

AGEs have been identified as a potent stimulus for the production of MCP-1 (Matsui et al. 2007; Giunti et al. 2008) and its secretion by mesangial, epithelial cells and podocytes (Giunti et al. 2008; Tesch 2008). MCP-1 is a chemokine which drives macrophage/monocyte infiltration into tissues and contributes to the progression of chronic disease. MCP-1 production and secretion from inflammatory or damaged cells in diabetes is activated by a number of signalling pathways. In an experimental model of type 1 diabetic nephropathy, a deficiency in MCP-1 resulted in a significant reduction in renal inflammatory cells and renoprotection (Kanamori et al. 2007). Many of the treatments which inhibit AGE accumulation or AGE-dependent signalling are also anti-inflammatory. A number of AGE inhibitors such as aminoguanidine (Lane et al. 1999) but also AT1R antagonists (Candido et al. 2004), aspirin (Makino et al. 2003; Zhang et al. 2008), sRAGE (Gu et al. 2006) and thiazolidinediones (Marx et al. 2004) have been shown to attenuate MCP-1 expression.

Downstream signalling of the AGE axis has been down to promote the action of pro-fibrotic molecules in the kidney and heart (Martin et al. 2005; Massague 1998; Wang et al. 2014). Transforming growth factor-β (TGF-β), a fibrogenic cytokine, has been implicated as a key effector molecule in renal fibrosis (Hill et al. 2000). A number of anti-AGE therapies including alagebrium (Forbes et al. 2003), aminoguanidine (Soulis et al. 1996, 1997b) and OPB-9195 (Tsuchida et al. 1999) but also AT1 antagonists (Cao et al. 2001) sRAGE (Wendt et al. 2003b) and aspirin (Makino et al. 2003) have been shown to attenuate the diabetes-induced increase in TGF-β1.

AGEs have also been demonstrated to modulate connective tissue growth factor (CTGF) which plays an important role in fibrosis but also in plaque remodelling in diabetes (Twigg et al. 2002a; Murphy et al. 1999; Riser et al. 2000). CTGF is elevated in both early and late diabetic nephropathy in humans (Ito et al. 1998). AGEs have been reported to increase CTGF expression in fibroblasts (Twigg et al. 2001) and renal cells including in mesangial cells (Twigg et al. 2001). A study in STZ-induced DN has demonstrated that the AGE inhibitor aminoguanidine attenuated the increased expression of renal cortical CTGF (Twigg et al. 2002b). Aspirin has also been shown to reduce increased expression of renal CTGF in diabetes in association with less mesangial expansion (Makino et al. 2003).

Vascular endothelial growth factor (VEGF) is a cytokine which plays in important role in diabetic retinopathy and nephropathy (Thallas-Bonke et al. 2004; Wendt et al. 2003b; Rizkalla et al. 2003; Wada et al. 2001; De Vriese et al. 2001). We and others have shown reduced VEGF expression with a number of AGE-lowering approaches including alagebrium (Thallas-Bonke et al. 2004), ACE inhibitors (Thallas-Bonke et al. 2004), sRAGE (Wendt et al. 2003b) and OPB-9195. In experimental models of DN, VEGF expression is decreased by treatment with an inhibitor of AGE formation (Tsuchida et al. 1999) and with the cross-link inhibitor alagebrium (ALA) (Thallas-Bonke et al. 2004).

4 AGE-Lowering Therapies

4.1 Direct Targeting of AGEs

A number of AGE-lowering compounds have been developed including aminoguanidine, and AGE formation inhibitor had progressed to clinical trials but had to be retracted due to severe side effects. ALA, a putative cross-link inhibitor did not progress into further clinical validation due to closure of the company. Inhibitors of AGE formation, including aminoguanidine (Soulis-Liparota et al. 1991) and OPB-9195 (Miyata et al. 2000a) have been shown to reduce AGE accumulation by scavenging free reactive carbonyl groups (Brownlee et al. 1986; Miyata et al. 2000b; Khalifah et al. 2005; Booth et al. 1997). The anti-hyperglycaemic agent metformin can also trap reactive carbonyl groups (Beisswenger et al. 1999). Aspirin also has also been shown to scavenging free carbonyls groups, as well as decreasing AGE levels by targeting preformed intermediates via chelation of copper and other transition metals which can contribute to ROS production (Urios et al. 2007).

LR-90 (methylene bis [4,4′-(2chloropheylureido phenoxysiobutyric acid)] is a compound proven to be effective in reducing renal and circulating AGE accumulation (Figarola et al. 2003, 2008). It attenuates AGE accumulation via its potent metal chelating abilities and its interaction with reactive carbonyl species (Figarola et al. 2003). LR-90 has shown renoprotective benefits in experimental models of diabetes and attenuated both glomerulosclerosis and albuminuria (Figarola et al. 2003, 2008). In addition, we have demonstrated that LR-90 attenuated diabetes-associated atherosclerosis in experimental models of diabetes (Watson et al. 2010).

Vitamin B derivatives such as benfotiamine, thiamine and pyridoxamine have been trialled and have shown initial beneficial AGE-lowering effects in the clinical context (Table 1). Pyridoxamine prevents the formation of AGEs from Amadori intermediates (Khalifah et al. 2005; Booth et al. 1997) and cleaves 3-deoxyglucosone reactive carbonyl intermediates (Chetyrkin et al. 2008). The inhibitory actions of pyridoxamine on AGE accumulation are associated with improvements in renal function in experimental models (Degenhardt et al. 2002) and a decrease in diabetes-associated atherosclerosis (Watson et al. 2011). In a phase II study in patients with diabetic renal disease, pyridoxamine has also been shown to be renoprotective (Williams et al. 2007).

The liposoluble derivatives of vitamin B1, benfotiamine and thiamine, also exhibit AGE-lowering properties. In contrast to pyridoxamine, benfotiamine and thiamine can decrease the formation of reducing sugars and polyol pathway intermediates (Berrone et al. 2006). Both have also been shown to be beneficial in experimental models of diabetic nephropathy (Babaei-Jadidi et al. 2003; Karachalias et al. 2003). Type 2 diabetic patients consuming a high-AGE diet treated with benfotiamine showed reduced levels of circulating AGE levels, as well as lower levels of ROS (Stirban et al. 2006). Another study failed to show a positive effect of benfotiamine in type 1 diabetic patients however (Du et al. 2008).

Carnosine, a naturally occurring dipeptide in the brain other tissues, is another antioxidant. Carnosine also reacts with aldehydes, including aldose and ketose

Table 1 Direct and indirect AGE-lowering therapies and their mechanism of action

Therapy	Mechanisms of actions
Direct	
Alagebrium chloride	Carbonyl scavenger, reduces cross-links, ROS inhibition
Aminoguanidine	Carbonyl and dicarbonyl scavenger, ROS inhibition
Benfotiamine	Reduce glycolysis and polyol pathway, transketolase activation, ROS inhibition
Carnosine	Reacts with aldehydes, antioxidant
OPB-9195	Carbonyl scavenger, ROS inhibition
Pyridoxamine	Reacts with carbonyl group in Amadori products, metal chelator, ROS scavenger
Thiamine	Reduce glycolysis and polyol pathway, transketolase activation, ROS inhibition
Indirect	
ACE inhibitors	Reduced formation of AII, anti-inflammatory, ROS inhibition
Aspirin	Thromboxane A_2, antiplatelet agent, anti-inflammatory, ROS inhibition
AT1R antagonists	Reduced signalling via AT1 receptor, reduced inflammation, ROS inhibition
Metformin	Reduced gluconeogenesis in the liver, glucose control, AMPK activation, reduction in cAMP and protein kinase A, ROS inhibition
NOX Inhibitors	ROS inhibition, anti-inflammatory
SGLT-2 Inhibitors	Glycaemic control, anti-inflammatory, ROS inhibition
sRAGE	Decoy receptor for circulating AGEs, reduced RAGE signalling
Thiazolidinediones	Glycaemic control, ROS inhibition, reduction in circulating AGEs

sugars, which attenuates AGE formation (Alhamdani et al. 2007; Hipkiss and Chana 1998; Hipkiss et al. 1998). It has been found to have renoprotective effects experimental diabetic nephropathy models (Janssen et al. 2005) (Table 1).

4.2 Cleavage of Preformed AGEs

AGEs form nonreversible covalent cross-links within and between tissue proteins and other organic compounds. N-Phenacylthiazolium bromide (N-PTB) (Vasan et al. 1996) and alagebrium (Forbes et al. 2003) novel therapies postulated to cleave or at last reduce cross-linking, allowing glycated proteins to be removed via scavenger receptors and renal excretion, although the exact mechanism is yet to be determined. Specifically, N-PTB cleaved α-dicarbonyl intermediates and reduced AGE formation (Vasan et al. 1996). Unexplainable increases in blood pressure and associated toxicity are seen with PTB; thus this was not translated into the clinic (Cooper et al. 2000). Alagebrium treatment has been shown to attenuate human-isolated systolic hypertension and symptoms of diastolic heart failure (Little et al. 2005). Our own studies have shown that alagebrium had both reno- and atheroprotective actions by reducing circulating and tissue AGE accumulation (Watson et al. 2011, 2012).

4.3 Targeting AGE Precursors: Methylglyoxal

In addition to advanced glycation end products (AGEs), more recently it has been shown that the precursors of AGEs, early reactive dicarbonyls, may also play a role in the development of diabetic complications but also in ageing, neurodegenerative diseases and Alzheimer's disease. There are a number of early or intermediate AGEs, known as α-dicarbonyls including, glyoxal, 3-deoxyglucosone and methylglyoxal. They are formed as by-products of glycolysis, lipid peroxidation and during the degradation of AGEs (Thomas 2011). These intermediates can be rapidly transformed into irreversible AGEs and AGE cross-links. Indeed, our studies have shown that increasing MGO levels independent of diabetes is associated with increased vascular inflammation and plaque development with plaque area similar to that observed in diabetes (Watson et al. 2012; Tikellis et al. 2014). Type 2 diabetic patients also have elevated levels of reactive dicarbonyls in both their serum and urine (Waris et al. 2015) (Fig. 3).

Glyoxolase-1 (Glo-1) is a naturally occurring detoxification mechanism which reduces MGO levels (Fig. 3). Glo-1 levels are depleted in diabetes and in human carotid atherosclerotic plaques (Hanssen et al. 2014a). Glyoxalase-1 catalyses the glutathione-dependent conversion of MGO into S-D-lactoylglutathione which is further metabolised to D-lactate by Glo-2 (Fig. 3). Given that methylglyoxal and other early dicarbonyls can be detoxified by glyoxalase-1 and glyoxalase activity is

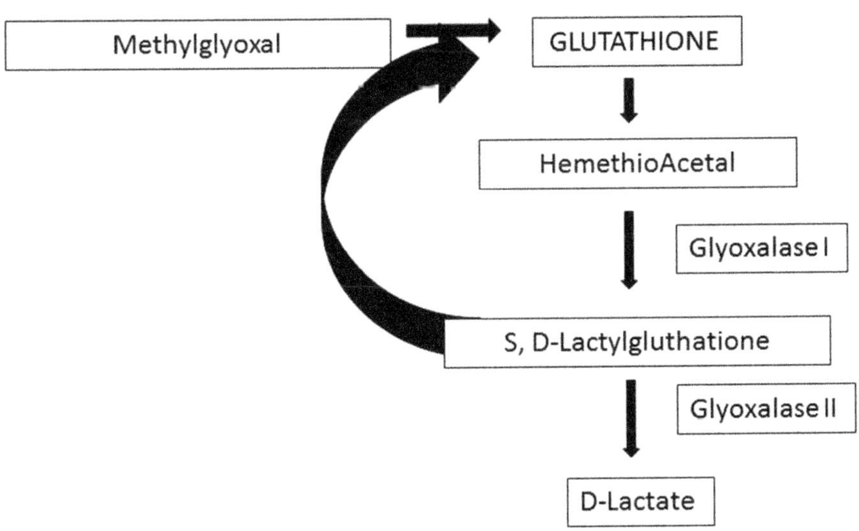

Fig. 3 Detoxification of methylglyoxal by glyoxalase I

deficient in diabetes, another approach to reduce dicarbonyl levels has been to increase glyoxalase expression or activity. Global Glo-1 KO is lethal, and partial Glo-1 knockdown in another study did not have effects on formation of atherosclerotic plaques in the short term (Wortmann et al. 2016). The overexpression of Glo-1 was associated with reduced methylglyoxal production and end-organ protection. However, Glo-1 transgenic/apoE−/− mice, a model of atherosclerosis, did not show reduced plaque area (Geoffrion et al. 2014; Hanssen et al. 2014b). However, Glo-1 transgenic mice were protected against the development of diabetic kidney disease with attenuation of albuminuria, mesangial area and a concurrent decrease in renal dicarbonyl levels (Geoffrion et al. 2014).

There is also an interaction between Glo-1 and Nrf2. Nrf2 is an essential transcription factor regulating the expression of genes containing an antioxidant response element (ARE) responsible for the protection against dysregulated ROS and glutathione (GSH) recycling. Nrf2 directly regulates the transcription of Glo-1 because there is a functional ARE in exon 1 of the Glo-1 gene. Nrf2 inducers such as sulphoraphane have shown to interact with the regulatory region of the Glo-1 gene. Another natural compound, mangiferin, has shown simultaneous activation of Glo-1 and Nrf2 (Liu et al. 2017).

Several Glo-1 activators have been tested including high-dose thiamine therapy, dicarbonyl scavenging with 3-deoxyglucosone, aminoguanidine (pimagedine), polypyrimidine tract-binding proteins (PTBP), arginine- or histidine-rich peptides, histidine glyoxalase mimetics and Nrf2 agonists as well as ψ-GSH, a synthetic cofactor of Glo-1, which is resistant to hydrolysis and is able to pass the blood-brain barrier (Table 2, (Mastrocola 2017)). ψ-GSH has been shown to reduce protein carbonyl content and Aβ plaque formation, resulting in a reduction in ROS formation and prevention of cognitive decline. A study combining trans resveratrol with hesperetin has been validated in clinical trials and shown to increase Glo-1 activity by 22%, reduced MGO levels and had beneficial effects on insulin resistance and fasting plasma glucose (Xue et al. 2016).

Further studies about the interaction between Nrf2 and Glo-1 are required, but there is emerging evidence that targeting the Nrf2/Glo-1 axis may be a promising

Table 2 Approaches to reduce dicarbonyl stress

1. Prevention of dicarbonyl formation	High-dose thiamine
2. Dicarbonyl scavengers	Aminoguanidine (pimagenidine) metformin Phenacylthiazolium bromide (PTB) arginine-rich peptides Histidine glyoxalase mimetics
3. Glo-1 inducers	Trans resveratrol-hesperetin Allyl isothiocyanate (ATTC) Sulphoraphane Butylated hydroxyamisole (BHA) n-3 polyunsaturated fatty acids (PUFA)

new approach to reduce dicarbonyl formation and AGE accumulation as well as to improve antioxidant defence.

5 Indirect Targeting of AGEs

5.1 Glucose-Lowering Approaches

Traditionally, glucose-lowering medications have shown AGE-lowering properties (Table 1). Some of these therapies used in the treatment of diabetes and insulin resistance have been found to have actions in addition to glucose lowering. Specifically, metformin has structural similarities to aminoguanidine, and it has been suggested that metformin detoxifies dicarbonyls. Importantly, metformin has been associated with reductions in CV events in the UKPDS study (Chalmers and Cooper 2008).

Some studies have demonstrated an AGE-lowering effect of metformin; however, most of these studies are observational and uncontrolled. In two randomised trials when metformin was compared to other glucose-lowering treatments, no difference was demonstrated. In a randomised clinical trial (RCT) comparing metformin with pioglitazone, metformin was equally effective with respect to pioglitazone in reducing AGE accumulation in type 2 diabetic patients (Esteghamati et al. 2013; Mirmiranpour et al. 2013) (Table 1).

Intensified glucose control has been shown to reduce AGE formation. However, in a substudy of the DCCT trial with intensive glycaemic control, no effects on skin fructosyllysine, CML or pentosidine were detected. In a larger substudy of the same trial, skin collagen glycation and skin CML, pentosidine and fluorescence decreased in the intensively treated group when compared to the conventionally treated group (Schiel et al. 2003, 2004). No difference was found when insulin was compared to other anti-hyperglycaemic agents in another RCT (Schiel et al. 2003, 2004). In general, effects of glucose-lowering agents on AGEs are inconsistent, and the type of AGEs effected varies from study to study. Moreover, in general no effects on serum CML or total AGEs have been demonstrated.

5.2 SGLT-2 Inhibitors

This new class of agents has recently been approved for treatment of type 2 diabetes. These drugs reduce blood glucose by increasing urinary glucose excretion. The anti-hyperglycaemic effect is usually mild and results in an HbA1c reduction of 0.6–0.8%. In the recent EMPA REG outcome study, cardiovascular and renal protection has been demonstrated (Cherney et al. 2017). In a type 2 diabetic cohort of approximately 7,000 patients, empagliflozin reduced UACR in normo-, micro- and macroalbuminuric patients relative to the placebo-treated group (Cherney et al. 2017). The effect of this type of therapeutic on AGEs has not been investigated. It is likely that, with the improvements in glycaemic control as well as restoration of renal and cardiovascular function, AGE formation and accumulation will be concomitantly reduced with these SGLT-2 inhibitors. SGLT-2 inhibitors have also been shown to reduce ROS production which further reduced AGE accumulation in chronic disease including diabetic complications (Tang et al. 2017; Ojima et al. 2015). Moreover, with improvements in glomerular filtration associated with SGLT-2 inhibition, it is likely that AGEs would be cleared more effectively (Table 1).

5.3 Inhibitors of the RAS

Inhibitors of the RAS are the standard treatment of care for hypertension in diabetes. The introduction of angiotensin-converting enzyme (ACE) inhibitors (ACEi) and angiotensin receptor blockers (ARBs) into the treatment regime of diabetic patients has significantly delayed the progression of diabetic nephropathy and subsequent end-stage renal disease. The interaction between the RAS and AGEs has been the focus of investigation for some time. It has been widely demonstrated that there are indeed important interactions between the RAS and AGEs and ACE inhibitors (Forbes et al. 2005; Miyata et al. 2002) as well as AT1 antagonists (Miyata et al. 2002; Nakamura et al. 2005; Forbes et al. 2004) were shown to be potent inhibitors of AGE accumulation. ACEi and AT1R antagonists reduce AGE accumulation by trapping reactive carbonyl groups, decreasing the formation of hydroxyl and carbon centred radicals, as well as chelating metal ions involved in AGE formation (Miyata et al. 2002).

It has been postulated that part of the renoprotective effect of RAS blockade is mediated via a reduction in serum and tissue AGEs. Inhibition of the renin angiotensin system attenuates the diabetes-induced increased formation of reactive oxygen species (Watson et al. 2014; Coughlan et al. 2007; Rosca et al. 2005). There are numerous studies demonstrating that ARBs and ACE inhibitors reduce AGE tissue formation (Watson et al. 2014; Coughlan et al. 2007; Rosca et al. 2005). Clinical studies have shown reductions in urinary (Ono et al. 2013) or in circulating AGEs (Matsui et al. 2007) in patients treated with ARBs. Larger RCTs with irbesartan failed to confirm this effect (Busch et al. 2008) (Table 1). The ARB olmesartan and the ACE inhibitor, temocaprilat, have been shown to significantly attenuate AGE (CML and pentosidine) formation (Miyata et al. 2002). The ARB telmisartan has been shown to reduce the expression of RAGE and sRAGE (Nakamura et al. 2005). In the KK/Ta diabetic mouse model, candesartan attenuated the formation of AGEs, reactive nitrogen-oxygen-species and RAGE expression (Fan et al. 2004). In type 2 diabetes, AT1R antagonists have been shown to reduce proteinuria in association with reduced renal AGE accumulation (Nangaku et al. 2003).

Importantly, a recent finding by Zheng et al. demonstrated that inhibition of AGEs and RAS led to a reduction in mortality and progression of diabetic nephropathy in an experimental model of type 2 diabetes (Zheng et al. 2006).

The interaction between AGEs and ROS is bidirectional (Fig. 2), and we have published data where administration of AGEs to rodents results in up-regulation of renal RAS, similar to that seen in diabetes. However, administration of angiotensin II to rats led to AGE accumulation (Thomas et al. 2005b). Furthermore, we have also shown benefits from ACE inhibition in diabetic nephropathy as soluble RAGE increased in both experimental models and in type 1 diabetic patients (Forbes et al. 2005). In addition to improvements in diabetic nephropathy, we have also demonstrated that the ACE inhibitor quinapril attenuated diabetes-associated atherosclerosis via a reduction in AGE accumulation (Watson et al. 2014). Only one small trial has investigated the effect of the ACE inhibitor ramipril (Sebekova et al. 2003). Studies using other antihypertensives such as the calcium channel blocker (CCB) amlodipine did not show a reduction in serum CML or pentosidine levels despite similar reductions in blood pressure (Busch et al. 2008). Thus, the AGE-lowering effects may be stronger with blockers of the RAS compared to other antihypertensives and may relate to their antioxidant and chelating characteristics. Importantly, a recent study by Pickering et al. demonstrated that the pro-inflammatory receptor AT1 complexes with the pro-inflammatory receptor for AGEs, RAGE. They demonstrated that the interaction between the two receptors enabled transactivation of the cytosolic tail of this receptor triggering pro-inflammatory NF-κB signalling. This interaction was found to occur independent of ligand binding in the extracellular region of RAGE. Treatment of diabetic mice with a RAGE peptide with a mutated cytosolic tail attenuated angiotensin II dependent inflammation and atherogenesis (Pickering et al. 2019). This further highlights the important interaction between these two pathways.

There may be synergistic benefits afforded by blockade of the RAS and by reducing AGE formation via a common pathway of attenuating aberrant ROS formation. Hence, the most rational therapeutic approach to treat chronic disease and diabetic complications is likely to be a combination of AGE lowering, RAS blockade and reduction in ROS formation (Watson et al. 2014; Coughlan et al. 2007; Davis et al. 2004) (Table 1).

5.4 Statins

In many preclinical trials, statin therapy has been associated with lower AGE formation. Statins reduce lipid oxidation, ROS formation and subsequently the formation of AGEs and ALEs. Statin treatment has been shown to be associated with reduced urinary AGEs (Vlad et al. 2017) serum AGEs (Ohsawa et al. 2015; Nakamura et al. 2010; Younis et al. 2010) and reduced accumulation of AGEs in plaques (Spadaccio et al. 2014). Furthermore, in animal models statins reduced not only AGE levels but also the expression of the receptor RAGE independent of their effects on lipids which may explain part of their cardiovascular protective effects (Feng et al. 2011).

5.5 Diet

Diets low in AGE content have been shown to provide beneficial effects on the kidney and vasculature in disease. It has been shown that a diet high in AGEs can mimic the effects of diabetes with respect to kidney disease and atherosclerosis (Gray et al. 2013). Methylglyoxal exposure to nondiabetic animals increased plaque formation in association with increased inflammation (Tikellis et al. 2014). We have recently shown in a randomised double-blind study that a low-AGE diet in overweight patients not only reduced AGE levels and glucose control but also improved renal function (Harcourt et al. 2011). Dietary interventions, to reduce AGE intake, have also been performed in patients with DN and have shown positive outcomes (Uribarri and Tuttle 2006; Uribarri et al. 2003a, b). However, the effects of a low-AGE diet on disease progression are not fully understood and need to be investigated in larger controlled trials.

5.6 Inhibitors of ROS Formation

5.6.1 Reactive Oxygen Species

Reactive oxygen species are important mediators in the formation of AGEs and are often generated in excessive in chronic disease, including diabetic complications (Forbes et al. 2008). There is a large body of evidence showing that AGEs may mediate ROS generation by both direct and indirect means. Concomitant dysregulation of antioxidant enzymes in diabetes leads to a state of ROS overproduction (Forbes et al. 2008). Most of the AGE inhibitors listed in Table 1, including AT1R antagonists (Miyata et al. 2002), vitamin B6 derivatives (Hammes et al. 2003; Endo et al. 2007), NOX inhibitors (Gray et al. 2013; Gray and Jandeleit-Dahm 2015; Gray et al. 2017; Jha et al. 2014) metformin (Rahbar et al. 2000), ACEi (Miyata et al. 2002; Coughlan et al. 2007), OPB-9195 (Wada et al. 2001; Mizutani et al. 2002), ALA (Coughlan et al. 2007) and sRAGE (Wautier et al. 1996) have shown to reduce superoxide generation within tissues in association with attenuation of end-organ injury.

Vitamin B-related therapies can effectively scavenge ROS intermediates. For example, pyridoxamine inhibits superoxide generation in association with attenuating neuropathy and retinopathy progression (Jain and Lim 2001). The vitamin B1 derivatives benfotiamine and thiamine have been shown to be beneficial by normalising ROS production (Berrone et al. 2006).

Induction of diabetes results in increased cytosolic and mitochondrial ROS production in the kidney. ALA treatment in an STZ rat model resulted in attenuation of both mitochondrial and cytosolic superoxide generation (Thallas-Bonke et al. 2004). Aspirin has been found to decrease reactive oxygen species production, in addition to increasing NO production (Dragomir et al. 2004, 2006). There is a close relationship between AGE formation and mitochondrial ROS production. Glycation of mitochondrial proteins in the diabetic rat kidneys has been associated with excess ROS production, which was associated with abnormalities in the mitochondrial respiratory chain complexes (Rosca et al. 2005). These changes were prevented with administration of aminoguanidine (Rosca et al. 2005).

5.6.2 NOX Inhibitors

There is increasing experimental evidence that inhibitors of ROS effectively reduce AGE formation and tissue accumulation. Furthermore, there is evidence that reduced expression of the receptor for AGEs, RAGE, is associated with reduced ROS generation. In recent preclinical studies using a first-in-class NOX inhibitor, GKT137831, in an animal model of diabetes-associated renal and vascular disease reduced ROS formation as well as RAGE expression in association with reno- and atheroprotection in this model (Watson et al. 2011, 2012) (Table 1).

5.7 Aldose Reductase Inhibitors

The AR inhibitor epalrestat has been investigated for effects on AGEs. One open-label randomised trial showed a decrease in serum CML in type 2 diabetic patients after 1 year of treatment when compared to control (Kawai et al. 2010). Some smaller studies have shown lower erythrocyte levels of CML, 3-deoxyguanidine (Hamada et al. 2000), sorbitol 3-phosphate and fructose 3-phosphate are reduced. Further RCTs are needed to investigate the potential of this therapeutic approach in diabetes.

6 Conclusion

AGE lowering has been shown to be an effective measure to reduce chronic disease, ageing and complications of diabetes via effects on ROS formation, PKC and NF-κB activation and inflammation. A number of currently used therapeutic agents such as blockers of the RAAS reduce AGEs as a side effect of their main action, in addition to BP reduction and attenuation of ROS dysregulation and inflammation. Although these agents lower AGEs, the AGE-lowering effect is usually modest. Thus there is an urgent need to develop more effective agents with potent AGE-lowering properties.

Only a handful of direct AGE inhibitors have been developed and are investigated at this stage in RCTs, such as pyridoxamine and vitamin B derivatives. Ongoing RCTs need to confirm their end-organ protection in chronic diseases including diabetes.

Other novel approaches include the inhibition of the formation of reactive dicarbonyls such as methylglyoxal as an important precursor of AGEs via increased activation of the detoxifying enzyme Glo-1. With the advent of specific and potent inhibitors of NOX-derived ROS as well as novel approaches to reduce the AGE/RAGE signalling, it may be possible to reduce the deleterious effects of AGEs in chronic diseases including diabetic complications.

References

Alhamdani MS, Al-Azzawie HF, Abbas FK (2007) Decreased formation of advanced glycation end-products in peritoneal fluid by carnosine and related peptides. Perit Dial Int 27(1):86–89

Alikhani Z et al (2005) Advanced glycation end products enhance expression of pro-apoptotic genes and stimulate fibroblast apoptosis through cytoplasmic and mitochondrial pathways. J Biol Chem 280(13):12087–12095

Babaei-Jadidi R et al (2003) Prevention of incipient diabetic nephropathy by high-dose thiamine and benfotiamine. Diabetes 52(8):2110–2120

Barnes PJ, Larin M (1997) Mechanisms of disease – nuclear factor-kappa-B – a pivotal transcription factor in chronic inflammatory diseases [review]. N Engl J Med 336(15):1066–1071

Baynes JW, Thorpe SR (1999) Role of oxidative stress in diabetic complications: a new perspective on an old paradigm. Diabetes 48(1):1–9

Beisswenger PJ et al (1995) Formation of immunochemical advanced glycosylation end products precedes and correlates with early manifestations of renal and retinal disease in diabetes. Diabetes 44(7):824–829

Beisswenger PJ et al (1999) Metformin reduces systemic methylglyoxal levels in type 2 diabetes. Diabetes 48(1):198–202

Berrone E et al (2006) Regulation of intracellular glucose and polyol pathway by thiamine and benfotiamine in vascular cells cultured in high glucose. J Biol Chem 281(14):9307–9313

Bierhaus A et al (2001) Diabetes-associated sustained activation of the transcription factor nuclear factor-kappaB. Diabetes 50(12):2792–2808

Bierhaus A et al (2005a) Understanding RAGE, the receptor for advanced glycation end products. J Mol Med 83(11):876–886

Bierhaus A et al (2005b) Advanced glycation end product receptor-mediated cellular dysfunction. Ann N Y Acad Sci 1043:676–680

Bohlender J et al (2005) Advanced glycation end products: a possible link to angiotensin in an animal model. Ann N Y Acad Sci 1043:681–684

Booth AA et al (1997) In vitro kinetic studies of formation of antigenic advanced glycation end products (AGEs). Novel inhibition of post-amadori glycation pathways. J Biol Chem 272 (9):5430–5437

Brownlee M et al (1986) Aminoguanidine prevents diabetes-induced arterial wall protein cross-linking. Science 232:1629–1632

Bucciarelli LG et al (2002) RAGE blockade stabilizes established atherosclerosis in diabetic apolipoprotein E-null mice. Circulation 106(22):2827–2835

Busch M et al (2008) Serum levels of the advanced glycation end products Nepsilon-carboxymethyllysine and pentosidine are not influenced by treatment with the angiotensin receptor II type 1 blocker irbesartan in patients with type 2 diabetic nephropathy and hypertension. Nephron Clin Pract 108(4):c291–c297

Candido R et al (2004) Irbesartan but not amlodipine suppresses diabetes-associated atherosclerosis. Circulation 109(12):1536–1542

Cao Z et al (2001) Additive hypotensive and anti-albuminuric effects of angiotensin-converting enzyme inhibition and angiotensin receptor antagonism in diabetic spontaneously hypertensive rats. Clin Sci (Colch) 100(6):591–599

Chalmers J, Cooper ME (2008) UKPDS and the legacy effect. N Engl J Med 359(15):1618–1620

Cherney DZI et al (2017) Effects of empagliflozin on the urinary albumin-to-creatinine ratio in patients with type 2 diabetes and established cardiovascular disease: an exploratory analysis from the EMPA-REG OUTCOME randomised, placebo-controlled trial. Lancet Diabetes Endocrinol 5(8):610–621

Chetyrkin SV et al (2008) Pyridoxamine protects proteins from functional damage by 3-deoxyglucosone: mechanism of action of pyridoxamine. Biochemistry 47(3):997–1006

Cooper ME et al (2000) The cross-link breaker, N-phenacylthiazolium bromide prevents vascular advanced glycation end-product accumulation. Diabetologia 43(5):660–664

Coughlan MT, Forbes JM (2011) Temporal increases in urinary carboxymethyllysine correlate with albuminuria development in diabetes. Am J Nephrol 34(1):9–17
Coughlan MT et al (2007) Combination therapy with the advanced glycation end product cross-link breaker, alagebrium, and angiotensin converting enzyme inhibitors in diabetes: synergy or redundancy? Endocrinology 148(2):886–895
Coughlan MT et al (2009) RAGE-induced cytosolic ROS promote mitochondrial superoxide generation in diabetes. J Am Soc Nephrol 20(4):742–752
Davis BJ et al (2004) Superior renoprotective effects of combination therapy with ACE and AGE inhibition in the diabetic spontaneously hypertensive rat. Diabetologia 47(1):89–97
De Vriese AS et al (2001) Vascular endothelial growth factor is essential for hyperglycemia-induced structural and functional alterations of the peritoneal membrane. J Am Soc Nephrol 12(8):1734–1741
Degenhardt TP et al (2002) Pyridoxamine inhibits early renal disease and dyslipidemia in the streptozotocin-diabetic rat. Kidney Int 61(3):939–950
Dragomir E et al (2004) Aspirin rectifies calcium homeostasis, decreases reactive oxygen species, and increases NO production in high glucose-exposed human endothelial cells. J Diabetes Complications 18(5):289–299
Dragomir E et al (2006) Aspirin and PPAR-alpha activators inhibit monocyte chemoattractant protein-1 expression induced by high glucose concentration in human endothelial cells. Vascul Pharmacol 44(6):440–449
Du X, Edelstein D, Brownlee M (2008) Oral benfotiamine plus alpha-lipoic acid normalises complication-causing pathways in type 1 diabetes. Diabetologia 51(10):1930–1932
Endo N et al (2007) Vitamin B6 suppresses apoptosis of NM-1 bovine endothelial cells induced by homocysteine and copper. Biochim Biophys Acta 1770(4):571–577
Esteghamati A et al (2013) Effects of metformin on markers of oxidative stress and antioxidant reserve in patients with newly diagnosed type 2 diabetes: a randomized clinical trial. Clin Nutr 32(2):179–185
Fan Q et al (2004) Candesartan reduced advanced glycation end-products accumulation and diminished nitro-oxidative stress in type 2 diabetic KK/ta mice. Nephrol Dial Transplant 19 (12):3012–3020
Febbraio M, Hajjar DP, Silverstein RL (2001) CD36: a class B scavenger receptor involved in angiogenesis, atherosclerosis, inflammation, and lipid metabolism. J Clin Invest 108 (6):785–791
Feng B et al (2011) Atorvastatin exerts its anti-atherosclerotic effects by targeting the receptor for advanced glycation end products. Biochim Biophys Acta 1812(9):1130–1137
Figarola JL et al (2003) LR-90 a new advanced glycation endproduct inhibitor prevents progression of diabetic nephropathy in streptozotocin-diabetic rats. Diabetologia 46(8):1140–1152
Figarola JL et al (2008) LR-90 prevents dyslipidaemia and diabetic nephropathy in the Zucker diabetic fatty rat. Diabetologia 51(5):882–891
Forbes JM, Cooper ME (2013) Mechanisms of diabetic complications. Physiol Rev 93(1):137–188
Forbes JM et al (2003) The breakdown of preexisting advanced glycation end products is associated with reduced renal fibrosis in experimental diabetes. FASEB J 17(12):1762–1764
Forbes JM et al (2004) The effects of valsartan on the accumulation of circulating and renal advanced glycation end products in experimental diabetes. Kidney Int Suppl 92:S105–S107
Forbes JM et al (2005) Modulation of soluble receptor for advanced glycation end products by angiotensin-converting enzyme-1 inhibition in diabetic nephropathy. J Am Soc Nephrol 16 (8):2363–2372
Forbes JM, Coughlan MT, Cooper ME (2008) Oxidative stress as a major culprit in kidney disease in diabetes. Diabetes 57(6):1446–1454
Gao ZQ et al (2008) RAGE upregulation and nuclear factor-kappaB activation associated with ageing rat cardiomyocyte dysfunction. Gen Physiol Biophys 27(3):152–158
Genuth S et al (2005) Glycation and carboxymethyllysine levels in skin collagen predict the risk of future 10-year progression of diabetic retinopathy and nephropathy in the diabetes control and

complications trial and epidemiology of diabetes interventions and complications participants with type 1 diabetes. Diabetes 54(11):3103–3111

Geoffrion M et al (2014) Differential effects of glyoxalase 1 overexpression on diabetic atherosclerosis and renal dysfunction in streptozotocin-treated, apolipoprotein E-deficient mice. Physiol Rep 2(6)

Giunti S et al (2008) Monocyte chemoattractant protein-1 has prosclerotic effects both in a mouse model of experimental diabetes and in vitro in human mesangial cells. Diabetologia 51 (1):198–207

Goldin A et al (2006) Advanced glycation end products: sparking the development of diabetic vascular injury. Circulation 114(6):597–605

Gray SP, Jandeleit-Dahm KA (2015) The role of NADPH oxidase in vascular disease – hypertension, atherosclerosis & stroke. Curr Pharm Des 21(41):5933–5944

Gray SP et al (2013) NADPH oxidase 1 plays a key role in diabetes mellitus-accelerated atherosclerosis. Circulation 127(18):1888–1902

Gray SP et al (2017) Combined NOX1/4 inhibition with GKT137831 in mice provides dose-dependent reno- and atheroprotection even in established micro- and macrovascular disease. Diabetologia 60(5):927–937

Gu L et al (2006) Role of receptor for advanced glycation end-products and signalling events in advanced glycation end-product-induced monocyte chemoattractant protein-1 expression in differentiated mouse podocytes. Nephrol Dial Transplant 21(2):299–313

Hamada Y et al (2000) Epalrestat, an aldose reductase ihibitor, reduces the levels of Nepsilon-(carboxymethyl)lysine protein adducts and their precursors in erythrocytes from diabetic patients. Diabetes Care 23(10):1539–1544

Hammes HP et al (2003) Benfotiamine blocks three major pathways of hyperglycemic damage and prevents experimental diabetic retinopathy. Nat Med 9(3):294–299

Hanssen NM et al (2014a) Higher levels of advanced glycation endproducts in human carotid atherosclerotic plaques are associated with a rupture-prone phenotype. Eur Heart J 35 (17):1137–1146

Hanssen NM et al (2014b) Glyoxalase 1 overexpression does not affect atherosclerotic lesion size and severity in ApoE−/− mice with or without diabetes. Cardiovasc Res 104(1):160–170

Harcourt BE et al (2011) Targeted reduction of advanced glycation improves renal function in obesity. Kidney Int 80(2):190–198

Hartog JW et al (2005) Accumulation of advanced glycation end products, measured as skin autofluorescence, in renal disease. Ann N Y Acad Sci 1043:299–307

Hill C et al (2000) The renal expression of transforming growth factor-beta isoforms and their receptors in acute and chronic experimental diabetes in rats. Endocrinology 141(3):1196–1208

Hipkiss AR, Chana H (1998) Carnosine protects proteins against methylglyoxal-mediated modifications. Biochem Biophys Res Commun 248(1):28–32

Hipkiss AR et al (1998) Pluripotent protective effects of carnosine, a naturally occurring dipeptide. Ann N Y Acad Sci 854:37–53

Horiuchi S et al (1996) Advanced glycation end products and their recognition by macrophage and macrophage-derived cells. Diabetes 45(Suppl 3):S73–S76

Horiuchi S et al (2005) Pathological roles of advanced glycation end product receptors SR-A and CD36. Ann N Y Acad Sci 1043:671–675

Humpert PM et al (2007) Soluble RAGE but not endogenous secretory RAGE is associated with albuminuria in patients with type 2 diabetes. Cardiovasc Diabetol 6:9

Iacobini C et al (2005) Development of age-dependent glomerular lesions in galectin-3/AGE-receptor-3 knockout mice. Am J Physiol Renal Physiol 289(3):F611–F621

Inagi R et al (2006) A severe diabetic nephropathy model with early development of nodule-like lesions induced by megsin overexpression in RAGE/iNOS transgenic mice. Diabetes 55 (2):356–366

Inoguchi T et al (1992) Preferential elevation of protein kinase C isoform beta II and diacylglycerol levels in the aorta and heart of diabetic rats: differential reversibility to glycemic control by islet cell transplantation. Proc Natl Acad Sci U S A 89(22):11059–11063

Isoda K et al (2006) Metformin inhibits proinflammatory responses and nuclear factor-kappaB in human vascular wall cells. Arterioscler Thromb Vasc Biol 26(3):611–617
Ito Y et al (1998) Expression of connective tissue growth factor in human renal fibrosis. Kidney Int 53(4):853–861
Jain SK, Lim G (2001) Pyridoxine and pyridoxamine inhibits superoxide radicals and prevents lipid peroxidation, protein glycosylation, and (Na+ + K+)-ATPase activity reduction in high glucose-treated human erythrocytes. Free Radic Biol Med 30(3):232–237
Janssen B et al (2005) Carnosine as a protective factor in diabetic nephropathy: association with a leucine repeat of the carnosinase gene CNDP1. Diabetes 54(8):2320–2327
Jha JC et al (2014) Genetic targeting or pharmacologic inhibition of NADPH oxidase nox4 provides renoprotection in long-term diabetic nephropathy. J Am Soc Nephrol 25(6):1237–1254
Jiang HB, Xu M, Wang XP (2008) Pancreatic stellate cells promote proliferation and invasiveness of human pancreatic cancer cells via galectin-3. World J Gastroenterol 14(13):2023–2028
Johnson KD et al (2007) Galectin-3 as a potential therapeutic target in tumors arising from malignant endothelia. Neoplasia 9(8):662–670
Kalousova M et al (2006) Soluble receptor for advanced glycation end products in patients with decreased renal function. Am J Kidney Dis 47(3):406–411
Kanamori H et al (2007) Inhibition of MCP-1/CCR2 pathway ameliorates the development of diabetic nephropathy. Biochem Biophys Res Commun 360(4):772–777
Karachalias N et al (2003) Accumulation of fructosyl-lysine and advanced glycation end products in the kidney, retina and peripheral nerve of streptozotocin-induced diabetic rats. Biochem Soc Trans 31(Pt 6):1423–1425
Kawai T et al (2010) Effects of epalrestat, an aldose reductase inhibitor, on diabetic peripheral neuropathy in patients with type 2 diabetes, in relation to suppression of N(varepsilon)-carboxymethyl lysine. J Diabetes Complications 24(6):424–432
Kellow NJ, Coughlan MT, Reid CM (2018) Association between habitual dietary and lifestyle behaviours and skin autofluorescence (SAF), a marker of tissue accumulation of advanced glycation endproducts (AGEs), in healthy adults. Eur J Nutr 57(6):2209–2216
Khalifah RG, Chen Y, Wassenberg JJ (2005) Post-Amadori AGE inhibition as a therapeutic target for diabetic complications: a rational approach to second-generation Amadorin design. Ann N Y Acad Sci 1043:793–806
Kikuchi Y et al (2004) Galectin-3-positive cell infiltration in human diabetic nephropathy. Nephrol Dial Transplant 19(3):602–607
Kikuchi Y et al (2005) Advanced glycation end-product induces fractalkine gene upregulation in normal rat glomeruli. Nephrol Dial Transplant 20(12):2690–2696
Kilhovd BK et al (2003) Increased serum levels of the specific AGE-compound methylglyoxal-derived hydroimidazolone in patients with type 2 diabetes. Metabolism 52(2):163–167
Koulis C et al (2014) Role of bone-marrow- and non-bone-marrow-derived receptor for advanced glycation end-products (RAGE) in a mouse model of diabetes-associated atherosclerosis. Clin Sci (Lond) 127(7):485–497
Koulis C et al (2015) Linking RAGE and Nox in diabetic micro- and macrovascular complications. Diabetes Metab 41(4):272–281
Kuniyasu A et al (2003) CD36-mediated endocytic uptake of advanced glycation end products (AGE) in mouse 3T3-L1 and human subcutaneous adipocytes. FEBS Lett 537(1–3):85–90
Lane TE, Fox HS, Buchmeier MJ (1999) Inhibition of nitric oxide synthase-2 reduces the severity of mouse hepatitis virus-induced demyelination: implications for NOS2/NO regulation of chemokine expression and inflammation. J Neurovirol 5(1):48–54
Lee FT et al (2004) Interactions between angiotensin II and NF-kappaB-dependent pathways in modulating macrophage infiltration in experimental diabetic nephropathy. J Am Soc Nephrol 15 (8):2139–2151
Little WC et al (2005) The effect of alagebrium chloride (ALT-711), a novel glucose cross-link breaker, in the treatment of elderly patients with diastolic heart failure. J Card Fail 11 (3):191–195

Liu SF, Ye X, Malik AB (1999a) Inhibition of NF-kappaB activation by pyrrolidine dithiocarbamate prevents in vivo expression of proinflammatory genes. Circulation 100(12):1330–1337
Liu SF, Ye X, Malik AB (1999b) Pyrrolidine dithiocarbamate prevents I-kappaB degradation and reduces microvascular injury induced by lipopolysaccharide in multiple organs. Mol Pharmacol 55(4):658–667
Liu H et al (2005) Overexpression of AGE-receptor-1 (AGE-R1) in mice prevent AGE accumulation and delays diabetic renal injury. Diabetes 54(Suppl):A21-B
Liu HQ et al (2006) Angiotensin II stimulates intercellular adhesion molecule-1 via an AT1 receptor/nuclear factor-kappaB pathway in brain microvascular endothelial cells. Life Sci 78 (12):1293–1298
Liu YW et al (2017) Mangiferin upregulates glyoxalase 1 through activation of Nrf2/ARE signaling in central neurons cultured with high glucose. Mol Neurobiol 54(6):4060–4070
Lutgers HL et al (2006) Skin autofluorescence as a noninvasive marker of vascular damage in patients with type 2 diabetes. Diabetes Care 29(12):2654–2659
Maillard L (1912) Action des acides amines sur les sucres: formation des melanoidines par voie methodique. C R Acad Sci 154:66–68
Makino H et al (2003) Roles of connective tissue growth factor and prostanoids in early streptozotocin-induced diabetic rat kidney: the effect of aspirin treatment. Clin Exp Nephrol 7 (1):33–40
Makita Z et al (1991) Advanced glycosylation end products in patients with diabetic nephropathy. N Engl J Med 325(12):836–842
Makita Z et al (1992) Hemoglobin-AGE: a circulating marker of advanced glycosylation. Science 258(5082):651–653
Makita Z et al (1994) Reactive glycosylation endproducts in diabetic uraemia and treatment of renal failure. Lancet 343(8912):1519–1522
Malhotra A et al (1997) Experimental diabetes is associated with functional activation of protein kinase C epsilon and phosphorylation of troponin I in the heart, which are prevented by angiotensin II receptor blockade. Circ Res 81(6):1027–1033
Martin J et al (2005) Tranilast attenuates cardiac matrix deposition in experimental diabetes: role of transforming growth factor-beta. Cardiovasc Res 65(3):694–701
Marx N et al (2004) Thiazolidinediones reduce endothelial expression of receptors for advanced glycation end products. Diabetes 53(10):2662–2668
Massague J (1998) TGF-beta signal transduction. Annu Rev Biochem 67:753–791
Mastrocola R (2017) AGEs and neurodegeneration: the Nrf2/glyoxalase-1 interaction. Oncotarget 8 (4):5645–5646
Matsui T et al (2007) Telmisartan, an angiotensin II type 1 receptor blocker, inhibits advanced glycation end-product (AGE)-induced monocyte chemoattractant protein-1 expression in mesangial cells through downregulation of receptor for AGEs via peroxisome proliferator-activated receptor-gamma activation. J Int Med Res 35(4):482–489
Matsumoto R et al (1998) Human ecalectin, a variant of human galectin-9, is a novel eosinophil chemoattractant produced by T lymphocytes. J Biol Chem 273(27):16976–16984
Mirmiranpour H et al (2013) Comparative effects of pioglitazone and metformin on oxidative stress markers in newly diagnosed type 2 diabetes patients: a randomized clinical trial. J Diabetes Complications 27(5):501–507
Miura J et al (2003) Serum levels of non-carboxymethyllysine advanced glycation endproducts are correlated to severity of microvascular complications in patients with type 1 diabetes. J Diabetes Complications 17(1):16–21
Miyata T et al (2000a) Mechanism of the inhibitory effect of OPB-9195 [(+/−)-2-isopropylidenehydrazono-4-oxo-thiazolidin-5-yla cetanilide] on advanced glycation end product and advanced lipoxidation end product formation. J Am Soc Nephrol 11(9):1719–1725
Miyata T, Kurokawa K, Van Ypersele De Strihou C (2000b) Advanced glycation and lipoxidation end products: role of reactive carbonyl compounds generated during carbohydrate and lipid metabolism. J Am Soc Nephrol 11(9):1744–1752

Miyata T et al (2002) Angiotensin II receptor antagonists and angiotensin-converting enzyme inhibitors lower in vitro the formation of advanced glycation end products: biochemical mechanisms. J Am Soc Nephrol 13(10):2478–2487

Mizutani K et al (2002) Inhibitor for advanced glycation end products formation attenuates hypertension and oxidative damage in genetic hypertensive rats. J Hypertens 20(8):1607–1614

Monnier VM et al (1999) Skin collagen glycation, glycoxidation, and crosslinking are lower in subjects with long-term intensive versus conventional therapy of type 1 diabetes: relevance of glycated collagen products versus HbA1c as markers of diabetic complications. DCCT Skin Collagen Ancillary Study Group. Diabetes control and complications trial. Diabetes 48 (4):870–880

Morcos M et al (2002) Activation of tubular epithelial cells in diabetic nephropathy. Diabetes 51 (12):3532–3544

Mulder DJ et al (2006) Skin autofluorescence, a novel marker for glycemic and oxidative stress-derived advanced glycation endproducts: an overview of current clinical studies, evidence, and limitations. Diabetes Technol Ther 8(5):523–535

Murphy M et al (1999) Suppression subtractive hybridization identifies high glucose levels as a stimulus for expression of connective tissue growth factor and other genes in human mesangial cells. J Biol Chem 274(9):5830–5834

Myint KM et al (2006) RAGE control of diabetic nephropathy in a mouse model: effects of RAGE gene disruption and administration of low-molecular weight heparin. Diabetes 55(9):2510–2522

Nakamura K et al (2005) Telmisartan inhibits expression of a receptor for advanced glycation end products (RAGE) in angiotensin-II-exposed endothelial cells and decreases serum levels of soluble RAGE in patients with essential hypertension. Microvasc Res 70(3):137–141

Nakamura T et al (2010) Atorvastatin reduces proteinuria in non-diabetic chronic kidney disease patients partly via lowering serum levels of advanced glycation end products (AGEs). Oxid Med Cell Longev 3(5):304–307

Nangaku M et al (2003) Anti-hypertensive agents inhibit in vivo the formation of advanced glycation end products and improve renal damage in a type 2 diabetic nephropathy rat model. J Am Soc Nephrol 14(5):1212–1222

Nangia-Makker P et al (2007) Galectin-3 in apoptosis, a novel therapeutic target. J Bioenerg Biomembr 39(1):79–84

Nathan DM et al (2003) Intensive diabetes therapy and carotid intima-media thickness in type 1 diabetes mellitus. N Engl J Med 348(23):2294–2303

Neeper M et al (1992) Cloning and expression of a cell surface receptor for advanced glycosylation end products of proteins. J Biol Chem 267(21):14998 15004

Nicholson AC et al (2001) Role of CD36, the macrophage class B scavenger receptor, in atherosclerosis. Ann N Y Acad Sci 947:224–228

Nishikawa T et al (2000) Normalizing mitochondrial superoxide production blocks three pathways of hyperglycaemic damage. Nature 404(6779):787–790

Odashima M et al (2006) Zinc L-carnosine protects colonic mucosal injury through induction of heat shock protein 72 and suppression of NF-kappaB activation. Life Sci 79(24):2245–2250

Ohgami N et al (2001a) CD36, a member of class B scavenger receptor family, is a receptor for advanced glycation end products. Ann N Y Acad Sci 947:350–355

Ohgami N et al (2001b) Scavenger receptor class B type I-mediated reverse cholesterol transport is inhibited by advanced glycation end products. J Biol Chem 276(16):13348–13355

Ohgami N et al (2003) Advanced glycation end products (AGE) inhibit scavenger receptor class B type I-mediated reverse cholesterol transport: a new crossroad of AGE to cholesterol metabolism. J Atheroscler Thromb 10(1):1–6

Ohsawa M et al (2015) Effects of pitavastatin add-on therapy on chronic kidney disease with albuminuria and dyslipidemia. Lipids Health Dis 14:161

Ojima A et al (2015) Empagliflozin, an inhibitor of sodium-glucose Cotransporter 2 exerts anti-inflammatory and antifibrotic effects on experimental diabetic nephropathy partly by suppressing AGEs-receptor Axis. Horm Metab Res 47(9):686–692

Ono Y et al (2013) Suppression of advanced glycation and lipoxidation end products by angiotensin II type-1 receptor blocker candesartan in type 2 diabetic patients with essential hypertension. Fukushima J Med Sci 59(2):69–75

Osicka TM et al (2000) Albuminuria in patients with type 1 diabetes is directly linked to changes in the lysosome-mediated degradation of albumin during renal passage. Diabetes 49(9):1579–1584

Penfold SA et al (2010) Circulating high-molecular-weight RAGE ligands activate pathways implicated in the development of diabetic nephropathy. Kidney Int 78(3):287–295

Pickering RJ et al (2019) Transactivation of RAGE mediates angiotensin-induced inflammation and atherogenesis. J Clin Invest 129(1):406–421

Pieper GM, Riazulhaq (1997) Activation of nuclear factor-kappa-B in cultured endothelial cells by increased glucose concentration – prevention by Calphostin C. J Cardiovasc Pharmacol 30 (4):528–532

Prevo R et al (2004) Rapid plasma membrane-endosomal trafficking of the lymph node sinus and high endothelial venule scavenger receptor/homing receptor stabilin-1 (FEEL-1/CLEVER-1). J Biol Chem 279(50):52580–52592

Prieto VG et al (2006) Galectin-3 expression is associated with tumor progression and pattern of sun exposure in melanoma. Clin Cancer Res 12(22):6709–6715

Rahbar S et al (2000) Evidence that pioglitazone, metformin and pentoxifylline are inhibitors of glycation. Clin Chim Acta 301(1–2):65–77

Rangan GK et al (1999) Inhibition of nuclear factor-kappa B activation reduces cortical tubulointerstitial injury in proteinuric rats. Kidney Int 56(1):118–134

Riser BL et al (2000) Regulation of connective tissue growth factor activity in cultured rat mesangial cells and its expression in experimental diabetic glomerulosclerosis. J Am Soc Nephrol 11(1):25–38

Rizkalla B et al (2003) Increased renal vascular endothelial growth factor and angiopoietins by angiotensin II infusion is mediated by both AT1 and AT2 receptors. J Am Soc Nephrol 14 (12):3061–3071

Rosca MG et al (2005) Glycation of mitochondrial proteins from diabetic rat kidney is associated with excess superoxide formation. Am J Physiol Renal Physiol 289(2):F420–F430

Schiel R et al (2003) Improvement in quality of diabetes control and concentrations of AGE-products in patients with type 1 and insulin-treated type 2 diabetes mellitus studied over a period of 10 years (JEVIN). J Diabetes Complications 17(2):90–97

Schiel R et al (2004) Improvement of the quality of diabetes control and decrease in the concentrations of AGE-products in patients with type 1 and insulin-treated type 2 diabetes mellitus: results from a 10 year-prospective, population-based survey on the quality of diabetes care in Germany (JEVIN). Eur J Med Res 9(8):391–399

Schlueter C et al (2003) Tissue-specific expression patterns of the RAGE receptor and its soluble forms – a result of regulated alternative splicing? Biochim Biophys Acta 1630(1):1–6

Schmidt AM et al (1994a) Cellular receptors for advanced glycation end products. Implications for induction of oxidant stress and cellular dysfunction in the pathogenesis of vascular lesions. Arterioscler Thromb 14(10):1521–1528

Schmidt AM et al (1994b) The endothelial cell binding site for advanced glycation end products consists of a complex: an integral membrane protein and a lactoferrin-like polypeptide. J Biol Chem 269(13):9882–9888

Schrijvers BF, De Vriese AS, Flyvbjerg A (2004) From hyperglycemia to diabetic kidney disease: the role of metabolic, hemodynamic, intracellular factors and growth factors/cytokines. Endocr Rev 25(6):971–1010

Sebekova K et al (2003) Effects of ramipril in nondiabetic nephropathy: improved parameters of oxidatives stress and potential modulation of advanced glycation end products. J Hum Hypertens 17(4):265–270

Sell DR et al (1992) Pentosidine formation in skin correlates with severity of complications in individuals with long-standing IDDM. Diabetes 41(10):1286–1292

Singh R et al (2001) Advanced glycation end-products: a review. Diabetologia 44(2):129–146

Son M et al (2017) Age dependent accumulation patterns of advanced glycation end product receptor (RAGE) ligands and binding intensities between RAGE and its ligands differ in the liver, kidney, and skeletal muscle. Immun Ageing 14:12

Soro-Paavonen A et al (2008) Receptor for advanced glycation end products (RAGE) deficiency attenuates the development of atherosclerosis in diabetes. Diabetes 57(9):2461–2469

Soulis T et al (1996) Effects of aminoguanidine in preventing experimental diabetic nephropathy are related to the duration of treatment. Kidney Int 50(2):627–634

Soulis T et al (1997a) Advanced glycation end products and their receptors co-localise in rat organs susceptible to diabetic microvascular injury. Diabetologia 40(6):619–628

Soulis T et al (1997b) Relative contributions of advanced glycation and nitric oxide synthase inhibition to aminoguanidine-mediated renoprotection in diabetic rats. Diabetologia 40 (10):1141–1151

Soulis-Liparota T et al (1991) Retardation by aminoguanidine of development of albuminuria, mesangial expansion, and tissue fluorescence in streptozocin-induced diabetic rat. Diabetes 40 (10):1328–1334

Soulis-Liparota T et al (1995) The relative roles of advanced glycation, oxidation and aldose reductase inhibition in the development of experimental diabetic nephropathy in the Sprague-Dawley rat. Diabetologia 38(4):387–394

Sourris KC, Forbes JM (2009) Interactions between advanced glycation end-products (AGE) and their receptors in the development and progression of diabetic nephropathy – are these receptors valid therapeutic targets. Curr Drug Targets 10(1):42–50

Sourris KC et al (2010) Receptor for AGEs (RAGE) blockade may exert its renoprotective effects in patients with diabetic nephropathy via induction of the angiotensin II type 2 (AT2) receptor. Diabetologia 53(11):2442–2451

Spadaccio C et al (2014) Simvastatin attenuates the endothelial pro-thrombotic shift in saphenous vein grafts induced by advanced glycation endproducts. Thromb Res 133(3):418–425

Steffes MW, Chavers BM, Molitch ME, Cleary PA, Lachin JM, Genuth S, Nathan DM (2003) Sustained effect of intensive treatment of type 1 diabetes mellitus on development and progression of diabetic nephropathy: the epidemiology of diabetes interventions and complications (EDIC) study. JAMA 290(16):2159–2167

Stirban A et al (2006) Benfotiamine prevents macro- and microvascular endothelial dysfunction and oxidative stress following a meal rich in advanced glycation end products in individuals with type 2 diabetes. Diabetes Care 29(9):2064–2071

Stitt AW, He C, Vlassara H (1999) Characterization of the advanced glycation end-product receptor complex in human vascular endothelial cells. Biochem Biophys Res Commun 256(3):549–556

Stitt A et al (2002) The AGE inhibitor pyridoxamine inhibits development of retinopathy in experimental diabetes. Diabetes 51(9):2826–2832

Sugiyama S et al (1998) Plasma levels of pentosidine in diabetic patients: an advanced glycation end product. J Am Soc Nephrol 9(9):1681–1688

Tamura Y et al (2003) FEEL-1 and FEEL-2 are endocytic receptors for advanced glycation end products. J Biol Chem 278(15):12613–12617

Tang L et al (2017) Dapagliflozin slows the progression of the renal and liver fibrosis associated with type 2 diabetes. Am J Physiol Endocrinol Metab 313(5):E563–E576

Tesch GH (2008) MCP-1/CCL2: a new diagnostic marker and therapeutic target for progressive renal injury in diabetic nephropathy. Am J Physiol Renal Physiol 294(4):F697–F701

Thallas-Bonke V et al (2004) Attenuation of extracellular matrix accumulation in diabetic nephropathy by the advanced glycation end product cross-link breaker ALT-711 via a protein kinase C-alpha-dependent pathway. Diabetes 53(11):2921–2930

Thallas-Bonke V et al (2014) Nox-4 deletion reduces oxidative stress and injury by PKC-alpha-associated mechanisms in diabetic nephropathy. Physiol Rep:2(11)

Thomas MC (2011) Advanced glycation end products. Contrib Nephrol 170:66–74

Thomas MC et al (2005a) Low-molecular weight advanced glycation end products: markers of tissue AGE accumulation and more? Ann N Y Acad Sci 1043:644–654

Thomas MC et al (2005b) Interactions between renin angiotensin system and advanced glycation in the kidney. J Am Soc Nephrol 16(10):2976–2984
Tikellis C et al (2014) Dicarbonyl stress in the absence of hyperglycemia increases endothelial inflammation and atherogenesis similar to that observed in diabetes. Diabetes 63 (11):3915–3925
Tsuchida K et al (1999) Suppression of transforming growth factor beta and vascular endothelial growth factor in diabetic nephropathy in rats by a novel advanced glycation end product inhibitor, OPB-9195. Diabetologia 42(5):579–588
Twigg SM et al (2001) Advanced glycosylation end products up-regulate connective tissue growth factor (insulin-like growth factor-binding protein-related protein 2) in human fibroblasts: a potential mechanism for expansion of extracellular matrix in diabetes mellitus. Endocrinology 142(5):1760–1769
Twigg SM et al (2002a) Renal connective tissue growth factor induction in experimental diabetes is prevented by aminoguanidine. Endocrinology 143(12):4907–4915
Twigg SM et al (2002b) Renal connective tissue growth factor induction in experimental diabetes is prevented by aminoguanidine. Endocrinology 143(12):4907–4915
Uribarri J, Tuttle KR (2006) Advanced glycation end products and nephrotoxicity of high-protein diets. Clin J Am Soc Nephrol 1(6):1293–1299
Uribarri J et al (2003a) Restriction of dietary glycotoxins reduces excessive advanced glycation end products in renal failure patients. J Am Soc Nephrol 14(3):728–731
Uribarri J et al (2003b) Dietary glycotoxins correlate with circulating advanced glycation end product levels in renal failure patients. Am J Kidney Dis 42(3):532–538
Urios P, Grigorova-Borsos AM, Sternberg M (2007) Aspirin inhibits the formation of pentosidine, a cross-linking advanced glycation end product, in collagen. Diabetes Res Clin Pract 77 (2):337–340
Vasan S et al (1996) An agent cleaving glucose-derived protein crosslinks in vitro and in vivo. Nature 382:275–278
Vlad A et al (2017) Therapy with atorvastatin versus rosuvastatin reduces urinary podocytes, podocyte-associated molecules, and proximal tubule dysfunction biomarkers in patients with type 2 diabetes mellitus: a pilot study. Ren Fail 39(1):112–119
Vlassara H (1995) Advanced glycation in diabetic renal and vascular disease. Kidney Int Suppl 51: S43–S44
Vlassara H (1997) Recent progress in advanced glycation end products and diabetic complications. Diabetes 46(Suppl 2):S19–S25
Vlassara H (2001) The AGE-receptor in the pathogenesis of diabetic complications. Diabetes Metab Res Rev 17(6):436–443
Vlassara H, Bucala R (1996) Recent progress in advanced glycation and diabetic vascular disease: role of advanced glycation end product receptors. Diabetes 45(Suppl 3):S65–S66
Wada R, Yagihashi S (2005) Role of advanced glycation end products and their receptors in development of diabetic neuropathy. Ann N Y Acad Sci 1043:598–604
Wada R et al (2001) Effects of OPB-9195, anti-glycation agent, on experimental diabetic neuropathy. Eur J Clin Invest 31(6):513–520
Wang B et al (2014) Transforming growth factor-beta1-mediated renal fibrosis is dependent on the regulation of transforming growth factor receptor 1 expression by let-7b. Kidney Int 85 (2):352–361
Waris S et al (2015) Increased DNA dicarbonyl glycation and oxidation markers in patients with type 2 diabetes and link to diabetic nephropathy. J Diabetes Res 2015:915486
Watson A, Thomas M, Koh P, Figarola J, Rahbar S, Jandeleit-Dahm K (2010) Attenuation of diabetes associated atherosclerosis with LR-90, a novel inhibitor of AGE formation. J R Soc Chem:137–143
Watson AM et al (2011) Delayed intervention with AGE inhibitors attenuates the progression of diabetes-accelerated atherosclerosis in diabetic apolipoprotein E knockout mice. Diabetologia 54(3):681–689

Watson AM et al (2012) Alagebrium reduces glomerular fibrogenesis and inflammation beyond preventing RAGE activation in diabetic apolipoprotein E knockout mice. Diabetes 61 (8):2105–2113

Watson AM et al (2014) Quinapril treatment abolishes diabetes-associated atherosclerosis in RAGE/apolipoprotein E double knockout mice. Atherosclerosis 235(2):444–448

Wautier JL, Guillausseau PJ (2001) Advanced glycation end products, their receptors and diabetic angiopathy. Diabetes Metab 27(5 Pt 1):535–542

Wautier JL et al (1996) Receptor-mediated endothelial cell dysfunction in diabetic vasculopathy. Soluble receptor for advanced glycation end products blocks hyperpermeability in diabetic rats. J Clin Invest 97(1):238–243

Wautier MP et al (2003) N(carboxymethyl)lysine as a biomarker for microvascular complications in type 2 diabetic patients. Diabetes Metab 29(1):44–52

Wendt T et al (2003a) Glucose, glycation, and RAGE: implications for amplification of cellular dysfunction in diabetic nephropathy. J Am Soc Nephrol 14(5):1383–1395

Wendt TM et al (2003b) RAGE drives the development of glomerulosclerosis and implicates podocyte activation in the pathogenesis of diabetic nephropathy. Am J Pathol 162 (4):1123–1137

Williams ME et al (2007) Effects of pyridoxamine in combined phase 2 studies of patients with type 1 and type 2 diabetes and overt nephropathy. Am J Nephrol 27(6):605–614

Wortmann M et al (2016) A Glyoxalase-1 knockdown does not have major short term effects on energy expenditure and atherosclerosis in mice. J Diabetes Res 2016:2981639

Xia P et al (1994) Characterization of the mechanism for the chronic activation of diacylglycerol-protein kinase C pathway in diabetes and hypergalactosemia. Diabetes 43(9):1122–1129

Xue M et al (2016) Improved glycemic control and vascular function in overweight and obese subjects by glyoxalase 1 inducer formulation. Diabetes 65(8):2282–2294

Yamamoto Y et al (2001) Development and prevention of advanced diabetic nephropathy in RAGE-overexpressing mice. J Clin Invest 108(2):261–268

Yan SD et al (1994) Enhanced cellular oxidant stress by the interaction of advanced glycation end products with their receptors/binding proteins. J Biol Chem 269(13):9889–9897

Yan SD, Stern D, Schmidt AM (1997) What's the RAGE? The receptor for advanced glycation end products (RAGE) and the dark side of glucose. Eur J Clin Invest 27(3):179–181

Yan SF et al (2006) Receptor for advanced glycation end products and the cardiovascular complications of diabetes and beyond: lessons from AGEing. Endocrinol Metab Clin North Am 35(3):511–524. viii

Yang Z et al (1991) Two novel rat liver membrane proteins that bind advanced glycosylation endproducts: relationship to macrophage receptor for glucose-modified proteins. J Exp Med 174 (3):515–524

Yang CW et al (1994) Advanced glycation end products up-regulate gene expression found in diabetic glomerular disease. Proc Natl Acad Sci U S A 91(20):9436–9440

Yonemura S, Tsukita S (1999) Direct involvement of ezrin/radixin/moesin (ERM)-binding membrane proteins in the organization of microvilli in collaboration with activated ERM proteins. J Cell Biol 145(7):1497–1509

Younis NN et al (2010) Small-dense LDL and LDL glycation in metabolic syndrome and in statin-treated and non-statin-treated type 2 diabetes. Diab Vasc Dis Res 7(4):289–295

Zhang Z et al (2008) Combination therapy with AT1 blocker and vitamin D analog markedly ameliorates diabetic nephropathy: blockade of compensatory renin increase. Proc Natl Acad Sci U S A 105(41):15896–15901

Zheng F, Guan Y (2007) Thiazolidinediones: a novel class of drugs for the prevention of diabetic nephropathy? Kidney Int 72(11):1301–1303

Zheng F et al (2006) Combined AGE inhibition and ACEi decreases the progression of established diabetic nephropathy in B6 db/db mice. Kidney Int 70(3):507–514

Zhuang A et al (2017) Increased liver AGEs induce hepatic injury mediated through an OST48 pathway. Sci Rep 7(1):12292

Correction to: Soluble Guanylate Cyclase Stimulators and Activators

Peter Sandner, Daniel P. Zimmer, G. Todd Milne, Markus Follmann, Adrian Hobbs, and Johannes-Peter Stasch

Correction to: Online First Chapter "Soluble Guanylate Cyclase Stimulators and Activators" in book series Handbook of Experimental Pharmacology, https://doi.org/10.1007/164_2018_197

The chapter 'Soluble Guanylate Cyclase Stimulators and Activators' has now been made available open access under a CC BY 4.0 license.

The updated online version of this chapter can be found at https://doi.org/10.1007/164_2018_197

H. H. H. W. Schmidt et al. (eds.), *Reactive Oxygen Species*, Handbook of Experimental Pharmacology 264, https://doi.org/10.1007/164_2019_249